RENEWALS 458-4574

DATE DUE

MAY 0 4			

Food Colloids
Self-Assembly and Material Science

Food Colloids
Self-Assembly and Material Science

Edited by

Eric Dickinson
Procter Department of Food Science, University of Leeds, Leeds, UK

Martin E. Leser
Nestlé Research Center, Lausanne, Switzerland

RSCPublishing

The proceedings of the Food Colloids 2006: Self-Assembly and Material Science conference held on 23-26 April 2006 in Montreux, Switzerland

Special Publication No. 302

ISBN 13: 978-0-85404-271-5

A catalogue record for this book is available from the British Library

©The Royal Society of Chemistry 2007

All rights reserved

Apart from any fair dealing for the purpose of research or private study for non-commercial purposes, or criticism or review as permitted under the terms of the UK Copyright, Designs and Patents Act, 1988 and the Copyright and Related Rights Regulations 2003, this publication may not be reproduced, stored or transmitted, in any form or by any means, without the prior permission in writing of The Royal Society of Chemistry, or in the case of reprographic reproduction only in accordance with the terms of the licences issued by the Copyright Licensing Agency in the UK, or in accordance with the terms of the licences issued by the appropriate Reproduction Rights Organization outside the UK. Enquiries concerning reproduction outside the terms stated here should be sent to The Royal Society of Chemistry at the address printed on this page.

Published by The Royal Society of Chemistry,
Thomas Graham House, Science Park, Milton Road,
Cambridge CB4 0WF, UK

Registered Charity Number 207890

For further information see our web site at www.rsc.org

Printed by Henry Ling Ltd, Dorchester, Dorset, UK

Preface

The subject of food colloids is the study of multiphase systems containing the main functional ingredients of food at the nanoscale or mesoscopic level. The scientific aim is to use the fundamental concepts of physical chemistry, material science and soft matter physics to understand how the properties of colloidal systems – emulsions, dispersions, gels and foams – are related to the interactions between the constituent macromolecules and particles, both in bulk solution and at fluid interfaces. The practical objective is to improve control of the taste, texture and shelf-life of existing foods, and to formulate new products of high quality using imaginative combinations of ingredients and processing methods. A major challenge emerging for the food industry is the design of novel nanoscale structures for delivery of physiologically active components, in order to provide additional specific health benefits over and above that expected from consuming the nutrients present in a normal diet.

This volume is based on the conference 'Food Colloids 2006' held in Montreux (Switzerland) from 23 to 26 April 2006. It was the 11th in the series of biennial European conferences on this subject organized under the auspices of the Food Group of the Royal Society of Chemistry, the first having been held 20 years ago at the University of Leeds. The conference programme was sub-titled 'Self-Assembly and Material Science', and there were five main themes: (i) material science concepts in food; (ii) self-assembly in food; (iii) colloidal aspects of nutrition; (iv) understanding electrostatic interactions in foods; and (v) particles at interfaces.

The Montreux conference was attended by more than 200 delegates from 26 different countries, including a large proportion of younger scientists and research students. The technical programme consisted of 42 lectures and around 125 posters. Some 34 of the conference lecture presentations are recorded as edited chapters in this book, beginning with the stimulating review article on 'Food Structure and Nutrition' based on the invited lecture of Professor Bruce German. A set of research papers based on selected poster presentations are to appear separately in a special issue of the journal *Food Hydrocolloids* edited by Christophe Schmitt and Raffaele Mezzenga.

For their advice and support, especially in the difficult task of choosing the presentations for the lecture programme from the large number of excellent abstracts submitted, we thank the other members of the International Organizing Committee: Prof. Björn Bergenståhl (University of Lund), Prof. David Horne (formerly of the Hannah Research Institute), Dr Reinhard Miller (Max-Planck-Institute, Golm), Prof. Juan Rodríguez Patino (University of

Seville), Dr Ton van Vliet (Wageningen University) and Prof. Pieter Walstra (formerly of Wageningen University). We also gratefully thank the members of the Local Organizing Committee: Dr A. Burbidge, Mr P. Frossard, Dr M. Michel, Dr C. Schmitt and Dr H. Watzke (Nestlé Research Center, Lausanne) and Prof. R. Mezzenga and Prof. P. Schurtenberger (University of Fribourg).

Eric Dickinson, Leeds
Martin Leser, Lausanne
July 2006

Contents

Chapter 1   **Food Structure for Nutrition**   1
D.G. Lemay, C.J. Dillard and J.B. German

     1.1   Introduction   1
     1.2   Food Structure and Nutrition – Then and Now   2
           1.2.1   The Past   2
           1.2.2   The Present   3
     1.3   Food Structure and Bioavailability   4
     1.4   Models of Food Components, Food Structure, and Health   5
     1.5   Milk as a Model of Food Structure and Nutrition   6
     1.6   Bioactive Molecules in Milk   8
           1.6.1   Milk and Glucose Bioavailability   9
           1.6.2   Milk and Protein Bioavailability   9
           1.6.3   Milk and Lipid Bioavailability   10
     1.7   Food Structure and Nutrition – The Future   11
     1.8   Conclusions   13
     References   13

PART I   **Self-Assembly and Encapsulation**   17

Chapter 2   **Self-Assembly in Food – A New Way to Make Nutritious Products**   19
M. Michel, H.J. Watzke, L. Sagalowicz, E. Kolodziejczyk and M.E. Leser

     2.1   Introduction   19
     2.2   Lipid Digestion – A Self-Assembly-Based Bioprocess   21
     2.3   Lipases: Transformation of Lipids into Amphiphiles   22
     2.4   Liquid Crystalline Phase Formation during Digestion   23
     2.5   Oil Droplet Digestion   25
     2.6   Emulsifier-Based Self-Assembly and Structuring of Oil Droplets   26
     2.7   Addition of Guest Molecules to Mesophases   27

2.8	Phytosterol Solubilization into Mesophases	29
2.9	Colloid Science and Nutrition	29
References		30

Chapter 3 Structure of Self-Assembled Globular Proteins   35
T. Nicolai

3.1	Introduction	35
3.2	Structure of Dilute Aggregates	38
3.3	Structure of Interacting Aggregates and Gels	43
3.4	Relationship between Structure and Linear Elasticity	51
3.5	Conclusions	53
Acknowledgement		54
References		54

Chapter 4 Similarities in Self-Assembly of Proteins and Surfactants: An Attempt to Bridge the Gap   57
E. van der Linden and P. Venema

4.1	Introduction	57
4.2	Surfactant Assembly	58
4.3	Spherical Protein Aggregates	61
4.4	Protein Assemblies of Arbitrary Morphology: A First Attempt	62
4.5	β-Lactoglobulin Fibrils: Equilibrium Assembly or Not?	64
References		66

Chapter 5 Self-Assembled Liquid Particles: How to Modulate their Internal Structure   69
S. Guillot, A. Yaghmur, L. de Campo, S. Salentinig, L. Sagalowicz, M.E. Leser, M. Michel, H.J. Watzke and O. Glatter

5.1	Introduction	69
5.2	Emulsified Liquid Crystals and Microemulsions	71
5.3	Solubilization of Oils in Self-Assembled Structured Particles	75
	5.3.1 Emulsification of Microemulsions	75
	5.3.2 Micellar Cubosomes and Transformation from Hexosomes to EME	79
	5.3.3 Control of the Internal Self-Assembled Structure	82
5.4	Conclusions and Outlook	83
References		84

Chapter 6 Synergistic Solubilization of Mixed Nutraceuticals in Modified Discontinuous Micellar Cubic Structures 87
R. Efrat, A. Aserin and N. Garti

    6.1 Introduction 87
    6.2 Experimental 89
    6.3 Results and Discussion 91
        6.3.1 Thermal Behaviour of the Q_L Phase 91
        6.3.2 Systems Loaded with Phytosterols 94
        6.3.3 Lycopene Solubilization 97
        6.3.4 Lycopene and Phytosterol Solubilization 97
    6.4 Conclusions 100
    References 101

Chapter 7 Scope and Limitations of Using Wax to Encapsulate Water-Soluble Compounds 103
M. Mellema

    7.1 Introduction 103
    7.2 Theory 106
    7.3 Materials and Methods 106
        7.3.1 Preparation of Wax Capsules 106
        7.3.2 Analysis of Wax Capsules 107
        7.3.3 Preparation and Analysis of Complex Coacervate Capsules 108
        7.3.4 Analysis of Wax Digestibility 109
    7.4 Results 110
        7.4.1 Wax Capsules Prepared by Technique 1 110
        7.4.2 Wax Capsules Prepared by Technique 2 110
        7.4.3 Complex Coacervate Capsules 112
        7.4.4 Digestibility of Wax 113
    7.5 Concluding Remarks 114
    Acknowledgements 114
    References 115

Chapter 8 Self-Assembly of Starch Spherulites as Induced by Inclusion Complexation with Small Ligands 117
B. Conde-Petit, S. Handschin, C. Heinemann and F. Escher

    8.1 Introduction 117
    8.2 Materials and Methods 118
    8.3 Results and Discussion 119
    8.4 Concluding Remarks 125
    References 125

PART II Biopolymer Interactions 127

Chapter 9 Electrostatics in Macromolecular Solutions 129
B. Jönsson, M. Lund and F.L.B. da Silva

9.1 Introduction – The Dielectric Continuum Model 129
9.2 The Simple Electrolyte Solution 130
9.3 A Charged Macromolecule in a Salt Solution 131
9.4 Interaction of Two Charged Macromolecules 136
    9.4.1 Interaction of Two Peptides 136
    9.4.2 Effect of Multivalent Ions 139
    9.4.3 Effect of Titrating Groups 140
    9.4.4 Bridging Attraction with Polyelectrolytes 145
9.5 Protein–Polyelectrolyte Complexation 148
9.6 Conclusions 153
References 153

Chapter 10 Casein Interactions: Does the Chemistry Really Matter? 155
D.S. Horne, J.A. Lucey and J.-W. Choi

10.1 Introduction 155
10.2 Caseins as Polymers 156
10.3 Introduction of Calcium Sequestrants 157
10.4 Casein Chemistry and its Role in Casein Micelle Structure 161
10.5 Concluding Remark 165
References 166

Chapter 11 Electrostatic Interactions between Lactoferrin and β-Lactoglobulin in Oil-in-Water Emulsions 167
A. Ye and H. Singh

11.1 Introduction 167
11.2 Experimental 168
    11.2.1 Preparation of Emulsions 168
    11.2.2 Characterization of Emulsions 168
11.3 Results and Discussion 169
    11.3.1 Emulsions formed with Mixture of Lactoferrin + β-Lactoglobulin 169
    11.3.2 Addition of β-Lactoglobulin to Lactoferrin-Stabilized Emulsions 171
    11.3.3 Addition of Lactoferrin to β-Lactoglobulin-Stabilized Emulsions 173
Acknowledgements 175
References 175

Chapter 12	β-Lactoglobulin Aggregates from Heating with Charged Cosolutes: Formation, Characterization and Foaming		177
	G. Unterhaslberger, C. Schmitt, S. Shojaei-Rami and C. Sanchez		
	12.1	Introduction	177
	12.2	Materials and Methods	178
		12.2.1 Materials	178
		12.2.2 Sample Preparation	178
		12.2.3 Determination of Denaturation Kinetics by RP-HPLC	179
		12.2.4 Determination of Stability Ratio	180
		12.2.5 Determination of Protein Aggregate Molecular Weight and Second Virial Coefficient	180
		12.2.6 Determination of Protein Aggregate Size and Electrophoretic Mobility	181
		12.2.7 Determination of Protein Aggregate Interfacial Properties	182
		12.2.8 Determination of Protein Aggregate Foaming and Foam-Stabilizing Properties	182
	12.3	Results and Discussion	183
		12.3.1 Protein Denaturation and Aggregation in Presence of Cosolutes	183
		12.3.2 Physicochemical Properties of Protein Aggregates in Relation to Interfacial and Foaming Properties	188
	12.4	Conclusions	192
	References		192
Chapter 13	Manipulation of Adsorption Behaviour at Liquid Interfaces by Changing Protein–Polysaccharide Electrostatic Interactions		195
	R.A. Ganzevles, T. van Vliet, M.A. Cohen Stuart and H.H.J. de Jongh		
	13.1	Introduction	195
	13.2	Experimental	198
	13.3	Adsorption Kinetics	199
	13.4	Surface Rheology and Structure of Adsorbed Layers	203
	13.5	Concluding Remarks	206
	Acknowledgements		206
	References		206

Chapter 14 Adsorption Experiments from Mixed Protein + Surfactant Solutions — 209
V.S. Alahverdjieva, D.O. Grigoriev, J.K. Ferri, V.B. Fainerman, E.V. Aksenenko, M.E. Leser, M. Michel and R. Miller

14.1	Introduction	209
14.2	Theoretical Approach	210
	14.2.1 Thermodynamic Model	210
	14.2.2 Adsorption Kinetics	212
14.3	Materials and Methods	213
14.4	Results and Discussion	214
14.5	Conclusions	223
	Acknowledgement	223
	References	223

Chapter 15 Role of Electrostatic Interactions on Molecular Self-Assembly of Protein + Phospholipid Films at the Air–Water Interface — 227
A. Lucero Caro, A.R. Mackie, A.P. Gunning, P.J. Wilde, V.J. Morris, M.R. Rodríguez Niño and J.M. Rodríguez Patino

15.1	Introduction	227
15.2	Experimental	228
15.3	Results and Discussion	229
	15.3.1 Effect of pH on Electrostatic Interactions	229
	15.3.2 DPPC Monolayers	231
	15.3.3 β-Casein Monolayers	234
	15.3.4 DPPC + β-Casein Monolayers	238
15.4	Conclusions	241
	Acknowledgements	241
	References	242

Chapter 16 Theoretical Study of Phase Transition Behaviour in Mixed Biopolymer + Surfactant Interfacial Layers Using the Self-Consistent-Field Approach — 245
R. Ettelaie, E. Dickinson, L. Cao and L.A. Pugnaloni

16.1	Introduction	245
16.2	Methodology and Model	247
	16.2.1 Self-Consistent-Field Calculations	247
	16.2.2 The Mixture Model	248
16.3	Results and Discussion	249

	16.3.1	Adsorption Isotherms	249
	16.3.2	Density Profiles	253
16.4	Conclusions		254
References			255

Chapter 17 Interactions during the Acidification of Native and Heated Milks Studied by Diffusing Wave Spectroscopy 257
M. Alexander, L. Donato and D.G. Dalgleish

17.1	Introduction	257
17.2	Diffusing Wave Spectroscopy	258
17.3	Materials and Methods	260
17.4	Results	261
	17.4.1 Unheated Milk	261
	17.4.2 Heated Milk	263
17.5	Discussion	265
References		266

Chapter 18 Computer Simulation of the Pre-heating, Gelation and Rheology of Acid Skim Milk Systems 269
J.H.J. van Opheusden

18.1	Introduction	269
18.2	Simulation Model	270
18.3	The Pre-Heating Phase	271
	18.3.1 Methodology	271
	18.3.2 Results and Discussion	273
18.4	The Gelation Phase	278
18.5	The Deformation Phase	283
18.6	Comparison with Experiment	285
Acknowledgement		286
References		286

Chapter 19 Xanthan Gum in Skim Milk: Designing Structure into Acid Milk Gels 289
P.-A. Aichinger, M.-L. Dillmann, S. Rami-Shojaei, A. Paterson, M. Michel and D.S. Horne

19.1	Introduction	289
19.2	Materials and Methods	290
	19.2.1 Solution Preparation	290
	19.2.2 Phase Separation	290
	19.2.3 Gel Formation	291
19.3	Manipulation of Particle Interactions	292
	19.3.1 Depletion-Induced Phase Separation	292

	19.3.2 Acid-Induced Aggregation and Gel Formation	293
	19.3.3 Effects of Pre-Heating of Skim Milk	294
19.4	Tailoring Gels from Particle Dispersions	296
19.5	Conclusions	300
Acknowledgement		301
References		301

PART III Particles, Droplets and Bubbles   303

Chapter 20 Particle Tracking as a Probe of Microrheology in Food Colloids   305
E. Dickinson, B.S. Murray and T. Moschakis

20.1	Introduction	305
20.2	Experimental	308
20.3	Results and Discussion	309
20.4	Concluding Remarks	316
Acknowledgements		317
References		317

Chapter 21 Optical Microrheology of Gelling Biopolymer Solutions Based on Diffusing Wave Spectroscopy   319
F. Cardinaux, H. Bissig, P. Schurtenberger and F. Scheffold

21.1	Introduction	319
21.2	Critical Gelation Microrheology	319
21.3	Sample Preparation	320
21.4	Diffusing Wave Spectroscopy Based on Optical Microrheology	321
21.5	Results and Discussion	323
21.6	Conclusions	325
Acknowledgements		325
References		325

Chapter 22 Gel and Glass Transitions in Short-range Attractive Colloidal Systems   327
G. Foffi, N. Dorsaz and C. De Michele

22.1	Introduction	327
22.2	The Gel Transition: Arrested Phase Separation	329
22.3	Dependence on Range of Attraction: Can an Arrested State Emerge from Equilibrium?	338
22.4	Conclusions	341
Acknowledgement		341
References		341

Chapter 23 Shape and Interfacial Viscoelastic Response of Emulsion Droplets in Shear Flow 343
P. Erni, V. Herle, E.J. Windhab and P. Fischer

- 23.1 Introduction 344
- 23.2 Experimental 344
  - 23.2.1 Materials 344
  - 23.2.2 Emulsion Preparation and Characterization 345
  - 23.2.3 Rheo-SALS and Rheology 345
  - 23.2.4 Single Drop Deformation Experiments 347
  - 23.2.5 Interfacial Characterization 347
- 23.3 Results 347
- Acknowledgements 353
- References 354

Chapter 24 Enhancement of Stability of Bubbles to Disproportionation Using Hydrophilic Silica Particles Mixed with Surfactants or Proteins 357
T. Kostakis, R. Ettelaie and B.S. Murray

- 24.1 Introduction 357
- 24.2 Materials and Methods 358
- 24.3 Results and Discussion 359
  - 24.3.1 Fumed Silica Particles + DDAB 359
  - 24.3.2 Fumed Silica Particles + Protein 361
  - 24.3.3 Fumed Silica Particles + Lecithin 362
  - 24.3.4 Colloidal Silica Particles + DDAB 363
  - 24.3.5 Colloidal Silica Particles + Protein 365
  - 24.3.6 Colloidal Silica Particles + Lecithin 367
- 24.4 Conclusions 367
- References 368

Chapter 25 Coalescence of Expanding Bubbles: Effects of Protein Type and Included Oil Droplets 369
B.S. Murray, A. Cox, E. Dickinson, P.V. Nelson and Y. Wang

- 25.1 Introduction 369
- 25.2 Materials and Methods 370
  - 25.2.1 Materials 370
  - 25.2.2 Emulsion Preparation and Characterization 370
  - 25.2.3 Single Bubble Layer Experiment 371
  - 25.2.4 Foam Stability Test 372
- 25.3 Results and Discussion 374
  - 25.3.1 Effect of Protein Type 374
  - 25.3.2 Effect of Oil Droplets 376

25.4	Conclusions	381
References		381

PART IV Emulsions 383

Chapter 26 Role of Protein-Stabilized Interfaces on the Microstructure and Rheology of Oil-in-Water Emulsions 385
P.J. Wilde, A.R. Mackie, M.J. Ridout, F.A. Husband, G.K. Moates and M.M. Robins

26.1	Introduction	385
26.2	Materials and Methods	387
26.3	Results	389
26.4	Discussion	394
26.5	Conclusions	396
Acknowledgement		396
References		396

Chapter 27 Crystallization in Monodisperse Emulsions with Particles in Size Range 20–200 nm 399
M.J.W. Povey, T.S. Awad, R. Huo and Y. Ding

27.1	Introduction	399
27.2	Materials and Methods	400
27.3	Crystal Nucleation Theory	403
27.4	Results and Discussion	404
27.5	Conclusions	411
References		411

Chapter 28 Instant Emulsions 413
T. Foster, A. Russell, D. Farrer, M. Golding, R. Finlayson, A. Thomas, D. Jarvis and E. Pelan

28.1	Introduction	413
28.2	Emulsification	415
28.3	Emulsion Stabilization	417
28.4	Rates of Hydrocolloid Hydration	418
28.5	Competitive Hydration	421
References		422

Chapter 29 Flavour Binding by Solid and Liquid Emulsion Droplets 423
S. Ghosh, D.G. Peterson and J.N. Coupland

29.1	Introduction	423
29.2	Experimental	424
29.3	Experimental Results and Preliminary Modelling	425
29.4	Nature of the Flavour–Binding Interactions	426

		29.5	Surface Binding and Droplet Dissolution Model	428
		29.6	Conclusions	431
		Acknowledgement		432
		References		432

Chapter 30 Adsorption of Macromolecules at Oil–Water Interfaces during Emulsification — 433

L. Nilsson, P. Osmark, M. Leeman, C. Fernandez, K.-G. Wahlund and B. Bergenståhl

30.1	Introduction		433
30.2	Experimental		434
30.3	Results		436
	30.3.1	Total Adsorption of OSA-Starch	436
	30.3.2	Selective Adsorption of OSA-Starch	439
	30.3.3	Protein Adsorption	440
30.4	Discussion		441
	30.4.1	Kinetically Controlled Adsorption	441
	30.4.2	Affinity-Controlled Adsorption	443
	30.4.3	Equilibrium-Controlled Adsorption	445
30.5	Conclusions		446
Acknowledgement			446
References			446

PART V Texture, Rheology and Sensory Perception — 449

Chapter 31 Tribology as a Tool to Study Emulsion Behaviour in the Mouth — 451

D.M. Dresselhuis, E.H.A. de Hoog, M.A. Cohen Stuart and G.A. van Aken

31.1	Introduction	451
31.2	Sensory Perception of Emulsions	452
31.3	Tribology and Sensory Science	453
31.4	Importance of Surface Characteristics	454
31.5	In-Mouth Emulsion Behaviour and Perception	458
31.6	Summary	460
Acknowledgements		460
References		461

Chapter 32 Saliva-Induced Emulsion Flocculation: Role of Droplet Charge — 463

E. Silletti, M.H. Vingerhoeds, W. Norde and G.A. van Aken

32.1	Introduction		463
32.2	Materials and Methods		464
	32.2.1	Materials	464

		32.2.2	Saliva Collection and Handling	464
		32.2.3	Preparation of Emulsions and their Mixtures with Saliva	465
		32.2.4	Characterization of the Flocculation	465
		32.2.5	Characterization of Complex Formation between Lysozyme and Saliva	466
	32.3	Results and Discussion		466
	32.4	Conclusions		470
	References			471

Chapter 33 Surface Topography of Heat-Set Whey Protein Gels: Effects of Added Salt and Xanthan Gum   473
J. Chen, E. Dickinson, T. Moschakis and K. Nayebzadeh

	33.1	Introduction		473
	33.2	Materials and Methods		474
	33.3	Results and Discussion		475
		33.3.1	Effect of Added Salt	475
		33.3.2	Effect of Added Xanthan Gum	478
	33.4	Concluding Remarks		482
	Acknowledgements			484
	References			484

Chapter 34 Mechanisms Determining Crispness and its Retention in Foods with a Dry Crust   485
T. van Vliet, J. Visser, W. Lichtendonk and H. Luyten

	34.1	Introduction		485
	34.2	Materials and Methods		486
	34.3	Mechanism Acting at the Molecular Scale		487
	34.4	Mechanism Acting at the Mesoscopic Scale		489
		34.4.1	Estimation of the Structural Length-Scale for Crispness	489
		34.4.2	Need for Crack Stoppers	491
		34.4.3	Need for Overlapping Sound Events	492
		34.4.4	Frequency Spectrum of Emitted Sound	493
		34.4.5	Effect of Oil Presence after Deep Frying	495
		34.4.6	Change in Mode of Fracturing with Increasing Water Content	497
	34.5	Effects at the Macroscopic Scale		498
	34.6	Conclusions		499
	Acknowledgements			499
	References			499

Subject Index   503

Chapter 1
Food Structure for Nutrition

Danielle G. Lemay,[1] Cora J. Dillard[1] and J. Bruce German[1,2]

[1] DEPARTMENT OF FOOD SCIENCE AND TECHNOLOGY, UNIVERSITY OF CALIFORNIA, DAVIS CA 95616, USA
[2] NESTLÉ RESEARCH CENTER, P.O. BOX 44, CH-1000 LAUSANNE 26, SWITZERLAND

1.1 Introduction

The great tradition of nutrition research has seen the creation of an unprecedented knowledge base of the essential nutrients, together with their absolute quantitative requirements at different life stages, and the pathological phenotypes experienced by populations who fail to consume sufficient quantities of them. The research that was necessary to assemble this knowledge base of essential molecules is one of the life science's great achievements. In retrospect, the achievement was made possible by some key strategic decisions by nutrition scientists. First, there was the critical decision for nutrition to become a molecular science. The object of the study of nutrition, namely food, was physically and conceptually disassembled into individual molecules. By eliminating food structure from nutrition research, it became possible to feed animals with purified diets in which specific suspected nutrients were explicitly included or assiduously removed. If the molecule was an essential nutrient, its elimination from a diet fed to a growing, reproducing animal model would produce overt deficiency symptoms. This so-called 'fault model' of nutrient discovery was critical to scientific studies designed to identify the essential nutrients. As a result of this very successful research strategy, all of the vitamins, minerals, amino acids, and fatty acids that are essential to the growth and reproduction of animals and humans are now known.

The knowledge of the nutrients that are necessary for humans shifted the public health emphasis to strategies designed to ensure that populations consume diets that achieve adequacy of all of the essential nutrients, and are thus sufficient to prevent deficiency diseases. An underlying assumption in this public health emphasis and focus on essential nutrients was that diets containing all of the essential nutrients in adequate amounts would largely eliminate diet-related diseases. But it has become disturbingly clear that it is possible to obtain all of the essential nutrients and still consume diets that are suboptimal for overall health. In fact, poorly balanced diets are now recognized to be one of the leading causes of metabolic diseases around the world.[1] Understanding how diet can produce such diseases is leading to a much more comprehensive understanding of the interactions between food components and health.

To understand food's broader role in nutrition, it will be necessary not only to go beyond essential nutrients, and to redesign nutritional models and experiments, but also to reconsider the meaning of 'health' itself. Health must be understood in terms of the wide diversity of normal metabolic states of humans. The recent epidemics in heart disease, obesity, diabetes, hypertension, and osteoporosis indicate that health can deteriorate in response to imbalanced diets almost as quickly as it can deteriorate in the absence of essential nutrients.[2] In particular, there is an immediate need to understand the role of diet and foods in managing food intake and in regulating whole body energy status. Most nutritional measures of essential nutrients and their status have been taken in the fasted condition to avoid confounding chronic status with acute food intake.[3] It is now becoming apparent[4] that variations in the dynamic or temporal responses to food components are also important to overall health. Nutrition research must, therefore, embrace the entire continuum of the fed state, leading to an understanding of how rapidly foods are digested, absorbed, and metabolized, and how these temporal and special events relate to overall health. With this perspective in mind, the influence of food structure on the dynamics of varying post-prandial metabolic states should be studied in mechanistic detail.

Food itself must also be understood comprehensively as multi-component, multi-phasic ensembles of biomaterials. Clearly, optimal diets are more than just the essential nutrients. Notably, several aspects of foods have emerged that were virtually ignored by nutritional research in the scientific pursuit of essential nutrients. Perhaps the most important omission was food structure itself. The science of food biomaterials must now deliver food structure as a fully controllable variable set to nutrition research. For nutrition researchers to understand the role of physical, colloidal, and macromolecular structure, these must be studied as independent variables in nutrition studies. Knowledge of food structure must, therefore, become highly predictive – not just descriptive.

1.2 Food Structure and Nutrition – Then and Now

1.2.1 The Past

The science of nutrition is the knowledge repository for the understanding of essential nutrients, their quantitative requirements, and the mechanisms behind their essentiality. Few achievements in science have been as complete, or as rapidly brought to public health practice, as the identification of nutrients essential for humans. The decision taken early in the 20th century to disassemble food commodities into molecules and study essentiality as a molecular phenomenon was the defining event in molecular nutrition, and few decisions in the history of science have been so successful. The concept of specific molecules as essential amines was initiated by Casimir Funk, who effectively spawned molecular nutrition with the isolation of vitamin B1 (thiamin) in 1912, as reviewed by Jukes;[5] and within 40 years the concept of essential nutrients was ostensibly complete.

This great legacy of discovery paved the way for the eradication of diseases caused by nutrient deficiencies. However, some of the important strategies of nutrient discovery are proving to be debilitating to the future growth of nutrition as a field. This is particularly true as nutrition emerges with the goal of guiding diets to improve overall health. For example, while disassembling foods into molecules enabled the differentiation of essential from nonessential nutrients, it left a prevailing assumption that food *per se* is immaterial to the provision of essentiality, and to health in general. Similarly, by establishing essential nutrients, the underlying assumption was that they were essential for *all* humans; thus, varying nutritional requirements within individuals among the population remained largely ignored. By defining bioavailability as the integration of the pharmacokinetic curve of an ingested nutrient in blood relative to the same nutrient when injected, the goal became a bioavailability of unity; hence, the foods themselves – in particular those with a degree of complexity of food matrices that may reduce bioavailability – were by definition deleterious. By defining an adequate diet as that producing adequate tissue levels of nutrients in the fasted condition, the relationship of the structure of foods and diet to the dynamics of the fed state has remained largely unstudied. Needless to say, if the dynamics of food and nutrient delivery remain unknown, they will continue to be unappreciated in terms of the value of food structure to its nutritive value.

1.2.2 The Present

The emergence of diseases of metabolic dysregulation associated with variations in diet is forcing scientists to refine the traditional views of nutrition. Diet affects health in more ways than simply through providing essential nutrients: it includes various aspects of chronic metabolic regulation. The science of nutrition has not yet built an understanding of this complexity. Furthermore, whereas recommendations for essential nutrients can be quite broad for the entire population, humans vary in their responses to diets as they affect metabolism, and the same diets fed to different individuals may develop quite distinctly different health consequences. Therefore, as the field of nutrition faces this new challenge of (re-)discovering the importance of food, nutrition scientists will also need to change their perspectives of health.

While food structure and nutrition have not been well studied, there are clear indications of the importance of the dynamics of food digestion to health from many aspects of ongoing scientific research. The most vivid is the variation in rate of absorption of the simple nutrient glucose. Although food structure has not been explicitly specified as an independent variable, for many years the rate of glucose delivery to blood has been included as an indicator of food as the

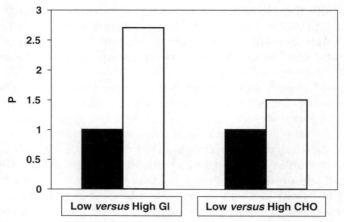

Figure 1 *Relative probability of developing premature macular degeneration (P) as affected by the glycemic index (GI) and the total carbohydrate (CHO) of reported habitual diets in the Nurses' Study population*[7]

glycemic index.[6] Numerous studies have documented the effect of glycemic index in accounting for variations in metabolic response to particular diets. Even more impressively, as glycemic index has been included as an independent variable in larger health issues, this simple surrogate reflection of food structure is being recognized as important to many end points that are not typically thought of as being related to food structure. For example, Taylor and co-workers[7] have shown that glycemic index as a dietary variable was associated with an almost threefold increase in risk of macular degeneration in the Nurses' Health Study population (Figure 1). Though many more studies have documented the effects on health of the rate of absorption of glucose as glycemic index, as reviewed by Dickinson and Brand-Miller,[8] little research has been pursued to extend these observations into a more precise understanding of the role of food structure in the dynamics of glucose delivery.

1.3 Food Structure and Bioavailability

Bioavailability is defined[9] as the difference in delivery to blood between an oral dose and an injected dose. It is determined experimentally by measuring the dose-corrected area under curve (AUC) of a substance administered orally (po) divided by the AUC intravenous (IV):

$$F = ([AUC]_{po} \times dose_{IV})/([AUC]_{IV} \times dose_{po}) \qquad (1)$$

If 100% of an ingested dose of a nutrient or a compound reaches the blood, the bioavailability (F) is considered to be 1. From the perspective of essential nutrients – especially when considering the risk of deficiency – the greater the bioavailability, the better the nutrient. Hence, food matrices promoting rapid and complete absorption were judged to be superior in their ability to deliver essential nutrients.[10] Food matrices that compromised rapid absorption, as a result of their structural complexity, were considered deleterious to bioavailability and hence to nutritional value. With these rather narrow but, in context, reasonable assumptions about food structure and nutrition, food processing steps that destroyed structure were advantageous and those that created structure were considered deleterious. Now that scientific considerations have begun to broaden beyond essential nutrients to nonessential food components and the food structure itself, the absolute dynamics of the delivery of food and food components is being studied. The key question then becomes: "what is the optimal situation – is it rapid complete absorption, or slow, perhaps even incomplete, absorption?" When such a question is asked, it becomes clear that there are no biological models of optimal food structure. This being the case, it is reasonable to ponder what is the preferable biological model one would wish to interrogate to answer questions about rate and extent of absorption of nonessential food components.

1.4 Models of Food Components, Food Structure, and Health

Nutrition has begun to address the scientific questions associated with nonessential food components in the diet. To date, most of these components have been derived from plants. The vast majority of small biological molecules as candidate bioactive food components are indeed derived from secondary metabolism in plants. The total number of secondary plant metabolites is estimated[11] to exceed 2×10^5. Nonetheless, in attempting to establish appropriate food components for nourishment beyond the essential nutrients, it is reasonable to ask what was the driving force for the emergence of these secondary metabolites within plants during evolution? For many of these compounds and their respective pathways, the divergence of secondary metabolism was driven by the competitive advantage for those plants to avoid predation. Because animals were a conspicuous source of predation of plants, much of the secondary metabolism of plants evolved explicitly by Darwinian selective pressure to avoid being eaten by animals. Thus, quite the opposite of providing compounds to improve animal health, plant secondary metabolism evolved to produce compounds that were explicitly toxic to animals. Examples of the success of this evolutionary pressure abound, from soybean trypsin inhibitors to alkaloids.[12,13] Thus, although the plant kingdom and the rapidly arriving plant genomes are interesting targets for the development of knowledge of toxicity and anti-nutritive components, it is difficult to defend plants and plant metabolites *per se* as targets for healthful food structures.

To understand how to design food structures that optimize nutrient delivery, it is necessary to examine a food system that ideally was under Darwinian selective pressure during evolution to be explicitly nourishing in all its connotations. Fortunately, there is such a model, with remarkable complexity and genetic diversity – mammalian milk.

1.5 Milk as a Model of Food Structure and Nutrition

Milk is the only biomaterial that has evolved for the purpose of nourishing growing mammals. Survival of mammalian offspring has exerted a strong selective pressure on the biochemical evolution of the lactation process. Just as for evolution of any biological process, the strong survive at the expense of the weak, which leads to the appearance of new traits that promote health, strength, and, ultimately, survival. The same is true of milk. The role of milk was to produce the molecules that would ensure the survival of the mammalian offspring. The various processes of lactation have been described: the immature gland, the developing gland,[14] colostrum formation,[15] secretory activation (lactogenesis),[16–18] and involution and senescence.[19] The overall process of lactation requires extensive physiological, structural, and metabolic remodelling of the mammary tissue to enable it to acquire milk-forming capabilities.

This evolutionary pressure led to the elaboration of a whole food that contains proteins, peptides, complex lipids, and oligosaccharides in higher order structures, and all coming together as a complex, multicomponent – yet highly organized – food. In simple terms, milk is an obligate liquid. The basic mammary tissue constraints and the early evolutionary direction led to a liquid emulsion that had to stay in the liquid form to exit the ductal tissue and flow to the infant. Table 1 shows that the variation in milk composition is significant across mammals.[20] The composition of all of the macronutrients (protein, fat, and carbohydrate) vary significantly, with the extremes of protein being from ~ 1 to ~ 11 g per 100 g of fresh milk.

As milk has been studied more intensively, it has become evident that it is more than a simple repository of essential nutrients in a protein/fat/lactose 'broth'. Milk contains many compounds that exhibit properties associated with biological functions beyond simple provision of essential building blocks. Milk compounds and their configurations act as growth factors, toxin-binding factors, antimicrobial peptides, prebiotics, and immune regulatory factors within the mammalian intestine.[21] Gradually, science has found that these macromolecules deliver biological advantages to the intestine and throughout

Table 1 *Protein, fat, and carbohydrate composition of milk from different mammalian species (per 100 g fresh milk) (data taken from Ref. 20)*

Species	Protein (g)	Fat (g)	Carbohydrate (g)
Human	1.1	4.2	7.0
Monkey (rhesus)	1.6	4.0	7.0
Donkey	1.9	0.6	6.1
Goat	2.9	3.8	4.7
Cow	3.2	3.7	4.6
Elephant	4.0	5.0	5.3
Water buffalo	4.1	9.0	4.8
Mouse	9.0	13.1	3.0
Seal	10.2	49.4	0.1
Whale	10.9	42.3	1.3

the body, ultimately contributing to the survival of neonatal mammals.[22] It should be emphasized that research on these compounds and their functions is proving to be more difficult than traditional nutritional investigations on essential nutrients. Because these compounds and their functions are not essential, they have discernible value only to infants under certain circumstances. Hence, their functions, actions, and overall benefits have proven exceedingly difficult to recognize, and much less to study, and it is generally agreed that only a very small subset of the total biological value of milk's components is known.[23] Nonetheless, in study after study, the identification of component after component is revealing the evolution of milk as a highly functional food with properties that may find multiple applications to the structure manipulation of other foods. The evolution of the mammary gland was under constant Darwinian selection for the successful survival of offspring.[24] Milk is indeed a genomic model for the multidimensional aspects of nourishment.[25]

The evolutionary origins of milk proteins and mammary regulation define the key functions of milk and the mammary gland. The emergence during evolution of the mammary gland likely involved adaptive recruitment of existing precursor genes through alteration of regulatory sequences to allow expression in primitive mammary glands, and duplication and mutation of structural sequences to acquire new functions from pre-existing primitive proteins. Evidence is consistent with early precursors to lactation being derived from an ancestral apocrine-like gland that was associated with hair follicles that led to secretions designed to nourish and protect the soft, rubbery shells of egg-laying monotrenes.[26] The earliest mammary function after provision of nutrition was possibly the passing of protective advantages onto offspring *via* immunoglobulins, and thus aiding selection for survival. This would have paved the way for the elaboration of myriad protective functions that are only now beginning to be appreciated.

The functions of milk generally can be considered supportive of both mammalian mothers and infants through several mechanisms. Milk provides nourishment to infant offspring; disease defence for the infant; disease defence for the mother; regulation or stimulation of infant development, growth, or function; regulation or stimulation of maternal mammary tissue development, growth, or function; inoculation, colonization, nourishment, regulation, and elimination of infant microflora; and inoculation, colonization, nourishment, regulation, and elimination of maternal mammary microflora.

Nourishing the mammalian neonate is the most obvious role of milk, and the success of mammals attests to the value of milk as an initial food source for the young of these species. The demands on milk as a sole source of nutrition are remarkable. All of the essential macronutrients – water, vitamins, minerals, amino acids, and fatty acids – plus the basic structural and energetic intermediates needed to sustain life, must be delivered to the neonate in a highly absorbable form that is appropriate to the species and the stage of development – all at minimal energy cost to the mother. Lactation research has illuminated many of the biological processes needed to mobilize the essential biomolecules from maternal stores and to convert them into dispersed, transportable, and bioavailable structures in milk.

Milk also provides myriad benefits to the growth, development, and health-supporting processes of infant and mother beyond those of the essential nutrients. The nonessential components of milk as well as those of essential nutrients are not understood, but research is now beginning to focus on their roles in the well-being of neonates.[23] The research strategies needed to discover these properties are different from those used to discover the properties and roles of essential nutrients. The latter can be studied with relative ease because their elimination from the diet of animals leads to overt signs of deficiency in each individual. Nonessential nutrients and their functions, however, are only valuable within a particular context, and thus investigations of benefits of nonessential nutrients must first recognize the context in which they are valuable. This has become the great challenge for discovery of nutritional value of milk's nonessential components: when and why are they beneficial? Anything that is added to milk – literally anything – costs the mammalian mother; so, in a highly competitive environment, if it does not profit the infant, it would put the mother at selective disadvantage. Hence, it can be assumed that there was a selective advantage to the incorporation of the components in milk, and it is an exciting pursuit of modern nutrition to discover the biological value of each.

1.6 Bioactive Molecules in Milk

Milk is an excellent source of nutrients: high-quality protein, water-soluble and fat-soluble vitamins, calcium, phosphorus, magnesium, *etc*. But it is more than this. It is also a source of proteins with biological activities that have been demonstrated *in vitro*, in animal models, and in infant and adult humans. The physiological activities provided by milk proteins in the gastrointestinal tract include enhancement of nutrient absorption, enzyme activity, inhibition of enzymes, growth stimulation, modulation of the immune system, and defence against pathogens.[23,27–31]

Milk provides a myriad of defensive strategies for the intestine of the infant against exogenous pathogens, including immunological factors (antibodies, cells, cytokines), proteins (lactoferrin, enzymes, *e.g.*, lysozyme), oligosaccharides, and glycoproteins, gut microflora (prebiotics), and nutrients to optimize the infant's immune system, as reviewed in the recent literature.[32–35] Lactoferrin, lysozyme, and haptocorrin (a vitamin B_{12}-binding protein in human milk) have been proposed[36] to influence the growth of bacteria and elimination of particular pathogens. Although just beginning as a discrete field of nutrition, many of the intact milk proteins contain peptide sequences that appear to be expressly liberated by the proteases of intestinal digestion.[37]

With these multiple confirmations of milk's discrete biological functions beyond delivering simple calories, essential nutrients, and building blocks, milk is unquestionably a knowledge resource for nutrition. Millennia of Darwinian selective pressure have been guiding its evolution in many ways. Therefore, it is reasonable to interrogate the basic structures of milk for lessons on how to design molecular food structures for optimal nutrition.

1.6.1 Milk and Glucose Bioavailability

The rate of delivery of glucose has been under considerable scrutiny for several years because of the implications for adverse health of rapidly absorbed glucose from foods of high glycemic index.[8] It is thus potentially instructive to examine how glucose is, in effect, delivered in milk.

The vast majority of glucose is present in milk not as glucose, but as the disaccharide lactose. This is formed from glucose and galactose as a final step in the assembly of milk in the lactating epithelial cell. Lactose is a disaccharide that is strikingly non-digestible by humans. Only a single enzyme – lactase – has emerged through mammalian evolution to digest lactose, and this enzyme exhibits an equally remarkable regulation. This enzyme is produced by mammalian infants during the suckling period, and then, in virtually all mammals, this activity is down-regulated on weaning. Thus, intriguingly, mammalian mothers are unable to digest the lactose that they produce. Although various functions have been suggested for lactose's unique structure, there is one indisputable consequence of this structure to digestion. By delivering glucose in the form of lactose, mammalian evolution has placed the delivery of glucose exclusively under control of the infant and its lactase activity. Glucose will be digested and absorbed only as rapidly as the activity of its endogenous intestinal lactase enzyme allows. So, through the evolution of lactation, glucose is delivered as lactose, and this lactose invariably causes slow delivery of glucose.

1.6.2 Milk and Protein Bioavailability

The rate of delivery of protein has not been as well studied as that of other macronutrients, although recent research[38,39] has begun to examine variations in proteins as delivery agents for amino acids. Casein is the most abundant single protein source in most mammalian milks, and it is the protein component that is uniquely associated with milk.[29] Thus, it is interesting to examine casein structure for its nutritional value in terms of structure, digestion, and rate of absorption.

Proteins in general are digested by proteases as a consequence of their structure. In fact, all levels of protein structure – primary, secondary, tertiary, and quarternary – are known to affect proteolysis. The greater the degree of native structure, the less digestible is the protein. In general, denaturation of a protein increases its digestion by proteases. Thus, if evolutionary pressure were to be actively providing a selective advantage to highly digestible proteins, then the less the structure, the more digestible a given protein would be expected to be. Consistent with this interpretation, casein is perhaps the least structured protein known.[40] There is remarkably less secondary structure within any of the casein subunits. Thus, from this criterion of structure, casein would appear to be unusually susceptible to rapid hydrolysis by intestinal proteases. *In vitro* studies of protein digestibility have confirmed[41] that caseins are highly susceptible to proteolysis. This would logically lead to rapid digestion and absorption of the amino acids of casein proteins. Nonetheless, the casein proteins evolved with a quite remarkably distinct additional property. Assembled as multi-subunit complexes termed casein micelles, the caseins are actually large protein aggregates that, while soluble, are dispersible only in water, thanks to an exterior of predominantly hydrophilic κ-casein subunits. When the casein micelle encounters the mammalian infant stomach, one of the most ingenious events in nutrition occurs – the enzyme chymosin, expressed in the stomach of the infant, and with an acidic activity range, hydrolyses a single peptide bond on κ-casein. This reaction splits off the hydrophilic peptide of κ-casein (glycomacropeptide), which contains the surface-stabilizing element of the entire casein micelle. As a result, the balance of casein proteins within the stomach rapidly aggregates into a large, insoluble curd. This association process has the net effect of slowing the release, digestion, and absorption of casein's amino acids. Thus, mammalian casein has, in fact, a unique structure and structuring process that, at least in part, *slows* the delivery of milk protein.

1.6.3 Milk and Lipid Bioavailability

The digestion and absorption of lipids has been well studied from a wide variety of sources, including milk. There is a considerable challenge to the process considering the inherent insolubility of most lipids; thus, the conspicuous success of fat digestion by animals has been studied extensively. Although normally quite successful, several genetic and physiological defects in the intestine can lead to failures in fat absorption, termed steatorrhea. The difficulties that are presented by absorbing fat are the likely reason that fat is the most common form of intestinal malabsorption. Invariably, this clinical condition is highly deleterious, because of both the failure to absorb essential fatty acids and fat-soluble vitamins, and the loss of the substantial caloric density of dietary lipids. Therefore, there is considerable medical importance to ensuring that fat is in a highly absorbable form in milk.

Given the basic requirement to produce a fat globule for export into milk, there has existed in place, within the ancestors of mammals, some genetic motifs that were ostensibly 'available' to evolution of a fat-globule delivery system. The synthesis of lipoproteins was a well-established biological process in early animal evolution, with highly effective transport of intact fat globules being critical to the physiology of reptiles, amphibia, and fish.[42–44] These phospholipid-bound and apoprotein-targeted lipoproteins deliver triglycerides from the liver through the blood to the surface of various peripheral tissues in these organisms, and lipoprotein lipase activity on the endothelial surfaces is capable of very rapidly transferring the lipid contents to the targeted tissues.[45] Therefore, producing lipoproteins as fat globules in milk was an available alternative.

This is not, however, the structure of milk fat globules. Milk fat globules are simple, lipoprotein-like particles, as they are produced in the epithelial cell; but they do not acquire an apoprotein on their surface; and further, on exiting from the cell, each globule is enrobed with an intact coating of plasma membrane bilayer.[46] It is still perplexing to scientists as to why precisely this unusual course to milk fat structure evolved, but one thing has been determined. The hydrolysis rate of the milk fat globule is dramatically slowed by the presence of this additional plasma membrane bilayer.[47]

1.7 Food Structure and Nutrition – The Future

As a speciality within food science, the structure–function analysis of genetically defined biopolymers and their complexes has enormous potential in the coming decades. Previously, each field within the life sciences made progress within the narrow constraints of its particular discipline. Whereas there was sharing of the final outcomes (*i.e.*, confirmed or refuted hypotheses), the detailed information that was responsible ultimately for building the knowledge that they generated was only usable within each field. This perspective is now changing dramatically. Many fields of scientific inquiry are beginning to make progress in areas that produce information that, in its breadth, can be leveraged for the structure–function analysis of food materials. Given these technological innovations, an overview of what goals may be achievable is presented in Table 2.

To date, the most notable progress in structure–function analysis has been made with proteins. In the 1970s, the Brookhaven National Laboratory established the Protein Data Bank (PDB) as a repository for three-dimensional structural data of biological macromolecules.[48] A key aspect of this endeavour

Table 2 *From food structure to nutrition function: a vision for future decades*

Goal	Key requirements	Fields to follow
Establish function of known structure set	1. Standardization (structure format, experimental conditions, measurable end points) 2. Database development (repository)	Biochemistry, pharmacology
Establish function of synthetic structure libraries	1. Combinatorial synthesis of structures 2. Development of assays amenable to high-throughput screening	Chemical genomics
Establish individual response to structure	1. High-throughput screening for relevant polymorphisms 2. Optimization of food structure for specific profiles	Pharmaco-genomics

was the use of a standard format, the PDB format, to represent structural data derived from X-ray diffraction and NMR studies. As of 13 June 2006, there were 37,136 protein and nucleic acid structures in the PDB database of the Research Collaboratory for Structural Bioinformatics (RCSB). Some more recently established databases, such as the Structure Function Linkage Database (SFLD), are integrating structural data with functional information.[49] The information in this database can be used for rule-based prediction of functional capabilities of new structures with unknown functions so long as the new structure is a member of a protein 'superfamily' in the database. The goal of food scientists must increasingly be to extend such annotations to include the structures and functions of proteins in food, and the diverse consequences of such structures when consumed for the relationship between diet and health.

While a database like the SFLD has had to be painstakingly assembled from the primary literature, the assembly of a food materials database could be accelerated if food scientists were to standardize experimental conditions and measurable end points at the outset. The integrity of inter-experimental analysis would also be improved. To some scientists, the need for a database may not seem immediately obvious. After all, new knowledge on the effect of a particular lipid structure on the glycemic index of a food bolus, for example, would be interesting in itself. But in the modern age of informatics, the data can be recycled from past experiments to generate and/or test hypotheses that could not have been conceived at the time of data collection. Furthermore, there remains the tantalizing proposal that the analysis of well-characterized known structures might help to predict the actions of unknown/unsynthesized structures.

Because it is unlikely that the predictive power of a few known structures will be sufficient to estimate the action of the infinite number of possible structures, food scientists may benefit from advances in chemical genomics. The objective of chemical genomics is to find those small molecules that interact with the genome. These molecules can then be used therapeutically as drugs or scientifically in experiments to better understand specific biological pathways.[50] The relevance to the science of food structure is that technologies for the combinatorial organic synthesis of small molecules may provide insight into methods for the combinatorial synthesis of food structures. Also, the application of creative labelling techniques, such as molecular tags,[51,52] could prove to be as useful in refining food biopolymer structures as they have proven to be for their biological counterparts.

To establish the functions of food structures in a fast and efficient manner, food scientists need to develop assembly systems and assays that are amenable to 'high-throughput' screening. At the cellular level, automated techniques have been developed for everything from microarrays to cell imaging. At the level of the organism, the means to measure metabolites in a high-throughput manner is already available.[53] Concentrations of various metabolites and fluxes between different body pools can be measured simultaneously. Elucidating the effect of food structures on the metabolome should lead to food products that have an influence on chronic diseases, and thus have a widespread impact on human health.

Finally, as the understanding of food structure–function relations in humans is improved, it is quite likely that responses to some food materials will be individual in character, *i.e.*, depending on a person's genotype. The emerging field of pharmacogenomics is likely to be the technology leader in this area. The goal of pharmacogenomics is to enable doctors to prescribe drugs on the basis of a person's genetic profile. To transfer this idea to food science, human genetic polymorphisms will need to be screened on a high-throughput basis to determine which of these may have an effect on a person's response to food materials. Then, ideally, food structures can be optimized for particular genetic profiles or for targeted nutritional goals.

1.8 Conclusions

The role of food structure in the nutritional value of diets is becoming recognized as being much more important than previously assumed. In fact, the examination of epidemiological associations between food structure and certain chronic and degenerative processes, from diabetes to macular degeneration, implicates the rate of nutrient delivery to be as important to health as the diet's overall macronutrient composition (protein, fat, and carbohydrate). Such provocative epidemiological results have prompted a re-examination of milk by asking the question: "how has the Darwinian selective pressure during evolution influenced the structure and delivery of milk components for infants?" Each component of milk – glucose, protein, and fat – although ultimately digestible and absorbable by the infant, is structured in such a way as to prolong its absorption rather than to accelerate it. These interrogations of milk are suggestive that the role of food structure should be an important part of all future diet and health research. Such a mandate would require that structural food chemists assemble the tools of food structure manipulation and examination expressly for use in nutrition-oriented studies. In essence, the fields of food structure, food chemistry, and nutrition must be reunited to consolidate the scientific depth and public health relevance of each.

References

1. G. Alberti, *Bull. World Health Organ.*, 2001, **79**, 907.
2. D. Yach, D. Stuckler and K.D. Brownell, *Nat. Med.*, 2006, **12**, 62.
3. E. Cahill, J. McPartlin and M. Gibney, *Int. J. Vitam. Nutr. Res.*, 1998, **68**, 142.
4. D.G. O'Donovan, S. Doran, C. Feinle-Bisset, K.L. Jones, J.H. Meyer, J.M. Wishart, H.A. Morris and M. Horowitz, *J. Clin. Endocrinol. Metab.*, 2004, **89**, 3431.
5. T.H. Jukes, *Prev. Med.*, 1989, **18**, 877.
6. D.J. Jenkins, T.M. Wolever and R.H. Taylor, *Am. J. Clin. Nutr.*, 1981 **34**, 362.

7. C.J. Chiu, L.D. Hubbard, J. Armstrong, G. Rogers, P.F. Jacques, L.T. Chylack, Jr., S.E. Hankinson, W.C. Willett and A. Taylor, *Am. J. Clin. Nutr.*, 2006, **83**, 880.
8. S. Dickinson and J. Brand-Miller, *Curr. Opin. Lipidol.*, 2005, **16**, 69.
9. K.J. Wienk, J.J. Marx and A.C. Beynen, *Eur. J. Nutr.*, 1999, **38**, 51.
10. R.M. Faulks and S. Southon, *Biochim. Biophys. Acta*, 2005, **1740**, 95.
11. E. Pichersky and D. Gang, *Trends Plant Sci.*, 2000, **5**, 439.
12. S.W. Applebaum, in *Comprehensive Insect Physiology, Biochemistry and Pharmacology*, Vol. 4, G.A. Kerkut and L. Gilbert (eds), Pergamon, New York, 1985, p. 279.
13. A.A. Seawright, *J. Nat. Toxins*, 1995, **3**, 227.
14. M.S. Holland and R.E. Holland, *J. Dairy Sci.*, 2005, **88**(Suppl. 1), E1.
15. C.M. Fetherston, C.S. Lee and P.E. Hartmann, *Adv. Nutr. Res.*, 2001, **10**, 67.
16. J.L. McManaman and M.C. Neville, *Adv. Drug Deliv. Rev.*, 2003, **55**, 629.
17. J.L. McManaman, C. Palmer, S. Anderson, K. Schwertfeger and M.C. Neville, *Adv. Exp. Med. Biol.*, 2004, **554**, 263.
18. M.C. Neville, T. McFadden and I.J. Forsyth, *J. Mammary Gland. Biol. Neoplasia*, 2002, **7**, 49.
19. B. Stefanon, M. Colitti, G. Gabai, C.H. Knight and C.J. Wilde, *J. Dairy Res.*, 2002, **69**, 37.
20. B. Webb, A.H. Johnson and J. Alford (eds), *Fundamentals of Dairy Chemistry*, 2nd edn, AVI, Westport, CT, 1974, Chap. 1.
21. R.L. Walzem, C.J. Dillard and J.B. German, *Crit. Rev. Food Sci. Nutr.*, 2002, **42**, 353.
22. D.S. Newburg (ed), *Advances in Experimental Medicine and Biology: Bioactive Components of Human Milk*, Kluwer Academic/Plenum, New York, 2001.
23. J.B. German, C.J. Dillard and R.E. Ward, *Curr. Opin. Clin. Nutr. Metab. Care*, 2002, **5**, 653.
24. F.L. Schanbacher, R.S. Talhouk and F.A. Murray, *Livestock Production Science*, 1997, **50**, 105.
25. R.E. Ward and J.B. German, *J. Nutr.*, 2004, **134**, 962S.
26. O.T. Oftedal, *J. Mammary Gland Biol. Neoplasia*, 2002, **7**, 225.
27. C. Castillo, E. Atalah, J. Riumallo and R. Castro, *Bull. Pan Am. Health Org.*, 1996, **30**, 125.
28. F.F. Rubaltelli, R. Biadaiol, P. Pecile and P. Nicoletti, *J. Perinat. Med.*, 1998, **26**, 186.
29. W.H. Oddy, *Breastfeed Rev.*, 2001, **9**, 11.
30. B. Lönnerdal, *Am. J. Clin. Nutr.*, 2003, **77**, 1537S.
31. B. Lönnerdal, *Adv. Exp. Biol. Med.*, 2004, **554**, 423.
32. A. Baldi, P. Ioannis, P. Chiara, F. Eleonora, C. Roubini and D. Vittorio, *J. Dairy Res.*, 2005, **72**(Special Issue), 66.
33. H. Meisel, *Curr. Med. Chem.*, 2005, **12**, 1905.
34. K.J. Rutherfurd-Markwick and P.J. Moughan, *J. AOAC Int.*, 2005, **88**, 955.

35. S. Severin and X. Wenshui, *Crit. Rev. Food Sci. Nutr.*, 2005, **45**, 645.
36. Y. Adkins and B. Lönnerdal, *Am. J. Clin. Nutr.*, 2003, **77**, 1234.
37. H. Meisel and R.J. FitzGerald, *Br. J. Nutr.*, 2000, **84**, S27.
38. Y. Boirie, M. Dangin, P. Gachon, M.P. Vasson, J.L. Maubois and B. Beaufrere, *Proc. Natl Acad. Sci. USA*, 1997, **94**, 14930.
39. M. Dangin, C. Guillet, C. Garcia-Rodenas, P. Gachon, C. Bouteloup-Demange, K. Reiffers-Magnani, J. Fauquant, O. Ballevre and B. Beaufrere, *J. Physiol.*, 2003, **549**, 635.
40. H.E. Swaisgood, in *Advanced Dairy Chemistry, Volume 1 – Proteins*, 3rd edn, P.F. Fox and P.L.H. McSweeney (eds), Kluwer Academic/Plenum, New York, 2003, Part A, p. 139.
41. A. Baglieri, S. Mahe, R. Benamouzig, L. Savoie and D. Tome, *J. Nutr.*, 1995, **125**, 1894.
42. W.R. Garstka, R.R. Tokarz, M. Diamond, A. Halpert and D. Crews, *Horm. Behav.*, 1985, **19**, 137.
43. J. Ndiaye and S. Hayashi, *Cell Struct. Funct.*, 1996, **21**, 307.
44. J.A. Sellers, L. Hou, D.R. Schoenberg, S.R. Batistuzzo de Medeiros, W. Wahli and G.S. Shelness, *J. Biol. Chem.*, 2005, **280**, 13902.
45. I.J. Goldberg, *J. Lipid Res.*, 1996, **37**, 693.
46. M. Hamosh, J.A. Peterson, T.R. Henderson, C.D. Scallan, R. Kiwan, R.L. Ceriani, M. Armand, N.R. Mehta and P. Hamosh, *Semin. Perinatol.*, 1999, **23**, 242.
47. J.B. German and C.J. Dillard, *Am. J. Clin. Nutr.*, 2004, **80**, 550.
48. F.C. Bernstein, T.F. Koetzle, G.J. Williams, E.F. Meyer, Jr., M.D. Brice, J.R. Rodgers, O. Kennard, T. Shimanouchi and M. Tasumi, *J. Mol. Biol.*, 1977, **112**, 535.
49. S.C. Pegg, S.D. Brown, S. Ojha, J. Seffernick, E.C. Meng, J.H. Morris, P.J. Chang, C.C. Huang, T.E. Ferrin and P.C. Babbitt, *Biochemistry*, 2006, **45**, 2545.
50. S.L. Schreiber, *Chem. Eng. News*, 2003, **81**, 51.
51. S. Brenner and R.A. Lerner, *Proc. Natl. Acad. Sci. USA*, 1992, **89**, 5381.
52. M.H. Ohlmeyer, R.N. Swanson, L.W. Dillard, J.C. Reader, G. Asouline, R. Kobayashi, M. Wigler and W.C. Still, *Proc. Natl. Acad. Sci. USA*, 1993, **90**, 10922.
53. S.M. Watkins and J.B. German, *Curr. Opin. Biotechnol.*, 2002, **13**, 512.

PART I
Self-Assembly and Encapsulation

Chapter 2

Self-Assembly in Food – A New Way to Make Nutritious Products

Martin Michel, Heribert J. Watzke, Laurent Sagalowicz, Eric Kolodziejczyk and Martin E. Leser

NESTLÉ RESEARCH CENTER, VERS-CHEZ-LES-BLANC, CH-1000 LAUSANNE 26, SWITZERLAND

2.1 Introduction

For many years food manufacturers have been trying to meet the rising expectations of consumers for nutritionally balanced and healthy foods. The current way to address this challenge is to 'enhance' nutritional functionality within a product by adding to a common food base various bioactives such as probiotics, sterols, flavones, carotenoids and polyphenols.[1] Such enrichments are often linked to a health claim, supported by a clinical study, and with the stated aim of preventing a potential health problem.[2] Recent examples of added bioactives are plant sterols to prevent cardiovascular diseases,[3-5] isoflavones for improving bone health[6] and lycopene for cancer prevention.[7-9]

The next major step in value addition to food will be to deliver food products adapted to the nutritional and health needs of an individual, because different people respond differently to similar diets and life styles.[10,11] This step requires new strategies to elucidate the causes of these differences and to establish how to address them *via* nutritional means. The current focus lies on nutrigenomics,[12,13] *i.e.*, the study of how genetics and metabolic processes relate to nutrition. This approach will help to identify more effectively those people who are statistically more likely to develop a particular disease and would benefit from a personalized diet. A major challenge will be the translation of the resulting data into adequate nutrition solutions.

The study and understanding of how the digestive system functions will play another important role in food personalization. After all, making an 'individual' means keeping apart what is 'in' from what is 'out'.[14] Humans (and most animals) digest their food extra-cellularly, *i.e.*, outside of the cells. The digestive space is delimited by the gastrointestinal tract, an approximately 10-m long tube from mouth to anus. The digestive system is strictly speaking 'outside the body'. It is divided into regions that specialize in the separate processes of digestion, nutrient absorption and waste elimination: the mouth, the pharynx, the oesophagus, the stomach, the small and large intestine, and the rectum. The digestive system can therefore be seen as the gatekeeper, controlling what is allowed to get into the body and what is kept out.

As a consequence we need to see food development in a broader context. Food scientists will have to provide food products adapted to the physiological, physico-chemical and colloidal processes involved in perception, as well as nutrient transformation, liberation and absorption during digestion.[15] In other words, it will be important to master the kinetics and thermodynamics of nutrient incorporation into a food product, liberation during digestion and absorption into the cells, making use of food structures that facilitate the realization of the desired nutritional benefits. It is worthwhile to learn from past experience of food–drug interactions, which represent a similar context to that mentioned above. Food–drug interactions can lead either to improvement of drug absorption or to treatment failure; they may even provoke serious toxicity.[16,17] The most important pharmacokinetic effects of food–drug interactions are caused by changes in drug absorption due to physico-chemical interactions (*e.g.*, chelation) between the drug and certain food components, and physiological responses such as changes in gastric acidity, bile secretion or gastrointestinal motility.[18] Effects of foods on drug absorption can be classified in terms of those causing decreased, delayed, increased or accelerated absorption, and those where the food has no significant effect. It has been shown[19] that negative food effects can be influenced by adapting the pharmaceutical formulation. Furthermore, food–drug interactions typically depend on the size and composition of the meal, as well as the exact timing of drug intake in relation to eating. Bioavailability of lipophilic drugs is often increased by a high fat content, either due to an increase in drug solubility or stimulation of bile secretion. A high fibre content can reduce bioavailability of some drugs (*e.g.*, digoxin) because of drug binding to the fibre. The complexity in foods with built-in health benefits will be even greater, where the food itself will take over the place of the pharmaceutical formulation. Instead of seeing it from the negative side, we can explore the different 'kinetic' effects on nutrient uptake – *i.e.*, reducing it ('keep something out'), sustaining it (slow release), increasing it (better utilization) or accelerating it ('getting a boost').

Of particular interest is the question as to how the dynamics of nutrient uptake is influenced by food structure and composition, *e.g.*, microstructural features such as particle size and shape, self-assembled structures and molecular structure. Figure 1 illustrates the complexity existing in foods in the form of a phase diagram. The macronutrients (carbohydrates, proteins and lipids) determine the dry matter of any food. Adding water to each binary composition will determine the phase behaviour in systems of lipid + protein + water, carbohydrate + protein + water and lipid + carbohydrate + water. These ternary systems can be used as a guide to get a better insight into the complexity of structure formation in complex foods. Of special interest to us is the spontaneous formation of

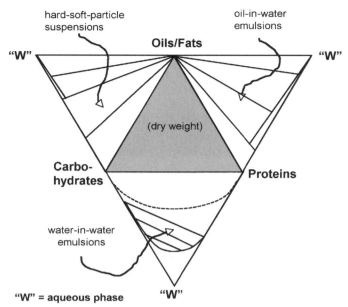

Figure 1 *Schematic phase diagram indicating the complexity of food structures based on the three macronutrients: proteins, lipids (oils/fats) and polysaccharides (carbohydrates). The centre triangle corresponds to the dry matter composition. In the case of emulsifiers and proteins, addition of water produces a very complex phase behaviour. On the same diagram, digestion behaviour can also be represented if the water corner is replaced by an artificial digestion medium* (Copyright PhysChem Consulting (2006))

self-assembled aggregates, and developing an understanding as to how these structures can be applied in form of colloids.

In this article we review important physico-chemical and colloidal aspects regarding the digestion of lipids in relation to emulsion structure. Lipid digestion is a rather complex process, resulting in different intermediate self-assembled structures. We may consider emulsions as ideal systems for studying the relationship between food structure formation and lipid digestion, because they are easy to produce with different droplet sizes, and they can be created with various internal self-assembled structures, resulting in cubosomes, hexosomes and isasomes.[20–25] Moreover, emulsions can be loaded with lipophilic and amphiphilic bioactives such as esterified and non-esterified plant sterols. Emulsions are also ideal model systems, as they can easily be exposed to artificial or native digestion media such as saliva, gastric juices, lipolytic enzymes and bile.

2.2 Lipid Digestion – A Self-Assembly-Based Bioprocess

On average, the daily adult Western diet contains approximately 100 g of lipids, of which 92–96% consist of long-chain triacylglycerols, also referred to as triglycerides.[26,27] Fats supply more than 40% of the calories ingested in the Western diet. The majority of triglycerides contain saturated and unsaturated long-chain fatty acids (length > C_{14}). The two unsaturated long-chain fatty acids, oleic and linoleic, account for 34% and 19%, respectively, of all the edible oils and fats produced industrially. The process of fat digestion and absorption is very efficient (95%) and there is no feedback regulation to reduce fat assimilation. In order to become absorbable, lipids undergo a number of biochemical and colloidal transformations during fat digestion. For a detailed description of this topic, the interested reader is referred to various literature reviews.[26–30] In summary, lipid digestion is a highly complex sequential process involving:

(i) enzymatic hydrolysis of triglycerides into partially ionized fatty acids and monoglycerides (ratio 2:1) catalysed by lipolytic enzymes;
(ii) emulsification of bulk fat;
(iii) micellar solubilization due to association of monoglycerides and fatty acids with themselves and with endogenous, biliary colloidal components such as bile salts, phosphatidylcholine and cholesterol; and
(iv) transport of digestion products through an unstirred, mucine-rich water layer into the enterocytes.

2.3 Lipases: Transformation of Lipids into Amphiphiles

The digestion of triglycerides in humans is done by gastric and pancreatic lipases, hydrolysing the *sn*-1 and *sn*-3 ester bonds of triglycerides, and producing *sn*-2 monoglycerides. Enzymatic hydrolysis of lipids can only be achieved at the interface between lipids and the surrounding aqueous medium. Lipid hydrolysis begins in the stomach with the action of gastric lipase, a product of the chief cells of the gastric mucosa. Around 30% of the total triglycerides may be digested by gastric lipases in the stomach.[31] Only the short-chain fatty acids can be absorbed into the bloodstream.

As it passes by peristaltic movements through the pylorus, the stomach digest undergoes shear forces sufficient for emulsification of bulk fats. The potential emulsifying agents are monoglycerides, phospholipids, polysaccharides and peptidic fragments originating from protein digestion. The presence of these surfactants can lead to a rather low interfacial tension and 'spontaneous emulsification'. Whereas pure bile salts are poor emulsifiers of fat, when present together with biliary lipids they facilitate fat emulsification.[32–34] The digestion of dietary fat emulsions in humans and rats has been studied by Armand *et al.*[35] They found that dietary lipids are emulsified in the duodenum as droplets of size 1–50 μm in humans[36] and 14–33 μm in rats.[37] These values are much larger than those that were reported 50 years earlier by Frazer *et al.*[38]

Pancreatic juice contains pancreatic lipase, phospholipase A2, colipase and carboxylic ester hydrolase (which needs bile salts for its activity). Enzymatic hydrolysis of lipids with pancreatic lipases can only occur at the oil–water interface in the presence of a ternary lipase–colipase–bile salt micelle complex.[39]

Bile salts are found at concentrations above the critical micellar concentration (CMC) in the small intestine during digestion. The pure pancreatic lipase is inhibited *in vitro* by the physiological concentrations of bile salts.[40] Colipase counteracts this effect, facilitating the adsorption of lipase to bile salt-covered lipid–water interfaces. Lipolysis is a very fast process: it is almost complete on moving from the duodenum into the jejenum where mucosal lipid uptake takes place. Phospholipase A2 hydrolyzes the *sn*-2 position of a variety of phospholipids producing lyso-phospholipids. This enzyme has an absolute requirement for Ca^{2+} which binds in a 1:1 stoichiometry to both substrate and enzyme.[41] It does not act on sphingolipids.

2.4 Liquid Crystalline Phase Formation during Digestion

Hofmann and Borgström[42–46] were the first to show the existence of mixed micelles in human intestinal lipid digestion products. They separated the lipid digestion products by ultracentrifugation into three parts: (a) an oil phase containing triglycerides, diglycerides and minor amounts of fatty acids; (b) an aqueous mixed micellar phase consisting of bile salts, *sn*-2 monoglycerides and dissociated fatty acids; and (c) a precipitated pellet. Patton *et al.*[47,48] showed by light microscopy that, during *in vitro* digestion of emulsion droplets, a lamellar liquid crystalline phase is formed containing calcium and ionized fatty acids, followed by the production of a viscous isotropic phase composed of monoglycerides and protonated fatty acids. Similar structural observations were obtained by Holt *et al.*[49] and Rigler *et al.*[50] Some examples are shown in Figure 2. The latter group identified lamellar and vesicular lipolytic product phases by freeze-fracture transmission electron microscopy (TEM). Using detergent removal techniques such as dialysis and dilution in model biles made of phospholipids or monoglycerides and native bile, the presence of micelle–vesicle transitions was demonstrated[52–54] by dynamic light-scattering and freeze-fracture TEM. All these later studies have revealed that the original mixed micelle model of Hofmann and Borgström is an oversimplification: some other phases do play a role in lipid digestion.

Lindström and co-workers[55] studied in more detail the *in vitro* transformation of aqueous self-assembled lipid phases into dilute micellar bile salt solutions. They showed that the cubic phase is easily solubilized into bile micelles. The reversed hexagonal phase is more difficult to solubilize, and the lamellar phase can be dispersed into liposomes even in the absence of bile salts. Their X-ray data indicate that a monoolein + oleic acid mixture of molar ratio of 1:2 at pH = 6.5 (the composition and conditions of intestinal contents) forms an inverse micellar L2-phase upon swelling with water. It can be suggested that the L2 phase (W/O microemulsion) plays an important role in increasing the efficiency of lipolysis. As cholesterol and phospholipids constitute only minor fractions in the digestion of a fatty meal, they assumed[55] that these components do not significantly influence the observed phase behaviour. Staggers *et al.*[28]

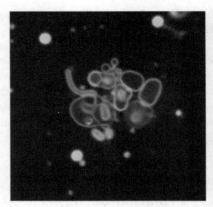 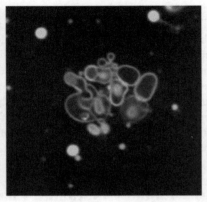

Figure 2 *Oil droplets and vesicles identified during in vitro digestion of lipids. Fat digestion (Pancreatin, Sigma) was simulated in an artificial bile at pH = 6.5, as described by Kossena et al.,[51] incubating 4 mg of soybean oil emulsified by 1% lecithin (fat/pancreatin ratio = 6). After incubation for 1 h at 37°C, 30 mL of digestion medium were collected and transferred into a 100-μm deep chamber covered with a coverslide coated with a film of Nile Red (Sigma) to stain lipids. Real-time imaging was achieved at room temperature using the 'time series' functions of the Zeiss LSM510 confocal microscope over periods of 4 min at ~0.8 s intervals*
(Copyright Nestec Ltd (2006))

performed a systematic phase equilibrium investigation of model lipid systems corresponding to lipid compositions typical of upper intestinal contents during lipid digestion in adult humans; the systems contained bile salts, mixed intestinal lipids, partially ionized fatty oleic acids, monooleylglycerol and cholesterol at 37°C, pH = 6.5 and ionic strength=0.15 M (Na^+). They identified a rich phase behaviour, particularly the formation of metastable unilamellar vesicles of variable sizes in coexistence with thermodynamically stable saturated mixed micelles. So it is very likely, as also suggested by Hernell and Blackberg,[56] that vesicles are directly involved in lipid absorption, since bile salt-deficient subjects can take up 50–75% of dietary lipids.

Kossena and co-workers[57] established phase diagrams to determine the behaviour of the digestion products of common formulation lipids (C8:0, C12:0 and C18:1) under model physiological conditions. Pseudo-ternary phase diagrams were constructed using varying proportions of simulated endogenous intestinal fluid (SEIF), fatty acids and monoglycerides (as representative of exogenous lipid digestion products). A change from liquid crystals to mixed micelles and vesicles was observed with decreasing fatty acid/monoglyceride concentrations. Large solubilization enhancement ratios for a series of poorly water-soluble compounds such as hydrocortisone and its esters were measured relative to the intrinsic solubility in buffer, resulting in an increased solubilization capacity in lamellar liquid crystal phases (a factor of 10–2000) and cubic liquid crystal phases (a factor of 10–30,000). Positive correlations were observed between the solubilization benefit provided by each phase and the drug lipophilicity. Kossena et al.[57] suggested that, in their in vitro model test, the action of SEIF might result in the formation of a lamellar droplet surface with C_8 and C_{12} lipids, whereas with C_{18} lipids an additional cubic phase layer might form, the obtained phase/solubility trends indicating that poorly water-soluble drugs are transported across the intestinal colloidal species that form during the digestion of lipid-based drug delivery systems. These observations are of importance for food, as the different mesophases formed could have an influence on the bioavailability of oil-soluble and amphiphilic bioactives present in chyme.

A comparison of in vitro equilibrium structures[28] with structures obtained from aspirated duodenal contents[29] shows that the sizes of the micelles are comparable, whereas the unilamellar vesicles in vivo are substantially smaller. Borné et al.[58–61] studied the effect of lipase on different liquid crystalline phases in water, such as lamellar (monoolein + sodium oleate), reversed bicontinuous cubic (monoolein + oleic acid) and reversed hexagonal (monoolein + diolein). Under their experimental conditions they could not detect any difference in enzyme activity when comparing cubic to hexagonal phases. They do not exclude the possibility, however, that other experimental conditions (different mesophases or sample compositions) could result in different enzyme activities.

2.5 Oil Droplet Digestion

The digestion and lipid absorption from emulsions with different droplet sizes has been investigated in humans by Armand and co-workers.[35,36] Healthy subjects received intragastrically in random order a coarse lipid emulsion (10 μm diameter) and a fine lipid emulsion (0.7 μm diameter); the emulsions were of identical composition. Gastric and duodenal aspirates were collected throughout digestion in order to measure changes in fat droplet size, gastric and pancreatic lipase activities and fat digestion. Despite an increase in droplet size in the stomach (2.75–6.20 μm), the fine emulsion retained droplets of smaller size and its degree of lipolysis was greater than that of the coarse emulsion (37% compared with 16%; $P < 0.05$). In the duodenum, the extent of lipolysis of the fine emulsion was on the whole higher (73% compared with 46%). The overall 0–7-h plasma and chylomicron responses given by the areas under the curve were not significantly different between the emulsions, but the triacylglycerol peak was delayed with the fine emulsion (4 h compared with <3 h), as previously shown in rat experiments.[37] These data suggest that emulsions with different particle sizes lead to different time courses of plasma lipids postprandially. Although lipid digestion was faster for the finer emulsions – as is to be expected, owing to the larger interface area – the appearance of triacylglycerol in the blood was found to be slower. The authors mention as possible causes for the latter a difference in the rate of gastric emptying, the formation of intermediate phases such as mixed micelles, and vesicles, and a difference in the diffusion of lipids through the unstirred water layer. It might also be that the two emulsions studied had different interfacial composition, as the total area was much higher in the fine emulsion.

All these results show that digestion is a complex biotransformation process involving enzymes, amphiphilic substances and colloids. The mechanism of final uptake of molecular species through the mucine-rich unstirred water layer and the cell membranes into the enterocytes remains unresolved. Both passive diffusion and carrier-mediated processes have been proposed.[62] For long-chain fatty acids and cholesterol, the passage across the unstirred water layer is rate-limiting, whereas the passage of short- and medium-chain fatty acids is limited by the brush border membrane. The observation that polystyrene latex particles with sizes of 200 nm and 2 μm can be absorbed in the small intestine indicates that absorption of colloidal particles even larger than micelles might be possible,[63–65] although Rigler and co-workers[50] did not provide any evidence for such a mechanism.

2.6 Emulsifier-Based Self-Assembly and Structuring of Oil Droplets

The digestion of triacylglycerols or other lipids leads to the production of numerous amphiphilic molecules such as emulsifiers, as discussed above. When such surfactants are exposed to water, they exhibit a rich phase behaviour. Especially of interest to us are self-assembled aggregates of mono- and diglycerides because these emulsifiers are specified as GRAS (generally recognized as safe), and so they can be used with confidence to study the phase behaviour in the presence of oil and water, mimicking the phases that occur during digestion. The glycerol fatty acid esters and their derivatives account for about 75% of the world production of food polar lipids, and they are considered to be the most important group of amphiphiles in food.[66] Their use dates back to the 1930s when they were first used in margarine production. Their major applications today are in bread, cakes, margarine, ice-cream and chewing gum. Bakery is by far the biggest application, with approximately 60% of all monoglycerides being used by this industry.[66] Luzzati[67] reported the first phase diagrams of monoglyceride + water systems. X-ray diffraction has allowed the unambiguous identification of the formation of the different mesophases. It was shown by Qui and Caffrey,[68] for instance, that the monoolein + water system forms an inverted bicontinuous cubic reversed hexagonal phase and W/O microemulsions.

Oil droplets in emulsions can be internally structured using self-assembly principles by the addition of appropriate amphiphiles.[20] Such internally self-assembled emulsion droplets are of interest in food technology as delivery system for oil-soluble and amphiphilic bioactives.[69] In the context of the subject of the present article, a deeper understanding of their phase behaviour in the presence of water, oil and fatty acids is essential for further elucidating the phase behaviour during lipid digestion. This knowledge will definitely be an advantage for the development of lipid-based delivery colloids with specific release functions for bioactives. The dispersion of reversed liquid crystalline mesophases, however, is a complex task. They show only a limited stability in water, and they quickly phase separate after dispersion. Intrinsic to the behaviour of these systems is the problem of lipophilicity of the surfaces formed after fragmentation of the mesophases. Since the inner self-assembled structure of the particle is bent towards the water phase (*i.e.*, it has a negative natural curvature), the lipophilic domains become exposed to the continuous water phase, which is energetically unfavourable. Therefore, an external stabilizer has to be used to protect these lipophilic particles against aggregation.[21,22] Most work relevant to food has been done on dispersions of reversed bicontinuous cubic phases, made of glyceryl monoolein (GMO). Landh[70] and Gustafsson *et al.*[21,22] have reported that amphiphilic block copolymers provide an exceptional stabilization capability for reversed bicontinuous cubic phase dispersions, most probably through steric stabilization. Poloxamer 407 (from BASF, also denoted as Pluronics F-127) is by far the most studied block copolymer in this context. The dispersed bicontinuous cubic particles are called 'cubosomes' and the dispersed reversed hexagonal particles 'hexosomes'.[21,22] Using the method of Gustafsson *et al.*,[22] it has recently been demonstrated[24] that it is indeed possible to disperse W/O microemulsions, cubic phases and hexagonal phases without losing the structural features of the bulk phase inside the particles after dispersion.

We have shown[71] that the use of cryo-TEM in combination with fast Fourier transform analysis and tilting experiments is an effective alternative to small-angle X-ray scattering for obtaining information on the crystallographic structure and space group of particles having reversed bicontinuous cubic and hexagonal structures. A major advantage of cryo-TEM is the possibility to analyse single particles. This allows identification of particles present at very low concentrations and the coexistence of particles having different internal self-assembly structures. Using this technique one can visualize a multitude of different self-assembly particles as indicated in Figure 3. Further details on structural and thermodynamic properties of emulsified mesophases can be found elsewhere.[72]

2.7 Addition of Guest Molecules to Mesophases

The addition of guest molecules to monoglyceride mesophases can significantly influence the phase behaviour of the binary monoglyceride + water system. After solubilization of a certain 'critical' amount of additive, a phase transition is induced. Each additive affects the interfacial curvature of the lipid bilayer in a slightly different way. Lipophilic additives such as oleic acid,[73,74] triglycerides[22,75] or *n*-tetradecane[24] induce a transition from a reversed bicontinuous cubic phase to a reversed hexagonal phase, resulting in a more negative mean interfacial curvature. More hydrophilic additives such as diglycerol monooleate[76,77] induce the transition from reversed bicontinuous cubic to lamellar, and the mean interfacial curvature changes towards zero. However, addition of hydrophilic molecules, such as polysaccharides, can also induce a transition towards more

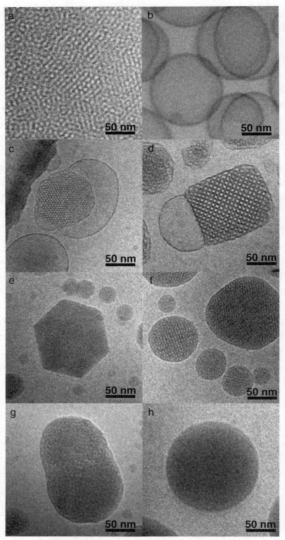

Figure 3 *Self-assembled structures formed by food amphiphiles: (a) micelles (2% polysorbate in water); (b) vesicles (Dimodan U, sodium stearoyl lactylate, water); (c) cubosomes Pn3m (Dimodan U and Pluronic F127 in water); (d) cubosomes Im3m (Dimodan U and protein);[71] (e) hexosomes (monolinolein, n-tetradecane, Pluronic F127, water); (f) emulsified micellar cubic Fd3m (monolinolein, higher n-tetradecane content than for image (e), Pluronic F127, water); (g) emulsified L2 (Dimodan U and soya oil in water); and (h) emulsion droplet (soya oil, Pluronic F127). (Images (c) and (e) reprinted with permission from refs. 20 and 24.)*
(Copyright Nestec Ltd (2006))

negatively curved phases. Mezzenga et al.[78] showed that the presence of monosaccharides, oligosaccharides or polysaccharides in the bicontinuous cubic mesophase results in a lowering of the cubic-to-hexagonal transition temperature. The authors state[78] that, in the presence of hydrophilic compounds with a strong hydrogen bond acceptor, the number of water molecules hydrating the monoglyceride polar head is reduced. Thus, provided that the polysaccharide is small enough to fit within the water channels of the cubic phase, the change in the critical packing parameter is proportional to the number of hydrophilic groups (i.e., the carbohydrate concentration) dissolved in the aqueous phase.

2.8 Phytosterol Solubilization into Mesophases

Phytosterols reduce the absorption of cholesterol and lower the serum concentrations of cholesterol and low-density lipoproteins (LDL). Their effectiveness is highly dependent on the molecular structure and their physical state. Enrichment of foods with natural, non-esterified (free) phytosterols is difficult, since plant sterols are insoluble in water and are also poorly soluble in oils. Lipid solubility can be enhanced by esterification of phytosterols with fatty acids. The incorporation of plant stanol esters into margarine is one of the first examples of a functional food with proven cholesterol-lowering effectiveness.[79,80] One mechanism suggested for the cholesterol-lowering effect of plant stanols/sterols is a reduction in intestinal cholesterol absorption, a process in which changes in micellar composition are thought to play a major role. Another mechanism is the process in which plant stanols/sterols actively influence cellular cholesterol metabolism within intestinal enterocytes.[81,82]

The food matrix has a strong influence on lowering the LDL cholesterol level via plant sterols, with milk and yoghurt being more effective than bread and cereals.[83] Microcrystalline-free plant sterols have been shown[84] to be effective for both serum and LDL cholesterol lowering. In the non-esterified form, the presence of lecithin- or monoglyceride-solubilized stanols incorporated in O/W emulsions massively reduces cholesterol absorption.[4,5] The amphiphilic character of non-esterified phytosterols facilitates their solubilization into mesophases (Figure 4), thereby considerably increasing the absorption capacity in the digestive tract.[5]

2.9 Colloid Science and Nutrition

We have reviewed some of the self-assembly processes of polar lipids involved in the digestion processes of dietary lipids, as well as the multitude of self-assembled structures that can be engineered into colloidal particles. Similar self-assembled colloidal structures that occur during digestion are currently being exploited to increase bioavailability of poorly water- and oil-soluble bioactives such as non-esterified plant esters. However, the development of self-assembly delivery systems is still mostly being done independently of the transformation processes occurring during digestion. Emerging colloidal

Figure 4 *Solubilization of free phytosterols in (a) an emulsion and (b) an emulsified L2 phase. Polarized light microscopy clearly indicates plant sterol crystals in the emulsion after a storage time of 4 weeks at 4°C, whereas the emulsified L2 phase does not indicate any traces of crystalline phytosterols*
(Copyright Nestec Ltd (2006))

concepts have to establish a solid link, on the one hand, between the biological mechanisms of perception and digestion, and, on the other, the engineering of food structures that can deliver the desired sensorial and nutritional impacts.

Analysing the processes involved in food production, consumption and digestion, it becomes evident that the essential properties of a food critically depend on phenomena or processes taking place over various spatial and temporal ranges. The length-scales involved are molecular, supra-molecular, colloidal and macroscopic, and the time-scales range from nanoseconds (enzyme reactions) to years (shelf-life). Key physico-chemical mechanisms, thermodynamic as well as kinetic, have to be identified where the control of events at the molecular and nanoscopic scales is critical for product performance, *e.g.*, the controlled delivery of a bioactive. An interesting colloidal concept is the so-called 'pro-colloid' approach, where the colloid is designed in a way that it delivers its benefits in the context of the involved bio-transformation, *e.g.*, during consumption or digestion. Possible built-in functionalities are time-controlled events (*e.g.*, retarded or sequential release), activation of enzymes or precursors, and inactivation or extraction of intermediates.

At present the design of delivery colloids is a very slow process because we lack efficient *in vitro* models for studying digestion and nutrient liberation. There is a strong need to set up reasonable methodologies for *in vitro* assessment of delivery colloid performance in the context of the food complexity and digestion. Once properly established, this should considerably increase the range of colloidal structures providing sensorial and nutritional benefits. Furthermore, modular concepts using a variety of colloids targeted to specific nutritional situations will then offer solutions for personalized nutrition. The successful realization of these concepts requires a strong interdisciplinary approach, involving physics, chemistry, colloid science, process-engineering, biology and nutrition.

References

1. P.R. Guesry, *Forum Nutr.*, 2005, **57**, 73.
2. M.B. Katan and N.M. de Roos, *Crit. Rev. Food Sci. Nutr.*, 2004, **44**, 369.

3. A. Berger, P.J.H. Jones and S.S. Abumweis, *Lipids Health Dis.*, 2004, **3**, 5.
4. G. Gremaud, E. Dalan, C. Piguet, M. Baumgartner, P. Ballabeni, B. Decarli, M.E. Leser, A. Berger and L.B. Fay, *Eur. J. Nutr.*, 2002 **41**, 54.
5. E.B. Pouteau, I.E. Monnard, C. Piguet-Welsch, M.J. Groux, L. Sagalowicz and A. Berger, *Eur. J. Nutr.*, 2003, **42**, 154.
6. Z.C. Dang and C. Lowik, *Trends Endocrinol. Metab.*, 2005, **16**, 207.
7. G. Banhegyi, *Orv. Hetil.*, 2005, **146**, 1621.
8. H. Nishino, H. Tokuda, Y. Satomi, M. Masuda, Y. Osaka, S. Yogosawa, S. Wada, X.Y. Mou, J. Takayasu, M. Murakoshi, K. Jinnno and M. Yano, *Biofactors*, 2004, **22**, 57.
9. M.P. Wirth and O.W. Hakenberg, *Dtsch. Med. Wochenschr.*, 2005, **130**, 2002.
10. J.B. German and H.J. Watzke, *Comp. Rev. Food Sci. Food Safety*, 2004, **3**(4), 145.
11. J.B. German, C. Yeretzian and H.J. Watzke, *Food Technol.*, 2004, **58**(12), 26.
12. I. Corthesy-Theulaz, J.T. den Dunnen, P. Ferre, J.M. Geurts, M. Muller, N. van Belzen and B. van Ommen, *Ann. Nutr. Metab.*, 2005, **49**, 355.
13. M. Muller and S. Kersten, *Nat. Rev. Genet.*, 2003, **4**, 315.
14. R. Piazza, *Curr. Opin. Colloid Interface Sci.*, 2006, **11**, 30.
15. P. Richmond, *Pure Appl. Chem.*, 1992, **64**, 1751.
16. D.A. Maka and L.K. Murphy, *AACN Clin. Issues*, 2000, **11**, 580.
17. L.E. Schmidt and K. Dalhoff, *Drugs*, 2002, **62**, 1481.
18. B.N. Singh, *Clin. Pharmacokinet.*, 1999, **37**, 213.
19. D. Fleisher, C. Li, Y. Zhou, L.H. Pao and A. Karim, *Clin. Pharmacokinet.*, 1999, **36**, 233.
20. L. de Campo, A. Yaghmur, L. Sagalowicz, M.E. Leser, H. Watzke and O. Glatter, *Langmuir*, 2004, **20**, 5254.
21. J. Gustafsson, H. Ljusberg-Wahren, M. Almgren and K. Larsson, *Langmuir*, 1996, **12**, 4611.
22. J. Gustafsson, H. Ljusberg-Wahren, M. Almgren and K. Larsson, *Langmuir*, 1997, **13**, 6964.
23. K. Larsson, *J. Dispersion Sci. Technol.*, 1999, **20**, 27.
24. A. Yaghmur, L. de Campo, L. Sagalowicz, M.E. Leser and O. Glatter, *Langmuir*, 2005, **21**, 569.
25. A. Yaghmur, L. de Campo, S. Salentinig, L. Sagalowicz, M.E. Leser and O. Glatter, *Langmuir*, 2006, **22**, 517.
26. M.C. Carey, D.M. Small and C.M. Bliss, *Ann. Rev. Physiol.*, 1983, **45**, 651.
27. H.I. Friedman and B. Nylund, *Am. J. Clin. Nutr.*, 1980, **33**, 1108.
28. J.E. Staggers, O. Hernell, R.J. Stafford and M.C. Carey, *Biochemistry*, 1990, **29**, 2028.
29. O. Hernell, J.E. Stagger and M.C. Carey, *Biochemistry*, 1990, **29**, 2041.
30. W.N. Charman, *J. Pharm. Sci.*, 2000, **89**, 967.
31. T.H. Liao, P. Hamosh and M. Hamosh, *Pediatr. Res.*, 1984, **18**, 402.
32. A.C. Frazer, J.H. Schulman and H.C. Stewart, *J. Physiol.*, 1964, **103**, 306.

33. J.M. Linthorst, S. Bennett-Clark and P.R. Holt, *J. Colloid Interface Sci.*, 1977, **60**, 1.
34. G.F. Dasher, *Science*, 1952, **116**, 660.
35. M. Armand, B. Pasquier, M. Andre, P. Borel, M. Senft, J. Peyrot, J. Salducci, H. Portugal, V. Jaussan and D. Lairon, *Am. J. Clin. Nutr.*, 1999, **70**, 1096.
36. M. Armand, P. Borel, B. Pasquier, C. Dubois, M. Senft, M. Andre, J. Peyrot, J. Salducci and D. Lairon, *Am. J. Physiol.*, 1996, **271**(G1), 72.
37. P. Borel, M. Armand, B. Pasquier, M. Senft, G. Dutot, C. Melin, H. Lafont and D. Lairon, *J. Parenter. Enteral Nutr.*, 1994, **18**, 534.
38. A.C. Frazer, J.H. Schulman and H.C. Stewart, *J. Physiol.*, 1944, **103**, 306.
39. M. Charles, H. Sari, B. Entressangles and P. Desmuelle, *Biochem. Biophys. Res. Commun.*, 1975, **65**, 740.
40. S. Labourdenne, O. Brass, M. Ivanova, A. Cagna and R. Verger, *Biochemistry*, 1997, **36**, 3423.
41. L.L.M. van Deenen and G.H. DeHass, *Adv. Lipid Res.*, 1964, **2**, 162.
42. B. Borgström, G. Lundh and A. Hofmann, *Gastroenterology*, 1963, **45**, 229.
43. B. Borgström, G. Lundh and A. Hofmann, *Gastroenterology*, 1968, **54**, S781.
44. A.F. Hofmann and B. Borgström, *J. Clin. Invest.*, 1964, **43**, 247.
45. A.F. Hofmann and B. Borgström, *Fed. Proc.*, 1962, **21**, 43.
46. A.F. Hofmann and B. Borgström, *Biochim. Biophys. Acta*, 1963, **70**, 317.
47. J.S. Patton and M.C. Carey, *Science*, 1979, **204**, 145.
48. J.S. Patton, R.D. Vetter, M. Hamosh, B. Borgström, M. Lindström and M.C. Carey, *Food Microstruct.*, 1985, **4**, 29.
49. P.R. Holt, B.M. Fairchild and J. Weiss, *Lipids*, 1986, **21**, 444.
50. M.W. Rigler, R.E. Honkanen and J.S. Patton, *J. Lipid Res.*, 1986, **27**, 836.
51. G.A. Kossena, B.J. Boyd, C.J. Porter and W.N. Charman, *J. Pharm. Sci.*, 2003, **92**, 634.
52. P. Schurtenberger, N. Mazer, S. Waldvogel and W. Kanzig, *Biochim. Biophys. Acta*, 1984, **775**, 111.
53. P. Schurtenberger, N.A. Mazer and W. Kanzig, *Hepatology*, 1984, **4**, 143S.
54. P. Schurtenberger, M. Svard, E. Wehrli and B. Lindman, *Biochim. Biophys. Acta*, 1986, **882**, 465.
55. M. Lindström, H. Ljusberg-Wahren, K. Larsson and B. Borgström, *Lipids*, 1981, **16**, 749.
56. O. Hernell and L. Blackberg, *J. Pediatr.*, 1994, **125**, S56.
57. G.A. Kossena, W.N. Charman, B.J. Boyd, D.E. Dunstan and C.J. Porter, *J. Pharm. Sci.*, 2004, **93**, 332.
58. J. Borné, T. Nylander and A. Khan, *J. Phys. Chem. B*, 2002, **106**, 10492.
59. J. Borné, T. Nylander and A. Khan, *Langmuir*, 2002, **18**, 8972.
60. J. Borné, T. Nylander and A. Khan, *J. Colloid Interface Sci.*, 2001, **257**, 310.
61. J. Borné, T. Nylander and A. Khan, *Langmuir*, 2001, **17**, 7742.
62. F.A. Wilson, *Am. J. Physiol.*, 1981, **241**, G83.

63. I. Behrens, A.I. Pena, M.J. Alonso and T. Kissel, *Pharm. Res.*, 2002, **19**, 1185.
64. G.M. Hodges, E.A. Carr, R.A. Hazzard and K.E. Carr, *Dig. Dis. Sci.*, 1995, **40**, 967.
65. E. Sanders and C.T. Ashworth, *Exp. Cell Res.*, 1961, **22**, 137.
66. N.J. Krog, in *Food Emulsions*, 3rd edn, S.E. Friberg and K. Larson (eds), Marcel Dekker, New York, 1997, 141.
67. V. Luzzati, in *Biological Membranes*, Vol. 1, D. Chapman (ed), Academic Press, New York, 1968, p. 71.
68. H. Qiu and M. Caffrey, *Biomaterials*, 2000, **21**, 223.
69. L. Sagalowicz, M.E. Leser, H.J. Watzke and M. Michel, *Trends Food Sci. Technol.*, 2006, **12**, 204.
70. T. Landh, *J. Phys. Chem.*, 1994, **98**, 8453.
71. L. Sagalowicz, M. Michel, M. Adrian, P. Frossard, M. Rouvet, H.J. Watzke, A. Yaghmur, L. de Campo, O. Glatter and M.E. Leser, *J. Microscopy*, 2006, **221**, 110.
72. S. Guillot, A. Yaghmur, L. de Campo, S. Salentinig, L. Sagalowicz, M.E. Leser, M. Michel, H.J. Watzke and O. Glatter, this volume, p. 69.
73. F. Caboi, G.S. Amico, P. Pitzalis, M. Monduzzi, T. Nylander and K. Larsson, *Chem. Phys. Lipids*, 2001, **109**, 47.
74. M. Nakano, T. Teshigawara, A. Sugita, W. Leesajakul, A. Taniguchi, T. Kamo, H. Matsuoka and T. Handa, *Langmuir*, 2002, **18**, 9283.
75. I. Amar-Yuli and N. Garti, *Colloids Surf. B*, 2005, **43**, 72.
76. P. Pitzalis, M. Monduzzi, N. Krog, K. Larson, H. Ljusberg-Wahren and T. Nylander, *Langmuir*, 2000, **16**, 6358.
77. A. Yaghmur, L. de Campo, L. Sagalowicz, M.E. Leser and O. Glatter, submitted for publication.
78. R. Mezzenga, M. Grigorov, Z. Zhang, C. Servais, L. Sagalowicz, A.I. Romoscanu, V. Khanna and C. Meyer, *Langmuir*, 2005, **21**, 6165.
79. H. Gylling and T.A. Miettinen, *Ann. Clin. Biochem.*, 2005, **42**, 254.
80. G.R. Thompson and S.M. Grundy, *Am. J. Cardiol.*, 2005, **96**, 3D.
81. J. Plat and R.P. Mensink, *Am. J. Cardiol.*, 2005, **96**, 15D.
82. K. von Bergmann, T. Sudhop and D. Lutjohann, *Am. J. Cardiol.*, 2005, **96**, 10D.
83. P.M. Clifton, M. Noakes, D. Sullivan, N. Erichsen, D. Ross, G. Annison, A. Fassoulakis, M. Cehun and P. Nestel, *Eur. J. Clin. Nutr.*, 2004, **58**, 503.
84. L.I. Christiansen, P.L. Lahteenmaki, M.R. Mannelin, T.E. Seppanen-Laakso, R.V. Hiltunen and J.K. Yliruusi, *Eur. J. Nutr.*, 2001, **40**, 66.

Chapter 3
Structure of Self-Assembled Globular Proteins

Taco Nicolai

POLYMÈRES, COLLOÏDES, INTERFACES, UMR CNRS, UNIVERSITÉ DU MAINE, 72085 LE MANS CEDEX 9, FRANCE

3.1 Introduction

Large structures can be created by assembling many smaller elements. The way we do this is determined by the properties that are sought after in the structure. So, for example, we build a bridge in such a way that it has the property of spanning a river. There is feedback from the overall structure to the assembling process, as the latter is adapted to obtain the desired structure. Similarly, biomolecular structures formed by an evolutionary process, such as globular proteins, also result from this feedback mechanism, because assembling processes that do not lead to successful structures are less frequently reproduced.

The elementary units – polymers, colloids, proteins, and droplets – may themselves spontaneously assemble to form larger structures that can be very complex. This self-assembly is caused by the interaction between the particles without feedback from the large-scale structure, which therefore may be considered epiphenomenal. The large-scale structure is entirely determined by interactions on small length-scales. Sometimes, the structures formed by self-assembly have useful properties that can be exploited to create new materials. If we are able to predict the large-scale structure on the basis of the local interaction, we can try to modify the latter to obtain desired properties for the former.

Unfortunately, it is generally far from straightforward to predict the large-scale structure even if we have detailed knowledge of the particles and their interaction. A system of hard Brownian spheres that stick irreversibly when they collide (diffusion-controlled aggregation) represents one of the simplest self-assembly processes; nevertheless, it leads to very complex structures,[1] as illustrated in Figure 1. These structures can only be understood by invoking the concepts of self-similarity and fractal scaling. The number m of particles in the aggregates increases with the radius of gyration R_g of the aggregates following a

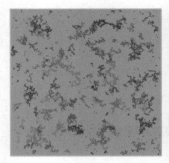 

Figure 1 *Computer simulation of irreversible aggregation of hard spheres (2 vol%) to form aggregates (left image) and a gel (right image). (Reproduced from ref. 1.)*

power law:

$$m = aR_g^{d_f}. \qquad (1)$$

The quantity d_f is the so-called fractal dimension(ality), and it can have any value between 1 and 3. If particles assemble into compact structures, we get $d_f = 3$; and if they assemble to form a rod, then we get $d_f = 1$. The pre-factor a is larger if the elementary unit of the self-similar structure is denser. The elementary unit is, of course, larger than the individual self-assembling particles. While the value of d_f cannot be predicted theoretically, it has been determined experimentally[2] and in computer simulations.[3,4] Irreversible diffusion-limited aggregation gives $d_f = 1.8$ as long as the aggregates are, on average, far away from each other.

It can be easily seen that the density ρ of aggregates with fractal dimension smaller than 3 decreases with increasing radius of gyration: $\rho \propto m/R^3 \propto R^{d_f-3}$. The implication is that, as the assembly process progresses, the average distance between the aggregates decreases until they finally fill up the whole space. The process by which close-packed aggregates join together is different from assembly in the dilute state, and so it results in the formation of aggregates characterized by a larger fractal dimension on longer length-scales, *i.e.*, $d_f = 2.5$.[5] This process is called as percolation and it leads to gelation. There is thus a transition between two types of assembly process, each giving rise to self-similar structures, but with different fractal dimensions. The complete process has only recently been elucidated in detail using computer simulation.[1] (Here we have ignored the possibility that the aggregates sediment when they become large, in which case a precipitate is formed instead of a gel.)

Random aggregation does not lead to monodisperse clusters, but to a distribution of sizes. The polydispersity may be characterized by the ratio of the weight-average molar mass M_w, and the number-average molar mass M_n. Dilute aggregation leads to a relatively narrow size distribution ($M_w/M_n = 2$), if the aggregation is diffusion-controlled. But, if it is reaction-controlled, the distribution is broad, and M_w/M_n increases with increasing M_w. The percolation process gives rise to such a broad distribution that the apparent fractal dimension determined by scattering techniques is lower, *i.e.*, d_f (apparent) ≈ 2.0.[5]

If the binding between the particles is reversible, the aggregates and gels that are formed are only transient. Because the assembly is reversible, it will yield configurations that minimize the free energy of the system. This can lead to phase separation between a dense phase of close-packed particles and a dilute phase of small transient aggregates. It is possible that during the phase separation a transient gel is formed by the aggregating particles.[6,7] Although phase separation will finally lead to two macroscopic phases, this may be a very slow process, and so the transient gel may be long-lived. When both reversible and irreversible binding occur together, transient structures formed by reversible binding may become permanently frozen-in by the formation of irreversible bonds.

Many other factors can modify the assembly process. Particles may feel both attractive and repulsive interactions or may contain only a limited number of specific binding sites. Furthermore, different kinds of particles may be involved, interacting differently with their own kind and with other kinds. The interaction itself may be time-dependent; *e.g.*, reversible bonds may become permanent as the assembly process progresses. Computer simulations are probably the only way to understand in detail how each of these factors influences the large-scale structure.[8] In real systems, various factors often play a role simultaneously and it is not always evident which one is predominant.

Self-assembly is frequently used to prepare food systems. For example, cheese and yoghurt are the result of self-assembly of casein; and sauces and puddings may be thickened by the self-assembly of polysaccharides or gelatin into a system spanning network. In all these cases the self-assembly is induced by a change in the medium which changes the interaction between the particles/polymers. This paper considers the structures formed by self-assembled globular proteins. Gelled egg-white is an obvious example of such a food system. Self-assembly of globular proteins can be induced in different ways, *e.g.*, by adding solvents or denaturants, heating, or applying pressure. The aggregation leads to gel formation if the concentration is sufficiently high.[9] Aggregation and gelation can also be induced by changing the pH or adding salt. In the latter case one often uses small aggregates formed by a pre-treatment of the native globular protein. This process is generally called as 'cold gelation' in the literature.[10] The structure of globular protein aggregates and gels has been investigated using the complementary techniques of microscopy and scattering.

Globular proteins have different sizes and shapes, and different secondary structures. Depending on the pH they have different charge densities and charge distributions. The types of interaction involved in the assembling include covalent bonding through the formation of disulfide bridges, charge interactions, hydrophobic interactions, and hydrogen bonding.[10] It would therefore be naïve to imagine that one can use the simple model of attractive hard spheres to explain the self-assembly of all globular proteins. Nevertheless, remarkably similar structures are formed by different globular proteins. The differences in the structure for a given protein under different conditions of pH and ionic strength are often larger than between different proteins under the same conditions. This observation justifies discussing the self-assembly of globular proteins in general terms.

3.2 Structure of Dilute Aggregates

Individual aggregates can be visualized using microscopy and hence their degree of connectivity (branching) can be appreciated. However, the established techniques of transmission electron microscopy (TEM) and scanning electron microscopy (SEM) necessitate treatments of the sample that can potentially perturb the structure. Cryo-TEM does not need such treatment, but for this technique very thin liquid films are formed on a grid, and the resulting confinement, shear stress, and interfacial effects could potentially modify the structure. Light microscopy and confocal laser scanning microscopy (CLSM) are less perturbing, but only features on length-scales larger than about a micrometre can be resolved. Atomic force microscopy (AFM) has also been used; it has very high resolution, but the aggregates need to be adsorbed on a solid surface, which can again potentially modify the structure. A disadvantage of microscopy in general is that a 2-D slice or projection of a 3-D system is obtained, although SEM does give some impression of the 3-D structure and CLSM could in principle yield the full 3-D structure.

Very different geometries have been observed using microscopy ranging from rigid rods to densely branched clusters. Figure 2 shows some examples of different kinds of β-lactoglobulin (β-LG) aggregates. Rigid rods are formed under certain conditions when the protein is highly charged and the ionic strength is low.[11-14] Nevertheless, the rigidity of these aggregates cannot be explained solely on the basis of electrostatic repulsion because the screening length of the electrostatic interaction is not much larger than the diameter of individual proteins. The rods have diameters of a few nanometres, *i.e.*, close to that of the individual protein molecules; and they may become micrometres long. Parallels have been drawn between these rod-like aggregates and the amyloid fibrils that are responsible for a range of diseases.[15] While the mechanism of formation is not yet elucidated, it has been speculated that it occurs by a nucleation and growth process, on the basis of the observation that a few long rods may be formed in the presence of a large majority of unaggregated protein.[14,16]

Under other conditions, where repulsive electrostatic interactions are still important, flexible linear aggregates are formed. In these cases the aggregation

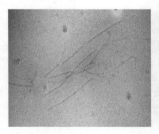 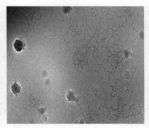 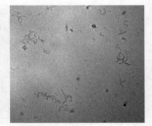

Figure 2 *Cryo-TEM images of β-LG aggregates formed under different conditions: (a) pH = 2, no added salt; (b) pH = 2, 0.1 M NaCl; and (c) pH = 7, 0.1 M NaCl. The total width of each image is 1 μm. (Taken from ref. 17.)*

appears more progressive since small aggregates are formed initially which grow in time.[14,18–20] When the electrostatic repulsion is weaker, either because the pH is closer to the isoelectric point or because salt is added, the aggregates appear more densely branched.[21,22]

Scattering measurements can be used to determine the average structure of aggregates over a range of length-scales between that of the size of an individual protein molecule up ~ 1 µm. Scattering techniques have the advantage that the samples do not need to be specially treated and that average properties of the 3-D structure are obtained. However, it is difficult to extract information on the variability of the structures and on the connectivity, e.g., the degree of branching. In addition, in order to eliminate the effect of interactions on the results, the aggregates need to be highly dilute. Fortunately, most globular protein aggregates are stable and can be diluted without modifying their structure. The properties that can be derived using scattering techniques are illustrated in Figure 3, where I/KC is plotted as function of the scattering wave vector q for a dilute solution of idealized globular protein aggregates. The quantity I is the excess scattering intensity over the solvent, C the concentration, and K an optical constant.[23,24]

The aggregates have a weight-average molar mass defined by

$$M_w = \int MC(M)\,\mathrm{d}M/C, \qquad (2)$$

where $C(M)$ is the weight concentration of aggregates with molar mass between M and $M + \mathrm{d}M$, and C the total concentration. They have a z-average radius of gyration defined by

$$R_{gz} = \left[\int MC(M)R_g^2\,\mathrm{d}M / \int MC(M)\,\mathrm{d}M\right]^{0.5}. \qquad (3)$$

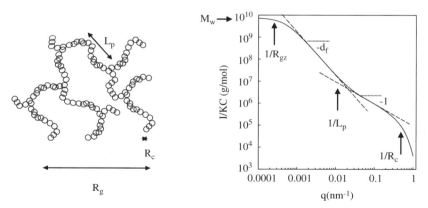

Figure 3 *Idealized structure of a globular protein aggregate (left) and the corresponding q-dependent scattering intensity I (right) indicating the various parameters that can be derived, as described in the text*

Four domains can be distinguished in the plot of I/KC versus q:

(i) At the smallest q-values, I is independent of q, and we have $I(0)/KC = M_w$.
(ii) For $q \approx R_{gz}^{-1}$, the q-dependence crosses over to a power-law behaviour, $I/KC = bq^{-d_f}$ characterizing the self-similar structure. The pre-factor b is larger if the local structure of the aggregates is denser, and it is directly related to the pre-factor a of Equation (1).[25] The value of R_{gz} can be determined from the initial q-dependence of I:

$$I = I(0)/[1 + (qR_{gz})^{2/3}]. \qquad (4)$$

(iii) For $q \approx L_p^{-1}$, the q-dependence crosses over to $I/KC = cq^{-1}$, because a characteristic structure of the rods is probed on length-scales below the persistence length (L_p). The parameter L_p characterizes either the average length-scale over which the aggregates do not bend or the average distance between branch points. The pre-factor c is determined by the molar mass per unit length of the rods (M_1), i.e., $c = \pi M_1$.
(iv) At $q > R_c^{-1}$, the q-dependence becomes steeper again because the internal structure of the rods (and branch points) is probed on length-scales smaller than their radius (R_c).

In order to cover all the relevant length-scales, light scattering has to be combined with neutron or X-ray scattering. As mentioned earlier, it is difficult to obtain information about the degree of branching using scattering techniques. One cannot distinguish between increased flexibility of linear aggregates and an increase in the branching density. Both lead to a decrease of L_p. Neither can one easily distinguish between a change in the structure of the rods and a change in the number of branch points; both will influence b and R_c.

Full characterization of aggregate structure by combined light scattering and X-ray scattering has so far been done only for ovalbumin (OA) and β-LG at pH = 7,[21,23,24] and for β-LG at pH = 2.[14] In each case the amount of added NaCl was varied. Light-scattering measurements have also been carried out for BSA aggregates formed by heating at pH = 7.[26] Extensive heating at different protein concentrations leads to the formation of clusters with values of M_w and R_{gz} that increase with increasing C. A gel is formed above a critical protein concentration C_g. The rate of aggregation increases strongly with increasing temperature, but the structure of the aggregates is independent of the heating temperature.[27] The temperature dependence of the growth rate, and thus the gel time t_g, is characterized by a large activation energy of ~ 300 kJ mol^{-1}, suggesting that it is determined by the protein denaturation step.

Figure 4 shows the q-dependence of I in the range covered by light scattering for β-LG aggregates formed at different concentrations in the presence of 100 mM NaCl at pH = 7.[28] The largest aggregates show a power-law behaviour over the whole q-range, and from the slope we can calculate $d_f = 2.0$. If the structure of the aggregates is self-similar, then the structure factor $S(q) = I(q)/I(0)$ is a

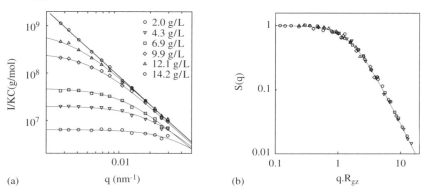

Figure 4 *The q-dependence of the light-scattering intensity of highly diluted β-LG aggregates formed after prolonged heating in aqueous solution (pH = 7, 0.1 M NaCl):[28] (a) I/KC versus q at different concentrations; and (b) same data after normalization of I/KC by M_w and q by R_{gz}*

universal function of qR_{gz} (Figure 4). This was found to be the case for aggregates formed (at pH = 7 for different ionic strengths) for OA,[29] β-LG,[28] or BSA.[26] The structure factor for all these systems is very similar at large length-scales and the inferred fractal dimension is close to 2, except at low ionic strength where it is slightly smaller ($d_f \approx 1.7$).

The structure on smaller length-scales has been investigated using small-angle X-ray scattering (SAXS). Figure 5 shows the q-dependence of I for large aggregates of OA and β-LG formed at different ionic strengths over a wide q-range combining light scattering and X-ray scattering.[21] The structure factor of aggregates formed at low ionic strength shows a cross-over from the fractal structure at low q to a rod-like structure for $q > L_p^{-1}$, so that both the persistence length and the molar mass per unit length can be calculated. The value of L_p decreases with increasing concentration of NaCl for both these proteins, indicating increased flexibility and/or branching density. Consequently, the elementary unit of the fractal structure at large length-scales is denser at higher ionic strength, and so the pre-factor b is larger. The q-dependence at large q can be used to calculate the radius of gyration of the cross-section of the rod-like structure (R_c). At higher ionic strength it is no longer possible to detect a rod-like local structure for β-LG clusters.

Although the values of the fractal dimensions of OA and β-LG aggregates are close, and the general features of the S(q) data are similar, the local structure is quite different. For OA, the behaviour at large q-values can be well described by a homogeneous rod-like structure for all the NaCl concentrations tested. The diameter (7 nm) is only slightly larger than that of monomeric OA (6 nm). For β-LG, the local structure is more complex. The diameter of the rod-like structure at low ionic strength (10 nm) is significantly larger than that of monomeric β-LG (4 nm). Furthermore, already at 100 mM NaCl a rod-like structure can no longer be detected and the local structure becomes denser with increasing ionic strength. It appears that β-LG clusters are

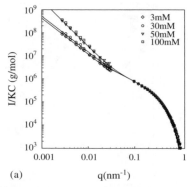 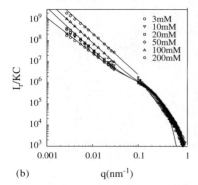

Figure 5 *The q-dependence of the intensities from light scattering (low q) and X-ray scattering (high q) of highly diluted very large aggregates formed by (a) OA and (b) β-LG close to the gel point at pH = 7 after prolonged heating at different NaCl concentrations.[21] The solid lines represent fits to the structure factor of semi-flexible chains*

formed in two steps at pH = 7: in the first step relatively small pre-aggregates are formed, which in a second step assemble into large fractal clusters.[30,31]

The increase in the density of the aggregates with increasing ionic strength has been confirmed by analysing the dependence of M_w on R_{gz}.[32] The results obtained for β-LG are plotted in Figure 6. The slopes are equal to the fractal dimension [see Equation (1)], while the pre-factor depends on the density of the elementary unit of the fractal structure. We have drawn solid lines through the data with slopes that are consistent with the structure factor of the large aggregates, i.e., $d_f = 2$ for NaCl concentrations > 30 mM and $d_f = 1.7$ in the absence of NaCl and at 30 mM. The lines are compatible with the data for large aggregates. However, a clear deviation can be found for the data obtained at 400 mM NaCl, for which M_w has a stronger dependence on R_{gz}. This can be interpreted by assuming either that the fractal dimension is larger at this ionic strength or that the elementary unit is larger so that the fractal structure is only revealed for large aggregates. In any case it is obvious that the density of the elementary unit increases with increasing ionic strength, especially for NaCl concentrations above 100 mM. Hagiwara et al.[26] observed that BSA aggregates formed close to pI were denser than at pH = 7, but the fractal dimension was close to 2 in both cases. Recently we have studied the structure of clusters formed at different values of pH. We have also found[33] that the fractal dimension remains close to 2, but the density increases when the pH approaches the isoelectric point.

The structure factors obtained for β-LG at pH = 2 indicate the formation of long rod-like structures at low ionic strength.[14] At 100 and 200 mM NaCl, the results are closer to those obtained for OA at pH = 7 than for β-LG at pH = 7. The aggregates globally have a self-similar structure and are locally rod-like with a diameter close to that of monomeric β-LG. The persistence length of the rod-like structure decreases with increasing ionic strength, but the diameter

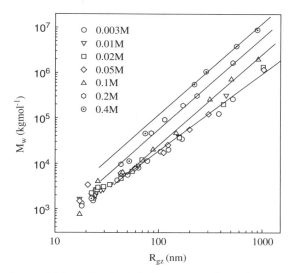

Figure 6 *Dependence of M_w on R_{gz} for β-LG aggregates formed at pH = 7 and different NaCl concentrations. (Taken from ref. 32.)*

remains the same. The difference between OA at pH = 7 and β-LG at pH = 2, on the one hand, and β-LG at pH = 7, on the other, is that in the former case the proteins form chains of monomers or dimers, while in latter they first form small pre-aggregates, which themselves assemble into more-or-less branched self-similar structures.

Before discussing the structure of interacting aggregates and gels, we mention a related light-scattering study on large β-LG clusters formed at room temperature.[34] In this work, small β-LG aggregates formed at pH = 7 in the absence of added salt were self-assembled by reducing the pH closer to the isoelectric point or by adding $CaCl_2$. Here large self-similar aggregates were also observed with fractal dimensions close to 2.

3.3 Structure of Interacting Aggregates and Gels

The scattering intensity of protein solutions increases during heating because both the numbers and the sizes of the aggregates grow. The rate of the initial increase can be used to estimate the aggregation rate.[35–37] However, the interaction between the aggregates influences the structure factor to an extent that depends on the protein concentration and the strength of the interactions. This is already evident for native proteins that show a peak in $S(q)$ at low ionic strength caused by correlations between the positions of the proteins.[38–42] The peak position q_p indicates the preferred distance between the proteins, and it increases approximately with the cube root of the concentration. The height of the peak increases with increasing protein concentration because the degree of order increases. The repulsion between the proteins decreases when salt is added, or the pH is adjusted closer to pI, and as a consequence the amplitude of the peak decreases.

When a protein solution is heated, the contribution of non-aggregated protein decreases and that of the aggregates increases, until finally all protein is transformed into aggregates – or, for $C > C_g$, into a gel. The value of C_g falls from approximately 100 g L^{-1} in the absence of NaCl and far from pI, to as low as 10 g L^{-1} at high ionic strength.[29,32,43,44] In the latter case a precipitate may be formed at even lower protein concentrations. An intriguing issue that has not yet been resolved is why below a critical concentration the aggregates stop growing. Apparently the aggregates cease to bind to each other in spite of the fact that in some cases they can be highly interpenetrated. This effect cannot be attributed to a dynamic equilibrium of bond formation and break-up, because generally the aggregates are stable to dilution. The stability of the aggregates is clearly related to electrostatic repulsion, because increasing the ionic strength, or reducing the pH towards pI, induces further assembly of the clusters even at room temperature.[10]

At low ionic strength, and away from pI, the heated systems remain transparent because the aggregate suspension and the gels are highly ordered and still produce a peak in the static structure factor.[38,42,45,46] This is illustrated in Figure 7, where the change of the structure factor is shown for β-LG heated at pH = 7 in the absence of added salt and at 100 mM NaCl.[45] Similar data were obtained for OA.[45] Initially at 0 mM one observes the interaction peak for native β-LG, and after extended heating one observes the interaction peak for the gel. The upturn at the lowest q-values is an artefact caused by the scattering cell. Surprisingly an iso-scattering point is observed during the gelation process, which suggests separation between the aggregates and the unaggregated protein during the process of self-assembly on the small length-scales covered by SAXS.

For the heated systems, the value of q_p increases with the square root of the protein concentration, because the aggregates are locally rod-like.[42] The peak height decreases with decreasing protein concentration and increasing salt concentration because there is a reduction in the degree of order in the structure of the gels. For the same reason, the apparent molar mass M_a calculated from $I(0)$ increases with decreasing protein concentration and increasing salt concentration. Above ∼30 mM NaCl, a scattering peak is no longer observed and the intensity at small q-values increases, as shown in Figure 7.

The ordered structure of concentrated aggregates of OA at low ionic strength has been visualized by cryo-TEM as indicated in Figure 8.[42] The typical distance between adjacent aggregate strands is close to the $2\pi/q_p$ obtained from the SAXS measurements on the same system. Gels cannot be investigated using cryo-TEM, and a highly perturbing treatment is necessary to observe them with TEM or SEM. This is probably the reason why the TEM micrographs of transparent gels do not clearly show the order that is implied by the scattering experiments. However, as shown in Figure 8, TEM does indicate cross-linked strands with diameters less than 10 nm. For this reason these ordered transparent gels are often called as fine-stranded gels.

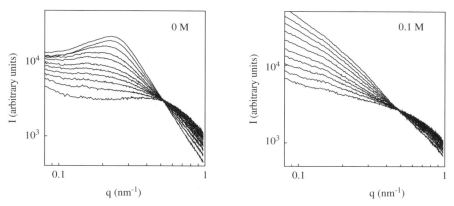

Figure 7 *The q-dependence of the X-ray scattering intensity during heating of β-LG solutions (C = 100 g L^{-1}, pH = 7) at two NaCl concentrations (0 and 0.1 M).[45] The intensity increases at low q and decreases at high q. A well-defined iso-scattering point is observed at all NaCl concentrations*

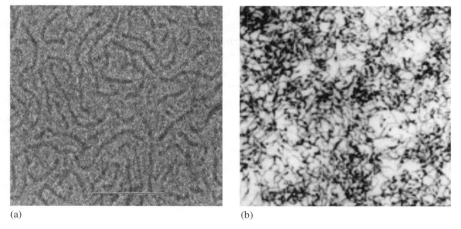

(a) (b)

Figure 8 *Electron micrographs of aggregates and gels formed at pH = 7 in the absence of added salt: (a) cryo-TEM image of concentrated OA aggregates;[42] and (b) TEM image of BSA gel.[16] The total width of each image is 0.7 μm*

The characteristic length-scale for gels that show a scattering peak is less than a few tens of nanometres. Therefore it can only be observed using small-angle neutron or X-ray scattering. When the degree of order decreases, the scattering peak disappears and the intensity increases with increasing q in the range covered by light scattering. The scattering intensity becomes important at smaller q-values, which means that more visible light is scattered at wide angles. Consequently less ordered gels are more turbid. The order can be reduced by adding more salt to the solutions or by setting the pH closer to the isoelectric point, which for many globular proteins used in food systems is close to pH=5, with the notable exception of lysozyme. The transition between transparent gels and very opaque gels occurs over a relatively narrow range of salt concentrations or pH. For instance, β-LG gels formed at pH = 7 are transparent if <50 mM NaCl is added, and are very opaque if 200 mM NaCl is added. In the absence of salt, the gels are transparent at pH = 6.2 and opaque at pH = 5.8. The transition depends mainly on the combination of ionic strength and pH, and it occurs approximately in the same range for gels formed by other globular proteins.[47]

The strong increase in turbidity is caused by the loss of the order due to the weakening of the repulsive interactions between the aggregates. The structure factor of more heterogeneous gels cannot be determined by standard light-scattering equipment because the contribution of multiple scattering, which gives the gels the turbid aspect, can no longer be neglected. Utilizing cross-correlation, dynamic light-scattering allows one to correct for the effect of multiple scattering and thus to obtain the true structure factor.[48,49] This technique has recently been used to obtain the structure factors of solutions and gels formed by heating β-LG,[28,50] OA,[19,29] or BSA[43] at pH = 7. Figure 9 shows examples of the q-dependence of the intensity for heated β-LG solutions at different protein concentrations. For heterogeneous gels we observe a power-law q-dependence indicating that the gels have a self-similar structure over a range of length-scales. The fractal dimension was found to be close to 2, *i.e.*, the same as that for diluted aggregates formed under the same conditions.

The scattering data for the gels were analysed using the same method as described above for the dilute aggregate solutions. Instead of M_w and R_{gz} one obtains an apparent molar mass M_a and apparent radius of gyration R_a. The

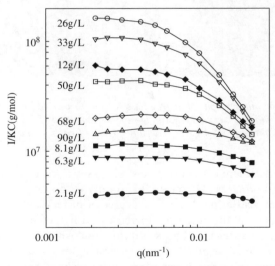

Figure 9 *The q-dependence of the light-scattering intensity of extensively heated β-LG solutions (pH = 7, 100 mM NaCl) at different protein concentrations.[28] Open points represent aggregated solutions; filled points represent gels ($C > C_g$)*

value of R_a is close to the correlation length of the concentration fluctuations and M_a is proportional to the osmotic compressibility.[25] The quantity R_a characterizes the length-scale above which the system is homogeneous. A simplified picture of the system is to consider it as an ensemble of close-packed 'blobs' with z-average radius of gyration R_a and weight-average molar mass M_a. This picture is helpful, but it should not be taken too literally. The centre of a blob can be chosen randomly in the system – so there is no distinction between bonds inside a blob and those between blobs. In addition, this representation fails completely to describe the ordered systems.

The quantities R_a and M_a are related in the same way as R_{gz} and M_w,[50] which implies that the blobs have the same structure as the aggregates. The concentration dependence of R_a for β-LG at 100 mM NaCl is compared with that of R_{gz} for diluted aggregates in Figure 10. As mentioned above, for $C < C_g$, aggregates are formed that increase in size with increasing concentration, which explains why R_a increases with increasing C. The aggregates are progressively more strongly interpenetrated with increasing C, so that R_a is smaller than R_{gz}. At some concentration close to C_g, the value of R_a reaches a maximum and subsequently decreases with increasing C, showing that denser gels are more homogeneous. Similar observations were made for OA[29] and BSA,[43] except that for very high concentrations R_a was found to increase again, especially for OA. Since for the latter system no densification was observed in the high q-range covered by SAXS, it was speculated that a very small fraction of the proteins formed large dense domains. Very recent results obtained for β-LG at lower pH also show a strong increase in R_a at high concentrations.[33]

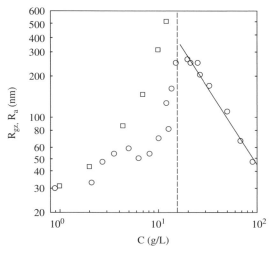

Figure 10 *Concentration dependence of apparent radius R_a for extensively heated β-LG solutions (○).[28] For comparison, the values of the radius of gyration (R_{gz}) of the aggregates (□) are shown for $C < C_g$. The solid line has slope of -1, as predicted for monodisperse blobs with $d_f = 2$ (see text). The dashed vertical line indicates the critical protein concentration C_g*

For a given protein concentration both R_a and M_a increase with increasing salt concentration. The range of salt concentrations that can be investigated with cross-correlation DLS is rather limited, however, because the gels become rapidly too turbid. The range can be extended using turbidimetry. The turbidity τ is directly related to the structure factor $S(q)$ by

$$\tau = K'CM_a \int_0^{2\pi} \int_0^{\pi} S(q)(1+\cos^2\theta)\sin\theta \, d\theta d\varphi, \quad (5)$$

where K' is an optical constant.[28] If $S(q)$ is known, M_a and R_a can be calculated from the wavelength dependence of τ. It turns out that the structure factor of globular protein gels in the q-range relevant for the turbidity can be well described by Equation (4) (replacing R_{gz} with R_a), so that an analytical expression for τ can be derived.[28]

We have recently carried out a systematic study of the turbidity for β-LG gels as a function of pH and ionic strength.[33] The values of M_a and R_a were derived from the wavelength dependence using Equations (4) and (5). Figure 11(a) shows M_a of β-LG gels formed at pH = 7 with $C = 100$ g L^{-1} as a function of the concentration of added NaCl. Small values of M_a characteristic of transparent samples were obtained using light-scattering measurements. The value of M_a was found to increase exponentially with the ionic strength between 0 and 200 mM. At higher salt concentrations, the transmission became less than a few percent even if we used path lengths of 1 mm, and M_a could no longer be determined accurately. (The upper limit of the turbidity that could be determined accurately was ~ 20 cm^{-1}.) The dependence of M_a on pH plotted in Figure 11(b) shows a strong increase of M_a with decreasing pH starting at pH = 6. We find that M_a increases slightly more strongly than R_a^2, which is to be expected because,

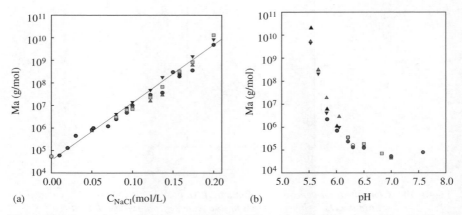

Figure 11 *Dependence of apparent molar mass M_a on (a) NaCl concentration at pH = 7 and (b) pH in absence of added salt, for gels formed by extensive heating of β-LG solutions at $C = 100$ g L^{-1}. (Taken from ref. 33.)*

although the fractal dimension remains at $d_f = 2$, the pre-factor a increases with increasing ionic strength and decreasing pH. The most turbid gels that could be studied had an average correlation length of ~ 1 μm.

Assuming that the system can be modelled as an ensemble of close-packed monodisperse blobs of size R_a and molar mass M_a, then the following equation relates the protein concentration to R_a and M_a:

$$C \approx 3M_a/(4N_A\pi R_a^3). \tag{6}$$

Utilizing the relationship between M_a and R_a, it follows that we have $R_a \propto a^{1.5} C^{-1}$ and $M_a \propto a^3 C^{-2}$ for $d_f = 2$. This result is compatible with the concentration dependence of M_a and R_a for β-LG at 100 mM NaCl.[28] However, the strong increase of M_a at a given protein concentration with increasing ionic strength cannot be explained by the weak increase of a. We conclude that the gels become increasingly heterogeneous; i.e., the concentration fluctuations are characterized by a range of correlation lengths. In terms of the blob picture, the gels consist of an ensemble of blobs with a size distribution that increases with decreasing electrostatic repulsion. The values of R_a and M_a determined by scattering techniques represent the z-average radius and the weight-average molar mass of the blobs, and therefore the largest blobs have a strong weight. If the size distribution of the concentration fluctuations varies, then R_a and M_a are no longer simply related to the protein concentration through the fractal dimension. Similar observations have been made for gels formed by small clay particles (Laponite) at different NaCl concentrations:[51] at a fixed concentration, the correlation length strongly increased with increasing ionic strength, while the local structure remained the same.

When the length-scale of the heterogeneity approaches 1 μm, it can be observed using CSLM,[29,43,50,52] as illustrated in Figure 12. The gels appear to consist of aggregated micrometre-sized particles. For this reason turbid globular protein gels are often called as particle gels or particulate gels. Typically, TEM and (particularly) SEM show a network of partly fused and roughly spherical particles with diameters of the order of 1 μm[16,53–55] (Figure 13).

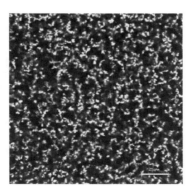

Figure 12 *CSLM image of a β-LG gel formed at $C = 50$ g L^{-1}, pH = 7, and 0.2 M NaCl.[50] The total width of the image is 160 μm*

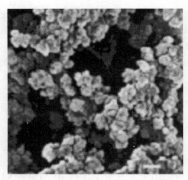

Figure 13 *SEM image of a β-LG gel formed at $C = 40$ g L^{-1}, $pH = 7$, and 0.6 M NaCl.[55] The total width of the image is 15 μm*

However, TEM has also shown[56] that the micrometre-size particles are clusters of smaller units. If the gels are truly made up of homogeneous domains with well-defined interfaces, one should observe a very steep decrease of the scattering intensity, i.e., $I \propto q^{-4}$.[23] This q-dependence has so far been observed in only one case for BSA gels formed at the isoelectric point using neutron scattering.[57] However, BSA aggregates formed under the same conditions were found to show a self-similar structure with $d_f = 1.8$–1.9 at the larger length-scales probed by light scattering.[26]

The formation of homogenous micrometre-size protein particles is not compatible with the scattering results discussed above. It is clear that, on the basis of the results shown in Figure 11, one cannot make an unambiguous distinction between fine-stranded gels and particulate gels in the pH and salt concentration range covered. Perhaps homogeneous micrometre-size particles are formed under conditions of higher salt concentrations, or closer to pI, but it is also possible that the structure is partly modified by the treatment (fixation, dehydration, embedding, and staining) that is necessary for TEM or SEM. A more systematic comparison of electron microscopy and scattering results for the same globular protein gels is needed to resolve this issue.

The principal origin of the strong, but gradual, increase of the turbidity with increasing ionic strength or pH is the gradually increased heterogeneity with relatively little change of the local structure. A possible origin for the increased heterogeneity is that the repulsive interaction between the aggregates decreases. This would lead to increased concentration fluctuations that are subsequently frozen in by gelation. One might even imagine that this process leads to micro-phase separation into well-defined spherical domains.[9,16] This latter mechanism has been clearly established for the gelation of β-LG at pH = 7 in the presence of anionic polysaccharides.

When protein aggregates are mixed at room temperature with polysaccharide, they phase separate into well-defined micrometre-sized domains that have a tendency to agglomerate into larger clusters[58,59] (Figure 14). Micro-phase separation is not observed for native proteins, and it occurs at lower polysaccharide concentrations when the aggregates are larger. Light scattering

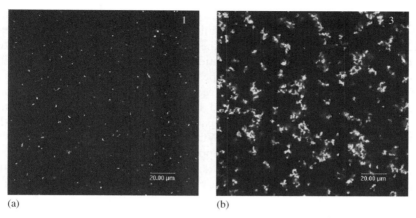

Figure 14 *CLSM images of mixtures of κ-carrageenan (10 g L^{-1}) with β-LG aggregates: (a) 0.2 g L^{-1} β-LG; and (b) 10 g L^{-1} β-LG[58]*

shows $I \propto q^{-4}$, which implies that the micro-domains are homogeneous and have a well-defined interface. Initially, the aggregates in the micro-domains are only loosely bound and can be dispersed by dilution so that the original aggregates are recovered. However, the domains transform into irreversibly bound micro-gels over a period of a day or two. Interestingly, smaller aggregates exit the liquid domains when the polysaccharide phase is gelled,[60] indicating less incompatibility of the aggregates with gelled polysaccharide. The structure of the agglomerated micro-domains formed by mixing pre-heated protein aggregates and polysaccharide at room temperature resembles that of the gels formed by heating mixtures of native protein and polysaccharide.

Confocal microscopy shows that the structure of these mixed gels is similar to that of heterogeneous gels formed by pure globular proteins in the presence of high salt concentrations or close to pI.[56,61,62] However, no micro-phase separation of protein aggregates is observed when salt is added to the solutions or the pH is lowered. Instead, a slow 'cold gelation' process is observed.[10] The structure of large clusters formed by the association of small aggregates at room temperature is similar to that of large clusters formed by heating native proteins. Interestingly, gels at high salt or close to pI are generally more homogeneous when they are formed from clusters made by pre-heating at conditions of low salt and far from pI than when they are formed by directly heating native protein.[55,63,64] This would suggest that micro-phase separation does not occur during the formation of pure globular protein gels.

3.4 Relationship between Structure and Linear Elasticity

One approach to calculate the shear modulus of globular protein gels is based on the hypothesis that the shear modulus is proportional to the concentration of independent elastic chains of the network formed by random aggregation of the globular proteins.[16] This concentration can be calculated using the mean-field approximation. In this approach there is no direct relationship between the structure and the elasticity; the contribution of an elastically active chain to the shear modulus is supposed to be independent of its size and structure.

The only model that involves a direct relationship between structure and mechanical properties is the so-called fractal gel model.[65–67] The system is considered as an ensemble of close-packed blobs connected to each other through a stress-bearing backbone. When all protein is incorporated in the gel, the size of the elastic blobs is approximately equal to the value of R_a determined with scattering techniques. Each blob acts as an elastic spring with a spring constant that depends on the size of the blob and the fractal dimension of the elastic backbone, d_b. If the elastic backbone is fully flexible, the elastic energy per blob in the linear response regime is proportional to the thermal energy, i.e., $E \propto kT$. If the elasticity is dominated by the bending enthalpy of the elastic backbone, then E is inversely proportional to the number of elementary units of the elastic backbone, N_b, which is related to the blob size R_b by $N_b \propto R_b^{d_b}$. The shear modulus is then E multiplied by the number concentration of the blobs (ν):

$$G \propto \nu kT, \quad \text{(entropy)}$$
$$G \propto \nu R_b^{-d_b}. \quad \text{(bending enthalpy)} \tag{7}$$

Here we have assumed that all links between elementary units are the same. Sometimes a distinction is made between links within blobs and those between blobs, but in the case of globular protein gelation this distinction does not apply. In fact, as mentioned above, the blobs are only a conceptual tool to describe the length-scale below which the system has a fractal structure and above which the system is homogeneous.

For globular protein gels, the shear modulus is determined by the enthalpy because G is much larger than expected for purely entropic elasticity.[28] When all the proteins are part of the gel, we have $\nu \propto R_b^{-3}$ and $R_b \propto C^{1/(d_f-3)}$; then G is given by

$$G \propto C^{(3+d_b)/(3-d_f)}. \tag{8}$$

At concentrations close to the gel point, not all the aggregated protein is part of the gel and Equation (8) is no longer valid, because only the gel fraction F_g contributes to the elastic modulus. However, the sol fraction does contribute to the scattering intensity; so R_b is not proportional to R_a unless $F_g = 1$. According to the percolation model,[5] the gel fraction (and thus the shear modulus) reach zero at the gel point concentration C_g or the gel time t_g following the power law $F_g \propto \varepsilon^{1.2}$, where ε can be $(C - C_g)/C_g$ for a given heating time much larger than t_g, or $(t - t_g)/t_g$ for a given protein concentration much larger than C_g. However, this limiting power-law dependence is only valid for $\varepsilon \ll 1$ and it is difficult to observe experimentally.

The model leading to Equation (8) is based on general properties of the gel structure. It does not give the pre-factor of the power-law increase; that depends on the strength of the individual bonds. Consequently, different structures can have similar absolute values of the shear modulus. Also different moduli can be found for systems with the same structure. For instance, simply reducing the temperature of globular protein gels from 80°C to 20°C increases the shear modulus reversibly by more than a factor of 3 without a significant change in the structure.[33,68]

The question is whether this relatively simple model can usefully be applied to globular protein gels. One difficulty is that oscillatory shear measurements show that the gels are not fully elastic.[69] The storage modulus G' has a weak power-law dependence on the frequency, and G'' is almost frequency-independent. Another difficulty is that the shear modulus continues to evolve logarithmically with time after all proteins have aggregated.[33,70] Even if we assume that the G' determined experimentally can be approximated as the elastic modulus of the gel, other conditions still have to be fulfilled before the fractal gel model may be applied. In the first place, the structure needs to be self-similar over a significant length-scale, implying that R_a has to be significantly larger than the elementary unit of the fractal structure. This means that the homogeneous (transparent) globular protein gels cannot be interpreted using the fractal gel model because R_a is close to the elementary unit of the aggregates. In fact, these gels are not self-similar on any length-scale. Secondly, Equation (8) is only valid if all blobs are connected to each other through their elastic backbone. This means that the protein concentration should not be close to C_g, so that the sol fraction is negligible. Thirdly, the gels do have some heterogeneity: they are characterized by a distribution of blob sizes. In the derivation of Equation (8) an average blob size is used, which is only valid if the polydispersity of the blobs is independent of the protein concentration. This is doubtful for very heterogeneous gels formed at high ionic strength or close to pI.

Clearly, one should not simply apply this model to globular protein gels in order to derive the fractal dimension of the blobs. The mere observation that G' increases with C following a power law generally over a limited concentration range is not sufficient proof that the structure of the gels is fractal over any significant length-scale. The concentration dependence of the blob size should be determined independently using scattering techniques. This has so far been done for one system: β-LG at pH = 7 and 0.1 M NaCl.[28] The structure and the mechanical properties of the extensively heated systems were found to be consistent with the fractal gel model for $C > 2C_g$ and $C < 90$ g L^{-1}. In this concentration range, the correlation length was significantly larger than the elementary unit of the gels, and at most less than a micrometre. Similar detailed comparison needs to be made for more homogeneous or heterogeneous gels in order to test whether the fractal gel model can be applied more generally.

3.5 Conclusions

It appears that globular proteins may form rigid rods when they are strongly charged and little salt is present. When salt is added or the charge density is reduced, flexible linear aggregates are formed. With increasing ionic strength and decreasing charge density, the degree of branching and flexibility increases, which enhances the local density of the aggregates. On length-scales larger than a few tens of nanometres, the aggregates have a self-similar structure characterized by a fractal dimension close to 2.

When repulsive charge interactions are important, the interpenetrating aggregates and gels are highly ordered and homogeneous on length-scales larger than a few tens of nanometres. The degree of order decreases with decreasing electrostatic interactions, which leads to an increase of the scattered light intensity. Beyond a certain degree of heterogeneity, the systems become turbid to the naked eye. The increase in the heterogeneity, and therefore in the turbidity, is strong over a narrow range of salt concentrations and pH, but the local arrangement of the proteins varies little. A clear distinction between the so-called fine-stranded gels and particulate gels is not justified – the difference simply represents a gradual transition from highly ordered to very heterogeneous gels.

Acknowledgement

Dominique Durand is thanked for fruitful discussions and a critical reading of the manuscript.

References

1. M. Rottereau, J.C. Gimel, T. Nicolai and D. Durand, *Eur. Phys. J. E*, 2004, **12**, 133.
2. M.Y. Lin, H.M. Lindsay, D.A. Weitz, R. Klein, R.C. Ball and P. Meakin, *J. Phys., Condens. Matter*, 1990, **2**, 3093.
3. M. Kolb, R. Botet and R. Jullien, *Phys. Rev. Lett.*, 1983, **51**, 1123.
4. P. Meakin, *Phys. Rev. Lett.*, 1983, **51**, 1119.
5. D. Stauffer and A. Aharony, *Introduction to Percolation Theory*, Taylor & Francis, London, 1992.
6. J.C. Gimel, T. Nicolai and D. Durand, *Eur. Phys. J. E*, 2001, **5**, 415.
7. N.A.M. Verhaegh, D. Asnaghi and H.N.W. Lekkerkerker, *Physica A*, 1999, **264**, 64.
8. E. Dickinson, *J. Colloid Interface Sci.*, 2000, **225**, 2.
9. A.H. Clark, in *Functional Properties of Food Macromolecules*, 2nd edn, S.E. Hill, D.A. Ledward and J.R. Mitchell (eds), Aspen, Gaithersburg, MD, 1998, p. 77.
10. M.C. Bryant and D.J. McClements, *Trends Food Sci. Technol.*, 1998, **9**, 143.
11. C. Veerman, H. Ruis, L.M.C. Sagis and E. van der Linden, *Biomacromolecules*, 2002, **3**, 869.
12. M. Weijers, L.M.C. Sagis, C. Veerman, B. Sperber and E. van der Linden, *Food Hydrocoll.*, 2002, **16**, 269.

13. W.S. Gosal, A.H. Clark and S.B. Ross-Murphy, *Biomacromolecules*, 2004, **5**, 2408.
14. P. Aymard, T. Nicolai and D. Durand, *Macromolecules*, 1999, **35**, 2542.
15. W.S. Gosal, A.H. Clark, P.D.A. Pudney and S.B. Ross-Murphy, *Langmuir*, 2002, **18**, 7174.
16. A.H. Clark, G.M. Kavanagh and S.B. Ross-Murphy, *Food Hydrocoll.*, 2001, **15**, 383.
17. D. Durand, J.C. Gimel and T. Nicolai, *Physica A*, 2002, **304**, 253.
18. C. Le Bon, T. Nicolai and D. Durand, *Macromolecules*, 1999, **32**, 6120.
19. M. Weijers, T. Nicolai and R.W. Visschers, *Macromolecules*, 2002, **35**, 4753.
20. T. Koseki, T. Fukuda, N. Kitabatake and E. Doi, *Food Hydrocoll.*, 1989, **3**, 135.
21. M. Pouzot, T. Nicolai, R.W. Visschers and M. Weijers, *Food Hydrocoll.*, 2005, **19**, 231.
22. T. Koseki, N. Kitabatake and E. Doi, *Food Hydrocoll.*, 1989, **3**, 123.
23. J.S. Higgins and K.C. Benoit, *Polymers and Neutron Scattering*, Clarendon, Oxford, 1994.
24. W. Brown, *Light Scattering: Principles and Development*, Clarendon, Oxford, 1996.
25. T. Nicolai, D. Durand and J.C. Gimel, in *Light Scattering: Principles and Development*, W. Brown (ed), Clarendon, Oxford, 1996, p. 201.
26. T. Hagiwara, H. Kumagai and K. Nakamura, *Biosci. Biotech. Biochem.*, 1996, **60**, 1757.
27. C. Le Bon, T. Nicolai and D. Durand, *Int. J. Food Sci. Technol.*, 1999, **34**, 451.
28. M. Pouzot, T. Nicolai, D. Durand and L. Benyahia, *Macromolecules*, 2004, **37**, 614.
29. M. Weijers, R.W. Visschers and T. Nicolai, *Macromolecules*, 2004, **37**, 8709.
30. P. Aymard, J.C. Gimel, T. Nicolai and D. Durand, *J. Chim. Phys.*, 1996, **93**, 987.
31. S. Ikeda, *Food Hydrocoll.*, 2003, **17**, 399.
32. K. Baussay, C. Le Bon, T. Nicolai, D. Durand and J. Busnel, *Int. J. Biol. Macromol.*, 2004, **34**, 21.
33. S. Mehalebi, D. Durand and T. Nicolai, unpublished results.
34. R. Vreeker, L.L. Hoekstra, D.C. Den Boer and W.G.M. Agterof, *Food Hydrocoll.*, 1992, **6**, 423.
35. U.M. Elofsson, P. Dejmek and M.A. Paulsson, *Int. Dairy J.*, 1996, **6**, 343.
36. M.A.M. Hoffmann, S.P.F.M. Roefs and M. Verheul, *J. Dairy Res.*, 1996, **63**, 423.
37. M. Verheul, S.P.F.M. Roefs and C.G. de Kruif, *J. Agric. Food Chem.*, 1998, **46**, 896.
38. A.H. Clark and C.D. Tuffnell, *Int. J. Peptide Protein Res.*, 1980, **16**, 339.
39. H. Matsuoka, N. Ise, T. Okubo, S. Kunugi, H. Tomiyama and Y. Yoshikawa, *J. Chem. Phys.*, 1984, **83**, 378.

40. R. Nossal, C.J. Glinka and S.-H. Chen, *Biopolymers*, 1986, **25**, 1157.
41. R. Giordano, A. Grasso, J. Teixeira, F. Wanderlingh and U. Wanderlingh, *Croatica Chem. Acta*, 1992, **65**, 411.
42. M. Weijers, E.H.A. de Hoog, M.A. Cohen Stuart, R.W. Visschers and P.A. Barneveld, *Colloids Surf. A*, 2005, **301**, 270.
43. L. Donato, C. Garnier, J.L. Doublier and T. Nicolai, *Biomacromolecules*, 2005, **6**, 2157.
44. D. Renard and J. Lefebvre, *Int. J. Biol. Macromol.*, 1992, **14**, 287.
45. T. Nicolai, M. Pouzot, D. Durand, M. Weijers and R.W. Visschers, *Europhys. Lett.*, 2006, **73**, 299.
46. D. Renard, M.A.V. Axelos, F. Boué and J. Lefebvre, *Biopolymers*, 1996, **39**, 149.
47. E. Doi and N. Kitabatake, *Food Hydrocoll.*, 1989, **3**, 327.
48. C. Urban and P. Schurtenberger, *J. Colloid Interface Sci.*, 1998, **207**, 150.
49. T. Nicolai, C. Urban and P. Schurtenberger, *J. Colloid Interface Sci.*, 2001, **240**, 419.
50. M. Pouzot, D. Durand and T. Nicolai, *Macromolecules*, 2004, **37**, 614.
51. T. Nicolai and S. Cocard, *Eur. Phys. J. E*, 2001, **5**, 221.
52. M. Verheul and S.P.F.M. Roefs, *J. Agric. Food Chem.*, 1998, **46**, 4909.
53. M. Langton and A.-M. Hermansson, *Food Hydrocoll.*, 1992, **5**, 523.
54. J.I. Boye, M. Kalab, I. Alli and C.Y. Ma, *Lebensm. Wiss. Technol.*, 2000, **33**, 165.
55. M. Vittayanont, J.F. Steffe, S.L. Flegler and D.M. Smit, *J. Agric. Food Chem.*, 2002, **50**, 2987.
56. C. Olsson, M. Langton and A.-M. Hermansson, *Food Hydrocoll.*, 2002 **16**, 111.
57. J. Lefebvre, D. Renard and A.C. Sanchez-Gimeno, *Rheol. Acta*, 1998 **37**, 345.
58. K. Baussay, T. Nicolai and D. Durand, *Biomacromolecules*, 2006, **7**, 304.
59. P. Croguennoc, D. Durand and T. Nicolai, *Langmuir*, 2001, **17**, 4372.
60. K. Baussay, D. Durand, T. Nicolai and *J. Colloid Interface Sci.,* 2006, **304**, 335.
61. P. Croguennoc, D. Durand and T. Nicolai, *Langmuir*, 2001, **17**, 4380.
62. L. Donato, C. Garnier, B. Novales and J.-L. Doublier, *Food Hydrocoll.*, 2005, **19**, 549.
63. E. Doi, *Trends Food Sci. Technol.*, 1993, **4**, 1.
64. D.J. McClements and M.K. Keogh, *J. Sci. Food Agric.*, 1995, **69**, 7.
65. Y. Kantor and I. Webman, *Phys. Rev. Lett.*, 1984, **52**, 1891.
66. M. Mellema, J.H.J. van Opheusden and T. van Vliet, *J. Rheol.*, 2002, **46**, 11.
67. W. Shih, W.Y. Shih, S. Kim, J. Liu and I.A. Aksay, *Phys. Rev. A*, 1990 **42**, 4772.
68. J.M. Aguilera, *Food Technol.*, 1995, **49**(10), 83.
69. M. Pouzot, T. Nicolai and L. Benyahia, *J. Rheol.*, 2004, **48**, 1123.
70. M. Paulsson, P. Dejmek and T. van Vliet, *J. Dairy Sci.*, 1990, **73**, 45.

Chapter 4

Similarities in Self-Assembly of Proteins and Surfactants: An Attempt to Bridge the Gap

Erik van der Linden and Paul Venema

FOOD PHYSICS GROUP, DEPARTMENT OF AGROTECHNOLOGY AND FOOD SCIENCES, WAGENINGEN UNIVERSITY, BOMENWEG 2, 6703 HD WAGENINGEN, THE NETHERLANDS

4.1 Introduction

The area of surfactant self-assembly has already received attention for more than half a century. Considerable progress has been made in connecting the molecular properties to the assembly morphology and the phase behaviour. A multitude of different (and rather exotic) types of mesophases with a large variation in topology can be distinguished. In addition, rheological response and even shear-induced transitions in morphology and phase behaviour have been reported. An important technological area is the assembly of surfactants at an interface.

The area of protein assembly and its associated phase behaviour has also received considerable attention. The main focus for many years has been on describing various properties – solubility, crystallization, interfacial assembly, phase separation and gel formation – in terms of the molecular properties of the protein or the protein mixture. Since assemblies of proteins are abundant in food products, the exploration of their morphology, physico-chemical characteristics and mutual interactions is relevant for the engineering of food materials. But it is only for the past few decades that the morphology of the assemblies has received more attention. This has mainly been explored in terms of fractal aggregates. The type of protein assembly that is usually found in foods (in 3 dimensions) exhibits a morphology that has many branches per unit length. An extreme case is an infinitely stiff aggregate with zero branches per unit length, *i.e.*, a rod (with a fractal dimension of 1). This fibrillar type has been the subject of intensive studies lately in various science areas, including material science, food science and medical sciences (β-amyloid diseases). Apart from the fact that the rod morphology is interesting as an extreme case of zero branches, it also has practical relevance, *e.g.*, because the fibrils act as a vehicle to form extremely low-weight fraction gels. Another example of a specific type of protein aggregate is a hollow capsule (serving as a protection device of many viruses).

In this article, the mesoscopic parameters relevant for describing the aggregate morphology, phase behaviour and topology of surfactant systems will first be addressed. Then, a similar scheme will be presented for the use of relevant mesoscopic parameters in describing the morphology and topology of protein aggregates, for fractal dimensions from 1 to 3. The influences of pH, salt concentration, salt type and temperature will be considered. Non-equilibrium effects regarding the particular assembly of proteins into fibrils will also be addressed. Finally, the similarities and differences regarding surfactant *versus* protein assembly will be discussed.

4.2 Surfactant Assembly

Assembly models of surfactants in solution that correctly describe existing experimental observations are based on a thermodynamic equilibrium description. The fluctuations in sizes of aggregates, the exchange rate of surfactants between molecules in solution and in the aggregates, as well as the rate of aggregate formation, are all assumed to be fast in comparison with the experimental time-scales. We follow here the scheme set out by Israelachvili.[1,2]

The first step in the scheme of any model of self-assembly is the distinction in solution between monomers, dimers, trimers, *etc.*, which are all in equilibrium with one another. Here we denote an aggregate containing a number of n monomers as an aggregate of size n – or, in short, an n-mer. The second step is to assume chemical equilibrium and hence to equate the chemical potential of the surfactant as monomer with the chemical potential of the surfactant within an n-mer. This leads to the following expression for all aggregates of size N or M (> 1):[1]

$$\mu_1 = \mu_1^0 + kT \ln X_1 = \mu_M = \mu_M^0 + \left(\frac{1}{M}\right)kT \ln\left(\frac{X_M}{M}\right)$$
$$= \mu_N = \mu_N^0 + \left(\frac{1}{N}\right)kT \ln\left(\frac{X_N}{N}\right). \tag{1}$$

In Equation (1), μ_1 denotes the chemical potential of the surfactants as monomer (*i.e.*, the surfactant in solution), M and N denote different sizes of aggregate, X_N is the mole fraction of surfactants that are present in aggregates of size $n = N$, μ_N the chemical potential of a surfactant within an aggregate of size N, and μ_N^0 the self-free energy of the surfactant within an aggregate of size N. As we assume a dilute solution here, the use of the term $kT \ln X_N$ is justified, and the factorization of $1/N$ ensures that the contribution of each surfactant within an aggregate of size N is only counted once. We note that mass balance yields $X_1 + X_2 + \cdots + X_N = X_{total}$, the total mole fraction of surfactant in solution. Equation (1) then yields[1]

$$X_N = N\{X_1 \exp[\mu_1^0 - \mu_N^0/kT]\}^N. \qquad (2)$$

Because the total mole fraction cannot exceed unity, there appears automatically an upper boundary for X_1, also known as a critical aggregation concentration, or critical micelle concentration, X_c, given by

$$X_c = \exp[-(\mu_1^0 - \mu_N^0/kT)]. \qquad (3)$$

We have not assumed anything about the type of surfactant, nor the morphology of the aggregates. So, for that matter, the above would be equally valid for proteins.

To advance further with the description set out above, one needs to include molecular information about the surfactant in order to calculate the interaction potential between surfactants within an aggregate of size n, and subsequently to arrive at an expression for the chemical potential of a surfactant in that aggregate of size N, i.e., μ_N^0. Such an expression should, in principle, also contain information on the shape of the aggregate, because aggregate shape also influences the distances between the hydrophilic and hydrophobic parts of the surfactants and thus their chemical potential. One should realise at this point the following general complicating factor. If one wants to arrive at a size distribution of aggregates of a particular surfactant on the basis of an equilibrium approach as above, one must include the shape of the aggregate in order to calculate the chemical potential. However, it is this shape itself that also should, in principle, be part of the minimization scheme of the free energy of the overall system. In fact, one should minimize the free energy on the basis of both the size distribution and the morphology of the system simultaneously. Setting aside for a moment this complication in variational calculus, one may alternatively set an *a priori* shape first, and put forward hydrophilic and hydrophobic interactions within such an aggregate, and hence determine the average size as a function of experimental parameters.

Israelachvili et al.[2] have argued that, despite the intrinsic difficulties of incorporating all the hydrophilic interaction contributions, the contribution from the repulsive interaction to the chemical potential per surfactant, within an aggregate of size N, can be assumed to adopt a simple form:

$$\mu_{N\ hydrophilic}^0 \propto De^2/\varepsilon a \qquad (4)$$

Equation (4) is based on modelling the energy contribution as the energy of a capacitor with a charge per unit area of e/a, and a separation D of the planes (resulting from the double-layer of charge with thickness D), where ε is the relative dielectric constant of the medium around the surfactant head-group. Indeed, Tanford[3] has shown that this $1/a$ dependence explains satisfactorily the micellar size and the critical micelle concentration.

Including hydrophobic interactions lead to an overall chemical potential of the form[1]

$$\mu_N^0 = \mu_N^0 \text{ hydrophobic} + \mu_N^0 \text{ hydrophilic} = \gamma a + c/a + g \qquad (5)$$

where γ denotes the surface tension of, and a is the area between, hydrophobic groups and water (a constant), and where the term g denotes constant surface area and/or bulk contributions. Combining Equations (4) and (5) yields an optimum area, $a_0 = (c/\gamma)^{1/2}$, for which the free energy is a minimum.[2] This relates to a specific number of surfactants per aggregate. Having a larger number of surfactants per aggregate yields a smaller area per surfactant, and *vice versa*. Both cases imply a higher free energy, leading to a size distribution of micelles.

The optimal area a_0 can be used to provide a practical criterion for the shape of the micelle *versus* the sizes of the head-group and tail of the surfactant,[2] assuming that the chemical potential is not influenced too much by the change of shape, *i.e.*, assuming that a_0 is not much affected by the choice of aggregate shape. Considering a spherical micelle of radius R, and denoting the volume of the hydrocarbon tail as v, and the surface area between hydrocarbon and water as a_0, we have $R = 3v/a_0$. Realising that there is an upper limit to the extension of the hydrocarbon chain given by the maximum tail length l_c, a spherical micelle will only be formed for $3v/a_0 < l_c$, leading to the condition for spherical micelle formation as $v/a_0 l_c < 1/3$.[2] Similarly, we have for cylinders the condition $v/a_0 l_c < 1/2$, and for planar objects $v/a_0 l_c < 1$.

We may investigate further, now assuming a certain shape, the effect of the number of surfactants on the chemical potential. Suppose one has an energy of binding between monomers within an aggregate of magnitude α. In the case of a rod-like structure, the endpoints are unbound, leading to[2]

$$N\mu_N^0 = -N\alpha kT + \alpha kT = -(N-1)\alpha kT \qquad (6)$$

or

$$\mu_N^0 = \mu_\infty^0 + \frac{\alpha kT}{N}, \qquad (7)$$

where μ_∞^0 is the energy of the surfactant within an infinite aggregate. Similarly, taking always into account the number of endpoints of the aggregate, we may derive[2] for planar aggregates:

$$\mu_N^0 = \mu_\infty^0 + \frac{\alpha kT}{N^{1/2}}. \qquad (8)$$

And for spherical aggregates we get

$$\mu_N^0 = \mu_\infty^0 + \frac{\alpha kT}{N^{1/3}}. \qquad (9)$$

Now that we know how the chemical potential depends on aggregate size, we may answer how the size depends on the value of α. Using Equations (7)–(9) we may rewrite Equation (2) as[2]

$$X_N = N\{X_1 \exp\alpha\}^N \exp(-\alpha) \qquad (10)$$

for rod-like structures, while for planar and spherical structures we have

$$X_N = N\{X_1 \exp\alpha\}^N \exp(-\alpha N^p), \qquad (11)$$

with p equal to 1/2 or 2/3 for discs and spheres, respectively. We can see from Equation (11) that, once α is of the order of unity, X_N becomes negligible for larger N. Indeed, a separate phase is then formed consisting of infinitely large aggregates. This is the case for a constant α, and where μ_N^0 is a constantly decreasing function of N, i.e., the chemical potential is smaller the larger the aggregate, so favouring infinite aggregates. It is only when μ_N^0 reaches a minimum value for finite N that we end up with finite aggregates – and more of them as the concentration of surfactants increases. This minimum in the chemical potential can be addressed by Equation (5), from which can be deduced an optimum head-group area of the surfactant and consequently an optimal number N per aggregate.

Once the concentration becomes high enough to induce interaction between aggregates, and as such to influence the chemical potential of the surfactants again, shape transitions may take place. This occurs, e.g., in the case of SDS exhibiting a spherical to rod transition at a certain concentration. The success of the above model can be attributed to an accurate enough description of the shape of the surfactant and the contributions to the chemical potential of the surfactant in various circumstances. Once α is large, but with $p = 1$ [i.e., referring to Equation (10)], X_N does not decrease rapidly to zero, and large aggregates of finite size may occur provided that the structure is rod-like.

4.3 Spherical Protein Aggregates

A class of spherical protein aggregates that are in equilibrium with their monomeric constituents, and so resemble micelles, are empty virus core shells (also called as capsids).[4-6] Rather detailed experiments have been undertaken by Ceres and Zlotnick,[7] and these have been analysed in terms of the theory of micellization by Kegel and van der Schoot.[4] This analysis successfully explains the influence of salt and temperature on the micellization of the capsids.

The expression for the chemical potential is based on the consideration that the hydrophobic patches account for a negative contribution to the chemical potential, while the electrostatic charge distribution accounts for a positive contribution. The number of units building up the spherical assembly is set as a constant. Kegel and van der Schoot[4] arrive at an energy of binding for the assembly which we rewrite as

$$N\mu_N^0 = -N\gamma a_h + N a_c k T \sigma^2 \lambda_B \kappa^{-1} + \text{constant} \qquad (12)$$

Here γ is the surface tension between the hydrophobic site of area a_h which is involved in self-assembly, a_c the area of interaction per protein involved in the Coulombic interaction, σ denotes the charge per surface area, λ_B the Bjerrum length and κ^{-1} the electrostatic Debye screening length. In fact, the electrostatic contribution, the second term in Equation (12), is based on similar considerations that led to Equation (5). The end effect that is taken into account in Equation (5) on the basis of the surface area of hydrophobic molecular parts that are exposed to water is omitted in Equation (12). It is assumed that the main contribution to the chemical potential due to hydrophobicity arises from the efficient shielding of hydrophobic sites, giving rise to a negative term.

The expression above can also explain effects of pH as observed with a closely related protein. The expression for σ originates from the Henderson–Hasselbalch equation[8] according to

$$\sigma = \frac{\sigma_b}{1 + 10^{pH-pK_b}} - \frac{\sigma_a}{1 + 10^{pK_a-pH}} \qquad (13)$$

where pK_b denotes an effective value for the basic groups of the protein sequence all taken together, and analogously for pK_a.

4.4 Protein Assemblies of Arbitrary Morphology: A First Attempt

From the surfactant point of view, we may determine a packing criterion originating from minimizing the sum of the two surface contributions to the chemical potential. The surfactant can be divided into hydrophobic and hydrophilic parts, which are clearly separated. The bulk contribution to the chemical potential due to the replacement of water molecules around hydrocarbon chains by its own hydrocarbon chains is the driving force for assembly. The packing criterion and the consequent optimal number of surfactants in a micelle is determined by the minimization of the surface contribution at finite N. The first step in testing the model of surfactant assembly is to test the size of spherical micelles as a function of experimental parameters like salt concentration and temperature.

By analogy with the surfactant approach, the assembly of proteins into empty virus core shells has been modelled as a case of spherical assembly, as described in Section 3. We propose, as a first ansatz, a simple extension to Equation (12) in order to account for non-spherical and/or branched structures, and at the same time to take into account endpoint defects (*i.e.*, the endpoints do not fully contribute to the chemical potential, compared to when an aggregate would be infinite, in the same manner as done in Equation (7) for aggregates of dimension 1). Taking into account an arbitrary shape, and/or a degree of branching, we may denote the dimensionality by its fractal dimension D, defined by

$$\phi = (R/r)^D. \qquad (14)$$

In Equation (14), ϕ is the volume fraction of material within the aggregate, r the size of the constituents building up the aggregate and R the size of the aggregate. In the case of surfactants, the dimensionality of a planar aggregate is 2, although there exists in this case a particular arrangement due to the anisotropic nature of the surfactant. In the case of proteins, there is usually less structural anisotropy present per molecule, which means that one expects arrangements that are more isotropic. In principle, D may range from 1 to 3, with $D = 1$ resembling rods, and $D = 3$ resembling a close-packed spherical assembly. The extent to which an aggregate exhibits linearity – and thus the less it exhibits branching – can be expressed in terms of the closeness to unity of the value of D. The number of endpoints, defined as the outermost points of a fractal aggregate, is

$$n_{\text{endpoints}} = DN^{(D-1)/D}. \tag{15}$$

So, taking into account also the endpoint contribution, we find

$$\mu_N^0 = (-\alpha + \beta)kT + \frac{(\alpha - \beta)kTD}{N^{1/D}}, \tag{16}$$

with $\alpha = \gamma a_h/kT$ and $\beta = a_c\sigma^2 \lambda_B \kappa^{-1}$; the term $(\alpha - \beta)kT$ denotes the bonding energy per particle. Again one finds that, if $(\alpha - \beta)$ is of the order of 1 or larger, there is phase separation for $N > 1$. If the bonding energy is higher than kT, only finite aggregates are formed for $D = 1$, i.e., rod-like structures.

The question remaining is whether we may expect equilibrium structures with a fractal dimension larger than 1. The idea behind surfactant assembly is that, also for $D > 1$, there can be finite aggregates once the bonding energy is of the order kT, since there is usually an optimum in the amount of surface area of the surfactant exposed to the water phase. This optimum in the case of surfactants is caused by two surface energy contributions per particle, which together can be minimized, yielding the optimal surface area per surfactant within the aggregate. Subsequently, it is possible to deduce minimal packing constraints for the surfactants, thus defining the preferred shape of the aggregate. We propose to follow an analogous procedure now for proteins.

We assume a number n_p of hydrophobic patches with equal surface area a_p; a patch may be 1–30 nm^2 in size.[9] We assume a number n_b of bonds per protein within the aggregate, i.e., neglecting the endpoints for the moment. The exposed area of the protein surface to the surrounding solution is then $a_{\text{exp}} = a - n_b a_p$, where a denotes the total surface area of the protein. There exist two contributions to the chemical potential arising from the exposed area of the protein within the aggregate to water. Their sum may be approximated by

$$\Delta\mu \approx \gamma(n_p a_p - a) + \gamma a_{\text{exp}} + 4\pi q^2 \lambda_B \kappa^{-1} \frac{kT}{a_{\text{exp}}} \tag{17}$$

with q being the number of unit charges. Minimization of Equation (17) with respect to a_{exp} leads to a minimal exposed surface area a_{exp}^0 given by

$$a_{\text{exp}}^0 = |q| \, (4\pi\lambda_B \kappa^{-1} kT/\gamma)^{1/2} \tag{18}$$

or

$$a^0_{\exp}/a = |\sigma|\,(4\pi\lambda_B\kappa^{-1}kT/\gamma)^{1/2}, \tag{19}$$

where we have introduced the surface charge density σ. In terms of the minimum number of neighbours, n_b^0, we find

$$n_b^0 = \frac{a - a^0_{\exp}}{a_p} = \frac{a}{a_p}\left[1 - |\sigma|(4\pi\lambda_B\kappa^{-1}kT/\gamma)^{1/2}\right]. \tag{20}$$

We assume $\lambda_B = 0.7$ nm, values of κ^{-1} in the range 0.25–1.75 nm (the experimentally existing window for a fibrillar β-lactoglobulin assembly at pH = 2), $kT = 4 \times 10^{-21}$ J, $\gamma = 50$ mN m^{-1} and $\sigma = 4.2 \times 10^{17}$ (using 21 unit charges at pH = 2, and a radius of 4 nm of the protein). We then arrive at a value for n_b^0 ranging between 0.4 a/a_p and 0.8 a/a_p for κ^{-1} ranging between 0.25 and 1.75 nm. Experimentally it is known that fibrils are being formed in the above-mentioned regime with almost no branches, i.e., $n_b = 2$, implying that a/a_p ranges between 2.5 and 5, i.e., 20–40% hydrophobic surface area per protein, a seemingly reasonable value. From Equation (13) it is clear that the charge density is also a strong function of pH, which in turn sharply affects the value of n_b^0, and thus the experimentally accessible n_b as is observed experimentally.[10-14]

It is clear that the above equation needs further refinement; it will be deferred to a future publication. The above represents a first crude estimate to predict equilibrium aggregate forms of proteins in solution. It needs to be emphasized that the description applies to an equilibrium aggregate morphology. This approach may also have merit for describing the behaviour within the so-called crystallization slot,[15] and beyond, as well as in describing the role of different salts via the effect on γ (the Hofmeister series).

One may take Equation (20) further by realizing that n_b^0 should not decrease below 1. If this should nonetheless happen, one may use Equation (20) as a criterion for aggregate formation where the single protein molecule is not the building block, but instead dimers or n-mers form the building blocks of the aggregate, since the presence of n-mers effectively decreases the exposed surface area. Thus, Equation (20) may also be used to identify transitions from single-stranded fibrils to fibrils having more than one strand wound around each another, or to fibrils with two strands having a twist with respect to one another (tubules or other such topologies).

4.5 β-Lactoglobulin Fibrils: Equilibrium Assembly or Not?

An example of fibrillar protein assembly is β-lactoglobulin assembling into fibrils from strands of one or two monomers thick, on heating above 80°C at pH = 2 and low ionic strength.[10-14] The length distribution of the mature fibrils is a polydisperse single-peaked distribution with a smooth linear decrease to zero. The initial reversibility of the fibril forming process becomes irreversible after some time, *i.e.*, the fibrils become stable towards dilution.[11] It is not obvious from our own measurements that there exists a critical concentration below which no fibrils are formed. If such a concentration would at all exist, it has to be substantially lower than 0.5 wt%, indicating a rather high bond energy between the proteins. From electric birefringence measurements, it has been concluded[16] that the proteins are ordered 'head-to-tail' within the fibril in a helical configuration.

The fact that a minimal temperature is needed in order to induce fibril formation is directly related to the fact that, at a certain elevated temperature, the protein will partially unfold. Since we have also observed the formation of fibrils at 4°C, after having applied this (partial) denaturation step, it seems that the elevated temperature is not essential during assembly. However, at the lower temperature, the assembly was found to be much slower, indicating that temperature affects the kinetics of the assembly process. The relation between the fibrillar type of assembly and the partially unfolded state also has been found for other proteins (ovalbumin, hen egg-white lysozyme, bovine serum albumin).[17] In the latter case it was found that, upon partial unfolding, some hydrophobic regions become exposed to the solvent. A requirement for obtaining long fibrils with few branches is maintenance of a low pH and a low ionic strength. A moderate ionic strength yields more flexible fibrils, while an even higher ionic strength and higher pH yields more condense spherical aggregates with a fractal dimension of around 2.[12–14]

The length distribution for β-lactoglobulin fibrils is not a sharply peaked Poisson distribution, in which case one would expect a 'full-width half-height' of around 100 nm. In the case of reversible aggregation, the equilibrium length distribution was predicted[1] using Equation (10), yielding a smooth and single-peaked distribution given by

$$C_L = L \exp(-aL) \qquad (21)$$

with the constant a being proportional to the chemical potential of the protein in the fibrils, and depending on concentration and temperature. The peak of the distribution is located at $1/a$. So, despite the more complex nature of proteins compared to simple surfactants, we do find a distribution with a shape that is at least similar to the theoretical prediction from Equation (21).

It is necessary to take into account two kinetic mechanisms that play a role in reaching the equilibrium distribution of fibrils. Firstly, there is a nucleation event, followed by subsequent growth of the fibril and possible redistribution of proteins between the fibrils. It is a reasonable assumption, when no shear is applied, that every fibril is extended at the same rate independent of its length. With this in mind, the resulting length distribution could never develop if there is an instantaneous nucleation of all the nuclei and if protein redistribution between the fibrils would be absent. Secondly, we also have to consider the influence of the (apparent) long-term irreversibility of the assembly. The fact that the attachment of proteins to the fibril becomes irreversible after some time will have a pronounced influence on the fibril length distribution. In the early stages of the self-assembly, when the typical time that the protein remains attached to the fibril's end is much longer than the time necessary to form a permanent bond, the fibril will grow and will be able to form irreversible bonds (β-sheet structures). The proteins in the fibril are in a 'trapped state' as a result of the high bond energy of the β-sheet. This, together with the high charge density at low pH, and the strong hydrophobic interactions at elevated temperature, may explain that, under these circumstances, one obtains fibrils as opposed to aggregates of fractal dimension much higher than 1.

The irreversibility does, in general, not apply to protein molecules attached to both ends of the fibril. If this were to be the case, all the proteins would eventually be incorporated in the fibrils – something contrary to our observations. At concentrations low enough, the typical time the average protein molecule remains attached to the fibril's end is much shorter than the time necessary to form a permanent bond; it is a two-step mechanism, where the protein first attaches to the fibril and subsequently becomes incorporated in the β-sheet of the fibril. Due to the partially irreversible assembly of most of the proteins in the fibril, only limited rearrangement of protein monomers can occur. Therefore we expect the final length distribution to be somewhat wider than the Poisson distribution, but narrower than predicted by Equation (21).

After a mild denaturation step, at an elevated temperature and pH = 2, the protein partially unfolds leading to an effective attraction between the proteins. As the protein molecules are in a head-to-tail arrangement within the fibril, this suggests that the attraction is of a dipolar nature. This attraction could very well be the sum of the different forces involved, *i.e.*, electrostatic, hydrophobic, *etc.* In order to initiate fibril growth, a nucleation event is necessary. Initially the assembly is reversible, but after some time (typically hours) the fibril is stabilized by a crossed β-sheet structure, running perpendicular to the long axis of the fibril. As a result of this irreversibility, thermodynamic equilibrium theories can only be applied with caution. The system might be kinetically trapped in the later stages of the assembly, and as such cannot reach its equilibrium by rearranging the proteins between different fibrils or even spherulites. It is unknown if the assembly is entropy-driven or enthalpy-driven.

It remains further to be tested whether at short times the above suggested attempt for an equilibrium approach to protein assembly into fibrils is applicable, and whether the equilibrium approach can be extended to regions of higher salt concentration and different pH.

References

1. J. Israelachvili, *Intermolecular and Surface Forces*, 2nd edn, Academic Press, London, 1992.
2. J.N. Israelachvili, D.J. Mitchell and B.W. Ninham, *J. Chem. Soc. Faraday Trans. II*, 1976, **72**, 1525.
3. C. Tanford, *J. Phys. Chem.*, 1974, **78**, 2469.

4. W.K. Kegel and P. van der Schoot, *Biophys. J.*, 2004, **86**, 3905.
5. R.F. Bruinsma, W.M. Gelbart, D. Reguera, J. Rudnick and R. Zandi, *Phys. Rev. Lett.*, 2003, **90**, 248101.
6. D.L.D. Caspar, *Biophys. J.*, 1980, **31**, 103.
7. P. Ceres and A. Zlotnick, *Biochemistry*, 2002, **41**, 11525.
8. L. Stryer, *Biochemistry*, Freeman, San Francisco, CA, 1980.
9. P. Lijnzaad, H.J.C. Berendsen and P. Argos, *Prot. Struct. Funct. Genet.*, 1996, **26**, 192.
10. A.M. Hermansson, O. Harbitz and M. Langton, *J. Sci. Food Agric.*, 1986, **37**, 69.
11. C. Veerman, H. Baptist, L.M.C. Sagis and E. van der Linden, *J. Agric. Food Chem.*, 2003, **51**, 3880.
12. P. Aymard, T. Nicolai and D. Durand, *Int. J. Polym. Anal. Charact.*, 1996, **2**, 115.
13. P. Aymard, T. Nicolai, D. Durand and A. Clark, *Macromolecules*, 1999, **32**, 2542.
14. E.P. Schokker, H. Singh, D.N. Pinder and L.K. Creamer, *Int. Dairy J.*, 2000, **10**, 233.
15. A. George and W.W. Wilson, *Acta Crystallogr. D*, 1994, **50**, 361.
16. S.S. Rogers, P. Venema, J.P.M. van der Ploeg, L.M.C. Sagis, A.M. Donald and E. van der Linden, *Eur. Phys. J. E*, 2005, **18**, 207.
17. F. Tani, M. Murata, T. Higasa, M. Goto, N. Kitabatake and E. Doi, *J. Agric. Food Chem.*, 1995, **43**, 2325.

Chapter 5
Self-Assembled Liquid Particles: How to Modulate their Internal Structure

Samuel Guillot,[1] Anan Yaghmur,[1] Liliana de Campo,[1] Stefan Salentinig,[1] Laurent Sagalowicz,[2] Martin E. Leser,[2] Martin Michel,[2] Heribert J. Watzke[2] and Otto Glatter[1]

[1] INSTITUTE OF CHEMISTRY, UNIVERSITY OF GRAZ, HEINRICHSTRASSE 28, A-8010 GRAZ, AUSTRIA
[2] NESTLÉ RESEARCH CENTER, VERS-CHEZ-LES-BLANC, CH-1000 LAUSANNE 26, SWITZERLAND

5.1 Introduction

In the modern type of food formulation, it is of increasing importance to have the essential components arranged in such a way that their functionality and bioavailability are optimized. This must be achieved and understood on the molecular level. Thus, self-assembly is a central mechanism for efficient structural design.[1] Moreover, for the successful incorporation of different functional molecules, it is also necessary to generate systems with hydrophilic and hydrophobic regions of large interfacial area. Microemulsions are self-assembling and thermodynamically stable systems that fulfil these conditions. The food-grade multicomponent microemulsions, which have been described recently,[2,3] are often applied in high concentrations, and so it is necessary to study the structure and the dynamics at high concentrations. This is possible with modern scattering methods.[4] Nevertheless, equilibrium systems such as micelles or microemulsions have one essential drawback: they may change dramatically, or even disassemble, if they are mixed with other systems or if just some parameters such as temperature or salinity are changed. Hierarchically organized systems can potentially avoid these difficulties.

Food emulsifiers such as monoglycerides or phospholipids are known to form self-assembly structures when added to water. The lamellar phase formed by phospholipids has attracted the most interest thus far due to its easy transformation into vesicles or liposomes. Unsaturated monoglycerides give rise to a series of reverse lyotropic liquid crystalline phases, such as the reverse bicontinuous, hexagonal, and microemulsion phases. However, the dispersibility of such phases is more complicated. In the pioneering work of Larsson and co-workers,[5–7] it was established that monoglycerides (*e.g.*, glycerol monooleate) can be dispersed in water using a dispersing agent such as a block copolymer, whereby sub-micrometre-sized particles with preserved internal self-assembly structures are formed. In analogy with liposomes, these dispersed particles were termed *cubosomes* when their internal structure consists of a cubic phase and *hexosomes* when it is a hexagonal phase (achieved by addition of a third component). These systems are important as membrane mimetic matrices,[8] as vehicles for the solubilization of active ingredients (*e.g.*, vitamins and enzymes),[8,9] and as unique microenvironments for the controlled release of additives (*e.g.*, drug delivery),[10,11] considering the hypothesis that the structural arrangement within the particle significantly determines the delivery properties.

An overview of such systems is provided in this article for monolinolein-based particles. Recently, we reported[12] that the internal structure of these particles is not only self-assembled but is also at equilibrium. The internal structure is in most cases a liquid–crystalline phase of either a cubic or hexagonal structure, but isotropic fluid structures can also exist at higher temperatures.[12] Moreover, we proved the reversible exchange of water from inside to outside the confined internal particle structures during the cooling and heating cycles. This can easily be modulated by either varying the temperature or solubilizing oil. The addition of oil at a constant temperature, or the increasing of temperature at a constant oil content, induces a transformation from cubosomes to hexosomes and dispersed water-in-oil (W/O) microemulsions.

We consider also the effect of variation of the lipid composition on the confined internal structures of the particles. The addition of oil can indeed lead to new systems: the emulsified microemulsions (EMEs). These are in a certain sense similar to double emulsions, *i.e.*, they are water-in-oil-in-water systems. However, they are extremely stable over time. This is caused by the fact that the dispersed phase is a W/O microemulsion that is dispersed in water using a high molecular weight hydrophilic secondary emulsifier.[13] We present evidence that particles having an internal structure of Fd3m (reverse discontinuous micellar cubic phase (MCP)) can also be created on addition of oil. Upon adding the oil, the interface is tuned to be more negative, *i.e.*, curved towards the aqueous phase. Nevertheless, this curvature effect can be tuned back by the addition of diglycerol monooleate (DGMO)[14] or other amphiphilic molecules, which then balances the structural effect of the added oils.

Small angle X-ray scattering (SAXS) and cryogenic transmission electron microscopy (Cryo-TEM) were used to analyse the internal structure of the particles. The results presented here mainly use *n*-tetradecane (TC) as the oil phase. All of the following internal structures are currently available for food-grade applications: cubic bicontinuous, hexagonal, discontinuous micellar cubic (all liquid–crystalline), and fluid isotropic (microemulsion).[15]

5.2 Emulsified Liquid Crystals and Microemulsions

The stabilized particles described here are based on monolinolein (MLO) + water mixtures. We aim to determine the role of temperature on their internal structure; but to understand such systems better, we need to first compare their internal structure to that of the corresponding bulk. So, the main goal of this part is to compare the phase transitions in dispersions that are induced by temperature variations and the behaviour of the nondispersed MLO + water binary system. Therefore, we determined the phase behaviour in terms of temperature and composition for the MLO + water system by investigating the internal structures using SAXS.

The binary phase diagram of MLO + water in Figure 1 comprises a variety of mesophases depending on temperature and water content. Left of the phase separation boundary (the thick line) the samples appear macroscopically as fairly transparent phase (single phase regions), whereas to the right the samples appear completely white (heterogeneous). In the latter case, the system separates into the mesophase and excess water. In this system, the bicontinuous cubic mesophase Pn3m can contain considerably more water than the reverse hexagonal phase H_2 or the L_2 phase formed at higher temperatures. At 20°C the MLO can solubilize up to 33 wt.% water, whereas at 94°C it can solubilize

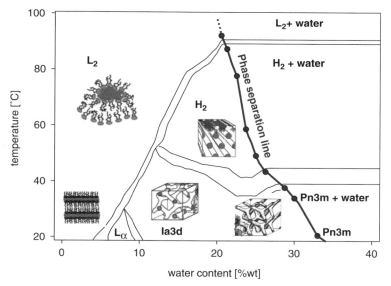

Figure 1 *Temperature-composition phase diagram of MLO + water determined by SAXS in the heating direction. The lines are guides for the eye to see the phase transitions and the regions of coexistence between the mesophases. The thick line corresponds to the phase separation boundary between the single-phase regions (left) and the excess water region (right)*
(Schematic representations reprinted with permission from Refs. 17,18.)

only 20.3 wt.% water. As soon as the water solubilization capacity is exceeded, *i.e.*, when the composition reaches the excess water region, the lattice parameters for the distinct symmetries coincide within experimental error:[12] above this point any additional water is then expelled from the mesophase (excess water). At 58°C, the inverse hexagonal phase can solubilize almost 23.8 wt.% water, as compared with just 21.3 wt.% at 87°C. Consequently, the lattice parameter of this mesophase (in the excess water) decreases from 5.54 to 5.10 nm. Thus, there exists a temperature-induced shrinking in the samples in excess water. This is important, since the most interesting of the bulk samples is the one that is already in equilibrium at room temperature with excess water (prepared with 40 wt.% water, for example), because this is the sample that can be best compared to the aqueous dispersions. In these bulk samples in excess of water, increasing temperature induces a phase transition from cubic Pn3m *via* H_2 to L_2, while greater amounts of water are expelled; for instance, from 20 to 94°C, 38% of the initial water content is expelled.

To produce an MLO + water system in its colloidal emulsified state, the MLO was dispersed in large amount of excess water by means of ultrasonication, using the Pluronic copolymer F127 as stabilizer. This led to the formation of sub-micrometre-sized particles with self-assembly structured interiors, which could be observed by cryo-TEM (Figure 2). Figure 3 shows the scattering curves of the dispersions (thin lines) and the corresponding nondispersed bulk phase in excess water (thick lines) at three different temperatures (25, 60, and 98°C). The resemblance of the corresponding scattering curves at each temperature is remarkable: the peak positions for bulk systems and dispersions practically coincide. The peaks of the dispersions are only broadened and lowered in the scattered intensity. At the same positions as in the bulk, we observe six peaks in the emulsified cubic phase (ECP) and three peaks in the emulsified hexagonal phase (EHP). This quantitative agreement between the structural parameters measured in the dispersed particles and the nondispersed bulk in excess water clearly demonstrates that the confined

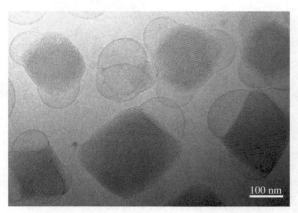

Figure 2 *Cryo-TEM image of a dispersion consisting of 4.625 wt.% MLO, 0.375 wt.% F127, and 95 wt.% water. Cubosomes coexist with vesicles at 25°C*

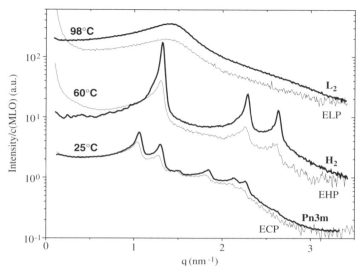

Figure 3 *Comparison of scattering curves from the same dispersion with those from the nondispersed bulk sample with excess water (40 wt.%) at three temperatures: thin lines correspond to the dispersion and thick lines to the bulk phase. Intensities were normalized by the respective MLO concentration*

interior of the emulsified particles is neither destroyed by the preparation method nor by the presence of the stabilizer F127. It also means that the interior of the dispersed particles is still highly ordered and only slightly affected by the dispersion procedure. The lower intensity and the slight broadening of the peaks for the dispersion can be attributed to an interplay of a limited size of the coherently scattering sub-volume (limited by the size of the dispersed droplets, and also indicated by the upturn of the scattering curves at low q-values) of a lower degree of order especially near the particle surface caused by incorporation of F127, and of the formation of vesicles from small parts of the MLO with F127 or possibly also small aggregates. Indeed, on almost all of the ECP particles, attached vesicles can be observed (Figure 2); this coexistence was also observed in previous studies on dispersed monoglycerides in which F127 was used as a stabilizer.[7,16] At 25°C the ECP of symmetry Pn3m has a mean lattice parameter of 8.47 ± 0.05 nm as determined by SAXS. The fast Fourier transform (FFT) of the internal structure of the particles observed by cryo-TEM (Figure 2) is compatible with the cubic Pn3m symmetry, and it gives a lattice parameter of ∼8.5 nm, which is in very good agreement with the SAXS analysis. At 60°C the internal structure of the particles was determined by SAXS to be hexagonal (EHP). This change in internal symmetry can also be seen in the respective cryo-TEM images.[12] The formation of an emulsified L_2 phase (ELP) is quite intriguing, but the data clearly show that the dispersion consists mainly of submicron-sized particles with a fluid-isotropic water + MLO-rich interior that has no long-range order. For all the investigated temperatures, therefore, the symmetries of the internal phase were preserved. Moreover, in all cases, a comparison of the temperature-dependent lattice parameters of the internal phase in the dispersion with the parameters of the nondispersed bulk phase with excess water gives identical values. This good agreement clearly shows that the internal structure of the dispersions at each temperature corresponds to the equilibrium structure of MLO with excess water.

To prove that the temperature-induced structural changes are reversible, a dispersion was measured at three different temperatures in a heating–cooling cycle (Figure 4). At 25°C the dispersed particles are ECP with an internal structure of Pn3m; after heating to 58°C they are transformed to EHP; and upon further heating to 87°C the transition to ELP takes place. On cooling the dispersion back to 58°C, we found an EHP again, and at 25°C the initial ECP. This means that, upon cooling the melted entrapped L_2 phase in the kinetically stabilized droplets of our dispersions, the ELP recrystallizes back to the original H_2 and finally to Pn3m at room temperature, *i.e.*, a reversible structural transformation. The respective scattering curves of the dispersion at 25°C (ECP) and 58°C (EHP) coincide. This shows that the internal structure of the dispersed droplets depends only on the current temperature, irrespective of whether it was reached by heating or cooling. This proves that the internal structure of the dispersed particles is a thermodynamic equilibrium structure. In addition, the mean radius of these particles was determined by dynamic light scattering to be ∼ 140 nm, with a rather polydisperse size-distribution width of *ca.* 25%.

In summary, aqueous submicrometre-sized dispersions of the binary MLO + water system, which are stabilized by means of the polymer, possess a distinct internal structure depending only on temperature and composition. We have demonstrated that the internal structure of the particles can be tuned with

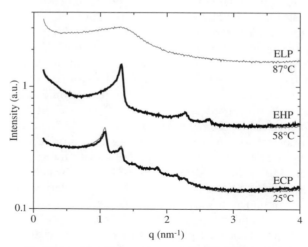

Figure 4 *Scattering curves of a dispersion, containing 8 wt.% MLO in a 1 wt.% F127 solution, measured in the heating direction at 25, 58, and 87°C (thin lines) and then in the cooling direction at 58 and 25°C (thick lines)*

temperature in a reversible way. Upon increasing the temperature, the internal structure undergoes a transition from cubic *via* hexagonal to fluid isotropic, and *vice versa*, in analogy to the binary MLO samples in equilibrium with excess water. This change in topology induces a different water intake in the particles. The internal structure expels water with increasing temperature in a reversible way. At each temperature, the internal structure of the dispersed particles corresponds very well to the structure observed for nondispersed bulk MLO with excess water. As this behaviour is independent of any thermal history, these particles are considered to be in thermodynamic equilibrium with the surrounding water phase.

5.3 Solubilization of Oils in Self-Assembled Structured Particles

5.3.1 Emulsification of Microemulsions

Incorporation of oil is of special interest from both the scientific and the industrial points of view. Indeed, one remaining question in the field is to know whether it would be possible to emulsify microemulsions. To study the possibility of the formation of what we shall refer to as an *EME*, we have studied the effect of the addition of oil on the confined internal structure of MLO-based aqueous solutions. In particular, we have investigated[13] the effect on the internal structure of the solubilization of *n*-tetradecane with different MLO/ (MLO + oil) ratios (denoted as δ%). To characterize better the self-assembly structure of the oil-loaded MLO-based aqueous dispersions, the corresponding bulk systems were also investigated.

Figure 5 shows the effect of TC on the internal structure at 25°C of the dispersions stabilized by F127, in comparison with the structure of the corresponding nondispersed, fully hydrated bulk systems of TC + MLO + water. In the absence of oil (Figure 5(a)), the scattering curve of the dispersion shows six peaks at the characteristic δ value for a cubic structure ($\delta = 100$) corresponding to the type Pn3m (cubosomes). As soon as TC is present ($\delta = 84$, Figure 5(b)), the scattering curve of the dispersion shows the three peaks characteristic of a hexagonal phase. A further increase of TC content in the dispersion ($\delta = 57.1$, Figure 5(c); $\delta = 47.6$, Figure 5(d)) leads to scattering curves with only one broad peak, as is typical for a concentrated microemulsion. The scattering curves of each of these dispersions with increasing TC content correspond very well to those of the nondispersed, fully hydrated bulk phases with the same δ values. This means that the internal structure of the particles is very similar to that of the nondispersed, fully hydrated bulk phase. This implies that, upon increasing the TC content, we actually observe a transition from an ECP, *via* an EHP, to an emulsified W/O microemulsion (EME).

It is important to note that this good structural agreement between the dispersed and the fully hydrated bulk phases implies that the TC/MLO mixing

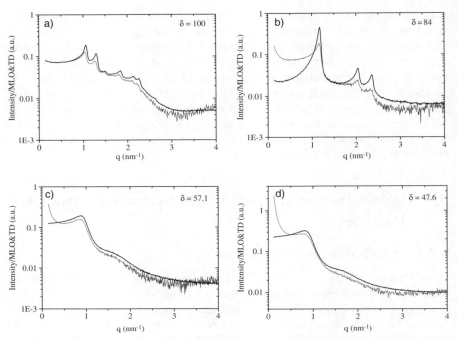

Figure 5 *Effect of solubilized n-tetradecane content on the scattering curves of MLO-based emulsified systems at 25°C (thin lines) and those of the nondispersed bulk sample with excess water (thick lines): (a) $\delta = 100$, (b) $\delta = 84$, (c) $\delta = 57.1$, and (d) $\delta = 47.6$. Intensities were normalized by the respective MLO + TC concentration*

ratio is maintained; that is, TC does not separate from MLO to form normal emulsion droplets in addition to the EME, despite the high-energy input of the dispersion procedure. Moreover, the polymer used to stabilize the particles does not disturb the internal structure. Stability tests of the dispersions do not show any change in the internal structure of the dispersions after four months storage at room temperature. Indeed, Figure 6 shows control of stability against ageing for two different oil-loaded dispersions (hexosomes and EME systems). This means that the addition of TC to monoglyceride + water mixtures leads to the ability to form confined nanoscaled tuneable hierarchical structures. The internal structure of the dispersed particles can then easily be tuned by adding a certain amount of TC at a certain temperature. As already found by SAXS, a low amount of added TC ($\delta = 84$) induces the formation of an H_2 internal particle structure (hexosomes). As shown in Figure 7, the particles exhibit internal arrangements of hexagonal symmetry and/or curved striations, which indicates the formation of hexosomes. As observed for hexosomes at high temperatures without oil, and in contrast to cubosomes, there are no vesicles or only very few vesicles coexisting with these hexosomes. The FFT of the internal H_2 phase reveals a characteristic spatial distance of ~ 5.6 nm, leading to a lattice parameter of ~ 6.5 nm, in good agreement with the SAXS analysis.

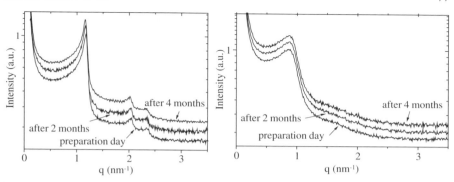

Figure 6 *Scattering curves of two different MLO-based dispersions at 25°C with δ = 84 (left) and δ = 57.1 (right). The curves for different ageing times are shifted vertically by a constant arbitrary factor for better visibility*

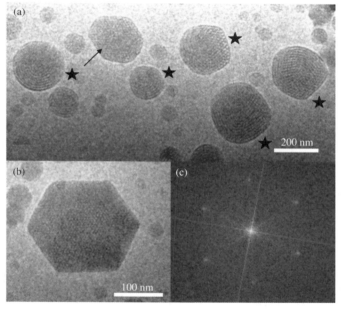

Figure 7 *Cryo-TEM observations of hexosomes (δ = 84). Indicated in (a) and (b) is the presence of curved striation (stars) and hexagonal internal symmetry (arrows). The FFT in (c) of the particle at the bottom left gives a lattice parameter of 6.5 nm*

We have shown, as for the case without added oil, that at each investigated temperature the structure of the entrapped internal phase is the same as that of the nondispersed samples coexisting with excess water. This again confirms that the internal structure of the dispersed particles at each temperature corresponds to an equilibrium mesophase. To elucidate further the structural changes of the

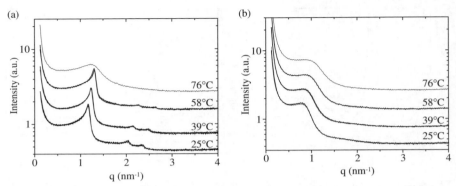

Figure 8 *Scattering curves of two dispersions with (a) $\delta = 84$ and (b) $\delta = 47.6$. These samples were measured while heating at 25, 39, 58, and 76°C (thin lines) and while cooling at 58, 39, and 25°C (thick lines). Curves are shifted vertically by a constant arbitrary factor for better visibility*

particle internal structure that occur during heating/cooling processes, we carried out an SAXS investigation on two aqueous dispersions (with $\delta = 84$ and 47.6). The scattering curves of these two dispersions, which are presented in Figure 8 as a function of temperature (from 25 to 76°C), reveal the structural reversibility of the confined structures. In both samples, the peaks shift in their dependence on temperature, and for $\delta = 84$ there is an additionally observed reversible transition of the type $H_2 \leftrightarrow L_2$. We note that, for each investigated temperature, the scattering curves are identical and are not dependent on whether the sample temperature was reached by heating or cooling. This indicates that the internal self-assembly structure in all of the samples containing oil, at a certain temperature, is independent of the thermal history: that is, heating or cooling to the required temperature leads to the same structure. This fact is clear evidence that the formed structures in the kinetically stabilized particles, in analogy to those in the nondispersed TC-rich bulk phases, are thermodynamic equilibrium structures.

Thus, when adding *n*-tetradecane to the MLO + water + F127 system at constant temperature the internal self-assembly structure of the kinetically stabilized particles transforms from Pn3m (cubosomes) to H_2 (hexosomes), and then to a W/O (L_2) microemulsion phase (EME). To our knowledge, this is the first time that the formation of stable EME systems has been properly described and proven to exist at room temperature. It thereafter becomes possible to form a W/O microemulsion-in-water emulsion system, which is different from the well-known double emulsion (emulsion-in-emulsion system) because the kinetically stabilized internal W/O emulsion in the double emulsion system is replaced by the thermodynamically stable W/O microemulsion as the 'inner' emulsion in the EME. Stable *EMEs* are superior to double emulsion systems, since they consist of stabilized droplets in an aqueous continuous phase, which contain in the inner part an entrapped equilibrium self-assembled structure.

5.3.2 Micellar Cubosomes and Transformation from Hexosomes to EME

This section focuses on the recent discovery of an Fd3m phase (reverse discontinuous micellar cubic), which is formed in the MLO + water + TC system at a specific TC/MLO weight ratio.

Figure 9 shows the SAXS scattering curves that were obtained from dispersions in which the δ value was varied from 100 down to 62.5 at 25°C. For $\delta = 100$ (without oil) a dispersed system is formed with a confined inverted bicontinuous cubic structure, belonging to the Pn3m space group. On increasing the oil content (decreasing δ), the internal ordered phase of the particles becomes an inverse hexagonal H_2. Intriguingly, a further increase of the TC concentration in the dispersion to $\delta = 71.4$ gives rise to a scattering curve that shows more than seven peaks. These peaks are in the characteristic ratio for a discontinuous micellar cubic structure of the type Fd3m (micellar cubosomes or emulsified MCP). Remarkably, this is not found in the absence of TC. At higher oil concentrations, the scattering curve shows only one broad peak, which is typical of that for a concentrated microemulsion phase. The present work proves that the structural transformation in the dispersed particles from H_2 (hexosomes) to the W/O microemulsion system (EME) may be indirect, occurring gradually *via* an emulsified intermediate phase (see Figure 9). We denote this new type of emulsified particles as *micellar cubosomes*. The Fd3m cubic phase consists of a three-dimensional periodically ordered complex packing of two different types of discrete inverse micelles, both quasispherical. In this phase, the unit cell contains 8 larger inverse micelles of symmetry $\bar{4}3\,m$

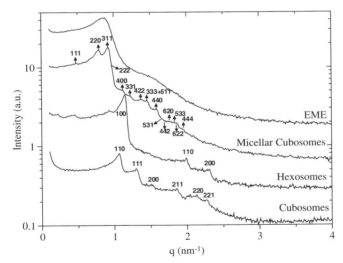

Figure 9 *Effect of n-tetradecane content on the scattering curves of MLO-based emulsified systems at 25°C: $\delta = 100$ (cubosomes), $\delta = 78.1$ (hexosomes), $\delta = 71.4$ (micellar cubosomes), and $\delta = 62.5$ (EME). Peaks are identified by their Miller indices*

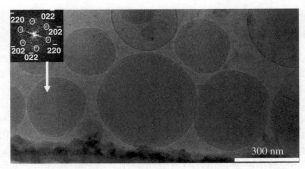

Figure 10 *Cryo-TEM image of MLO-based aqueous TC-loaded dispersion with δ = 71.4. Here, cubosomes with internal Fd3m phase are observed. Inset is an FFT of the arrowed particle viewed along [111]. A hexagonal motif formed by the {220} reflections is observed*

arranged tetrahedrally on a diamond lattice, and 16 smaller reverse micelles of symmetry $\bar{3}$ m. These discrete aggregates, whose core consists of hydrophilic aqueous domains, are embedded in a continuous three-dimensional hydrophobic matrix (the oil phase). The Fd3m phase appears between the H_2 and W/O microemulsion phases. For the TC-loaded dispersion with $\delta = 71.4$, the cryo-TEM image shown in Figure 10 reveals the formation of particles with an internal motif. The FFT (inset of Figure 10) shows a sixfold symmetry, and therefore the electron beam is very likely parallel to the [111] direction. The interplanar distance corresponding to the intensity peak in the FFT is ~8 nm, which is in agreement with the presence of {220} reflections leading to a lattice parameter of ~22 nm confirming the SAXS analysis.

These particles with self-assembled structured interiors can also be produced in food-grade systems. We present here a complete temperature–composition phase diagram for R-(+)-limonene-loaded Dimodan-based particles stabilized by F127 over the range 1–90°C. The phases identified by SAXS and their location in the temperature–δ representation are shown in Figure 11. We clearly see this new phase inserted between the inverse hexagonal and the L_2 phases. Before eventually reaching the MCP, the particles always first transform into hexosomes, but in the end, above a certain oil content, and depending on the temperature, they all turn into EME particles. In the case of pure phases, the hexosome to EME transformation can then occur directly (between 50 and 70°C) or indirectly through the MCP (between 1 and 40°C) by increasing the oil content. It is interesting to note that all the particles' internal structures melt into an L_2 phase at a certain temperature, which is strongly dependent on the oil content (80°C at $\delta = 100$, but only 40°C at $\delta = 70$). Above $\delta = 60$, the more oil we add, the lower is the transition. Below this value, we have EME particles at any temperature. Thus, in any case, increasing the temperature or oil content will lead in turn to EMEs. The eventual existence of the MCP is consistent with bulk studies where the location of the inverse cubic I_2 phase was thus far only found between the H_2 phase and an inverse micellar solution L_2. This internal phase behaviour, upon changing the oil/monoglyceride balance, can be

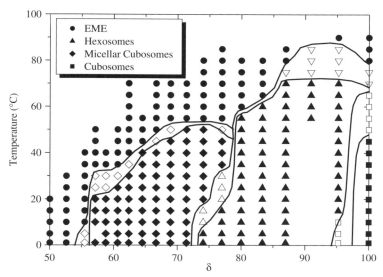

Figure 11 *Temperature–δ phase diagram of R-(+)-limonene-loaded Dimodan-based dispersions (4.625 wt.% Dimodan U + oil, 0.375 wt.% F127, 95.0 wt.% water). Dimodan U and MLO contain 59.5% and 86.1% MLO, respectively*

explained in terms of an increase of the negative mean curvature of the interface. The oil is a hydrophobic molecule that is not surface-active, and it is located around the fatty tails of the monoglycerides and not at the oil–water interfaces. Thus, the oil increases the hydrophobic volume of the chains compared to their head group area. The presence of oil will then tend to bend the interfaces towards a higher negative mean curvature, favouring the formation of spherical micelles. Nevertheless, this cannot be a *sufficient* condition to explain why we can obtain the MCP phase rather than an L_2, as many systems just show a direct transition from inverse hexagonal to L_2.

Figure 12 shows SAXS curves from R-(+)-limonene + Dimodan U dispersions stabilized by F127 at δ = 77. By decreasing the temperature from 90 to 1°C, we show not only a transformation from EME to micellar cubosomes at 45°C, but also a transition from Fd3m cubosomes to hexosomes at a rather low temperature (15°C). The latter transition happens over a broad temperature range (35–15°C) as can be seen from the first appearance of the (100) hexagonal peak at $q \approx 1.3$ nm^{-1} at 30°C and the vanishing of the last contribution of the MCP signal at 15°C. As can be seen in Figure 11, the transition line between the inverse hexagonal H_2 and MCP seems to be nearly vertical above 20°C. But it appears that below this temperature all transition lines are strongly tilted so that the phases are shifted to lower δ values. This enables the system to display an indirect transition from H_2 to L_2 *via* MCP upon varying the temperature. Nevertheless, even if this indirect transformation becomes possible by increasing either the oil content or the temperature, it is available only over a narrow range of oil content (22–27%). It is also interesting to note that, in all the

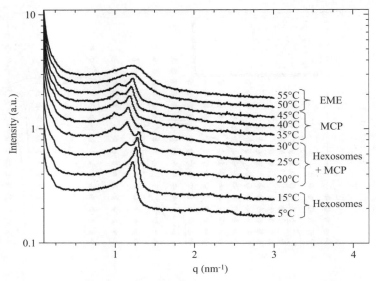

Figure 12 *Thermal transformation at $\delta = 77$ from hexosomes to EME via micellar cubosomes for R-(+)-limonene-loaded Dimodan-based dispersions. For clarity the SAXS curves are shifted vertically*

systems at very low temperatures, lamellar phases (zero curvature) are not observed; at 1°C only phases with negative curvature are observed – cubosomes, hexosomes, micellar cubosomes, and EME particles. Thus, the particles do maintain the self-assembled structures found at higher temperatures. This is important for storage reasons.

Here, we have reported on the effect of varying temperature and the solubilized amount of oil on the structural transitions observed in MLO-based dispersions. At a given temperature, the addition of oil induces a transition of the internal structure from the bicontinuous cubic phase (Pn3m) to the reversed hexagonal (H_2) and then to the isotropic liquid phase (W/O microemulsions). We have found an Fd3m phase (reverse discontinuous micellar cubic), which is formed at a specific oil/monoglyceride weight ratio. It is situated between the H_2 phase and the isotropic liquid phase (W/O microemulsion). The present work proves that the structural transformation in the dispersed particles from H_2 (hexosomes) to the W/O microemulsion system (EME) may occur directly or indirectly *via* an emulsified intermediate phase.

5.3.3 Control of the Internal Self-Assembled Structure

We showed above that we can tune the internal structure of monoglyceride-based particles upon addition of a certain amount of oil. Here we show that we can reverse the structural change using an additional surface-active component. In particular, DGMO has a counter effect to that of the oil (TC), in that it enables us to 'tune back' the self-assembled structure in the oil-loaded dispersions from hexosomes to cubosomes of Im3m symmetry.

The effect of DGMO on the structure of the MLO + TC + water system in its colloidal emulsified state is shown in Figure 13. The MLO + DGMO + TC mixtures, with different DGMO/MLO ratios (denoted as β in %) and a fixed δ value of 94.3, were dispersed in large amount of excess water using the block copolymer F127 as the stabilizer. At 25°C, in the absence of DGMO ($\beta = 0$), the scattering curve displays three peaks in the characteristic ratio for a hexagonal phase (hexosomes). When part of the MLO is replaced by DGMO ($\beta = 25$), the observed three peaks in the scattering curve move to lower angles, corresponding to larger scale structures. A further increase of the DGMO concentration in the dispersion ($\beta = 50$ and 100) leads to a scattering curve which shows the characteristic ratio for a cubic structure of the type Im3m. Previously, we found that the replacement of MLO by TC induces a transition from cubosomes *via* hexosomes to EME, while as shown here we see that DGMO promotes a reversed transition from hexosomes back to cubosomes. This means that this surfactant has a counter effect on the internal confined structure of the particles to that of the oil.

5.4 Conclusions and Outlook

Self-assembled structures, such as inverted mesophases of the cubic or hexagonal geometry, or reverse microemulsion phases, can be readily dispersed using a polymeric stabilizer. They are interesting for many reasons. They provide the

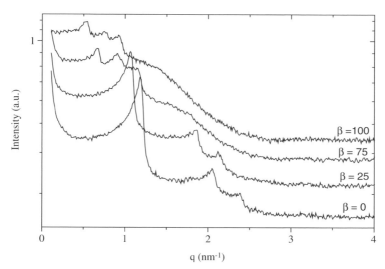

Figure 13 *Effect of DGMO content on the scattering curves of TC-loaded aqueous dispersions at 25°C for $\delta = 94.3$, and with different β values in the range of 0–100. Intensities are shifted vertically by a constant arbitrary factor for better visibility*

possibility to create hierarchically structured particles by using, for instance, the convenience of ultrasonication and the ability to tune their confined structures reversibly in a dispersion that is formed with kinetic stability with or without incorporated oil. The dispersed particles are in real thermodynamic equilibrium with the corresponding continuous aqueous phase.

We have also demonstrated that a microemulsion can be emulsified through addition of oil to the monoglyceride + water system. This is possible since the addition of n-tetradecane induces a transition of the internal particle structure from Pn3m to H_2 and L_2 at a given temperature. The same phase transition can be induced by increasing the temperature at a constant content of oil. In fact, both the increase of temperature at a constant oil content or the increase of the oil content in the particles at a constant temperature allow the spontaneous film curvature of the self-assembled structure formed inside the particles to become more negative. We also found that the temperature increase induces the microemulsion droplets to pack first on an Fd3m cubic lattice prior to the isotropic microemulsion phase being formed. The inner particle structure is easy to control by altering the composition of the system and/or varying the temperature. The use of surface-active molecules may tune the internal structure back to a less negative curvature of the interfaces so as then to recover bicontinuous cubic structures.

These self-assembled structured aqueous dispersions seem to offer a useful approach in the development of novel food. Indeed, these types of particles have a great potential to host lipophilic, hydrophilic, and amphiphilic functional molecules, which can be important for complex formulations in food–drug delivery, solubilization of active molecules, selective molecular transport, or the utilization of these systems as nano-reactors. They are stable over periods of months, if not years, when stored in aqueous solution. But as soon as they are mixed with other solutions, particles, *etc.*, they will start to exchange material for entropic reasons. In the future, we plan to study such transfer phenomena and their kinetics.

References

1. B.H. Robinson (ed), *Self-Assembly*, IOS Press, Amsterdam, 2003.
2. L. de Campo, A. Yaghmur, N. Garti, M.E. Leser, B. Folmer and O. Glatter, *J. Colloid Interface Sci.*, 2004, **274**, 251.
3. A. Yaghmur, L. de Campo, A. Aserin, N. Garti and O. Glatter, *Phys. Chem. Chem. Phys.*, 2004, **6**, 1524.
4. O. Glatter and L. de Campo, in *Self-Assembly*, B.H. Robinson (ed), IOS Press, Amsterdam, 2003, p. 284.
5. K. Larsson, *Curr. Opin. Colloid Interface Sci.*, 2000, **5**, 64.
6. J. Gustafsson, H. Ljusberg-Wahren, M. Almgren and K. Larsson, *Langmuir*, 1996, **12**, 4611.
7. J. Gustafsson, H. Ljusberg-Wahren, M. Almgren and K. Larsson, *Langmuir*, 1997, **13**, 6964.

8. S.M. Masum, S.J. Li, Y. Tamba, Y. Yamashita, T. Tanaka and M. Yamazaki, *Langmuir*, 2003, **19**, 4745.
9. T. Nylander, C. Mattisson, V. Razumas, Y. Miezis and B. Håkansson, *Colloids Surf. A*, 1996, **114**, 311.
10. K. Larsson, *J. Phys. Chem.*, 1989, **93**, 7304.
11. B.J. Boyd, *Int. J. Pharm.*, 2003, **260**, 239.
12. L. de Campo, A. Yaghmur, L. Sagalowicz, M.E. Leser, H. Watzke and O. Glatter, *Langmuir*, 2004, **20**, 5254.
13. A. Yaghmur, L. de Campo, L. Sagalowicz, M.E. Leser and O. Glatter, *Langmuir*, 2005, **21**, 569.
14. A. Yaghmur, L. de Campo, L. Sagalowicz, M.E. Leser and O. Glatter, *Langmuir*, 2006, **22**, 9919.
15. S. Guillot, C. Moitzi, S. Salentinig, L. Sagalowicz, M.E. Leser and O. Glatter, *Colloids Surf. A*, 2006, **291**, 78.
16. P.T. Spicer, K.L. Hayden, M.L. Lynch, A. Ofori-Boateng and J.L. Burns, *Langmuir*, 2001, **17**, 5748.
17. H. Qiu and H. Caffrey, *Biomaterials*, 2000, **21**, 223.
18. J.M. Seddon, *Biochim. Biophys. Acta*, 1990, **1031**, 1.

Chapter 6

Synergistic Solubilization of Mixed Nutraceuticals in Modified Discontinuous Micellar Cubic Structures

Rivka Efrat, Abraham Aserin and Nissim Garti

CASALI INSTITUTE OF APPLIED CHEMISTRY, INSTITUTE OF CHEMISTRY, THE HEBREW UNIVERSITY OF JERUSALEM, JERUSALEM 91904, ISRAEL

6.1 Introduction

Lyotropic liquid crystals are known to self-assemble in lamellar, hexagonal and cubic symmetries. These are mostly bicontinuous, but they can become discontinuous in certain cases and under specific conditions.[1] Several symmetries of cubic phases have been described in the literature. A discontinuous cubic structure with space group of Pm3n (Q^{223}) was first discovered by Balmbra *et al.*[2] in 1969 and further studied by Tardieu and Luzzati[3] in a variety of lipid + water systems. The space group of Fd3m (Q^{227}) was first described by Tardieu[4] and observed in a variety of lipid + water systems, in all of which the lipid was heterogeneous.[5-11] The space group of P4$_3$32 (Q^{212}) has, so far, been reported in a system composed of protein + lipid + water, as in monoolein + cytochrome C + water.[12,13] The space group Fm3n (Q^{225}) was first observed in a surfactant + water system[13] and later in several ganglioside + water systems.[14]

Once the domain of a liquid crystalline phase becomes discontinuous, micellar systems with similar structures can be formed. Only a few reports have dealt with the discontinuous structures and all of them are related to cubic symmetries. Micellar discontinuous cubic phases are composed of discrete aggregates (micelles) ordered in a three-dimensional array.[1,7] These structures have been recognized as a separate and unique form of a micellar solution. Topologically, micellar cubic systems can be structured as discrete aggregates of well-organized oil-in-water (type I) micelles or inverse water-in-oil (type II) micelles.[1,12]

Recently, we detected[15,16] a new and modified organization of a self-assembled discontinuous micellar cubic phase in systems composed of glycerol monooleate (GMO) + ethanol + water, which was designated as a Q_L phase. We describe this type of system as 'modified' because, unlike previously reported cubic structures that are semi-solid or gel-like, our mesophase is characterized by its low viscosity (~ 40 Pa s), high fluidity and transparency. This micellar cubic phase has a body-centred cubic (bcc) symmetry and it seems to belong to $P4_232$ (Q^{208}) space group. The $P4_232$ space group belongs to incommensurate crystals, *i.e.*, crystals consisting of a basic ordered structure that is perturbed periodically and 'modulated' by a sub-system.[17,18] The structure with the $P4_232$ symmetry is characterized by at least four periodicities, three of which represent known cubic symmetries of crystalline structures.

In water, the monoglyceride GMO is ordered in various liquid crystalline phases depending on the water content, temperature and other physical conditions.[19,20] We are concentrating our efforts in revealing the possible types of self-assembly. We have found that, once a controlled quantity of a 'lamellar structure-breaker' (such as short-chain alcohol) is carefully added to the binary mixture, some unique discontinuous structures are formed. We recently described and partially characterized one such mesophase denoted as the Q_L phase.[15,16] GMO is widely used as a food emulsifier,[21] and its liquid crystalline phases have the potential of serving as sustained release carriers of nutraceuticals, aromas and drugs.[19–23] In this work we report on the solubilization patterns of two of these nutraceuticals (lycopene and phytosterols), which are practically insoluble in water but which have been solubilized in the Q_L mesophases.

Phytosterols (plant sterols) belong to the tri-terpene family of natural products which includes over 200 different sterols.[24–26] Phytosterols are a mixture of sterols that are structurally similar to cholesterol, but with the inclusion of an extra hydrophobic carbon chain at the C-24 position (each phytosterol has an additional side chain) [Figure 1(a)]. Phytosterols in their alcoholic-'free' form are non-esterified sterols. Sterols can be esterified to form the corresponding fatty acid esters (phytosterol esters). Sterols and sterol esters are recognized as cholesterol reducing agents. Phytosterols act in the digestive track and are not transported to the blood stream, and therefore they exhibit virtually no side effects and no mutagenic activity or sub-chronic toxicity in animals. Phytosterols and phytosterol esters are hydrophobic structures poorly soluble in food-grade oils and in water, which limits their application in water-based applications such as clear beverages.

Lycopene is a unique carotenoid known to contribute to a lower risk factor for coronary heart disease and some types of cancer.[27] Lycopene is a straight chain hydrocarbon,[28] highly unsaturated, consisting of all-*trans* conjugated double-bonds[28–31] [Figure 1(b)]. Lycopene is an extremely insoluble compound; its solubility in both water and food-grade solvents is extremely low (several hundred parts per million). Lycopene exists in tomatoes in its crystalline form and as such it has very low bioavailability. For nutraceutical applications, raw powdered tomatoes or lycopene powdered matter containing large crystals cannot be used because the bioavailability is so very low. It has therefore been recognized by formulators and manufacturers that, for improved bioavailability, the crystalline lycopene must be 'micronized' or reduced to nanometre-sized

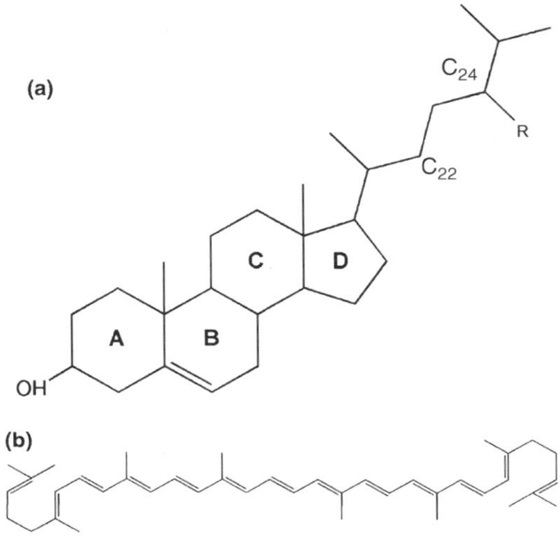

Figure 1 Chemical structures of (a) various phytosterols and (b) lycopene ($C_{40}H_{56}$). Phytosterol key: cholesterol, R=H; β–sitosterol, R=CH_2CH_3; stigmasterol, R=CH_2CH_3 and additional double bond at C_{22}; campasterol, R=CH_3; brassicasterol, R=CH_3 and additional double bond at C_{22}

particles – or, better still, solubilized in a proper liquid vehicle. A complication, however, is that lycopene in its molecular form is highly sensitive to oxidation and so requires oxidative protection.

The aim of this study was to develop technology for making water-based concentrates enriched with free sterols (and/or lycopene). The new formulations (nano-sized vehicles) are aimed at being more efficient in reducing cholesterol levels in humans since the vehicles onto which they are loaded are excellent carriers for molecularly solubilized nutraceuticals.[31] Lyotropic liquid crystals have been proposed as potential carriers for drugs and nutraceuticals. Lyotropic cubic phases have a high solubilization capacity, yet they form relatively large microcrystalline domains. The guest molecules are easily entrapped within the domains; but, depending on their nature and concentrations, they may induce mesophase transformations from cubic to hexagonal or lamellar, or even to micellar structures.

We report here on the solubilization capacity of lycopene and phytosterols into our newly discovered Q_L phase, both alone and together, and including any synergistic effects. We also report on the phase transformation that may occur in the presence of these solubilizates.

6.2 Experimental

GMO was purchased from Riken Vitamin (Japan); it contained > 98.5% monoesters of oleic acid. Crystalline lycopene was obtained from Lycored

(Beer-Sheva, Israel) as powdered material (all *trans*) containing 67.4% lycopene and the rest being tomato oleoresins. The phytosterols ($\geq$ 97%-free phytosterols, extracted from soybeans) were a gift from ADM Nutraceuticals (Decatur, IL, USA). The lycopene and phytosterols were used without any further purification. Analytical grade ethanol (EtOH, purity > 99%) was purchased from Frutarom (Israel). The water was double distilled.

The samples of Q_L mesophase were prepared by weighing appropriate amounts of molten GMO into culture tubes sealed with Viton-lined screw caps. The lycopene and the phytosterols were melted prior to being added to the GMO. Once the samples had reached room temperature, the ethanol and water were added. The clear samples were stored for several days at 25°C to make sure that no precipitation of the solubilizates occurred. The transparent samples were analysed by small-angle X-ray scattering (SAXS), synchrotron-SAXS and differential scanning calorimetry (DSC) as described below.

SAXS measurements were performed using Ni-filtered Cu Kα radiation (1.540 Å) from an Elliott GX6 rotating anode operating at 1.2 kW. X-ray radiation was further monochromated and collimated by a single Franks mirror and then collimated by a series of slits and height limiters. The scattering profile was measured by a linear position-sensitive detector of the delay line type; it was stored on a PC as a 256-channel histogram. The sample-to-detector distance was approximately 48 cm. Each sample was sealed in a glass capillary tube (diameter 1.5 mm) placed in a thermostatic bath. Measurements were carried out at 25 ± 0.5°C with an exposure time of 1–3 h.

Temperature-dependent synchrotron measurements of SAXS and wide-angle X-ray scattering (WAXS) combined with DSC were carried out on a modified Mettler model FP99A-V4.0 (SAXS–WAXS–DSC) at BL-15A Photon Factory (KEK, Tsukuba, Japan). The X-ray wavelength was 0.15 nm and the beam size was 500×500 μm at the sample position. The images of SAXS and WAXS were measured with an X-ray CCD detector coupled with a 230 mm diameter X-ray image intensifier (XRII), which was set at a distance of 160 mm from the sample. In addition to a usual beam stop, a small beam stop of 3 mm in diameter was set at a distance of 35 mm from the sample in order to reduce air scattering. The samples were sealed in the same cell, and were heated to 40°C and cooled at a rate of 2°C min^{-1} to -20°C for 15 min, and then heated to 25°C at a rate of 1°C min^{-1}. The SAXS–WAXS data were recorded at 10 s intervals with an exposure time of 1.5 s.

Some DSC measurements were carried out on a Mettler Toledo DSC822 instrument. Samples (5–15 mg) were weighed in standard 40 μL aluminium pans using a Mettler M3 microbalance and immediately sealed by a press. The samples were rapidly cooled to -10°C at pre-determined quenching rates from ambient to -10°C. Each sample remained at this temperature for 20 min and was then heated (from *ca.* -10°C) at 2°C min^{-1} to 40°C. The cooling/heating cycle was repeated twice. An empty pan was used as reference. The fusion temperatures of the solid components and the total heat transferred in any of the observed thermal processes were determined. The enthalpy changes, associated with thermal transition, were obtained by integrating the area of each pertinent DSC peak. The DSC temperatures reported here were reproducible to ± 0.5°C.

6.3 Results and Discussion

6.3.1 Thermal Behaviour of the Q_L Phase

A typical DSC thermogram of the first heating scan of the frozen GMO + water + ethanol mixture (pre-cooled to −10°C) reveals the existence of three endothermic events at 0.26, 6.0 and 19.32°C (Figure 2). The endothermic peak at 0.26°C is partially overlapped by a large endothermic peak (as indicated by an asterisk in the thermogram) and so is difficult to resolve. This endotherm corresponds to the thawing of free water. The existence of this free water was confirmed by replacing H_2O with D_2O, which resulted in an endotherm shift of 2.74°C (not shown). The endotherm is small, even at 50-wt% water content, with a melting enthalpy of only -5.16×10^{-2} J g^{-1}. The small enthalpy involved in the thawing process means that most of the water is bound to the lipid and alcohol, and that only minor amounts remain as free water.

The second event, appearing at 6.0°C, has a ΔH value of -15.91 J g^{-1}. This endothermic event corresponds to a typical L_β–L_α phase transition. Usually, the L_β phase is formed under intermediate conditions between the crystalline state and the more fluid mesophases, particularly where the crystal structure itself consists of parallel stacks of bilayers.[1] Formation of the L_β phase can be further confirmed by the SAXS and WAXS synchrotron measurements. The analysis of the reciprocal spacing of SAXS data in this temperature range (0–6°C) reveals diffractogram peaks typical of L_β with a 1:2 ratio of first and second diffraction orders. The d-spacing of L_β is 49.6 Å, in good agreement with the previously reported values.[32,33] The WAXS results show five sharp peaks, cooperative with a broad peak centred at 4.611 Å^{-1} [Figure 3(a)]. These

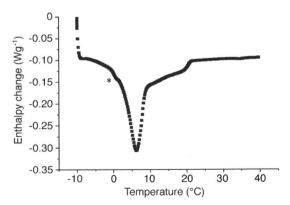

Figure 2 *Typical DSC thermogram of the first scan following heating of the Q_L phase. Enthalpy change ΔH is plotted against temperature. The shoulder at position * is explained in the text*

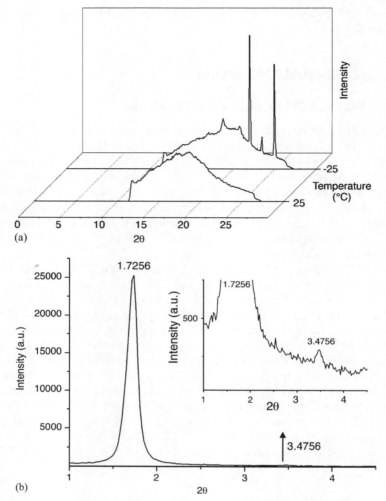

Figure 3 *X-ray scattering patterns of the Q_L phase: (a) WAXS diffractions at $-25°C$ and at $+25°C$; (b) SAXS diffractions at $5°C$. Intensity is plotted against scattering angle 2θ*

sharp WAXS and SAXS peaks indicate that the sample has a lamellar structure, characterized by a crystalline packing of the amphiphile chains (as evidenced by sharp Bragg diffraction peaks in the WAXS scattering regime), which are ordered in parallel stacks of bilayers. The high-range order is evidenced by sharp Bragg diffraction peaks in the SAXS scattering regime in the ratio 1:2 [Figure 3(b)].

The third endothermic event occurs at $19.32°C$ ($\Delta H = -1.16$ J g^{-1}); it is attributed to the L_α–Q_L phase transition. The analysis of the reciprocal spacing of the SAXS and WAXS synchrotron measurements in the temperature range from -10 to $+10°C$ shows the existence of 1:2 reciprocal ratios of the first and the second diffraction orders; the intensity decreases with increase in the temperature, but at 10 °C the second peak could not be seen. Close examination of the SAXS diffractions in the temperature range 8–22°C reveals the coexistence of two mesophases, the Q_L phase and the lamellar phase. The Q_L phase was discussed in our earlier report.[16]

Once the sample reaches 23°C, the ratio of the reciprocal spacing changes in good agreement with a typical Q_L phase. The WAXS diffractogram exhibits a broad peak centred at 4.6Å [Figure 3(a)]. The lamellar and Q_L phases were further analysed by the SAXS synchrotron between 0 and 40°C. It was found that, from 8 to 20°C, the intensity from the lamellar mesophases becomes gradually reduced and a transformation to the Q_L phase occurs. The Q_L mesophase peaks progressively become more pronounced and they have higher intensities, as shown in Figure 4. The progressive thermal transformation is strongly consistent with a gradual disordering of the L_α phase. This can be seen from the disappearance of the peaks related to the L_α phase, which is accompanied by the appearance of the Q_L peaks.

These overall findings are in good agreement with the study of Cherezov et al.[34] on the phase transition from L_α to the inverse bicontinuous cubic phase

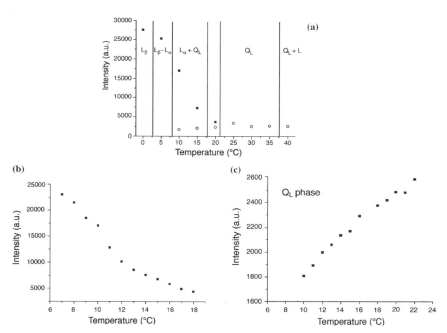

Figure 4 *Synchrotron SAXS intensity as function of temperature. (a) L_β and L_α (■) and Q_L (○) mesophases at 0–40°C. (b) Expanded plot for L_α mesophases. (c) Expanded plot for Q_L mesophases*

based on a hydrated N-monomethylated dioleoylphosphatidyl ethanolamine system.

6.3.2 Systems Loaded with Phytosterols

The main purpose of this set of experiments was to determine the effect of added guest nutraceutical molecules such as phytosterols on the integrity of the Q_L mesophase, and to detect any phase transformation that might occur as a result of the incorporation of the lipophilic guest molecules.

Samples loaded with phytosterols at several solubilization capacities were tested for their thermal behaviour. Figure 5 shows DSC measurements carried out using the same thermal protocol on samples containing 0.2, 1.4 and 2.5 wt% phytosterols. All the samples were liquid, optically isotropic and completely transparent; and no phytosterol crystals were detected, meaning that all the added phytosterol was fully solubilized within the mesophase. The DSC measurements were carried out with the same heating protocol as that of the unloaded (control) Q_L sample. The enthalpy values of all the endothermic events occurring in the control sample and in the presence of increasing amounts of solubilized phytosterols are summarized in Table 1. The thermograms of the heating scans revealed that the temperature of the L_β–L_α phase transition almost did not shift: the endothermic events occurred at 6.04, 5.66, 5.63 and 5.68°C for increasing amounts of solubilized phytosterols. Yet the magnitude of the enthalpy of transformation was found to decrease substantially with increase in the phytosterol content. The effect of phytosterols on the L_α–Q_L transition enthalpies is also significant, although the transition temperature did not change. The enthalpy change was larger for 0.2-wt% phytosterols (-1.24 J g^{-1}), but it decreased with increased content of phytosterols above 0.2 wt%. The sharp drop in the transition entropy indicates that the phase is almost completely disordered in the presence of a high concentration of solubilized phytosterols.

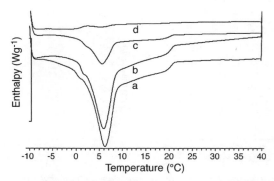

Figure 5 *The heating curve from the DSC thermograms carried out on samples containing various concentrations of phytosterols: (a) 0 wt%, (b) 0.2 wt%, (c) 1.4 wt% and (d) 2.5 wt%. Enthalpy change ΔH is plotted against temperature*

Table 1 *Transition enthalpy change ΔH, transition point temperature, transition temperature range and transition entropy change ΔS of the endothermic events occurring on heating of the control samples and in the presence of increasing quantities of solubilized phytosterols*

	Temperature (°C) $L_\beta - L_\alpha$	ΔH (J g^{-1}) $L_\beta - L_\alpha$	Temperature (°C) $L_\alpha - Q_L$	ΔH (J g^{-1}) $L_\alpha - Q_L$	ΔS (J g^{-1} K^{-1}) $L_\alpha - Q_L$
Q_L	2.79–8.53^a 6.04^b	−15.91	17.33–21.12^a 19.32^b	−1.16	−0.06
Q_L + 0.2-wt% phytosterols	3.03–8.08^a 5.66^b	−13.63	14.94–20.99^a 19.14^b	−1.24	−0.0645
Q_L + 1.4-wt% phytosterols	2.98–8.16^a 5.63^b	−4.17	16.92–20.61^a 18.74^b	-5.6×10^{-2}	−0.0030
Q_L + 2.5-wt% phytosterols	2.55–8.17^a 5.68^b	−1.09	Not observed	−^{c}	>0.003^c

^{a} Range of peak.
^{b} Centre of peak.
^{c} Difficult to measure.

The cooling stage (from 40 to −10°C) exhibits only one exothermic event, corresponding to the crystallization from liquid crystal phase to gel lamellar phase. The formation of the gel lamellar phase occurred at a lower temperature than during the heating scan (Figure 6 and Table 2). The lower crystallization point reflects the supersaturation effect, which is common in all fats and lipids. The transition temperatures were even lower in the presence of increasing contents of solubilized phytosterols. Similarly, the enthalpy values were found to decrease for > 0.2 wt% of phytosterols.

It is thus understood that a moderately high level (> 0.2 wt%) of solubilized phytosterol lowers the order in the Q_L phase. Phytosterols at high content destabilize the Q_L phase, and so the transition is observed at lower temperatures.

The SAXS measurements reveal that phytosterols solubilized into the Q_L phase have a progressive concentration-dependent effect on the nanostructure of the phase. At low solubilizate concentration, the guest molecules exhibit an ordering effect, so that the SAXS diffractions became sharper, better resolved and the peaks are intensified (Figure 7). The diffractions reveal the existence of a series of six Bragg peaks with calculated reciprocal spacing ratios of $1:\sqrt{2}:\sqrt{3}:\sqrt{4}:\sqrt{5}:\sqrt{6}$. The reciprocal d space ($1/d_{hkl}$) of the six reflections *versus* $(h^2 + k^2 + l^2)^{1/2}$ intercepts the axis of the origin with a very small deviation of 2.7×10^{-4} and high linearity of 0.9991. The indexation indicates the existence of the P4$_2$32 space group and the lattice periodicity spacing is 94.9 Å. Therefore we can conclude that a solubilization level of 0.2 wt% of phytosterols enhances the order in the structure, since a very good matching of the actual diffractions and the permitted reflections was clearly

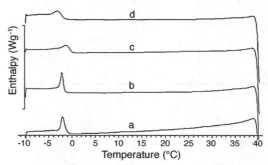

Figure 6 *The cooling curve from the DSC thermograms carried out on samples containing various concentrations of phytosterols: (a) 0 wt%, (b) 0.2 wt%, (c) 1.4 wt% and (d) 2.5 wt%. Enthalpy change ΔH is plotted against temperature*

Table 2 *Transition enthalpy change ΔH, transition point temperature and transition temperature range of the endothermic events occurring on cooling of the control samples and in the presence of increasing quantities of solubilized phytosterols*

	Temperature (°C) $L_\alpha - L_\beta$	ΔH (J g^{-1}) $L_\alpha - L_\beta$
Q_L	$-(1.14-2.70)^a$ -2.03^b	1.03
Q_L + 0.2-wt% phytosterols	$-(1.63-2.81)^a$ -2.10^b	1.18
Q_L + 1.4-wt% phytosterols	$-(1.68-3.91)^a$ -2.94^b	0.91
Q_L + 2.5-wt% phytosterols	$-(1.77-4.62)^a$ -3.09^b	0.70

^{a} Range of peak.
^{b} Centre of peak.

detected. It should be also stressed that the intensities of the SAXS diffractions were more pronounced.

Once a higher quantity (2.5 wt%) of phytosterols had been solubilized, the SAXS diffraction again became less pronounced and a clear loss of peak resolution was observed. Figure 7 shows that the peaks have profiles similar to those of an empty blank sample of the Q_L phase, but with lower intensity (64% for the blank system of the Q_L phase and 53% for the enriched Q_L phase with 2.5 wt% solubilized phytosterols). Four diffraction peaks with *d*-spacing ratios of $\sqrt{2}:\sqrt{5}:\sqrt{8}:\sqrt{9}$ can be attributed to a cubic symmetry, but the peaks are insufficiently well resolved to identify precisely the symmetry of the cubic phase. We can provide a possible explanation for this behaviour: high levels of solubilized phytosterols cause structural defects in the cubic phase due to the differences in molecular size and packing between GMO and phytosterols which affects the inner order once the levels of added phytosterol become significant.

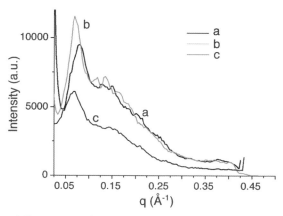

Figure 7 *X-ray diffraction profiles of samples containing various concentrations of phytosterols: (a) 0 wt%, (b) 0.2 wt% and (c) 2.5 wt%. Intensity is plotted against wave number q*

It should be noted that, for > 2.5 wt% of solubilized phytosterol, a rapid precipitation of the phytosterol is detected, and the overall structure seems to decompose. Therefore, the maximum solubilization capacity of the sterols within the Q_L mesophase was set at ~2.5 wt%.

6.3.3 Lycopene Solubilization

Lycopene could be solubilized in the Q_L phase in only minor quantities (*ca.* 100 ppm); above these levels the lycopene sedimented immediately as small crystals. The amount of solubilized lycopene was so small that it practically could not be quantitatively measured. At any measurable levels of added lycopene, the red crystals that migrated out of the microemulsion were easily detected by the naked eye.

The compound is known to be insoluble in most of the food-grade solvents. The low solubilization capacities in water are not completely surprising, but somewhat disappointing, since in our previous studies[30,31] we managed to solubilize the lycopene in water-in-oil and oil-in-water microemulsions at higher levels.

In preliminary work on synergistic drug solubilization in the Q_L phase (unpublished results), we have detected that combinations of two solubilized molecules with very different nature could have a very significant complementary effect on the nanostructure, leading to synergistic solubilization. Therefore, an attempt was made to solubilize the two nutraceuticals in the Q_L mesophase.

6.3.4 Lycopene and Phytosterol Solubilization

Lycopene and phytosterol do demonstrate a synergistic solubilization. At low phytosterol solubilization levels (< 2.5 wt%), the lycopene was found to precipitate immediately even at very low levels (<100 ppm). But once 2.5 wt% of phytosterols had been solubilized within the Q_L phase, the lycopene remained molecularly solubilized even at levels exceeding 400 ppm. We believe that the co-solubilization of these two nutraceuticals is a worthwhile achievement which may have some application advantages.

The microstructural behaviour in the presence of the two guest molecules was studied using DSC and SAXS measurements with the same thermal protocols as for the previous systems. Samples containing the two solubilized molecules exhibited thermal behaviour similar to the Q_L phase (Figure 8). However, close inspection of the effect of lycopene content (100–400 ppm) on the Q_L phase containing 2.5-wt% phytosterols reveals a small temperature shift in the phase formation upon heating towards higher temperatures with increase in lycopene content, and a large effect on the transition enthalpy. The trend in the enthalpy change for the synergistic samples was quite complex and unexpected. With 2.5 wt% of phytosterols and all three levels of solubilized lycopene (100, 200 and 400 ppm), the ΔH values were higher than for the systems containing 2.5-wt% phytosterol alone. This means that the lycopene contributes to improving the internal order of the Q_L system. This same trend was also observed in the cooling scans (results not shown).

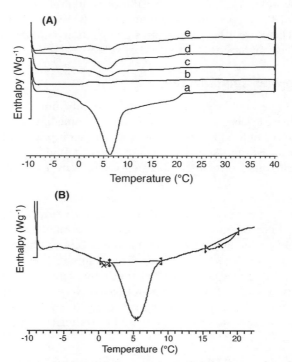

Figure 8 *(A) The heating curve from the DSC carried out on various samples: (a) control (without phytosterols or lycopene), (b) 2.5-wt% phytosterols, (c) 2.5-wt% phytosterols + 100-ppm lycopene, (d) 2.5-wt% phytosterols + 200-ppm lycopene and (e) 2.5-wt% phytosterols + 400-ppm lycopene. Enthalpy change ΔH is plotted against temperature. (B) Expanded plot for heating curve of 2.5 wt% + 100-ppm lycopene*

The maximum synergistic effect on the solubilization, and on the inner micellar order, was reached at 2.5-wt% phytosterol and ca. 200-ppm lycopene. At 2.5-wt% phytosterols + 400-ppm lycopene, an inner order begins and the solubilizates cause interfacial damage. As a result, the enthalpies decrease and the transitions occur with lower transformation energy (Table 3). The Q_L phase entrapping 2.5-wt% phytosterols + 400-ppm lycopene shows the smallest values of ΔH, while that entrapping 2.5-wt% phytosterols + 200-ppm lycopene shows the largest ΔH values.

The results indicate that the solubilized molecules can undergo a certain level of entrapment with an internal ordering effect, while above certain levels they move from the interface and cause a disordering effect which limits solubilization and affects the mesophase (causing a phase transformation). The structural change reduces the solubilization capacity of the structures, and the excess solubilizate migrates to the continuous phase and precipitates or separates.

The SAXS diffractograms reveal (similar to the effect exhibited by the thermal behaviour) that the solubilized lycopene added to the Q_L phase containing 2.5 wt% of phytosterol causes a progressive concentration-dependent effect on the nanostructure of the Q_L phase (Figure 9). The maximum effect on the inner order of the mesophases was observed at 200 ppm of solubilized lycopene. The diffractions reveal the existence of a series of seven Bragg peaks with calculated reciprocal spacing ratios of $\sqrt{2}:\sqrt{4}:\sqrt{5}:\sqrt{7}:\sqrt{9}:\sqrt{11}:\sqrt{12}$. The reciprocal d space ($1/d_{hkl}$) of the seven reflections $versus$ $(h^2 + k^2 + l^2)^{1/2}$ intercepts the axis of the origin with a very small deviation of 3.0×10^{-4} and a high linearity of 0.99975. The indexation indicates the existence of a $P4_232$ space group. Several Bragg peaks ($\sqrt{3},\sqrt{6},\sqrt{8},\sqrt{10}$) are missing. The diffraction peak $\sqrt{3}$ is not seen because the first peak in the SAXS diffractogram is wider

Table 3 *Transition enthalpy change ΔH, transition point temperature and transition temperature range of the endothermic events occurring on heating in the presence of 0.25 wt% of phytosterols and increasing amounts of lycopene*

	Temperature (°C) $L_\beta - L_\alpha$	ΔH (J g^{-1}) $L_\beta - L_\alpha$	Temperature (°C) $L_\alpha - Q_L$	ΔH (J g^{-1}) $L_\alpha - Q_L$
Q_L + 2.5-wt% phytosterols + 100-ppm lycopene	2.39–7.98[a] 5.42[b]	−1.98	15.18–19.94[a] 17.12[b]	−0.28
Q_L + 2.5-wt% phytosterols + 200-ppm lycopene	2.01–7.88[a] 5.57[b]	−3.71	13.33–20.66[a] 18.00[b]	−0.53
Q_L + 2.5-wt% phytosterols + 400-ppm lycopene	2.38–7.98[a] 5.88[b]	−1.59	13.63–19.82[a] 18.58[b]	−0.19

[a] Range of peak.
[b] Centre of peak.

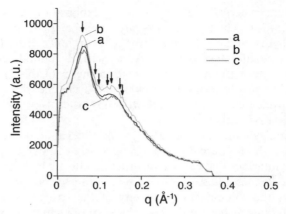

Figure 9 *X-ray diffraction profiles from samples containing 2.5-wt% phytosterol + various concentrations of lycopene: (a) 100 ppm, (b) 200 ppm or (c) 400 ppm. Intensity is plotted against wave number q*

than this diffraction peak. The other missing Bragg diffraction peaks could not be identified because of the low intensity and the high noise level.

At lower and higher concentrations of lycopene (100 and 400 ppm), the SAXS diffractions are similar to those observed for the pure Q_L system, but with lower intensities and lower resolutions. The Q_L system containing 2.5-wt% phytosterol + 100-ppm lycopene produced five Bragg peaks ($\sqrt{2}:\sqrt{5}:\sqrt{6}:\sqrt{8}:\sqrt{9}$). A sample containing 2.5-wt% phytosterol + 400-ppm lycopene also had five Bragg peaks, but with different relations ($\sqrt{2}:\sqrt{6}:\sqrt{7}:\sqrt{9}:\sqrt{10}$). This means that the initial permitted Bragg peaks of the $P4_232$ space group in these systems was reduced and long-range order also was reduced. It should therefore be stressed that, while at 200 ppm of lycopene the only Bragg peak missing is the one at $\sqrt{3}$, at 100 ppm of lycopene the peaks at $\sqrt{3}$ and $\sqrt{4}$ are missing and at 400 ppm of lycopene the ones at $\sqrt{3}$, $\sqrt{4}$ and $\sqrt{5}$ are missing. The pattern of SAXS scattering indicates clear evidence of some long-range order; therefore, the structures may have an inner order somewhere between cubic and sponge assemblies.

From the trend of the missing peaks we can also deduce that, at low and medium levels of lycopene (up to 200 ppm), the presence of the guest molecule adds to the internal order, while at more elevated levels (*ca.* 400 ppm) the presence of the guest molecule has a destructive effect, and the system symmetries are minor or small.

6.4 Conclusions

The scientific value of these results is threefold.

 (i) We have established that the Q_L phase has the capability and capacity to solubilize guest molecules of different natures.

(ii) We have developed an analytical method to detect at what concentration levels the guest molecules contribute to the inner order and symmetry of the mesophase, and at what concentration levels they destroy the inner symmetry and cause phase transformations.

(iii) We have found that certain guest molecules can complement each other structurally at the interface and exhibit synergistic solubilization.

References

1. S.T. Hyde, in *Handbook of Applied Surface and Colloid Chemistry*, K. Holmberg (ed), Wiley, New York, 2001.
2. R.R. Balmbra, J.S. Clunie and J.F. Goodman, *Nature*, 1969, **222**, 1959.
3. A. Tardieu and V. Luzzati, *Biophys. Acta Biomembr.*, 1970, **219**, 11.
4. A. Tardieu, Ph.D. Dissertation, Université de Paris-Sud, 1972.
5. V. Luzzati, R. Vargas, A. Gulik, P. Mariani and J.M. Seddon, *Biochemistry*, 1992, **31**, 279.
6. H. Delacroix, T. Gulik-Krzywicki, P. Mariani and V. Luzzati, *J. Mol. Biol.*, 1993, **229**, 526.
7. M. Clerc, A.M. Kevelut and J.F. Sadoc, *J. Phys. II (Paris)*, 1991, **1**, 1263.
8. P. Alexandridis, U. Olsson and B. Lindman, *Langmuir*, 1996, **12**, 1419.
9. J.M. Seddon, *Biochemistry*, 1990, **29**, 7997.
10. J.M. Seddon, E.A. Bartle and J. Mingins, *J. Phys., Condens. Matter*, 1990, **2**, 285.
11. P. Sakya, J.M. Seddon, R.H. Templer, R.J. Mirkin and G.J.T. Tiddy, *Langmuir*, 1997, **13**, 3706.
12. V. Luzzati, H. Delacroix and A. Gulik, *J. Phys. II (Paris)*, 1996, **6**, 405.
13. P. Mariani, B. Paci, P. Bösecke, C. Ferrero, M. Lorenzen and R. Caciuffo, *Phys. Rev. E*, 1996, **54**, 5840.
14. A. Gulik, H. Delacroix, G. Kirschner and V. Luzzati, *J. Phys. II (Paris)*, 1995, **5**, 445.
15. R. Efrat, A. Aserin, D. Danino, E.J. Wachtel and N. Garti, *Aust. J. Chem.*, 2005, **58**, 762.
16. R. Efrat, A. Aserin, E. Kesselman, D. Danino, E.J. Wachtel and N. Garti, *Colloids Surf. A*, in press.
17. R. Lifshitz, *Physica A*, 1996, **232**, 633.
18. R. Lifshitz, *Found. Phys.*, 2003, **33**, 1703.
19. J.C. Shah, Y. Sadhale and D.M. Chilukuri, *Adv. Drug Deliv. Rev.*, 2001, **47**, 229.
20. K. Kumar, M.H. Shah, A. Ketkar, K.R. Mahadik and A. Paradkar, *Int. J. Pharm.*, 2004, **272**, 151.
21. K. Larsson, *J. Phys. Chem.*, 1989, **93**, 7304.
22. M.G. Lara, M. Vitória, L.B. Bentley and J.H. Collett, *Int. J. Pharm.*, 2005, **293**, 241.
23. G.A. Kossena, W.N. Charman, B.J. Boyd and C.J.H. Porter, *J. Control. Release*, 2004, **99**, 217.

24. S. Rozner and N. Garti, *Colloids Surf. A*, 2006, **282–283**, 435.
25. A. Spernath, A. Yaghmur, A. Aserin, R. Hoffman and N. Garti, *J. Agric. Food Sci.*, 2003, **51**, 2359.
26. N. Garti, I. Amar-Yuli, A. Spernath and R. Hoffman, *Phys. Chem. Chem. Phys.*, 2004, **6**, 2968.
27. E. Giovannucci, E.B. Rimm, Y. Liu, M.J. Stampfer and W.C. Willett, *J. Natl. Cancer Inst.*, 2002, **94**, 391.
28. S. Agarwal and A.V. Rao, *Can. Med. Assoc. J.*, 2000, **19**, 739.
29. W. Stahl and H. Sies, *Arch. Biochem. Biophys.*, 1996, **336**, 1.
30. N.I. Krinsky and K.J. Yeum, *Biochem. Biophys. Res. Commun.*, 2003 **305**, 754.
31. A. Spernath, A. Yaghmur, A. Aserin, R.E. Hoffman and N. Garti, *J. Agric. Food Sci.*, 2002, **50**, 6917.
32. J. Briggs, H. Chung, and M. Caffrey, *J. Phys. II (Paris)*, 1996, **6**, 723.
33. H. Qiu and M. Caffrey, *Biomaterials*, 2000, **21**, 223.
34. V. Cherezov, D.P. Siegel, W. Shaw, S.W. Burgess and M. Caffrey, *J. Membr. Biol.*, 2003, **195**, 165.

Chapter 7
Scope and Limitations of Using Wax to Encapsulate Water-Soluble Compounds

Michel Mellema

UNILEVER R&D VLAARDINGEN, P.O. BOX 114, 3130 AC VLAARDINGEN, THE NETHERLANDS

7.1 Introduction

One of the challenges in developing functional foods is to achieve incorporation of the functional ingredient with acceptable bioavailability, but without interfering with product quality in terms of taste (bitterness, oxidation) or texture (insolubility, phase separation). Microencapsulation can be a suitable technique for incorporation of some types of functional ingredients. The crucial criterion is that the encapsulated compound should not leak out of the capsules during the storage period (weeks or months).

The following parameters are important in determining leakage from a (solid) microcapsule:

(i) Partition coefficient of the compound between the (wall) material of the capsule and the surrounding food matrix
(ii) Diffusion coefficient of the compound in the (wall) material of the capsule
(iii) Size of the capsule (thickness of the capsule wall)
(iv) Ductility of the capsule (wall)
(v) Mesh size of the capsule (wall)
(vi) Osmotic difference between internal capsule and the surrounding food matrix[1]
(vii) Molecular size of the compound

Assuming that the mesh size of the wall material is much larger than the molecular size of the compound, it is commonly accepted that the most crucial parameter determining leakage upon storage is the partition coefficient. The partition coefficient is often identified by the symbol $K_{O/W}$, representing the equilibrium distribution of the compound between octanol and water. For practical purposes this is considered similar to the relative solubility of a compound between oil and water.

Incorporation of lipids or lipophilic compounds has typically been achieved by encapsulation with a wall material of complex coacervates.[2,3] The long-term stability of such oily encapsulates is excellent (> weeks or months) for compounds like carotenoids,[4] which have an extremely high partition coefficient of $K_{O/W} \sim 10^{18}$. Similarly, compounds with an extremely low partition coefficient can, in principle, be retained by a hydrophobic wall material in an aqueous environment. The most water-soluble compounds that could be of interest to encapsulate in foods are compounds like the vitamins B and C, which have partition coefficients around 10^{-2}.

An example of a technique for encapsulating water-soluble compounds is the use of duplex or water-in-oil-in-water (W/O/W) emulsions. This technique has shown its potential in controlled release of drugs.[5,6] But the experience with duplex emulsions is that even the most extremely water-soluble compounds leak out within a few minutes (or even faster). For instance, Rayner et al.[7] reported the encapsulation of dextran blue (a hydrophilic colourant) in a W/O/W emulsion, made with sunflower oil, polyglycerol polyricinoleate (PGPR), and caseinate. They succeeded in entrapping dextran blue for 16 s in the droplets in an aqueous environment; but, due to osmotic bursting, the droplets exploded and released the dye. Instead of using liquid wall materials, one could use solid fat instead. However, this increases the complexity of the preparation considerably, and it also increases the chances of crack formation triggered by osmotic differences. We therefore conclude that existing encapsulation techniques cannot be used for the purpose of long-term retention of water-soluble compounds in aqueous applications like foods. This calls for the use of a drastically different type of wall material.

Waxes are often mentioned as excellent moisture barriers. They are esters of long-chain alcohols and fatty acids, and are found naturally in fruits, seeds, and petroleum. The most commonly used wax in foods is produced by insects – namely beeswax.[8] The most abundant ester in beeswax is C_{30}–C_{16} (Figure 1). Beeswax is used as glazing agent on candy or as coating on fruits.

Waxes are known for their hydrophobicity. At room temperature a wax is ductile without giving cracks[9]; generally waxes are softer if their chains are of unequal length. Also there are indications that the plate-like crystals are efficient in hampering the diffusion of small compounds.[10,11] Bodmeier et al.[12] explored the use of wax for encapsulation of a water-soluble compound using the W/O/W technique. Unfortunately, they found that the drug release (in this case pseudoephedrine) was much faster than from traditional polymeric microspheres, possibly because wall thicknesses were not large enough. In some recent applications,[13,14] it has also been found that for small wax microcapsules (~ 20 μm size) containing hydrophilic compounds, approximately 25–50% of the compounds leak out within half an hour.

In theory larger capsules should give better retention. Indeed it has been shown[14] that large particles (from 50 μm to 10 mm) can give excellent stability compared to similar particles prepared from solid fat. Unfortunately, even the smallest have been shown to be too large for application in foods because they gave a sandy texture.[16,17] Table 1, taken from Heith and Prinz,[17] can be used as

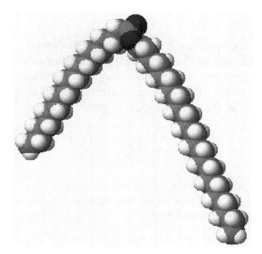

Figure 1 *Schematic representation of the main ester of the long-chain alcohol (C_{30}) and fatty acid (C_{16}) found in beeswax. The angle between the alcohol and the fatty acid chains is approximately 90°*

Table 1 *The detection threshold of various particles in foods*

Detecting Agent	Particle and Medium	Detection Threshold (μm)
Teeth	Aluminium trihydrate	5–15
Tongue	Separate particles	13
Palate	Hard, angular particles	11–22
Teeth	Not stated	15
Palate	Crystals	22
'Oral'	Sugar crystals	22
Tongue	Chocolate crystals	>20% are 22 μm
Tongue	Margarine fat (β' crystals)	25
Tongue/palate	Soft, round, biopolymer, 'flat' particles	80–100
'Oral'	Average food particles at end of mastication	1–2

a guideline to define which range of particle sizes are needed to prevent such texture defects.

We can conclude from Table 1 that (wax) particles need to be smaller than ~25 μm. To reach such small sizes a preparation technique has been developed. Results of this preparation technique are presented in this paper. Furthermore, we will compare experimental data obtained using a technique described previously,[15] and experimental data of the leakage of oil-soluble model compounds from complex coacervate capsules, with model calculations. Finally, we shall discuss the issues of wax digestibility, which could limit the acceptability of wax as a wall material.

7.2 Theory

The calculation of the release rate of an ingredient from a capsule or sphere is based on the Crank model.[18] The underlying assumption is that ingredients that diffuse out of the capsule are instantaneously removed, and so there is no external resistance in the surrounding phase to the mass transfer. The short-time approximation of the Crank model is given by Baker:[19]

$$\frac{M_t}{M_0} = 6\sqrt{\frac{D_{ep}t}{\pi R^2}} - 3\frac{D_{ep}t}{R^2} \tag{1}$$

where t is the time, M_0 the initial total mass of compound in the capsule, M_t the mass of compound diffused out of the capsule, R the diameter of the capsule, and the quantity D_{ep} is defined by

$$D_{ep} = \frac{D_e}{1 - \varphi + \varphi K_{dc}} \tag{2}$$

The effective diffusion coefficient D_e depends on the microstructure of the capsule material. It may be estimated by empirical correlation, e.g., by the general Maxwell equation[20] for spheres,

$$D_e = D_c \frac{2D_c(1 - \varphi) + K_{dc}D_d(1 + 2\varphi)}{D_c(2 + \varphi) + K_{dc}D_d(1 - \varphi)} \tag{3}$$

where φ is the volume fraction of the inner phase of the capsule, K_{dc} the partition coefficient, and D_d and D_c are diffusion coefficients of the compound in, respectively, the capsule (wall) material and the continuous phase.

7.3 Materials and Methods

7.3.1 Preparation of Wax Capsules

The following ingredients were used in the making of the various wax-based encapsulates: beeswax (Fagron, Cera Alba #7323), polyglycerol polyricinoleate (PGPR, Admul WOL, Quest International), acids (Merck), fluorescein isothiocyanate (FITC, Janssen Chimica), rhodamine B (Aldrich), and refined sunflower oil from Albert Heijn ('Zonnebloemolie').

Two preparation techniques were employed as illustrated schematically in Figure 2. Method 1, the 'solid' preparation technique, is elaborately explained elsewhere.[15] In preparation technique 2, water containing FITC (total volume 20%) was added to a solution of oil containing 1.6 wt% PGPR and 2 wt% beeswax. The mixture was gently stirred at 65°C for 10 min, and then was cooled down to room temperature or 5°C within 1 h while still stirring. The resulting liquid-like viscous W/O emulsion consisted of small water droplets (with dye) of size 1–10 μm stabilized by Admul WOL and wax.

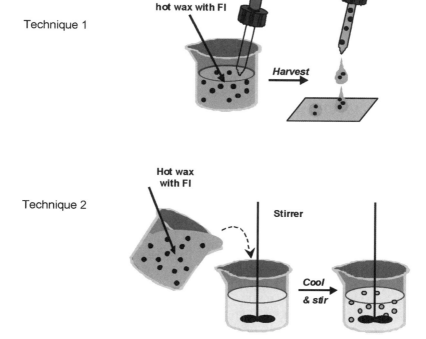

Figure 2 *Schematic representation of techniques for preparing the wax capsules. Technique 1 involves deposition on a plate of hot wax with the encapsulated compound (a 33% solution of citric acid), leading to large particles (1–10 mm). Technique 2 involves stirring in hot vegetable oil of hot wax with the encapsulated compound (a solution of FITC), followed by cooling while stirring, giving small particles (1–10 μm)*

7.3.2 Analysis of Wax Capsules

The leakage test of capsules made by preparation technique 1 involves storage of the capsules in demineralized water at room temperature in the dark. A quantity of capsules (2.5 g) was added to 100 mL water. Leakage of citric acid was quantified by measuring the pH of the outer water phase.

The leakage test of the capsules made by technique 2 basically involves mixing of freshly made wax 'capsules' (basically a W/O emulsion made from water + oil + FITC + wax + PGPR) with a W/O emulsion made from water + oil + PGPR at a volume ratio of 1:1. After this, some rhodamine was added to colour the wax and PGPR. Measurement of the leakage rate was performed by monitoring the concentration of FITC in the 'empty' water droplets of the second emulsion. This was done using a confocal laser scanning microscope (CLSM), a Bio-Rad MRC1024 equipped with an Ar/Kr laser combined with a

Zeiss Axiovert (inverted) microscope. Images were obtained at two different magnifications: objective 40× (zoom 2) and objective 63× (zoom 2.5), resulting in image widths of 130 and 65 μm, respectively.

To visualize the water droplets in the oil phase the following excitation/emission wavelengths were used: water droplets with rhodamine B (excitation, 568 nm; emission, ≥ 585 nm); water droplets with FITC (excitation, 488 nm; emission, 522 ± 16 nm); water droplets with rhodamine B; and water droplets with FITC in one W/O emulsion (excitation, 488 + 568 nm; emission, ≥ 585 nm for water droplets with rhodamine B, channel A; and 522 ± 16 nm for water droplets with FITC, channel B).

7.3.3 Preparation and Analysis of Complex Coacervate Capsules

The following materials were used to make the gelatin/arabic gum complex coacervate capsules: beef gelatin 250 bloom (Geltec), gum arabic (Merck), 50.3% glutamic dialdehyde solution in water (Acros), ethylbenzene, n-hexylbenzene, n-decylbenzene, n-tetradecylbenzene, n-octadecylbenzene (Aldrich), and refined sunflower oil from the Unilever Vlaardingen pilot plant. Table 2 gives the $K_{O/W}$ value of the encapsulated oil-soluble alkylbenzenes.

A 2-wt% solution of gum arabic in 1 L of demineralized water was prepared, with the gum added while stirring at 60°C. A 2-wt% solution of gelatin in 1 L of demineralized water was prepared, with the gelatin added while stirring at 60°C. The gum arabic and gelatin solutions were mixed together at 60°C while stirring at 150 rpm. To this aqueous mixture of gelatine + gum arabic, 100 g of sunflower oil was added, together with 3 mmol of the specific alkylbenzene. The mixture was stirred with an Ultraturrax mixer for 1 min at 1.35×10^4 rpm, with 1.0 M hydrochloric acid added while stirring until a pH of 4.2 was reached. The vessel was then slowly cooled to 5°C, the solution was decanted, and 2 L of fresh water was added. Some 0.25 g of glutaric dialdehyde solution (50%) was added to 180 mL water containing 100 g of coacervate, and mixed for 18 h at approximately 20°C. Water was removed by paper filtration, and the capsules were washed several times to reduce the amount of cross-linking agent. In the last washing step 0.1% potassium sorbate was added. The capsules were mixed into a 70% fat margarine spread ('Bona') to a concentration of 5%, and stored at 5°C.

Table 2 *Octanol–water partition coefficients and molecular weights of oil-soluble alkylbenzenes encapsulated in complex coacervate capsules*

Component	Log $(K_{O/W})$	Molecular weight (Da)
Ethylbenzene	3.03	106.2
n-Hexylbenzene	5.00	162.3
n-Decylbenzene	6.96	218.4
n-Tetradecylbenzene	8.92	274.5
n-Octadecylbenzene	10.89	330.6

At regular time intervals upon storage, the fat phase of the spreads was analysed using gas chromatography (GC) for the presence of alkylbenzenes. Fat phases of the spreads were extracted by melting, centrifugation, and filtration. Samples were saponified and extracted with n-hexane. The sensitivity of the GC analysis was 1–3 mg kg^{-1}.

7.3.4 Analysis of Wax Digestibility

To estimate the ease of digestibility of the wax, an *in vitro* lipolysis test was performed. As a reference substance we used Tween 20, which is assumed to hydrolyse completely. The waxes used were: beeswax, candelilla wax, ricebran wax, and carnauba wax, all obtained from Koster Keunen (the Netherlands). Sunflower wax slurry was obtained from the Unilever refinery at Inverno, Italy. Final sunflower wax was obtained after further fractionation at the Unilever Vlaardingen pilot plant.

To prepare the wax emulsions, 88.0 g of demineralized water was placed into a double-walled vessel, connected to a water bath at a temperature of 85°C, and 150 mg of xanthan gum (Keltrol RD) was mixed into the water with an Ultraturrax. A sample of 10.0 g of wax or 2.0 g of Tween 20 was heated in a glass beaker until melted and then mixed. The liquid lipid phase was added to the water phase while mixing with the Ultraturrax at 1.5×10^4 rpm for 5 min at 80–85°C. By replacing the content of the water bath with ice water, the emulsion was cooled within 5 min to 20°C while stirring at 1×10^4 rpm.

To carry out the lipolysis reaction, an incubation buffer of 150 mM sodium chloride, 2 mM tris(hydroxymethyl)aminomethane maleate, 95 mM calcium chloride, and 0.04-wt% Tween 20 was prepared. The pH of the buffer was adjusted with 1.0 M sodium hydroxide solution to pH = 6.5. For each lipolysis assay 40 mL of incubation buffer, 10 mL of wax emulsion, and 280 mg of bile extract were mixed. The lipolysis reaction was started with the addition of 1.80 g of pancreatin powder. During the incubation period of 30 min, automatic addition of sodium hydroxide solution was performed to maintain a pH value of 6.5 (pH stat).

The molar amount of sodium hydroxide solution added is equal to the molar amount of fatty acids formed, which are present after the hydrolysis. For the wax hydrolysis, it was assumed that pancreatic lipase hydrolyses 1 mol of wax into 1 mol of fatty acid and 1 mol of fatty alcohol. For Tween 20 hydrolysis, it was assumed that pancreatic lipase hydrolyses 1 mol of Tween 20 into 1 mol of polyoxyethylene sorbitan and 1 mol of lauric acid. The degree of hydrolysis of the wax was calculated from the molar amount of added sodium hydroxide divided by the molar amount of wax present in the incubation medium. The assumed average molecular masses of the different compounds were 1227 Da (Tween 20), 674 Da (beeswax, candelilla wax, carnauba wax, and ricebran wax), and 760 Da (sunflower wax).

7.4 Results

7.4.1 Wax Capsules Prepared by Technique 1

Some large capsules of varying size were made by preparation technique 1. They were loaded with a 33% solution of citric acid. We monitored the leakage of citric acid from these capsules as a function of time.[15] The experimental data presented in Figure 3 clearly show the effect of particle size. The same graph shows the results of model calculations using a short-time approximation of the Crank model. The parameter values of the model were chosen as follows: $D_d = 4.5 \times 10^{-15}$ m^2 s^{-1}, as taken from Scheiber and Riederer,[11] $\varphi = 0.4$, and $K_{O/W} = 22$.

From Figure 3 we can see that the leakage rate is unacceptable for the 1-mm size capsules over a typical storage period of a couple of weeks, whereas the larger capsules are considerably more stable. The parameter values of the model calculations are realistic for the given situation, as they lead to a reasonable fit to the experimental data, which confirms the reliability of the theoretical approach.

7.4.2 Wax Capsules Prepared by Technique 2

A 20% W/O emulsion was made by preparation technique 2 using PGPR and beeswax as 'emulsifiers'. This technique was developed to prepare the smallest possible wax capsules in order to overcome any textural defects in foods associated with the other preparation techniques.[15] The water-soluble fluorescent dye FITC was used as the model encapsulated compound.

The previously described emulsion was mixed with another emulsion of similar composition, except that no wax and FITC were included. We used confocal microscopy to monitor the leakage from the FITC-loaded droplets into the 'empty' ones. The results are presented in Figure 4.

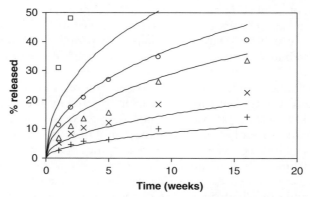

Figure 3 *Leakage test results of wax capsules. Percentage of maximum citric acid leaked out is shown as a function of time for various capsule sizes: □, 1 mm; ○, 1.5 mm; △, 2 mm; ×, 4 mm; and +, 7 mm. Lines show results of model calculations using a short-time approximation of the Crank solution for the corresponding capsule sizes*

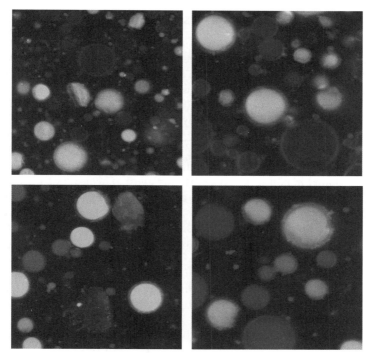

Figure 4 *Confocal micrographs of a mixture of FITC-loaded and wax-stabilized water droplets (whitish on the inside) and 'empty' water droplets (dark on the inside). Rhodamine B colours the wax and PGPR is situated at the interface. Top images are for samples one-hour old. Bottom images are for samples stored chilled for one week. Image widths: 130 µm (left-hand side) and 65 µm (right-hand side)*

In Figure 4 we see that the FITC-filled droplets have an irregular structure at the interface, originating from the presence of the wax particles. The 'empty' droplets have a smooth interface because they are stabilized by PGPR. We can speculate that there is Pickering stabilization by the wax particles[21] on top of the molecular steric stabilizing effect of the PGPR.

In the one-hour-old samples we clearly notice two types of droplets: one type containing FITC (white) and the other type not containing FITC (black). We conclude from this that the FITC does not leak from one droplet to the other over a time period of an hour. In the one-week-old samples, however, we notice an increase in FITC content of the droplets with the smooth interfaces, *i.e.*, those droplets which were depleted in FITC and wax. This shows that wax-stabilized water droplets in an oily environment cannot retain water-soluble compounds over a storage period of a week or longer. The retention is here nevertheless better than described previously by Bodmeier *et al.*,[12] probably because the waxy water droplets in this study are immersed in oil.

7.4.3 Complex Coacervate Capsules

The previous sections give a picture of the scope and limitations of the use of wax as a wall material for encapsulation of water-soluble compounds. We argue that, although wax is a strong barrier, it is not effective at the small particle sizes required to prevent negative mouthfeel issues. Of course, the problem could also be approached differently: that is, it could be argued that the compounds we are trying to encapsulate are not sufficiently water-soluble to be held inside simply because the mesh size of any digestible wall material will always be too large. This hypothesis is also supported by the observation that only oil-soluble compounds with extreme partition coefficients (*e.g.*, carotenes) can be successfully encapsulated in complex coacervate capsules.[4] This leaves us with one key question: how water-soluble should a compound be in order to allow stable encapsulation? While the requirements of the wall material and the internal phase are opposing in terms of hydrophobicity, the example of complex coacervate capsules could contribute towards answering this question. Hence we have tested a range of oil-soluble compounds with varying partition coefficient in systems of complex coacervate capsules. The capsules were added to a 70% fat margarine spread (Bona). Leakage was monitored by measuring the concentrations of oil-soluble compounds (alkylbenzenes) in the Bona fat phase at various time intervals.

We discussed above the relationship between leakage data from wax capsules with the simple model predictions. We conclude from this that the model used gives a reasonable prediction of the particle-size dependency of the leakage. Now we use the same model to predict leakage data for the complex coacervate capsules.

Figure 5 shows the experimental data and Figure 6 gives the results from the calculations assuming $D_d = 10^{-9}$ m^2 s^{-1}, $\varphi = 0.35$, and $R = 100$ μm. We

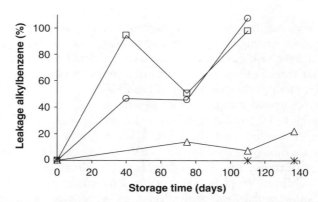

Figure 5 *Experimental data for leakage of alkylbenzenes from gelatin–arabic gum complex coacervate capsules during storage in spreads at 5°C:* □, *ethylbenzene* ($log_{10}K_{O/W} = 3$); ○, *hexylbenzene* ($log_{10}K_{O/W} = 5$); △, *decylbenzene* ($log_{10}K_{O/W} = 7$); ×, *tetradecylbenzene* ($log_{10}K_{O/W} = 9$); +, *octadecylbenzene* ($log_{10}K_{O/W} = 11$)

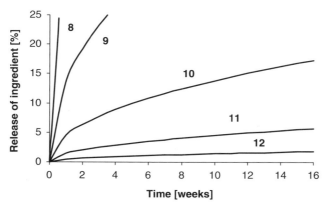

Figure 6 *Results of model calculations on the leakage of oil-soluble ingredients with log value of partition coefficient $K_{O/W}$ as indicated (ranging from 8 to 12) from capsules with a hydrophilic barrier (like complex coacervate capsules)*

conclude from Figure 5 that a partition coefficient of at least 10^9 is required to give stable encapsulation. Figure 6 points in the direction of even higher partition coefficients ($K_{O/W} > 10^{10}$). It could be argued that the effective diffusion coefficient D_e for wax wall materials is much smaller than for complex coacervate wall materials. However, the effect of this parameter in the model is much less than, for instance, the effect of the partition coefficient or of the mean particle size.

We conclude that, for stable encapsulation of water-soluble compounds in wax-based capsules of a size smaller than what is perceived in the mouth as sandiness, we need partition coefficients many orders of magnitude smaller than those of existing compounds of interest like water-soluble vitamins. Hence, with existing food-grade digestible wall materials, we are far away from being able to encapsulate water-soluble compounds into aqueous foods.

7.4.4 Digestibility of Wax

Waxes are common compounds in the everyday food of various animals.[22,23] Most mammals are able to digest waxes only to a limited extent, however, because they do not produce sufficient bile to facilitate the digestion process. To determine whether some are more digestible than others, we have performed hydrolysis tests on a range of waxes.

The data in Figure 7 confirm our expectations that waxes are indeed hydrolysed, but very slowly. Part of the slow rate originates from the fact that the material to be hydrolysed is solid: from other experiments (results not shown here) we know that sunflower oil has a similar hydrolysis profile to Tween 20, but that hardened vegetable fats like fully hardened rapeseed oil are hydrolysed more slowly, though still twice as fast as waxes.

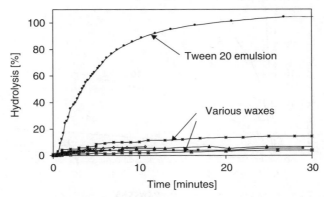

Figure 7 *Time-dependent hydrolysis of solutions of various types of waxes in comparison with a solution of Tween 20*

Whether the poor digestibility of waxes affects their potential as wall materials is not fully answered by our data. It appears that waxes are probably sufficiently digested to release a portion of an encapsulated ingredient, but probably not all of the ingredient. Differences between the various waxes are rather small. One type of wax was hydrolysed faster than the others because it contained a higher proportion of vegetable oil. This indicates that digestibility could be further improved by mixing the wax with liquid triglyceride oil.

7.5 Concluding Remarks

We have presented results of leakage tests for two types of wax capsules. The particles are either too large, and so give rise to textural defects, or are much smaller and then leakage occurs within hours to days. We conclude from this that the expected shelf-life of most foods cannot be reached using wax as a wall material. Moreover, wax is rather poorly digested, which could further limit its potential use.

Using model calculations and comparing with encapsulation techniques developed for oil-soluble compounds, we have shown theoretically that the oil/water partition coefficient of a water-soluble compound needs to be many orders of magnitude lower than that of existing functional compounds of interest. This is the case when the mesh size of the wall material is much larger than the size of the encapsulated compound. We therefore need digestible food-grade wall materials that have a much smaller mesh size. Such materials are, however, not currently available.

Acknowledgements

The following people contributed to this study: Ruud den Adel, Christiaan Beindorff, Wim van Benthum, Brigitta Boer, Guus Duchateau, Henk Husken, Guoping Lian, Rob Vreeker, and Aad Visser.

References

1. R. Mezzenga, B.M. Folmer and E. Hughes, *Langmuir*, 2004, **20**, 3574.
2. H. Jozomoto, E. Kanaoka, K. Sugita and K. Hirano, *Pharm. Res.*, 1993, **10**, 1115.
3. F. Weinbreck, M. Minor and C.G. de Kruif, *J. Microencaps.*, 2004, **21**, 667.
4. J. Bodor, M. Gude, E.G. Pelan and A. Visser. Patent WO 02 41711, 2002.
5. S.K. Das, *Drug Dev. Ind. Pharm.*, 1991, **17**, 2521.
6. M. Moroshita, A. Matsuzawa, K. Takayama, K. Isowa and T. Nagai, *Int. J. Pharm.*, 1998, **172**, 189.
7. M. Rayner, B. Bergenståhl, L. Massarelli and G. Tragaradh, conference abstract, *Food Colloids 2004: Interactions, Microstructure and Processing*, Harrogate, UK, April 2004.
8. S. Patel, D.R. Nelson and A.G. Hibs, *J. Insect Sci.*, 2001, **1.4**, 1.
9. L.C. McMillan and B.W. Darvell, *Dental Mater.*, 2000, **16**, 337.
10. I.G. Donhowe and O. Fennema, *J. Am. Oil Chem. Soc.*, 1993, **70**, 867.
11. L. Scheiber and M. Riederer, *Plant Cell Environ.*, 1996, **19**, 1075.
12. R. Bodmeier, J. Wang and H. Bhagwatwar, *J. Microencaps.*, 1992, **9**, 99.
13. F. Quaglia, F. Barbato, G. de Rosa, E. Granata, A. Miro and M.I. la Rotonda, *J. Agric. Food Chem.*, 2001, **49**, 4808.
14. M.S. Uddin, M.N.A. Hawlader and H.J. Zhu, *J. Microencaps.*, 2001, **18**, 199.
15. M. Mellema, W.A.J. van Benthum, B. Boer, J. von Harras and A. Visser, *J. Microencaps.*, 2006, **23**, 729.
16. P. Tyle, *Acta Psychol.*, 1993, **84**, 111.
17. M.R. Heath and J.F. Prinz, in *Food Texture: Measurement and Perception*, A.J. Rosenthal (ed), Gaithersburg, MD, 1999, p. 18.
18. J. Crank, *The Mathematics of Diffusion*, Oxford University Press, Oxford, 1956.
19. R. Baker, *Controlled Release of Biologically Active Agents*, Wiley, New York, 1987.
20. J.C.A. Maxwell, in *Treatise on Electricity and Magnetism*, 2nd edn, Clarendon Press, Oxford, 1881, p. 435.
21. N.X. Yan, M.R. Gray and J.H. Masliyah, *Colloids Surf. A*, 2001, **193**, 97.
22. A.R. Place, *Am. J. Physiol.*, 1992, **263**, R464.
23. A.R. Place, *Poultry Avian. Biol. Rev.*, 1996, **7**, 127.

Chapter 8

Self-Assembly of Starch Spherulites as Induced by Inclusion Complexation with Small Ligands

Béatrice Conde-Petit,[1] Stephan Handschin,[1] Cornelia Heinemann[2] and Felix Escher[1]

[1] INSTITUTE OF FOOD SCIENCE AND NUTRITION, ETH ZURICH, CH-8092 ZURICH, SWITZERLAND
[2] NESTLE PTC KONOLFINGEN, NESTLESTRASSE 3, CH-3510 KONOLFINGEN, SWITZERLAND

8.1 Introduction

Starch is synthesized in higher plants in the form of partly crystalline granules with different morphologies and sizes between 1 and 100 µm. The linear amylose and the branched amylopectin are the starch polymers, and most common starches are composed of 25% amylose and 75% amylopectin. The formation of starch granules in the plant is a process governed by genetics[1] and thermodynamics.[2] The process yields starch assemblies with a more or less spherical morphology and with alternating shells of amorphous and crystalline material.

To a certain extent native starch granules can be considered as natural spherulites. Starch granule assembly is thought to start from the hilum. This 'nucleus' of the granule and its surrounding core contain less well-organized polymers and sometimes an amylose-rich core. Spherulitic starch crystallization has been proposed as a model for starch granule initiation *in vivo*.[3,4] Ziegler *et al.*[4] have demonstrated the spontaneous formation of spherulites in aqueous mung bean starch dispersions at concentrations between 10% and 20%. This type of spherulitic assembly presents a B-type crystallinity, which corresponds to a double-helical arrangement of starch in the crystals. The spherulites are mainly composed of amylose or slightly branched material, and their crystallization is preceded by a phase-separation process.

Aqueous solutions of maltodextrins obtained by enzymatic degradation and debranching may also present spherulitic crystallization on holding at moderately elevated temperature (52°C).[5] In the early stages of the reaction the crystalline material presents B-type crystallinity, while the material shifts to an A-type crystalline packing in the later stages of the process. Both A- and B-type starch crystals are based on double-helical starch assemblies, with the difference being that the A-type crystal has a denser packing and only a few water molecules in the monoclinic cell.[6] The process of split crystallization yields highly crystalline material with spherical morphology which is resistant towards enzymatic degradation in the small intestine. This type of 'resistant starch' has found application as functional fibre in food because of promising physiological effects in humans.[7]

An alternative mechanism by which spherulitic crystallization can be initiated is the formation of helical amylose inclusion complexes in the presence of suitable ligands that induce the helix formation of amylose. Among the first reports on starch spherulites were the investigations of Bear[8] and Schoch[9] published in 1942. It was observed that addition of butanol to starch dispersions leads to a selective precipitation of amylose in the form of six-segmented spherulites with a diameter of 15–50 μm. The spherulites were found to be birefringent under polarized light and showed an interference pattern similar to the well-known Maltese cross of native starches. In contrast to the native starch granules, the *de novo* formed spherulites are composed of amylose only, this starch fraction being particularly prone to crystallization. The partly crystalline amylose structures formed by amylose inclusion complexation are also known as 'reformed amylose particles' and the process has been termed 'high-temperature retrogradation'.[10] In cereal starches, endogenous lipids may promote the formation of spherulites.[11,12] Besides linear alcohols and monoacyl lipids, other ligands suitable for inducing the formation of starch spherulites are terpenes and lactones. The present contribution reports on the spherulitic crystallization of potato starch as induced by inclusion complexation with different lactones. The latter molecules are aroma active and are relevant as naturally occurring flavouring agents in foods.

8.2 Materials and Methods

Potato starch was obtained from Blattmann (Wädenswil, Switzerland). The flavour compounds, γ-heptalactone, γ-nonalactone, γ-decalactone, γ-dodecalactone, δ-decalactone and δ-dodecalactone, all of *purum* quality, were supplied by Fluka (Buchs, Switzerland), as was the α-amylase, isolated from pig pancreas with an activity of 20 U mg^{-1}.

Starch dispersions (2 wt%) were prepared by heating native starch suspensions in cans at 121°C for 30 min or in beaker at 95°C for 30 min. In the latter case, the remnants of swollen starch granules were dispersed by treating the dispersion for 1 min with a benchtop homogenizer (Polytron, Kinematica, Littau, Switzerland). The lactones were added to the starch dispersions at room temperature at concentrations between 10 and 50 mmol mol^{-1} of anhydroglucose (2% starch ≡ 122.6 mmol mol^{-1} anhydroglucose) so that the extent of amylose complexation was rather low. The complexation index of amylose in the range 5–30% was measured by iodine titration.[13,14] Selected samples were complexed at higher lactone concentration to saturate the amylose-binding sites.[13] The samples were mixed for 20 s and kept at 25°C for 24 h before analysis. In one case a sample was annealed by heating the suspension of starch spherulites to 55°C for 60 min before further analysis by differential scanning calorimetry (DSC).

Spherulites were characterized by light microscopy by placing on a slide a drop of the starch dispersion, preferably from the precipitate. The sample was observed with an Axioskop microscope (Zeiss, Germany) equipped with a Hamamatsu 58103 CCD camera under polarized light, in phase contrast, or in the differential interference contrast (DIC) mode.

The dynamics of spherulitic starch crystallization was also observed *in situ* by light microscopy. For this purpose a potato starch dispersion was mixed with δ-dodecalactone in a syringe and injected onto the sample holder of a hot stage at 35°C (Linkam LTS 350 hot stage controlled by the LinkSys32 software, Linkam, Surrey, UK). The same system was also used to follow the structural changes of spherulites upon heating from 30 to 115°C with a heating rate of 5°C min^{-1}.

DSC and wide-angle X-ray diffraction measurements were performed on samples of extensively complexed starch at high lactone concentration. The samples were freeze-dried and measured as described elsewhere.[13] For DSC measurements the rehydrated samples (30% dry matter) were weighed (~ 40 mg) into pressure pans (Perkin Elmer, Norwalk, CT, USA) and measured in a DSC instrument (TA Instruments 2920, USA) at 10°C min^{-1} from 4 to 160°C. After cooling the sample to 4°C with an average cooling rate of ~ 20°C min^{-1}, a second run was performed. For the wide-angle X-ray diffraction measurements, the moisture content of freeze-dried samples was adjusted to 10–15%. The samples were compressed into tablets and measured in the transmission mode on a powder diffractometer (Siemens Kristalloflex D500, Karlsruhe, Germany) using CuKα radiation (1.54 Å) with 35 mA and 40 kV. A divergence slit of 2 mm and a receiving slit of 1° were chosen.

The susceptibility of spherulites towards α-amylase was assessed by treating a suspension of spherulites with α-amylase (activity 100 U g^{-1} dry starch) for 24 h at 25°C. The changes at the molecular level were followed by measuring the iodine-binding capacity by amperometric iodine titration.[14] The thermal stability and the microstructure of the spherulites were followed by DSC and light microscopy, respectively.

8.3 Results and Discussion

These experiments were carried out with potato starch because it is a lipid-free starch, which therefore does not contain pre-existing complexes. Upon addition of lactones at concentrations leading to a partial complexation of starch dispersions,[13,14] the spherulites could be observed by light microscopy after 24 h as shown in Figure 1. Spherocrystalline structures with diameters ranging from 10 to 90 μm were found. Three different morphologies are present: small

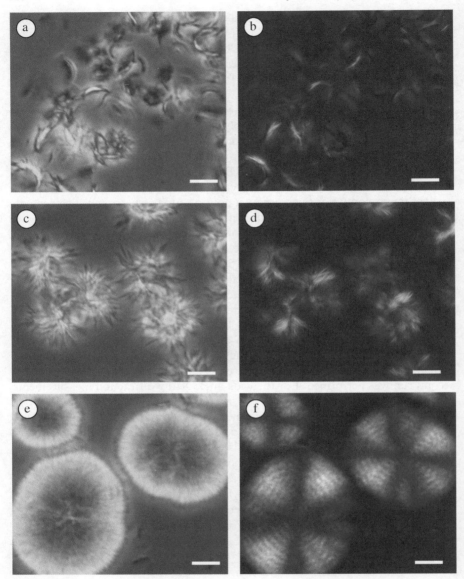

Figure 1 Polarized light microscopy [images (a), (c) and (e)] and phase contrast light microscopy [images (b), (d) and (f)] of amylose−lactone spherulites, 24 h after complexation (scale bar = 10 μm). Different spherulitic morphologies were obtained depending on the type of lactone: rod-like spherulites with γ-dodecalactone (a,b), irregular spherulites with γ-nonalactone (c,d) and spherulites with a Maltese cross with δ-dodecalactone (e,f)

rod-like spherulites with dimensions from 10 to 15 μm, irregular spherulites with diameters from 20 to 50 μm and rather large spherulites from 20 to 90 μm in size, which presented a Maltese cross and a banded extinction pattern when viewed under polarized light. Most likely the spherulites shown in Figure 1 correspond to different stages of spherulitic growth, as also shown for chitosan crystals.[15] Late stages of spherulitic growth lead to spherical structures that display a distinct Maltese cross.

The appearance of spherulites is linked to the helix-forming tendency of starch in the presence of lactones and the subsequent formation of ordered structures. The complexation of amylose can be viewed as an interaction that drives this polymer out of solution, due to the strong association tendency of the helical amylose segments, in order to escape the unfavourable interaction with water. Spherulites grown by the addition of lactones to starch dispersions present the typical V_h-type crystalline packing of amylose inclusion complexes as characterized by wide-angle X-ray diffraction (Figure 2). The reflections at 7.5°, 13.0° and 20.0° correspond to crystalline assemblies of single-left-handed amylose helices with six glucose units per helical turn. The X-ray diffractogram

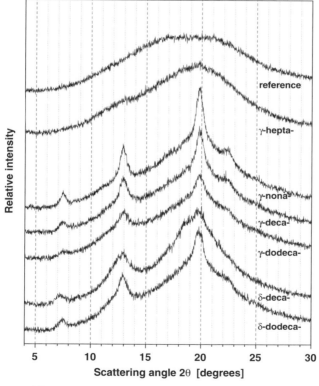

Figure 2 *X-ray diffraction diagrams of freeze-dried potato starch dispersions without addition (reference) and with addition of various lactones. Relative intensity is plotted against scattering angle*

of potato starch in the presence of γ-heptalactone suggests that the ligand was not able to form stable complexes, whereas all other lactones were good complexing partners. Depending on the rate of complexation, the formation of amylose complexes leads to the formation of a network gel or spherulitic crystallization accompanied by solid−liquid bulk phase separation.[14] The complexation rate is influenced by the type of ligand, the dispersion state and concentration of the ligand, and the temperature and shearing conditions. It can be assumed that starch spherulites are primarily composed of amylose as suggested by blue staining with iodine (results not shown). This implies that the formation of spherulites is the result of an amylase + amylopectin phase-separation process. Amylose and amylopectin possess limited miscibility in aqueous systems and the spontaneous phase-separation process is promoted by the formation of amylose complexes. Spherulitic crystallization is favourable if the continuity of the amylose phase falls below a critical level before extensive amylose complexation is reached. On the other hand, if the spontaneous amylase + amylopectin phase separation progresses faster than amylose aggregation, the formation of a kinetically stabilized amylose network is promoted.[14]

The dynamics of spherulite formation upon complexation of potato starch with δ-dodecalactone was followed with light microscopy. Like most good complexing partners, δ-dodecalactone possesses a low solubility in water which is revealed by the emulsion character of the starch-δ-dodecalactone system. The micrographs in Figure 3(a–d) show that the growth of spherulites starts in the vicinity of a drop of δ-dodecalactone, where a ring-like structure consisting of small crystals or a gel phase appears after a few minutes. This ring forms a kind of envelope around the ligand droplet and serves as nucleation point for the formation of spherulites. Large spherulites that present birefringence are visible after 24 h at the interface of the lactone droplet [Figure 3(e,f)]. Numerous spherulites also develop in the bulk aqueous phase, which implies that lactone molecules diffuse into the aqueous phase. The size and the degree of perfection of the spherulites increase with the distance to the lactone droplets. This is in line with the hypothesis that a low complexation rate − in this case as a consequence of the low ligand concentration − favours self-assembly of the complexes into well-ordered structures.[14,16]

The thermal stability of amylose−lactone spherulites was analysed by DSC and hot-stage microscopy. The results are presented in Figure 4. Spherulites formed with δ-decalactone exhibited a broad multiple melting transitions with a peak temperature around 80°C and a melting range of ∼40°C. The annealing of spherulites at 55°C for 60 min did not lead to significant changes in the melting range, but the curve shape suggests that more thermostable crystals are formed on annealing at the expense of less stable ones.[17] The existence of different thermodynamic states of complexes with different thermal stabilities has also been described for amylose−monoglyceride complexes.[18] Similar to the melting of native starch granules, the melting of amylose spherulites promotes the disappearance of the birefringence of spherulites. At the end of the melting process, remnants of spherulitic structures could be seen in phase

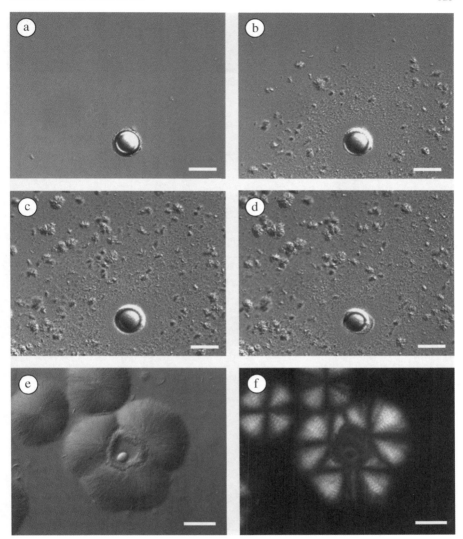

Figure 3 *Monitoring spherulitic crystallization in a potato starch dispersion in presence of δ-dodecalactone by DIC and polarized light micrographs at 35°C. Images were taken after various times in DIC mode: (a) 3 min; (b) 10 min; (c) 20 min; (d) 60 min; and (e) 24 h. Image (f) was taken after 24 h in polarized light mode. Scale bars: (a–d), 50 μm; (e,f), 20 μm*

contrast mode (results not shown). These structures were not birefringent, but were similar to the 'ghosts' that are observed after heating of native starch granules.[19]

Finally, the structural reorganization of amylose–γ-nonalactone spherulites upon enzymatic degradation by α-amylase at rather low enzyme activity was investigated. Figure 5 presents the structural changes as followed by iodine

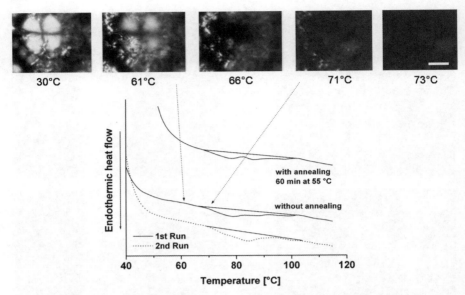

Figure 4 *Melting behaviour of amylose + δ-decalactone spherulites as assessed by hot-stage polarized light microscopy and DSC. The latter measurements were carried out without and with previous annealing at 55°C for 60 min. Scale bar = 20 µm*

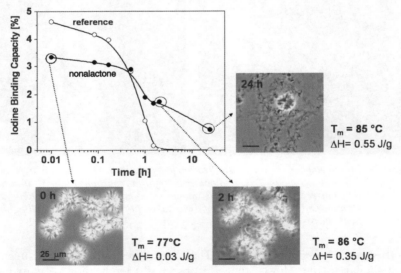

Figure 5 *Time-dependent structural changes of starch–γ-nonalactone spherulites at different length scales upon incubation with α-amylase of rather low enzyme activity at 25°C, as assessed by iodine-binding capacity, DSC and light microscopy. Values are given of the melting temperature T_m and melting enthalpy ΔH*

titration, DSC and light microscopy. The degradation of starch at the molecular level led to a decrease of the iodine-binding capacity, and the rate of starch degradation was lower for complexed starch compared with the reference. In the micrometre range, the disintegration of the spherulites was seen by microscopy, although remnants of the supramolecular assemblies could still be recognized after an incubation time of 24 h. Interestingly, the melting temperature T_m and the melting enthalpy ΔH of amylose–γ-nonalactone complexes were found to increase with enzymatic treatment. These results indicate that at low α-amylase activity the complexes are not completely degraded. By analogy with amylose–lipid complexes, it can be assumed that ordered helical segments are less susceptible towards enzymatic degradation than the amorphous regions.[20] The DSC results suggest that a partial degradation of starch contributes to a higher mobility of the polymer, which in turn contributes to the formation of more stable crystallites at the expense of less stable ones. This means that a slight enzymatic hydrolysis of starch complexes may increase the thermal stability of the remaining complexes, but at the same time it induces partial disassembly of the spherulites.[16] It can be hypothesized that an enzymatic annealing of spherulites in combination with a thermal annealing treatment is a promising route for preparing spherulites with increased molecular order.

8.4 Concluding Remarks

The specific interaction between starch and small molecules capable of inducing the formation of helical amylose complexes can promote the self-assembly of complexed amylose into spherulitic structures. The formation of spherulites can be viewed as an interaction that drives amylose out of solution, promoting solid–liquid bulk phase separation, since the helical segments have a strong tendency to associate. The formation of partly crystalline spherulites can be controlled by the type and the concentration of the ligands, the temperature/time conditions and the presence of starch-degrading enzymes.

Establishing phase and state diagrams of starch in terms of amylose + amylopectin miscibility and complex formation would allow better control of the self-assembly rate, and the size and crystallinity of spherulites. This approach offers a route to tailoring the structure-related properties of starch, such as solubility, thermal stability, and retention and release of small ligands from starch matrices. One application could be the encapsulation of active compounds into partly crystalline starch material *via* a solid–liquid separation process.

References

1. C. Martin and A.M. Smith, *Plant Cell*, 1995, **7**, 971.
2. V.I. Kinseleva, N.K. Genkina, R. Tester, L.A. Wassermann, A.A. Popov and V.P. Yurev, *Carbohydr. Polym.*, 2004, **56**, 157.
3. S. Fujimoto, T. Nagahama and M. Kanie, *Die Stärke*, 1972, **24**, 363.

4. G.R. Ziegler, J.A. Creek and J. Runt, *Biomacromolecules*, 2005, **6**, 1547.
5. A. Pohu, V. Planchot, J.L. Putaux, P. Colonna and A. Buléon, *Biomacromolecules*, 2004, **5**, 1792.
6. A. Buléon, P. Colonna, V. Planchot and S. Ball, *Int. J. Biol. Macromol.*, 1998, **23**, 85.
7. M.G. Sajilata, R.S. Singhai and P.R. Kulkarni, *Comp. Rev. Food Sci. Food Safety*, 2006, **5**, 1.
8. R.S. Bear, *J. Am. Chem. Soc.*, 1942, **64**, 1388.
9. T.J. Schoch, *J. Am. Chem. Soc.*, 1942, **64**, 2957.
10. T. Davies, D.C. Miller and A.A. Procter, *Starch/Stärke*, 1980, **32**, 149.
11. G.F. Fanta, F.C. Felker and R.L. Shogren, *Carbohydr. Polym.*, 2002, **48**, 161.
12. T.S. Nordmark and G.R. Ziegler, *Carbohydr. Polym.*, 2002, **49**, 439.
13. C. Heinemann, B. Conde-Petit, J. Nuessli and F. Escher, *J. Agric. Food Chem.*, 2001, **49**, 1370.
14. C. Heinemann, F. Escher and B. Conde-Petit, *Carbohydr. Polym.*, 2003, **51**, 159.
15. E. Belamie, A. Dormard, H. Chanzy and M.-M. Giraud-Guille, *Langmuir*, 1999, **15**, 1549.
16. C. Heinemann, M. Zinsli, A. Renggli, F. Escher and B. Conde-Petit, *LWT Food Sci. Technol.*, 2005, **38**, 885.
17. C. Heinemann, Ph.D. Thesis, No. 14958, ETH Zurich, 2002.
18. C.G. Biliaderis and G. Galloway, *Carbohydr. Res.*, 1989, **189**, 31.
19. N.J. Atkin, R.M. Abeysekera, S.L. Cheng and A.W. Robards, *Carbohydr. Polym.*, 1998, **36**, 173.
20. H.D. Senevirante and C.G. Biliaderis, *J. Cereal Sci.*, 1991, **13**, 129.

PART II
Biopolymer Interactions

Chapter 9

Electrostatics in Macromolecular Solutions

Bo Jönsson, Mikael Lund and Fernando L. Barroso da Silva

THEORETICAL CHEMISTRY, LUND UNIVERSITY, P.O. BOX 124, S-221 00 LUND, SWEDEN

9.1 Introduction – The Dielectric Continuum Model

An aqueous solution containing biological molecules can be described in a general sense as an electrolyte solution. That is, it contains simple ions such as Na^+, K^+, Cl^-, *etc.*, but it can also include macromolecules with a net charge significantly different from unity. Proteins, polysaccharides and DNA are important examples of macromolecules of natural origin, but different synthetic additives can also be described as charged macromolecules, sometimes collectively refered to as *polyelectrolytes*. It is our intention to discuss the interactions and stability of biological polyelectrolytes in a few generic situations, some of which could be of interest for a food chemist.

Despite the progress in computer technology and numerical algorithms during the last decades, it is still not feasible to treat a general solution of charged macromolecules in an atomistic model. This becomes especially clear when we are trying to calculate the interaction between macromolecules and how the interaction can be modulated by other charged species. The alternative at hand is to use the dielectric continuum model – crudely refered to as the *primitive model*. The solvent is then described as a structureless medium solely characterized by its relative dielectric permittivity ε_r. This simplification facilitates both the theoretical treatment and the conceptual understanding of electrostatic interactions in solution. In contrast to its name, it is a very sophisticated approximation, which allows an almost quantitative description of widely different phenomena involving such diverse materials as sea water and cement paste!

In the primitive model we treat all charged species as charged hard spheres. The interaction between two charges i and j separated by distance r can be formally described as

$$u(r) = \frac{Z_i Z_j e^2}{4\pi\varepsilon_0 \varepsilon_r r} \quad (r > d_{hc}), \tag{1}$$

$$u(r) = \infty \quad (r < d_{hc}), \tag{2}$$

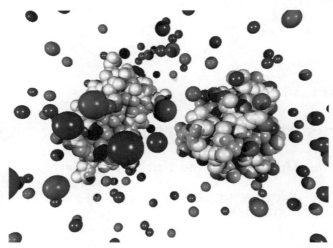

Figure 1 *Snapshot from an MC simulation of two proteins. The black and grey spheres illustrate mobile cations and anions. Amino acids are depicted as white spheres, clustered to form the two proteins. In the simulation, the proteins are displaced along a line and rotated independently. Ions are displaced in all three directions and the whole system is enclosed in a sphere of appropriate radius*

where Z_i is the ion valency, e the elementary charge, ε_0 the dielectric permittivity of vacuum and d_{hc} the hard sphere diameter of an ion. For simplicity, we will in this communication mostly assume d_{hc} to be the same for all ionic species and equal to 4 Å.

While these charges can be the small mobile ions in a salt solution, they can also be the charged groups on a protein or some other macromolecule. The model is schematically depicted in Figure 1 in terms of two macromolecules in a salt solution. We will solve this model exactly using Monte Carlo (MC) simulations, or in an approximate way with either the Poisson–Boltzmann (PB) equation or its linearized version, the Debye–Hückel (DH) equation. For an introduction to the DH theory, the reader is recommended to consult the excellent textbook of Hill.[1] Engström and Wennerström[2] have solved the PB equation for a charged surface with neutralizing counter-ions, and their paper is a good starting point on this subject. Monte Carlo and other simulations are well described in the textbooks by Allen and Tildesley[3] and Frenkel and Smit.[4] The MC simulations allow us to emphasize where the simple theory is applicable and where a more accurate treatment is needed. The simulations also give an opportunity to clarify certain physical mechanisms, providing a deeper understanding of the system at hand.

9.2 The Simple Electrolyte Solution

An important property of an electrolyte solution is the activity factor (activity coefficient) γ, or excess chemical potential μ_{ex}, which is a part of the total chemical potential μ, i.e.,

$$\mu = \mu_0 + \overbrace{kT \ln c}^{\mu_{\text{ideal}}} + \overbrace{kT \ln \gamma}^{\mu_{\text{excess}}}. \tag{3}$$

The quantity μ_0 is an uninteresting reference chemical potential and c the concentration. It is straightforward to calculate γ in an MC simulation, and we can also obtain it from the DH approximation,

$$kT \ln \gamma^{\text{DH}} = -\frac{Z^2 e^2 \kappa}{8\pi \varepsilon_0 \varepsilon_r (1 + \kappa d_{\text{hc}})}. \tag{4}$$

The important quantity in Equation (4) is the inverse screening length κ, which is related to the ionic strength:

$$\kappa^2 = \frac{e^2}{\varepsilon_0 \varepsilon_r kT} \sum_i c_i z_i^2. \tag{5}$$

Figure 2 shows how γ varies as a function of salt concentration for two different salts. The accuracy of the simple DH theory is surprisingly good. The main discrepancy comes from the too approximate treatment of the excluded volume effect, *i.e.*, the hard-core interaction.

A knowledge of γ allows us to calculate a number of interesting quantities. For example, we can calculate the dissolution of carbon dioxide in the ocean. The high salt content of the oceans increases the solubility of CO_2, which is apparent from the equilibrium relations,

$$H_2CO_3 \rightleftharpoons HCO_3^- + H^+, \quad K_1 = \underbrace{\frac{C_{HCO_3^-} C_{H^+}}{C_{H_2CO_3}}}_{K_1^s} \cdot \frac{\gamma_{HCO_3^-} \gamma_{H^+}}{\gamma_{H_2CO_3}}, \tag{6}$$

$$HCO_3^- \rightleftharpoons CO_3^{2-} + H^+, \quad K_2 = \underbrace{\frac{C_{CO_3^{2-}} C_{H^+}}{C_{HCO_3^-}}}_{K_2^s} \cdot \frac{\gamma_{CO_3^{2-}} \gamma_{H^+}}{\gamma_{HCO_3^-}}. \tag{7}$$

Note that the *thermodynamic* equilibrium constants, K_1 and K_2, are true constants, in contrast to the *stoichiometric* ones, K_1^s and K_2^s. Table 1 presents experimental and simulated activity factors for some salts relevant to sea water. The departure from ideality ($\gamma = 1$) is non-negligible, and as a consequence the dissolution of CO_2 in sea water is significantly larger than in fresh water. The excellent agreement between measured and simulated activity factors in Table 1 gives a strong support for the primitive model.

9.3 A Charged Macromolecule in a Salt Solution

We can use the activity factors in order to study how the binding of a charged ligand to a charged macromolecule is affected by addition of salt or by changes

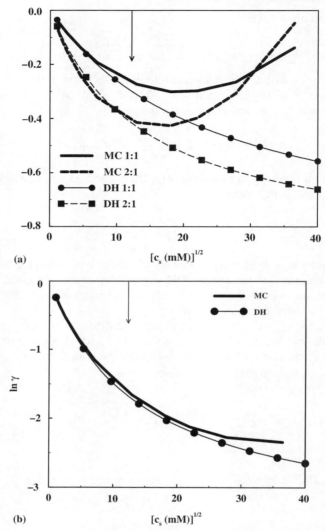

Figure 2 *Individual activity factors from MC simulations and from DH theory for a hard-core diameter of 4 Å. The logarithm of the activity factor γ is plotted against salt concentration c_s: (a) monovalent ion in 1:1 and 2:1 electrolytes; and (b) divalent ion in a 2:1 electrolyte. The arrows indicate the physiological salt condition*

in pH (meaning that the net charge on both ligand and macromolecule can vary). These changes will affect the electrostatic interactions and they are almost quantitatively captured by the activity factors. The simplest approach would then be to treat the macromolecule as a charged spherical object and directly apply Equation (4).

Table 1 *Mean activity factors in sea water (3.5% salinity) at 298 K from experiment[5,6] and simulation[7]*

Salt	γ (experimental)	γ (simulated)
Na_2SO_4	0.37	0.37
K_2SO_4	0.35	0.36
NaCl	0.67	0.67
KCl	0.66	0.66
$CaSO_4$	0.14	0.15

Let us take the calcium ion binding to the small chelator 5,5'-Br2BAPTA as an example:[8]

$$Ch + Ca^{2+} \rightleftharpoons ChCa, \qquad K = \overbrace{\frac{c_{ChCa}}{c_{Ch} c_{Ca^{2+}}}}^{K_s} \frac{\gamma_{ChCa}}{\gamma_{Ch} \gamma_{Ca^{2+}}}. \qquad (8)$$

Since K is a true constant, we can write a relation between the stoichiometric binding constants at two different salt concentrations as

$$K_s^I \frac{\gamma_{ChCa}^I}{\gamma_{Ch}^I \gamma_{Ca^{2+}}^I} = K_s^{II} \frac{\gamma_{ChCa}^{II}}{\gamma_{Ch}^{II} \gamma_{Ca^{2+}}^{II}}. \qquad (9)$$

The charge of the chelator (Ch) is $-4e$ at neutral pH, and it is assumed to have a radius of 7 Å. When calcium ion is bound to the chelator it is simply modelled by a reduction of the chelator charge from $-4e$ to $-2e$. This simple model captures the salt dependence from 1 mM to 1 M. Table 2 shows how the stoichiometric binding constant K^s varies with salt concentration. Both the simulated and the DH results are in excellent agreement with experiment.

A quantitatively more correct alternative is to use the so-called Tanford–Kirkwood (TK) model.[9] This takes the detailed charge distribution into account and solves the electrostatic problem using a variant of the DH approximation. The final result is the free energy for the macromolecule in a salt solution. For a not too highly charged macromolecule, this is usually a very efficient and reliable approach and the relevant equations are easily evaluated numerically. Figure 3 shows how the calcium-binding constant to the small protein calbindin D_{9k} varies with salt concentration.[10] Both simulated and TK results are based on the detailed charge distribution of the protein with the calbindin structure obtained from an X-ray study.[11] The agreement between the two theoretical approaches is excellent and so is the comparison with the experimental results.

It is worthwhile to investigate the limitations of the TK approach, as one should expect deviations from the simulated values for a really highly charged protein. This is indeed the case, and Figure 4 reveals typical behaviour for the binding of a charged ligand to an oppositely charged macromolecule or

Table 2 *Shift in the stoichiometric calcium-binding constant K_s as a function of salt concentration c_s for the chelator BAPTA. The case of 2 mM salt has been taken as the reference point, and the shifts are calculated relative this value*

c_s (mM)	ΔpK_s^{Exp}	ΔpK_s^{Sim}	ΔpK_s^{DH}
2	0.00	0.00	0.00
10	0.26	0.32	0.32
25	0.64	0.60	0.59
50	0.89	0.85	0.84
100	1.20	1.12	1.11
300	1.58	1.58	1.59
500	1.77	1.79	1.81
1000	1.97	2.05	2.09

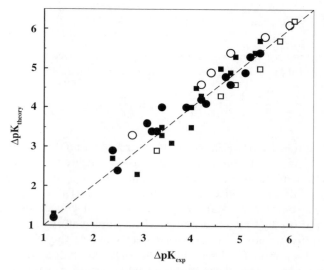

Figure 3 *Comparison of experimental and theoretical-binding constant shifts for the calcium-binding protein calbindin D_{9k}. The electrostatic interactions have been modified by adding salt in the range 2–150 mM and by mutating (neutralizing) charge residues in the protein.[10] Spheres are simulated data and squares are calculated using the TK approach. Filled symbols denote the addition of KCl and open symbols the addition of K_2SO_4. The dashed line corresponds to perfect agreement. The shifts are calculated relative to the native protein at 2 mM salt concentration*

particle. That is, when the charge reaches a certain level, the electrostatic response is no longer linear, but rather it approaches an asymptotic value. This means that the binding becomes 'saturated', and, *e.g.*, a further increase of negatively charged residues in the protein does not lead to an increased binding of calcium.

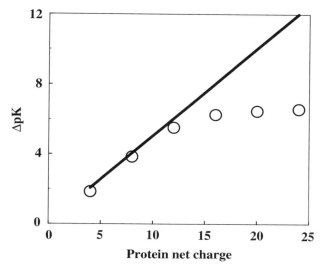

Figure 4 *Binding constant shifts as a function of protein net charge: ———, DH theory; O, MC simulations. The protein is modelled as a sphere of radius 14 Å with two binding sites close to the surface. The shift refers to a change in salt concentration from 1 to 500 mM. The protein concentration is 20 μM and the binding process involves two divalent ions*

The electrostatic model in colloid chemistry has always been one with a uniform dielectric permittivity for the whole system, typically chosen to be equal to that of water. In the calculations reported above we have followed this tradition. Obviously, the dielectric permittivity of a protein is different from that of bulk water; but we do not know its exact value; and, to be more formal, it is not a well-defined quantity. We also note that charged species prefer the high dielectric region – ions dissolve in water and not in oil! Another way to express it is to say that the electric field lines remain in the aqueous phase. Hence a small body of low dielectric material has only a marginal effect on the electrostatic interactions. These conclusions are supported by a wealth of experimental results on colloidal systems.

In biophysics, the opposite paradigm prevails: the low dielectric interior of a protein is usually assumed to be the clue to many properties of biochemical interest. The electrostatic approach is based on the PB or DH equation. A technical feature with the 'low dielectric' assumption is that the calculations contain a divergence, which can cause numerical problems. Or it can be used as a 'fitting parameter'. The divergence in electrostatic calculations invoking the low dielectric region is apparent in many applications. One such clear example relates to the apparent pK_a values in the protein calbindin (Table 3), which have been determined experimentally by Kesvatera *et al.*[12] and theoretically by Spassov and Bashford[13] assuming a low dielectric response for the protein. Juffer and Vogel[14] have extended the DH calculations of Spassov and Bashford and allowed for a high dielectric response from the protein. The paper by

Table 3 *The apparent pK_a of titrating acidic groups in calbindin D_{9k}. Experimental and various theoretical results. The 'low dielectric' results of Spassov and Bashford are highlighted. Both Spassov and Bashford[13] and Juffer and Vogel[14] have used the DH approximation, but the latter authors have assumed a uniformly high dielectric permittivity in the same way as Kesvatera et al.[12] The r.m.s. deviations are given in units of pK_a*

Residue	pK_a (experiment)	pK_a (theory, ref. 13)	pK_a (theory, ref. 14)	pK_a (theory, ref. 12)
Glu-27	6.5	**21.8**	5.2	4.7
Asp-54	3.6	**16.9**	4.8	4.4
Asp-58	4.4	**9.1**	4.8	4.8
Glu-60	6.2	**13.2**	5.6	6.0
Glu-65	5.4	**12.7**	4.6	5.0
r.m.s.	–	**10.6**	0.93	0.92

Kesvatera et al.[12] also contains results from MC simulations using a uniform dielectric response equal to that of water. Obviously, the calculations using a low dielectric interior containing charged groups are unable to describe the electrostatic interactions in calbindin and the results are unphysical.

9.4 Interaction of Two Charged Macromolecules

9.4.1 Interaction of Two Peptides

Calmodulin binds to myosin light chain kinase (MLCK) *via* a small peptide that is rich in basic residues. Calmodulin and the peptide form a complex which has been isolated and crystallized. Let us remove from this complex the peptide, smooth muscle MLCK (= smMLCK), and study the interaction between the pair. The net charge of smMLCK at neutral pH is close to $+7e$ and the two peptides repel each other, *i.e.*, the *free* energy of interaction is positive [Figure 5(a)]. The unscreened direct electrostatic interaction between the peptides is, of course, strongly repulsive, but the *total* electrostatic energy, including the background electrolyte, is essentially zero or sligthly attractive for all separations. Thus, the repulsion between the equally charged peptides is totally dominated by the entropy – the entropy of the salt and counter-ions. An increase in salt concentration from 4 to 100 mM does not change this picture, as shown in Figure 5(b).

A different picture emerges for the interaction of two *oppositely* charged peptides. Figure 6 shows the free energy of interaction between smMLCK and a peptide section from calmodulin comprising Glu45–Glu67 with a net charge of $-8e$. The interaction free energy is strongly attractive and so is the total energy. Thus, the attraction is energy driven and the entropy change is in this case only marginal.

The interaction between two charged macromolecules in a salt solution is screened by salt particles. One can derive an expression for the free energy of

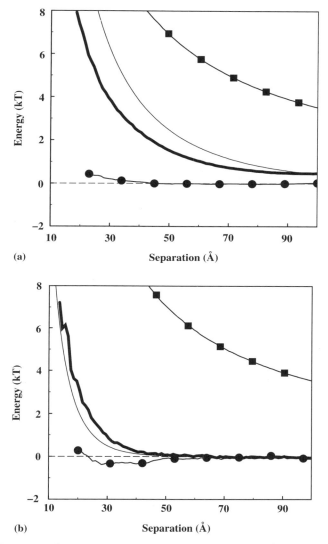

Figure 5 *Interaction between two smMLCK peptides at two different monovalent salt concentrations: (a) 4 mM and (b) 100 mM. The smMLCK peptide consists of 15 amino acids and its net charge is +7e. The thick lines show the simulated free energy of interaction. The solid thin line is from the screened Coulomb interaction, Equation (10). The thin line with filled circles is the simulated total energy of interaction. The line marked with filled squares is the electrostatic interaction between the charges on the two peptides only*

interaction, $A(r)$, based on the DH approximation

$$A(r)/kT = l_B Z_1 Z_2 \frac{\exp(-\kappa r)}{r_{12}}, \qquad (10)$$

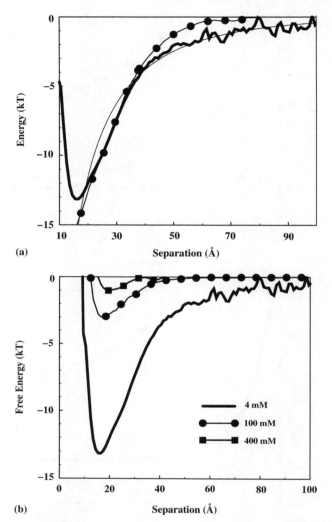

Figure 6 *Interaction between an smMLCK peptide and a fragment of calmodulin. The net charge of smMLCK is +7e and the calmodulin fragment has a charge of −8e. (a) The salt concentration is 4 mM. The solid thick line is the simulated-free energy of interaction, while the thin line is the corresponding screened Coulomb interaction. The thin line with filled circles is the simulated total energy of interaction. (b) The effect of added salt on the free energy of interaction*

where for convenience we have introduced the Bjerrum length, $l_B = e^2/4\pi\varepsilon_0\varepsilon_r kT$. Note that $A(r)$ is a *free* energy. Figure 5 shows that the screened Coulomb potential is a good approximation, and it is semi-quantitatively correct at both salt concentrations.

The results presented here for these particular peptides is generic; it is found in many cases with charged macromolecules or particles. The geometry is not crucial and the same qualitative behaviour is found for both interacting planes and interacting spheres. The screened Coulomb potential captures the change in free energy when the two macromolecules approach each other. It is, however, questionable to partition the screened Coulomb interaction into energy and entropy terms. More elaborate forms of the screened Coulomb potential can be derived,[15] where the macromolecular size is taken into account. The comparison in this section has been limited to a 1:1 electrolyte and to situations where κ^{-1} is of the same order (or greater) than the macromolecular dimension. To extend the use of the screened Coulomb potential to multivalent electrolytes usually leads to qualitatively incorrect results, as will be discussed below.

9.4.2 Effect of Multivalent Ions

We have shown above how a simple theory, the screened Coulomb potential, is capable of an almost quantitative description of the interaction between two charged proteins. This good agreement is limited to systems containing only monovalent counter-ions. There is a qualitative difference between the interaction of two charged macromolecules in the presence of monovalent counter-ions and in the presence of multivalent counter-ions. In the latter case the mean-field approximation behind the DH equation breaks down and one has to rely on simulations or more accurate theories like the hypernetted chain equation.[16,17] The deviation from the mean-field description due to *ion–ion correlations* has such a physical origin that the effect should be independent of the particular geometry of the charged aggregates. Clearly there are quantitative differences between cylindrical, spherical, irregularly shaped or flexible charged colloidal species, but the basic mechanism operates in the same way. The importance of ion–ion correlations can be seen in Figure 7, where the free energy of interaction for two charged spherical aggregates has been calculated by MC simulation. For monovalent counter-ions there is a monotonic repulsion in accordance with the screened Coulomb equation [Equation (10)]; but with multivalent counter-ions, or with a solvent of low dielectric permittivity, the entropic double-layer repulsion decreases and eventually the correlation term starts to dominate.

This phenomenon can be seen as a balance between entropy and energy. For two weakly or moderately charged macromolecules with monovalent counter-ions, the dominant contribution to the free energy of interaction comes, as we have seen in Figure 5, from a reduction in entropy when the two counter-ion clouds start to overlap. The energy of interaction is always attractive and is only weakly dependent on the counter-ion valency. The important difference between a system with monovalent or divalent counter-ions is the reduced entropy of the latter due to a lower number density of counter-ions. Thus, any change that reduces the entropy and/or increases the electrostatic interactions will eventually lead to a net attractive interaction. This is true for a model system of two spheres, each with a single net charge, but the same mechanism is also operating between two protein molecules with discrete charge distributions of irregular form,[18] or between two DNA molecules.[19]

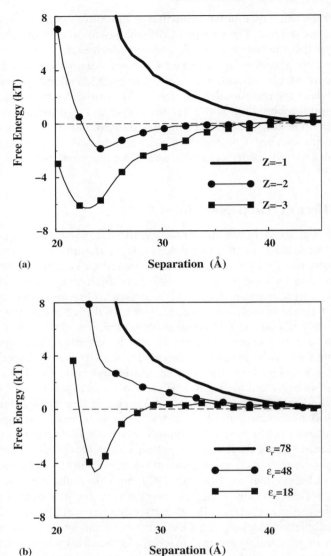

Figure 7 *Free energy of interaction between two spherical aggregates of radius 10 Å and net charge 24e. (a) The system contains no salt but only counterions of different valency (dielectric permittivity = 78, temperature = 298 K). (b) As (a), but with monovalent counter-ions and variation of the relative dielectric permittivity*

9.4.3 Effect of Titrating Groups

All proteins (and many other macromolecules) contain ionizable residues whose ionization status depends on the interaction with other molecules. This means that the electrostatic interaction between two proteins, besides involving the interaction between their average charges, also will contain terms originating from induced charges. These interactions can be formalized in a statistical mechanical perturbation approach,[20,21] so that a protein is characterized not only by its average net charge, but also by its *capacitance*. The induction interaction is important for the interaction of an approximately neutral protein with another charged macromolecule. The protein capacitance is a function of the number of titrating residues, and it will display maxima close to the pK_a of the titrating amino acids. In what follows, we derive a formal expression for the capacitance.

Consider the macromolecules A and B, described by two sets of charges $[\mathbf{r}_i, z_i]$ and $[\mathbf{r}_j, z_j]$, respectively. Their centres of mass are separated by $\mathbf{R}$, which means that the distance between two charges i and j is given by $r_{ij} = |\mathbf{R} + \mathbf{r}_j - \mathbf{r}_i|$. The average net charge of the distributions need not be zero. The free energy of interaction can be written as

$$A(R)/kT = -\ln\langle\exp(-U(R)/kT)\rangle_0 \approx \langle U(R)/kT\rangle_0 \\ -\frac{1}{2}\left\langle(U(R)/kT)^2\right\rangle_0 \\ +\frac{1}{2}\left[\langle U(R)/kT\rangle_0 + \frac{1}{2}\left\langle(U(R)/kT)^2\right\rangle_0\right]^2, \quad (11)$$

where $U(R)$ is the interaction between the two charge distributions and $\langle \ldots \rangle_0$ denotes an average over the unperturbed system, which in the present case is the single isolated protein in solution. The interaction energy is simply the direct Coulomb interaction between the two charge distributions, *i.e.*,

$$U(R)/kT = \sum_i \sum_j \frac{l_B z_i z_j}{r_{ij}}. \quad (12)$$

We can make a Taylor series expansion of U, assuming that $R \gg r_i$. This expansion includes ion–ion interaction, ion–dipole interaction, dipole–dipole interaction, *etc*. It also includes charge–induced-charge and induced-charge–induced-charge interactions. Thus, we can write an approximation to the free energy including all terms of order up to $1/R^2$. Note that the ion–dipole interaction disappears at first order and that the first non-vanishing dipole term, $-l_B^2 Z^2 \mu^2/6R^4$, is of order $1/R^4$:

$$A(R)/kT \approx \frac{l_B \langle Z_A \rangle \langle Z_A \rangle}{R} \\ -\frac{l_B^2}{2R^2}\left(\langle Z_A^2\rangle - \langle Z_A\rangle^2\right)\left(\langle Z_B^2\rangle - \langle Z_B\rangle^2\right) \\ -\frac{l_B^2}{2R^2}\left[\left(\langle Z_A^2\rangle - \langle Z_A\rangle^2\right)\langle Z_B\rangle^2 \\ +\left(\langle Z_B^2\rangle - \langle Z_B\rangle^2\right)\langle Z_A\rangle^2\right]. \quad (13)$$

The first term in Equation (13) is the direct Coulomb term, and the following term is the *induced-charge–induced-charge* term, and the final terms are the *charge–induced-charge* interactions. Note also that $\langle Z^2 \rangle \neq \langle Z \rangle^2$. If the molecules are identical, *i.e.*, $\langle Z_A \rangle = \langle Z_B \rangle = \langle Z \rangle$, then the expression simplifies to

$$A(R)/kT \approx \frac{l_B \langle Z \rangle^2}{R} - \frac{l_B^2}{2R^2} \left[\left(\langle Z^2 \rangle - \langle Z \rangle^2 \right)^2 + 2 \langle Z \rangle^2 \left(\langle Z^2 \rangle - \langle Z \rangle^2 \right) \right]. \quad (14)$$

If pH = pI, then we have $\langle Z \rangle = 0$, and the induced charge–induced charge interaction becomes the leading term,

$$A(R) \approx -\frac{l_B^2 \langle Z^2 \rangle^2}{2R^2}. \quad (15)$$

The above equations show that the fluctuating charge of a protein or macromolecule may under certain circumstances contribute significantly to the net interaction. We can define a 'charge polarizability' or a capacitance, C, as

$$C = \langle Z^2 \rangle - \langle Z \rangle^2. \quad (16)$$

With this definition of the capacitance, Equation (13) can be rewritten in a more compact form,

$$A(R)/kT \approx \frac{l_B \langle Z_A \rangle \langle Z_B \rangle}{R} - \frac{l_B^2}{2R^2} \left(C_A C_B + C_A \langle Z_B \rangle^2 + C_B \langle Z_A \rangle^2 \right). \quad (17)$$

We can use general electrostatic equations and relate the capacitance to the charge induced by a potential $\Delta\Phi$, *i.e.*,

$$Z_{\text{ind}} = \frac{C e \Delta \Phi}{kT}. \quad (18)$$

The capacitance C can also be derived from the experimental titration curve. The ionization degree α for a single titrating acid can be found in any elementary physical chemistry textbook:

$$\log K = \log \frac{\alpha}{1 - \alpha} - \text{pH}. \quad (19)$$

Taking the derivative of α with respect to pH gives

$$\frac{d\alpha}{d(\text{pH})} = \alpha(1 - \alpha) = C \ln 10, \quad (20)$$

where in the second step we have identified the capacitance defined in Equation (16). We can obtain an approximate value for the capacitance in a protein assuming that there is no interaction between the titrating sites. Any protein contains several titrating groups, like aspartic acid, glutamic acid, histidine, *etc.*, each with an ideal pK value. Denoting different titrating groups with subscript γ and their number with n_γ, the total capacitance can be approximated by

$$C^{ideal} \approx \frac{1}{\ln 10} \sum_{\gamma} n_{\gamma} \frac{10^{pH-pK_{\gamma}}}{(1+10^{pH-pK_{\gamma}})^2}. \quad (21)$$

We have calculated the capacitance for a number of proteins with different characteristics in terms of the number and type of residues. An MC simulation has to be performed at each pH for given salt and protein concentrations. Unless otherwise stated, we have used a salt concentration of 70 mM and a protein concentration of 0.7 mM. Figure 8(a) shows the capacitance *versus* pH for calbindin. The main difference from the ideal capacitance curve is a strong broadening of the two peaks corresponding to the responses from acidic and basic residues, respectively. If the protein has a significant net charge, the true curve will also shift away from the ideal one, as is seen for calbindin at high pH.

The protein hisactophilin is of the same size as calbindin, but it has a slightly different capacitance–pH curve [Figure 8(b)]. The protein contains 31 histidine residues, which is reflected in a large maximum in C at pH ≈ 5. The downward shift of the maximum is due to the high positive charge of hisactophilin at low pH. The net charge is $+28$ at pH $= 3$ and $+23$ at pH $= 4$. The isoelectric point found from the simulations is $pI = 7.3$, which is in good agreement with experimental estimates.

The electrostatic interaction between two proteins will be dominated by the direct Coulomb interaction provided that the net charge Z is sufficiently different from zero. The induced interactions will only play an important role at pH values close to the isoelectric point of one of the proteins – this can be seen from Equation (17). Figure 9(a) shows the free energy of interaction between the two proteins calbindin and lysozyme at pH $= 4$, which is close to the isoelectric point for calbindin. At contact there is a significant difference in interaction energy for a model with fixed charges as compared with a situation where the proteins are free to adjust their charges. The difference in free energy between the two models is mainly due to the interaction between the induced charge in calbindin and the permanent charge in lysozyme. This is a typical result, and significant effects from charge regulation can be expected when one of the interacting proteins has a large net charge and the other a large capacitance. Following Equation (17), we can approximate the difference as

$$\Delta A(R)/kT = (A_{reg} - A_{fix})/kT$$
$$= -\frac{l_B^2}{2R^2}\left(C_{calb}C_{lys} + C_{calb}Z_{lys}^2 + C_{lys}Z_{calb}^2\right). \quad (22)$$

Figure 10 shows almost perfect agreement between the simulated free energy difference and the one calculated according to Equation (22).

A noteworthy finding is that, despite calbindin and lysozyme both being positively charged at pH $= 4$, there is still an attractive electrostatic interaction between the two molecules. Such an attraction could, of course, be due

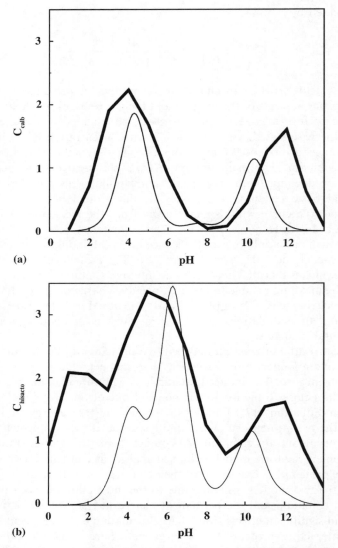

Figure 8 *Capacitance of proteins as a function of pH. (a) Calbindin D_{9k} ($pI \approx 4.2$): the thick solid curve is from MC simulation, while the thin solid line is the ideal capacitance calculated from Equation (21). (b) Hisactophilin ($pI = 7.3$)*

to charge–dipole and/or dipole–dipole interactions, but they do not seem to be important in the present case: the main contribution to the interaction-free energy comes from the induced charges. This is further demonstrated in Figure 9(b), where one can follow how the net charge of calbindin goes from around +1.4 at infinite separation to around −0.5 at contact. We will return to this issue when discussing protein–polyelectrolyte complexation.

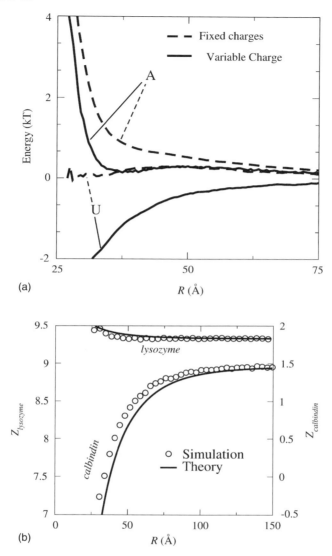

Figure 9 *(a) The energy and free energy of interaction between calbindin and lysozyme at pH = 4 for a protein model with fixed charges (dashed lines) and one with charge regulation (solid lines). The amino-acid model is used and the salt concentration is 5 mM. (b) The variation of net charge of calbindin and lysozyme as a function of their separation. The simulations are based on the amino-acid model (pH = 4, salt concentration = 5 mM)*

9.4.4 Bridging Attraction with Polyelectrolytes

The adsorption of a polyelectrolyte to an aggregate is a necessary, but not sufficient, condition to attain a modulation of the free energy. It actually has to

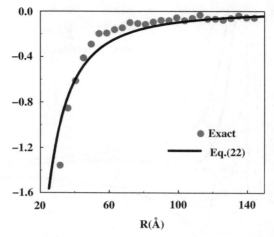

Figure 10 *The difference in free energy of interaction (ΔA) between calbindin and lysozyme at pH = 4 for a protein model with regulation and one with fixed charges, where R is the separation between the mass centres of the two proteins. Symbols denote the simulated difference (see Figure 9), and the solid line is obtained from Equation (22) with $Z_{calb} = 1.16$, $C_{calb} = 2.23$, $Z_{lys} = 10.2$ and $C_{lys} = 0.88$*

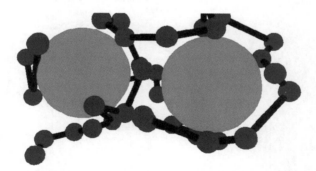

Figure 11 *Snapshot from an MC simulation of system containing two charged macromolecules and an oppositely charged polyelectrolyte*

adsorb to both aggregates in order to form *bridges* (Figure 11) leading to attractive interactions. For highly charged polyelectrolytes and oppositely charged macromolecules, bridge formation is usually a very effective way of destabilization. From simulations and mean field theories, we know that the attraction is rather short ranged and that it typically only extends over distances of the order of the monomer–monomer separation.[22–25] Figure 12(a) shows what happens if a polyelectrolyte salt is added to a solution of two charged macromolecules. The double-layer repulsion is replaced by a short-range attraction with a minimum at a surface-to-surface separation of approximately the monomer–monomer distance.

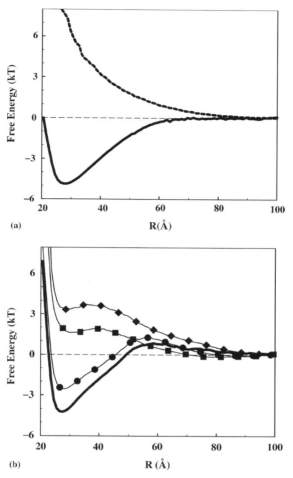

Figure 12 *Free energy of interaction between two charged spheres as a function of separation. (a) Situations in the presence of a polyelectrolyte salt (solid line) and in the presence of a 1:1 salt (dashed line). The charge of the aggregates is 10e and the radius is 10 Å. The freely jointed polyelectrolyte chain contains 10 charged monomers separated by a distance of 6 Å. (b) Two negatively charged spheres in the presence of a single neutral polyampholyte chain with 40 monomers. The charge topology has been varied and the following notation is used: diblock (solid line with no symbols), triblock (+10,−20,+10) (circles), tetrablock (+10,−10,+10,−10) (squares) and 'reversed' triblock (−10,+20,−10) (diamonds). Each macromolecule has a charge of +20e and the radius is 10 Å*

A polyelectrolyte adsorbs readily to an oppositely charged macromolecule, and in the presence of several charged spheres it is, of course, entropically favourable for the chain to adsorb to more than one sphere. This can only be accomplished at short separations, since the chain tries to avoid placing charges far from the charged aggregates, where the potential is high. Thus, a weakly charged chain – a chain with large separations between the charged monomers – will lead to a more long-ranged but weaker attraction. In general, one finds that highly charged systems give rise to fewer, but stronger 'bridges', and there is an optimal choice of polyelectrolyte structure for maximizing the attraction between the colloids.

The interaction between charged macromolecules is, from an electrostatic point of view, rather insensitive to the addition of neutral *random* polyampholytes. It is only with block polyampholytes that the normal double-layer repulsion can be reduced in the same way as with oppositely charged polyelectrolytes. The oppositely charged block acts like an oppositely charged polyelectrolyte. The only complication or constraint is that the equally charged blocks should avoid the aggregates. If the polyampholyte has a *net* charge, then it behaves qualitatively like a weakly charged polyelectrolyte. A mixing of positively and negatively charged monomers allows a tailoring of the range and magnitude of the attraction. Figure 12(b) shows the free energy of interaction between two charged macromolecules with different types of polyampholytes. A naïve picture of a triblock between two adsorbing macromolecules, which seems to be true for neutral block copolymers, is one where the two ends of the polyampholyte chain adsorb to one aggregate each and 'pull' them together. Such a structure is quite common in a simulation, but it does not lead to a significant 'pulling' force due to the weak force constant of a long segment of negatively charged monomers. Another way to express this is that the free energy gain of adsorbing a polyampholyte chain is approximately distance-independent for a tri-block of the type $(-10, +20, -10)$. Figure 13 is a snapshot from a simulation illustrating this conformation.

9.5 Protein–Polyelectrolyte Complexation

The complexation of polyelectrolytes and proteins is extensively exploited in pharmaceutics, foods and cosmetics.[26–33] The subject has been addressed by a

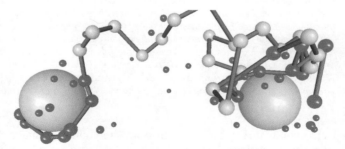

Figure 13 *Simulated picture of a triblock copolymer $(+10, -20, +10)$ adsorbing onto two negatively charged macroions $(Z = -20)$. Counter-ions and positively charged monomers are shown in grey and negatively charged monomers in white*

number of authors through experiment[32–35] and theoretical modelling.[36–38] The strength of the interaction is to a large extent regulated by electrostatic interactions, as governed by key parameters such as pH and salt concentration.

A particularly interesting observation is the apparently paradoxical formation of soluble complexes under conditions where the net charges of the protein and the polyelectrolyte have the same sign. The experimental studies of Dubin, de Kruif, and their co-workers,[33,36,39] have demonstrated this special feature of polymer–protein complexation. The expression 'complexation on the wrong side' has sometimes been used, meaning that a polyanion forms a complex with a protein at a pH above the isoelectric point of the protein. The molecular interpretation of such studies has focused on the assumption of 'charged patches' on the protein surface.[30,37,40,43]

A formal way to describe the interaction between oppositely charged patches on two macromolecules is in terms of a multipole expansion. That is, for two neutral protein molecules, the leading terms would be dipole–dipole, dipole–quadrupole, *etc.* Other electrostatic properties of the protein, however, may be more important, and already in 1952 Kirkwood and Shumaker[20] demonstrated theoretically that fluctuations of residue charges in two proteins can result in an attractive force. Recently, we have used MC simulations and the charge regulation theory described in the previous section in order to explain protein–polyelectrolyte association in a purely electrostatic model.[41] A charge regulation mechanism was also suggested by Cohen Stuart and Biesheuvel.[42]

We can use simulated capacitances and dipole moments in order to calculate analytically the ion–induced-charge and ion–dipole contributions to the interaction-free energy according to Equation (17). The results indicate that the regulation term is by far the most important term for lysozyme, while for α-lactalbumin and β-lactoglobulin the two terms are of comparable magnitude. The curves in Figure 14 should, of course, be regarded as qualitative and not quantitative. However, they still give, as will be seen below, a correct picture of the behaviour of the three proteins. The contact separation has been defined as protein radius + polyelectrolyte radius, $R_p + R_{pe}$. The latter has been chosen as half of the end-to-end separation of the corresponding neutral ideal polymer. Both protein and polyelectrolyte radii are approximate, but even with a rather generous variation of these values the general picture of Figure 14 will remain the same. The regulation term decays slower than the ion–dipole term, which means that it will gain in relative importance at larger separations, as shown in Figure 14. This means that, even if the two terms are comparable at contact, the regulation term can still dominate the contribution to the second virial coefficient.

We have performed four different simulations for each protein–polyelectrolyte complex:

(A) the 'neutral' protein (all charges set to zero),
(B) the protein with fixed charges at each amino-acid residue,
(C) the protein with an ideal dipole at its centre of mass, and
(D) the protein with titrating amino-acid residues.

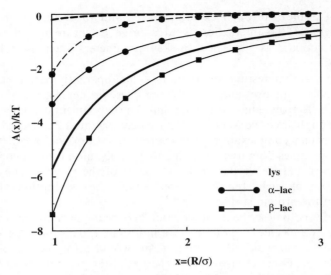

Figure 14 *The contribution to the free energy of interaction, $A(x)$, as a function of reduced separation $x = R/\sigma$ from the charge-induced–charge term (solid lines) and the ion–dipole term (dashed lines). Lines without symbols denote lysozyme, filled circles denote α-lactalbumin and filled squares denote β-lactoglobulin. The free energies are calculated from Equation (17) using simulated capacitances and dipole moments. Note that the ion–dipole terms for α-lactalbumin and β-lactoglobulin coincide*

The first set of simulations (A) describes only the shape of the protein, and so the free energy of interaction is, of course, everywhere repulsive. The second set (B) uses fixed fractional charges on all residues, which has been determined in a separate simulation of the isolated protein at the appropriate pH. In the next set (C), the charge distribution of the protein is replaced by an ideal dipole. And, in the fourth and final set (D), the amino acids are allowed to titrate, and so this simulation contains all electrostatic contributions including the ion–induced-charge term. The difference between the results from sets B and C indicates the importance of higher order electrostatic moments (quadrupole, octupole, *etc.*) in the protein, while the comparison of set C with set D reveals the importance of the regulation mechanism.

The calculated free energy of interaction, $A(R)$, for the three proteins at their respective p*I* values all show a clear minimum (Figure 15). The relative depths of the minima are in qualitative agreement with perturbation calculations (see Figure 14), though the actual numbers are approximately half the values predicted by second-order perturbation theory. The minima appear at roughly the same separation despite the fact that the β-lactoglobulin molecule is more than twice as big as the other two. This can be explained by the elongated shape of the former, which also results in a more long-ranged attraction. The separation R can approach zero, which corresponds to a situation where the polyelectrolyte wraps around the protein. We note here, however, that $A(0)$ is

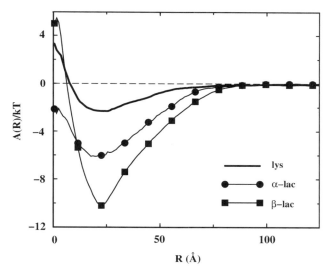

Figure 15 *The free energy of interaction $A(R)$ between the centres of mass of the protein and the polyelectrolyte at low salt concentration obtained from MC simulations with titrating amino acids (model D). The curves have been calculated at the respective isoelectric points for lysozyme (no symbols), α-lactalbumin (filled circles) and β-lactoglobulin (filled squares)*

repulsive, indicating that the 'wrapping' of the chain around the proteins is an entropically unfavourable structure.

The attractive minimum in the protein–polyelectrolyte potential is reduced upon addition of salt[37] and we can use the minima of $A(R)$ in Figure 15 in order to estimate the critical ionic strength. Assuming that the salt screening can be described by simple DH theory, and that the complex can be defined as 'dissolved' when the interaction is less than kT, we get the following relation:

$$\exp(-2\kappa R_{\min})|A(R_{\min})| \leq kT. \tag{23}$$

The factor of two in the exponent of Equation (23) comes from the fact that the second-order terms dominate the interaction. Following this recipe, we find that approximately 10–20 mM salt is sufficient to dissociate the α-lactalbumin–polymer or β-lactoglobulin–polymer complexes, respectively.

Thus, we have shown that a polyanion can form a complex with a neutral protein molecule. Next we make a numerically more rigorous partitioning of the contributions to the free energy of interaction shown in Figure 15. We can deduce that the minimum for lysozyme is solely due to charge regulation [Figure 16(a)]. If the charge distribution on lysozyme is considered fixed, then the polyanion–lysozyme interaction is essentially everywhere repulsive. Replacing the detailed charge distribution with an ideal dipole at the mass centre has a small effect on the free energy. This means that the ion–dipole interaction gives a very small attractive contribution, while the effect from higher order moments is negligible.

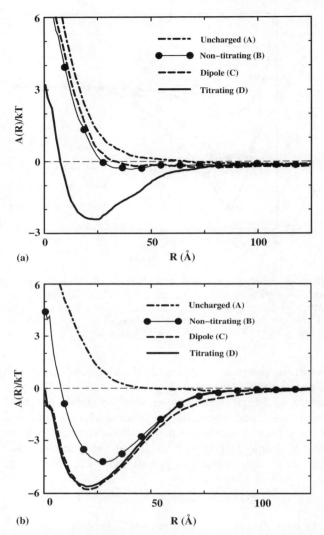

Figure 16 *The free energy of interaction $A(R)$ between the centres of mass of lysozyme and the polyanion at $pH = pI$. The four curves correspond to the different sets of simulations mentioned in the text: (a) lysozyme + polyanion; and (b) α-lactalbumin + polyanion*

Figure 16(b) shows that the polyanion interacts more strongly with α-lactalbumin than with lysozyme. For α-lactalbumin the regulation term increases the depth of the minimum from ∼4 to ∼6 kT. We note that the dipolar protein (C) shows a stronger interaction than the protein with a detailed but fixed charge distribution (B). This means that the ion–quadrupole interactions, *etc.*, add repulsive contributions to the overall interaction potential.

9.6 Conclusions

We have described a few generic situations in which electrostatic interactions between charged macromolecules seem to play an important role. With MC simulations we can obtain the exact answer within the given interaction model, which allows us to test the validity of approximate theories. Many biochemical systems are comparatively weakly charged, in contrast to many inorganic systems, and so simple theories based on the DH approximation give accurate answers. The long-range character of the Coulomb interaction often means that the geometry and detailed distribution of the charged groups are less important for the interaction of two charged macromolecules.

References

1. T.L. Hill, *An Introduction to Statistical Thermodynamics*, Dover, New York, 1986.
2. S. Engström and H. Wennerström, *J. Phys. Chem.*, 1978, **82**, 2711.
3. M.P. Allen and D.J. Tildesley, *Computer Simulation of Liquids*, Oxford University Press, Oxford, 1989.
4. D. Frenkel and B. Smit, *Understanding Molecular Simulation*, Academic Press, San Diego, 1996.
5. R.F. Platford and T. Dafoe, *J. Mar. Res.*, 1965, **23**, 63.
6. R.F. Platford, *J. Mar. Res.*, 1965, **23**, 55.
7. M. Lund, B. Jönsson and T. Pedersen, *Mar. Chem.*, 2003, **80**, 95.
8. L.A. Svensson, E. Thulin and S. Forsén, *J. Mol. Biol.*, 1992, **223**, 601.
9. C. Tanford and J.G. Kirkwood, *J. Am. Chem. Soc.*, 1957, **79**, 5333.
10. B. Svensson, B. Jönsson, C.E. Woodward and S. Linse, *Biochemistry*, 1991, **30**, 5209.
11. D.M.E. Szebenyi and K. Moffat, *J. Biol. Chem.*, 1986, **261**, 8761.
12. T. Kesvatera, B. Jönsson, E. Thulin and S. Linse, *Proteins*, 2001, **45**, 129.
13. V. Spassov and D. Bashford, *Protein Sci.*, 1998, **7**, 2012.
14. A.H. Juffer and H.J. Vogel, *Proteins*, 2000, **41**, 554.
15. B. Beresford-Smith, Some aspects of strongly interacting colloidal interactions, Ph.D. Thesis, Australian National University, Canberra, 1985.
16. R. Kjellander and S. Marelja, *Chem. Phys. Lett.*, 1984, **112**, 49.
17. R. Kjellander and S. Marelja, *J. Chem. Phys.*, 1985, **82**, 2122.
18. M. Lund and B. Jönsson, *Biophys. J.*, 2003, **85**, 2940.
19. S.M. Mel'nikov, M.O. Khan, B. Lindman and B. Jönsson, *J. Am. Chem. Soc.*, 1999, **121**, 1130.
20. J.G. Kirkwood and J.B. Shumaker, *Proc. Natl. Acad. Sci. USA*, 1952, **38**, 863.
21. M. Lund and B. Jönsson, *Biochemistry*, 2005, **44**, 5722.
22. C.E. Woodward, B. Jönsson and T. Åkesson, *J. Chem. Phys.*, 1988, **89**, 5145.
23. R. Podgornik, *J. Phys. Chem.*, 1991, **95**, 5249.
24. J. Ennis, L. Sjöström, T. Åkesson and B. Jönsson, *Langmuir*, 2000, **16**, 7116.

25. L. Sjöström and T. Åkesson, *J. Colloid Interface Sci.*, 1996, **181**, 645.
26. C. Schmitt, C. Sanchez, S. Desobry-Banon and J. Hardy, *Crit. Rev. Food Sci. Nutr.*, 1998, **38**, 689.
27. J.L. Doublier, C. Garnier, D. Renard and C. Sanchez, *Curr. Opin. Colloid Interface Sci.*, 2000, **5**, 202.
28. S. Zancong and S. Mitragotri, *Pharm. Res.*, 2002, **19**, 391.
29. G. Jiang, B.H. Woo, F.R. Kang, J. Singh and P.P. DeLuca, *J. Control. Release*, 2002, **79**, 137.
30. M. Simon, M. Wittmar, U. Bakowsky and T. Kissel, *Bioconjugate Chem.*, 2004, **15**, 841.
31. J. A. Hubbell, *Science*, 2003, **300**, 595.
32. M. Girard, S.L. Turgeon and S.F. Gauthier, *J. Agric. Food Chem.*, 2003, **51**, 6043.
33. C.G. de Kruif, F. Weinbreck and R. de Vries, *Curr. Opin. Colloid Interface Sci.*, 2004, **9**, 340.
34. E. Seyrek, P.L. Dubin, C. Tribet and E.A. Gamble, *Biomacromolecules*, 2003, **4**, 273.
35. R. Hallberg and P.L. Dubin, *J. Phys. Chem. B*, 1998, **102**, 8629.
36. K.R. Grymonpré, B.A. Staggemeier, P.L. Dubin and K.W. Mattison, *Biomacromolecules*, 2001, **2**, 422.
37. R. de Vries, *J. Chem. Phys.*, 2004, **120**, 3475.
38. F. Carlsson, P. Linse and M. Malmsten, *J. Phys. Chem. B*, 2001, **105**, 9040.
39. R. de Vries, F. Weinbreck and C.G. de Kruif, *J. Chem. Phys.*, 2003, **118**, 4649.
40. T. Hattori, R. Hallberg and P.L. Dubin, *Langmuir*, 2000, **16**, 9738.
41. F.L.B. da Silva, M. Lund, B. Jönsson and T. Åkesson, *J. Phys. Chem. B*, 2006, **110**, 4459.
42. M.A. Cohen Stuart and P.M. Biesheuvel, *Langmuir*, 2004, **20**, 2785.
43. R. de Vries, *J. Chem. Phys.*, 2004, **120**, 3475.

Chapter 10

Casein Interactions: Does the Chemistry Really Matter?

David S. Horne,[1] John A. Lucey[2] and Jong-Woo Choi[2]

[1] HANNAH RESEARCH INSTITUTE, AYR, SCOTLAND KA6 5HL, UK
[2] DEPARTMENT OF FOOD SCIENCE, UNIVERSITY OF WISCONSIN, 1605 LINDEN DRIVE, MADISON, WI 53706, USA

10.1 Introduction

'Soft matter' was a term first used in 1991 by Pierre-Gilles de Gennes in his acceptance address of the Nobel Prize for Physics. Since then the study of soft condensed matter has become a major growth area in physics, attracting significant funding worldwide. This sub-discipline encompasses the topics of colloids, polymers and surfactants, mainly in the high concentration region. While research activity in the area has undoubtedly grown in the intervening years, it would probably not be unfair to say that some research groups already working on these topics simply 're-badged' themselves to jump on the soft matter bandwagon.

Physics tends to be a subject where generic, unifying and often simplifying principles are sought to describe the properties and behaviour of the systems under study. The physics of soft condensed matter is no exception to this rule. In this article we revisit the subject of casein interactions, and challenge the validity of extending some of the simplifying assumptions of soft matter physics to the description of the behaviour of casein-containing colloidal systems.

In reality, we demonstrate here that there is a higher level of complexity to the interactions of the caseins in the casein micelle than a naïve application of any individual soft condensed matter theory would allow. In the development of any model there should arise questions as to how robust are its theoretical underpinnings and how can its premises and predictions be challenged. Two alternative models of the casein micelle provide a significant cross-linking role for calcium phosphate. Here, using the known stoichiometry of micellar calcium phosphate and some arguments as to interaction motifs in the caseins, we estimate the number of casein molecules interacting with each calcium phosphate nanocluster in the casein micelle.

10.2 Caseins as Polymers

Many of the properties of polymers can be described by considering each molecule as a wriggly piece of string in constant Brownian motion. The mean-square end-to-end distance of the polymer molecule is a good descriptor of its size in solution, and size variation with solvent quality can be captured by the Flory–Huggins χ-parameter. In effect, this energy term expresses the preferences of the monomers making up the polymer for locating themselves randomly within the solvent or close to other polymer segments. In the ideal solvent, there is no clear preference, and so the polymer component remains completely dispersed. In a non-ideal solvent, the polymer string collapses upon itself, and polymer aggregation and precipitation may follow.

Proteins, including the caseins, are polyelectrolytes. The solvent quality in a protein solution can be altered by adjusting the pH or ionic strength. In particular, we know that the caseins can be precipitated by lowering the pH to their isoelectric point, and that the heavily phosphorylated members of the family (α_{s1}-, α_{s2}- and β-caseins) can be precipitated by calcium addition. So, are these simply manifestations of the caseins being transferred into a poor solvent environment? Attractive as this polymer physics representation may appear as a unifying approach, it is clear that, in the case of calcium addition, at least, additional mechanisms are at work. It has been readily demonstrated[1] by simple equilibrium dialysis experiments that calcium ions bind weakly to the calcium-sensitive caseins. Though speculation still exists over the precise identity of the binding sites, the number of calcium ions bound as a function of the free ion concentration can be readily quantified.

By way of contrast, the completely different approach of applying the classical Derjaguin–Landau–Verwey–Overbeek (DLVO) theory of colloid stability[2] to the calcium-induced precipitation of the caseins has been more successful.[3,4] Colloid stabilization according to the DLVO theory is achieved by the domination of repulsive terms over attractive components in the pair potential existing between charged colloidal particles, which has the consequence of generating an energy barrier to aggregation.[2] The electrostatic repulsive contribution to the pair potential of mean force is given by:

$$U_E = \frac{\varepsilon a \Psi_0^2}{2} \ln\{1 + \exp(-\kappa r)\}, \qquad (1)$$

where a is the particle radius, r the inter-particle separation distance, ε the dielectric constant, and κ the Debye–Hückel parameter. The electrical potential Ψ_0 can be further related to the net protein charge Q by:

$$\Psi_0 = \frac{Q}{a\varepsilon(1 + \kappa a)}. \qquad (2)$$

Colloid stability is thus controlled by the magnitude of the charge carried by the particle (protein), and so anything which modifies that charge impacts on the predicted stability. Since the repulsion energy is proportional to Q^2, it is relatively simple to demonstrate[3] that the rate constant for aggregation will be an inverse exponential function of the square of the net charge.

In the case of the calcium-induced precipitation of α_{s1}-casein, the net protein charge – the algebraic sum of charges from lysine, arginine and histidine (all positive) and aspartic, glutamic and phosphoserine (all negative) – is reduced by the binding of ionic calcium. From the published calcium-binding isotherm,[1] we know the number of calcium ions bound per protein molecule in any calcium ion solution. The rate of precipitation can be measured by monitoring the increase in solution turbidity with time, and a straight line plot of the logarithm of this rate *versus* Q^2 testifies to the validity of this approach.[3,4]

Changes in the net charge carried by the protein can be produced by chemical modification of the amino acid residues.[5,6] Positively charged lysine residues can be converted to neutral residues at the pH of the kinetic experiments by the reaction with an acyl chloride (*e.g.*, dansyl chloride), or they are made negative by the reaction with fluorescent labelling reagent, fluorescamine. Negative charge can also be introduced by iodinating tyrosine residues, the di-iodo tyrosine hydroxyl group having a value of $pK_a = 6.36$ for the free amino acid. We select this particular set of reagents because in each case the extent of modification can be measured by spectrophotometry, an important consideration where accuracy and reliability in determining this parameter is essential, due to the extreme sensitivity of the precipitation reaction to the net protein charge. Each of these modifications effectively increases the negative charge of the protein, thereby reducing its propensity for calcium-induced precipitation and slowing down the rate of aggregation. However, once the protein charge is corrected for the measured extent of modification (from the number of lysine residues changed from +1 to 0 or from +1 to −1, or the number of di-iodotyrosines introduced), the logarithm of the rate of precipitation has been shown to remain linear in Q^2, the same relationship as followed by the unmodified protein (see Figure 1). That the precipitation reaction follows this simple DLVO-based model attests to the flexibility and mobility of the casein polymer; these characteristics allow it to respond in this non-specific fashion, where just the overall average charge dominates its stability behaviour.

10.3 Introduction of Calcium Sequestrants

The preceding studies of the calcium-induced precipitation of α_{s1}-casein were initiated in the late 1970s as part of a programme investigating casein micelle formation and structure. At that time, questions focused on the role of calcium phosphate in the casein micelle. One possible mechanism recognized the sequestrant action of phosphate and viewed the calcium phosphate as a 'sink' for ionic calcium, controlling its level and through that the amount of casein-bound calcium and the strength of the casein–casein interactions. A test of this hypothesis has been provided by introducing phosphate anions into a solution of Ca^{2+} + α_{s1}-casein and observing the effect on the rate of precipitation. A parallel series of experiments[7] monitored the influence of addition of citrate,

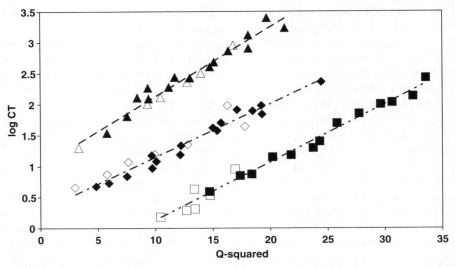

Figure 1 *Logarithm of the coagulation time CT as a function of the square of the net protein charge Q for chemically modified α_{s1}-caseins. Closed symbols distinguish the different modifications:* ■, *lysine residues reacted with fluorescamine;* ◆, *lysine residues reacted with dansyl chloride; and* ▲, *tyrosine residues converted to di-iodotyrosine. In all cases coagulation times were varied by combining various calcium solutions with proteins having differing extents of modification. Open symbols refer to native protein data at varying Ca^{2+} concentrations. Dashed lines are linear regression fits through all points in each data set. For clarity the data sets are shifted along the log (CT) axis (+1 for dansyl, and +2 for di-iodotyrosine)*

which is another calcium sequestrant, but one for which the complex is fully soluble in water.

The results for citrate addition are shown in Figure 2. The reaction was monitored by following the increase in solution turbidity with time. When Ca^{2+} was added to an α_{s1}-casein + citrate mixture, similar behaviour to that in the absence of citrate was observed – a lag phase followed by a linear increase in turbidity with time – except that the lag period increased and the slope of the linear phase decreased with increasing concentrations of citrate added. The sequestering action of citrate at pH = 7.0 is almost stoichiometric with one Ca^{2+} bound for every citrate anion introduced. Taking this reduction in available free calcium into account in the calculation of casein-bound calcium, we were able to recover the linear log (rate) *versus* Q^2 behaviour seen in the absence of citrate.[7] So it appears that citrate acts simply in a passive sequestrant role.

Inclusion of phosphate in the α_{s1}-casein solution produces a different effect when calcium is added. Unlike citrate, the coagulation time (CT) for the aggregation reaction is shortened (Figure 3). Moreover, in the presence of phosphate, precipitation is induced at Ca^{2+} levels where such mixtures do not normally precipitate under the same conditions of pH and temperature. Also,

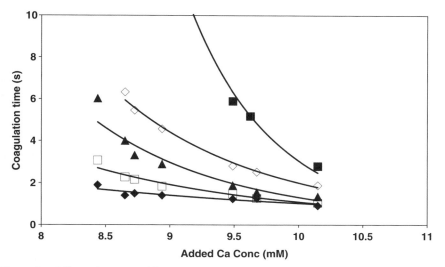

Figure 2 *Effect of citrate addition on calcium-induced aggregation of α_{s1}-casein. Coagulation time is plotted against the concentration of ionic calcium included for various citrate concentrations:* ◆, *no added citrate;* □, *1 mM;* ▲, *2 mM;* ◇, *3 mM and* ■, *5 mM. Lines are drawn to guide the eye*

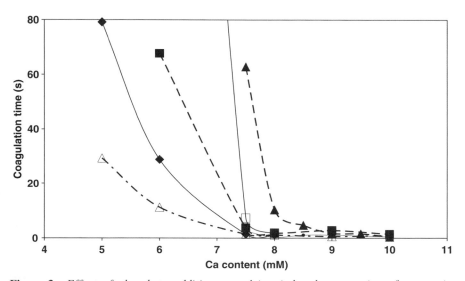

Figure 3 *Effect of phosphate addition on calcium-induced aggregation of α_{s1}-casein. Coagulation time is plotted against the concentration of ionic calcium included for various phosphate concentrations:* ▲, *no added phosphate;* □, *0.5 mM;* ■, *1.0 mM;* ◆, *1.5 mM and* △, *2.0 mM. Lines are drawn to guide the eye*

unlike the usual α_{s1}-casein + Ca^{2+} system behaviour, where the CT and the slope of the linear growth phase are inversely related, the slope in the presence of phosphate first increases and then decreases as the included phosphate concentration is increased.

Before considering possible explanations for the above behaviour, we must explore the possibility that the precipitation of calcium phosphate alone (independent of α_{s1}-casein) was causing the increase in turbidity. Experiments carried out in the absence of α_{s1}-casein showed measurable precipitation of calcium phosphate at higher concentrations of phosphate (P_i) and calcium ions. Taking the slope of the linear growth in turbidity of these solutions as a measure of the rate of aggregation, the results are conveniently plotted as log (rate) *versus* the ion product $[Ca^{2+}][P_i]$, as this representation shows linear behaviour (Figure 4). Inclusion of α_{s1}-casein produces precipitation at low ion-product values, lying well away from this linear behaviour and at rates approximately two orders of magnitude greater than predicted by extrapolation of the line into the same low ion-product range (Figure 4). The lowering of CT cannot be due to some independent precipitation of calcium phosphate because the inclusion of phosphate in the α_{s1}-casein + Ca^{2+} mixture speeds up the precipitation of this metastable system. There is nothing to suggest that the phosphate will precipitate the protein on its own. The mixture of α_{s1}-casein + P_i remains perfectly clear in the absence of ionic calcium. The three component mixture, α_{s1}-casein + Ca^{2+} + P_i, is destabilized by the inclusion of both protein and phosphate. Remove either protein or phosphate in this limited

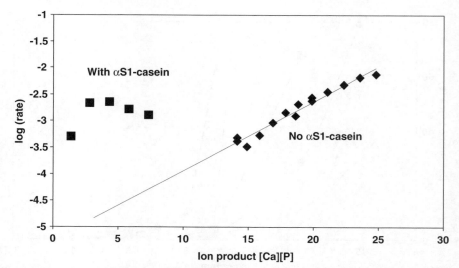

Figure 4 *Effect of inclusion of α_{s1}-casein on the rate of precipitation of calcium phosphate mixtures. Semi-log plot of dependence of rate of reaction (given by rate of increase of solution turbidity with time) on the value of the input ion product $[Ca][P_i]$ (mM^2). Data points (♦) obtained in the absence of protein, showing a linear dependence, appear to the right. Data points (■) obtained in the presence of protein appear to the left (well above the extrapolated line)*

concentration range and the rate of precipitation of the remaining pair of reagents is severely retarded. The phosphate anion is not functioning simply as a calcium chelator, but it is performing some active primary function in the precipitation process in partnership with the protein and calcium ions.

So, how successful are generic physical principles in explaining these phenomena? For simple calcium-induced precipitation of α_{s1}-casein, the simple model of the polymer in poor solvent is a non-starter because the cation binds to the protein rather than simply modifying solvent quality. But, for this reaction, the classical DLVO model works very well (indeed, possibly better than could realistically be hoped for), which points to a high degree of flexibility and mobility of the casein molecules in solution. When phosphate is included in the mixture, the mechanism of the precipitation process changes, and adherence to the simple DLVO-type concept is lost. This is where the special chemistry of the caseins emerges as a factor of major importance in understanding the stability behaviour.

10.4 Casein Chemistry and its Role in Casein Micelle Structure

The caseins are multi-phosphorylated, and the level of phosphorylation of the individual casein monomers parallels their sensitivity to calcium-induced precipitation. In particular, κ-casein, which generally carries only one phosphate residue, is not precipitated by ionic calcium.[8] In the other caseins (α_{s1}, α_{s2} and β), all of which are precipitated by Ca^{2+}, the majority of their phosphoserines are clustered in groups of 3 or more. Figure 5 identifies the positions of these cluster motifs in their respective sequences. It is a basic tenet of structural biology – the Anfinsen hypothesis – that primary structure and biological function are related. This hypothesis was enunciated to express the belief that primary sequence structure defines secondary structure and conformation, but a natural extension is that particular motifs in primary structure, particularly those preserved across species and families of proteins, play a major role in the biological function of those proteins. The phosphoserines of the caseins are clustered for a purpose, and it is our contention, and also that of other workers,[8–10] that the primary purpose of these motifs is to link within the casein micelle with the calcium phosphate microcrystallites, also known as 'nanoclusters'. This interaction forms the basis of the cross-linking mechanism in Holt's model of micellar assembly;[11] in fact, this is the only type of bonding specifically mentioned in Holt's model, but it is only one of the ways of linking casein molecules in the dual-binding model of Horne.[12] The other polymerization linkage in the dual-binding model is *via* interaction of the segregated hydrophobic regions found in all of the caseins. Both kinds of linkages allow the development of a three-dimensional network extending through space, the chains being terminated in the dual-binding model by κ-casein, which thus controls the casein micellar size.

For both of these models, an important question is this – how many casein molecules interact with the calcium phosphate nanocluster? Although Holt first estimated[13] a figure of 5 casein molecules, in more recent and substantially more detailed descriptions of his model he proposed[11,14,15] that around 48 casein molecules surround every calcium phosphate nanocluster in the casein micelle. This estimate was based on the belief that the *in vitro* entity stabilized by a preparation of β-casein phosphopeptide would be identical to the microcrystalline nodes in the casein micelle. Arguments against this assumption have been put forward elsewhere[16] and will not be repeated here. Instead we propose a simple method of estimating the number of phosphoserine cluster motifs interacting with the nanocluster, based on micellar composition and reasonable assumptions on motif numbers and sizes, and their role in controlling nanocluster size. Primarily, the phosphoserine cluster is seen as initiating the growth of the nanocluster, but it is also seen as terminating that growth by capping the facets of the growing microcrystallite. These capping motifs are not envisaged as adsorbing or coating the surface; rather, like the initiating cluster, their phosphate groups are considered to be integrated into, and forming part of, that surface layer.

Consideration of the casein sequences (Figure 5) indicates that there can be a maximum of 3 cluster motifs along α_{s2}-casein, 2 along α_{s1}-casein, 1 on β-casein and none on κ-casein. Figure 6(a) illustrates schematically the presence of a phosphoserine cluster on a calcium-sensitive casein. Bovine casein is generally accepted to consist of 40% α_{s1}, 40% β, 10% α_{s2} and 10% κ.[17] The weight

α_{S1}-casein (8P)

- - - -**SerP$_{46}$**-Glu-**SerP$_{48}$**-Thr-Glu- - - - - - **SerP$_{64}$**-Ile-**SerP$_{66}$**-**SerP$_{67}$**-**SerP$_{68}$**-Glu-Glu-
- - - - -**SerP$_{75}$**-Val-Glu- - - - - -**SerP$_{115}$**-Ala-Glu- - - - -

α_{S2}-casein (11 or 12P)

- - - - -**SerP$_8$**-**SerP$_9$**-**SerP$_{10}$**-Glu-Glu- - - - -**SerP$_{16}$**-Gln-Glu- - - - - - - - - - - - - -
-**SerP$_{56}$**-**SerP$_{57}$**-**SerP$_{58}$**-Glu-Glu-**SerP$_{61}$**-Ala-Glu- - - - -**SerP$_{129}$**-Thr-**SerP$_{131}$**-Glu-Glu-
- - - - -**SerP$_{143}$**-Thr-Glu-

β-casein (5P)

- - - -**SerP$_{15}$**-Leu-**SerP$_{17}$**-**SerP$_{18}$**-**SerP$_{19}$**-Glu-Glu- - - - -**SerP$_{35}$**-Glu-Glu- - - - -

K-casein (1P)

- - - - - - - - - -**SerP$_{149}$**-Pro-Glu- - - - - -

Figure 5 *Positions of phosphorylated residues (bold typeface) in the bovine caseins. Also indicated are the positions on the C-terminal side of each SerP group of the negatively charged residues (exclusively glutamic acid here) required by the casein kinase template*
Source: from ref. 8.

average of the number of cluster motifs is thus 1.5 per protein molecule. Or, alternatively expressed, there are 6 cluster motifs available for every 4 casein molecules.

The anhydrous composition of the casein is usually taken as 93% protein and 7% mineral ash.[11] Arguments exist as to the amounts of citrate or magnesium found in this mineral component, but here it is assumed to represent calcium phosphate. Thus, for every 100 kDa of molecular weight in the casein micelle, 93 kDa represents protein and 7 kDa represents calcium phosphate. We note that 93 kDa is approximately the combined molecular weight of 4 casein molecules, which, as calculated above, will have 6 phosphoserine cluster motifs available for association with the 7 kDa of calcium phosphate. Analysis of the mineral content of the casein micelle shows that the ratio of calcium to inorganic phosphate is ~ 1.5, close to that of the hydroxyapatite salt. Including the organic ester phosphate in the analysis, the ratio drops to ~ 1.0, close to the di-calcium salt figure.[13] This provides us with the necessary information to delineate the average composition of the minimum sized nanocluster. So, for every cluster motif of 4 ester phosphates, there are 8 inorganic phosphate anions and 12 matching calcium ions associated with the motif. The inorganic portion here has a molecular weight of 1.24 kDa. The available 7 kDa is thus 'in balance' with 5 or 6 of the available phosphoserine cluster motifs, indicating that these are probably satiated in their associations with the calcium phosphate nanocluster.

Though not considered to be adsorbed to the surface of the nanocluster, the phosphoserine cluster motifs control the surface area of the nanocluster by contributing an outer layer of phosphate groups. Considering these as a 2×2 rhombus [Figure 6(b)], we could then have the 8 inorganic phosphates required by stoichiometry forming a further two layers inwards, where (unless we have a hollow shell) they must meet the phosphates proceeding inwards from the opposite face of the nanocluster. Extending such groupings sideways in two dimensions would give platelets whose existence has never been reported. What this stoichiometry is really limiting, however, is the size of the nanocluster. The number of phosphoserine cluster motifs controls the surface area; the stoichiometry dictates the volume; the two together dictate the minimum size of the nanocluster, but only if all phosphoserine cluster motifs are involved and associated with nanoclusters.

There has been some controversy over the physical state of the calcium phosphate – whether it is crystalline or amorphous. While current opinion seems to favour the latter,[18] one might reasonably ask what is actually meant by 'amorphous calcium phosphate'? It seems to be generally considered as small clumps of calcium and phosphate (up to possibly 9 pairings), the clumps being arranged in the solid in a random haphazard fashion with no regular long-range periodic structure and no fingerprint radial distribution function with distinct spectral lines. This is thought also to be the situation for micellar calcium phosphate, but is the case clear-cut? While having close-range ordered structure, the small microcrystallite cannot have a long-range extended periodicity, as is the case for macroscopic crystal, because any spatial structure it

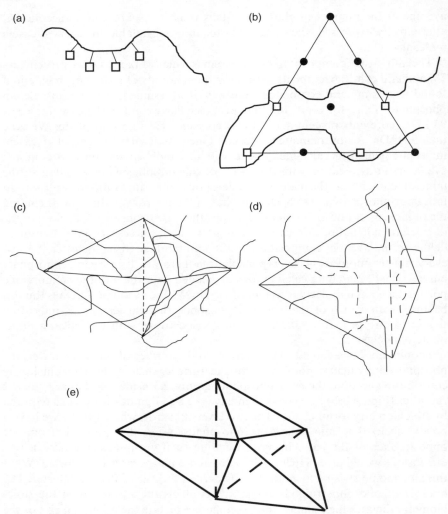

Figure 6 Schematic illustrations of putative chemical structures arising from interactions of calcium phosphate with casein: (a) phosphoserine cluster on a calcium-sensitive casein; (b) docking of protein cluster motif onto a facet of a calcium phosphate lattice, showing the ester phosphate (square) replacing the inorganic phosphate (circle); (c) crowding around the six facets of the bi-pyramid, showing only the four proteins docked on front faces; (d) situation with four proteins around the four faces of the tetrahedron; and (e) three tetrahedra in chain formation, with eight faces available for protein docking

possesses individually could not extend beyond a few peaks in the radial distribution function. As the nanoclusters are considered to be spread randomly within the space commanded by the casein micelle, their contributions to a scattering pattern would smear out and reduce the discreteness of any radial distribution function, perhaps making it appear close to the fewer broad features characterizing the radial distribution function of the amorphous bulk solid. Cross et al.[18] have recently reported a relatively featureless radial distribution function for a phosphopeptide-stabilized calcium phosphate dispersion. This contrasts with the results of Holt et al.[19] who found that the spectrum of brushite obtained by X-ray fine-structure absorption was close to their spectrum from milk calcium phosphate. This proposal of a brushite-like structure for micellar calcium phosphate was criticized by McGann et al.[20] on the basis of its neglect of citrate involvement in the casein micelle. They concluded[20] that micellar calcium phosphate was amorphous based on a comparison with bone hydroxyapatite, although it should be noted that their milk sample eventually subjected to X-ray diffraction had been freeze-dried and ground prior to examination. Accommodating the serine phosphates of the cluster motif may also produce strains in the lattice structure which could cause it to deviate slightly from the crystalline. Whether this would be enough to indicate a more amorphous nature may be in the eyes of the interpreter.

It is likely that the rigid entity that is the nanocluster will not be spherical, but rather it will have the shape of a multi-faceted solid. The tetrahedron is the solid geometrical object with the least number of faces, namely 4. As shown in Figure 6(c), the bi-pyramid, composed of two identical touching tetrahedra, has 6 faces. If each of these were to be occupied by a phosphoserine cluster motif, as illustrated in Figure 6(b), the bi-pyramid would contain 72 calcium ions and 72 phosphates made up of 48 inorganic and 24 ester phosphates from the 6 cluster motifs. The dicalcium phosphate unit cell (brushite if hydrated, monetite if anhydrous) has $Z = 4$, meaning the volume of the bi-pyramid would be that of 18 unit cells or equivalent to 9 per tetrahedron. A tetrahedron drawn internal to a cube has one-third the volume of that cube, meaning that our cube would have a total volume of 27 unit cells, or a side length of 3 unit cells. This gives the edge length of our tetrahedron as $3\sqrt{2}$ times the unit cell length, or approximately 2.5 nm, which is of the order of the size estimated by McGann et al.[20]

The cessation of growth of the calcium phosphate phase depends on the serendipitous arrival of a closure motif. Thus some nanoclusters could be closed off as tetrahedral, as shown in Figure 6(d). While this would distort the stoichiometry, it would be compensated if larger nanoclusters formed from chains of tetrahedra were also created, as illustrated in Figure 6(e) for 3 tetrahedra in the chain. Such entities would scatter X-rays or neutrons as short cylinders or needles, their distribution of sizes being such as to give the average observed stoichiometry.

10.5 Concluding Remark

What has emerged from our observation of an active role for phosphate in the precipitation of α_{S1}-casein + Ca^{2+} + phosphate mixtures is a speculative, but nonetheless plausible, estimate of the size of casein micelle nanoclusters and the number of casein phosphoserine cluster motifs involved with each nanocluster.

Confirmation of these ideas awaits the direct measurement of the size and molecular mass of nanoclusters excised from casein micelles with the minimum disturbance of pH and other milk serum conditions.

References

1. D.G. Dalgleish and T.G. Parker, *J. Dairy Res.*, 1980, **47**, 113.
2. E.J.W. Verwey and J.Th.G. Overbeek, *Theory of Stability of Lyophobic Colloids*, Elsevier, Amsterdam, 1948.
3. D.S. Horne and D.G. Dalgleish, *Int. J. Biol. Macromol.*, 1980, **2**, 154.
4. D.G. Dalgleish, E. Paterson and D.S. Horne, *Biophys. Chem.*, 1981, **13**, 307.
5. D.S. Horne, *Int. J. Biol. Macromol.*, 1983, **5**, 296.
6. D.S. Horne and P.D. Moir, *Int. J. Biol. Macromol.*, 1984, **6**, 316.
7. D.S. Horne, *J. Dairy Res.*, 1982, **49**, 107.
8. H.E. Swaisgood, in *Advanced Dairy Chemistry, Vol 1 – Proteins*, 3rd edn, part A, P.F. Fox and P.L.H. McSweeney (eds), Kluwer Academic/Plenum, New York, 2003, p. 139.
9. C. Holt and L. Sawyer, *Protein Eng.*, 1988, **2**, 251.
10. J.-C. Mercier, *Biochimie*, 1981, **63**, 1.
11. C.G. de Kruif and C. Holt, in *Advanced Dairy Chemistry, Vol 1 – Proteins*, 3rd edn, part A, P.F. Fox and P.L.H. McSweeney (eds), Kluwer Academic/Plenum, New York, 2003, p. 233.
12. D.S. Horne, *Int. Dairy J.*, 1998, **8**, 171.
13. C. Holt, in *Advanced Dairy Chemistry, Vol 3 – Lactose, Water, Salts and Vitamins*, P.F. Fox (ed), Chapman & Hall, London, 1997, p. 233.
14. C. Holt, *Eur. Biophys. J.*, 2004, **33**, 421.
15. C. Holt, *Mater. Res. Soc. Symp. Proc.*, 2004, **283**, w.7.1.1.
16. D.S. Horne, *Curr. Opin. Colloid Interface Sci.*, 2006, **11**, 148.
17. D.T. Davies and A.J.R. Law, *J. Dairy Res.*, 1977, **44**, 447.
18. K.J. Cross, N.L. Huq, J.E. Palamara, J.W. Perich and E.C. Reynolds, *J. Biol. Chem.*, 2005, **280**, 15362.
19. C. Holt, S.S. Hasnain and D.W.L. Hukins, *Biochim. Biophys. Acta*, 1982, **719**, 299.
20. T.C.A. McGann, R.D. Kearney, W. Buchheim, A.S. Posner, F. Betts and N.C. Blumenthal, *Calcif. Tissue Int.*, 1983, **35**, 821.

Chapter 11

Electrostatic Interactions between Lactoferrin and β-Lactoglobulin in Oil-in-Water Emulsions

Aiqian Ye and Harjinder Singh

RIDDET CENTRE, MASSEY UNIVERSITY, PRIVATE BAG 11 222, PALMERSTON NORTH, NEW ZEALAND

11.1 Introduction

Lactoferrin is a glycoprotein with around 700 amino acid residues and a molecular weight of $\sim 8 \times 10^4$ Da.[1] The polypeptide is folded into two globular lobes, representing its N- and C-terminal halves, which are commonly referred to as the N lobe and the C lobe. These two lobes are linked together by a short α-helix peptide and there is some 40% amino acid sequence identity between the N and C lobes.[1-3] The most important feature of lactoferrin is its very high affinity for iron; consequently, its biological functions with respect to its strong bacteriostatic properties depend on the iron-binding properties. The surface of the lactoferrin molecule has several regions with high concentrations of positive charge, giving it a high isoelectric point ($pI \approx 9$). This positive charge is one of the features that distinguishes lactoferrin from other milk proteins, such as β-lactoglobulin, which have isoelectric points in the range 4.5–5.5 and are negatively charged at neutral pH. Consequently, lactoferrin has been shown to interact with other milk proteins. This interaction appears to have a certain degree of specificity, as it occurs with β-lactoglobulin but not with α-lactalbumin.[4]

The adsorption of lactoferrin has been studied onto solid surfaces[5,6] and at air–water interfaces.[7] Wahlgren et al.[5] showed that electrostatic interactions between lactoferrin and β-lactoglobulin caused an increase in the measured amount of protein adsorbed on a hydrophilic silica surface, as compared with single proteins. Neutron reflectivity measurements have revealed[7] a strong structural unfolding of the lactoferrin molecule when adsorbed at the air–water interface from a neutral pH buffer solution over a wide concentration range. Two distinct regions, a top dense layer of 15–20 Å on the air side and a bottom diffuse layer of some 50 Å into the aqueous phase, characterizes the unfolded interfacial layer.

We recently examined the adsorption behaviour of lactoferrin in oil-in-water emulsions at pH 3.0 and 7.0. It was shown[8] that lactoferrin, like casein and whey proteins, is an excellent emulsifying agent. But, in contrast to the other milk proteins, cationic emulsion droplets can be formed with lactoferrin at neutral pH. For emulsions prepared under the same conditions, the droplet sizes in lactoferrin-stabilized emulsions were found to be similar to those in β-lactoglobulin-stabilized emulsions, but the surface protein coverage was higher in lactoferrin emulsions, possibly because of its higher molecular weight. The adsorption of lactoferrin from a binary mixture of lactoferrin + β-lactoglobulin at pH 7.0 was affected by electrostatic interactions, leading to higher amounts of adsorbed proteins at the droplet surface. Competitive adsorption was observed at pH 3.0, where both proteins carry net positive charges: β-lactoglobulin was adsorbed in preference to lactoferrin in emulsions made using low protein concentrations ($\leq 1\%$), whereas lactoferrin appeared to be adsorbed in preference to β-lactoglobulin in emulsions made using high protein concentrations.

The objectives of this study were to investigate further the interactions between lactoferrin and β-lactoglobulin in oil-in-water emulsions and to explore the possibility of making multilayered emulsions using these two proteins.

11.2 Experimental

11.2.1 Preparation of Emulsions

Protein solutions were prepared by dispersing lactoferrin (a gift from Tatua Co-operative Dairy Company, New Zealand), β-lactoglobulin (Sigma Chemical Co., St. Louis, MO, USA) or a mixture of powders of lactoferrin + β-lactoglobulin into Milli-Q water (water purified by treatment with a Milli-Q apparatus, Millipore, Bedford, MA, USA), and then stirring for 2 h at room temperature. The pH of the protein solution was adjusted to 7.0 using 1 M NaOH or 1 M HCl. Appropriate quantities of soya oil were mixed with the protein solution to give 30 wt.% oil in the final emulsion. The mixture was then heated to 55°C and homogenized in a two-stage homogenizer (APV 2000, Copenhagen, Denmark) at a first-stage pressure of 250 bar and a second-stage pressure of 40 bar. The emulsions were homogenized twice for more effective mixing of the oil phase and stored at 20°C.

In some cases, 50 g of emulsion sample prepared using 1 wt.% protein (lactoferrin or β-lactoglobulin) was mixed with 35 g of aqueous solution containing different concentrations of protein (lactoferrin or β-lactoglobulin) and gently stirred for 2 h at room temperature.

11.2.2 Characterization of Emulsions

A Mastersizer MSE (Malvern Instruments, Worcestershire, UK) was used to determine the average diameter of the emulsion droplets. The surface protein concentration of the emulsion droplets (milligrams of protein per square metre of oil surface) was determined using the method described by Ye and Singh.[8]

Sodium dodecyl sulfate polyacrylamide gel electrophoresis (SDS-PAGE), as described by Srinivasan et al.,[9] was used to determine the composition of the protein adsorbed at the surface of the emulsion droplets. The SDS gels were prepared on a Miniprotean II system (Bio-Rad Laboratories, Richmond, CA, USA). After destaining, the gels were scanned on a laser densitometer (LKB Ultroscan XL, LKB Produkter AB, Bromma, Sweden). The percentage composition of each sample was determined by scanning the areas for lactoferrin and β-lactoglobulin and expressing the individual whey protein peaks as fractions of the total.

The electrophoretic mobilities, and hence the calculated ζ-potentials of the emulsion droplets, were determined using a Malvern Zetasizer 4 instrument, the associated Malvern Multi-8 64 channel correlator and the AZ4 standard electrophoresis cell, incorporating a 4-mm diameter quartz capillary. The temperature of the electrophoresis cell was maintained at $25 \pm 0.5°C$ using a water jacket that was temperature controlled by the Peltier system. An applied voltage of 80 V (corresponding to a sensed voltage of approximately 60 V across the capillary, i.e., approximately 12 V cm^{-1}) and a modulation frequency of 250 Hz were used in all experiments.

11.3 Results and Discussion

11.3.1 Emulsions formed with Mixture of Lactoferrin + β-Lactoglobulin

The values of the average droplet diameters d_{32} of the emulsions formed with various concentrations of pure β-lactoglobulin, pure lactoferrin or a mixture of lactoferrin + β-lactoglobulin are shown in Figure 1. For all the emulsions, the droplet sizes were almost independent of the protein concentration in the range 1–3 wt.%, with the β-lactoglobulin emulsions showing slightly smaller droplets. At protein concentrations below 1 wt.%, the average droplet size of the emulsions formed with a mixture of lactoferrin + β-lactoglobulin was much larger than that of the emulsions formed with either single protein. The emulsions formed with a mixture of lactoferrin + β-lactoglobulin at <1 wt.% total protein concentration at pH 7.0 gave bimodal size distributions, whereas all other emulsions had monomodal distributions.

Figure 2 shows values of the ζ-potential of the emulsion droplets stabilized with β-lactoglobulin, lactoferrin or lactoferrin + β-lactoglobulin as a function of protein concentration. The ζ-potential of the emulsion droplets made with β-lactoglobulin alone was negative, and it became more negative with increase in protein concentration from 0.3 to 1 wt.%; but it did not change at higher protein concentrations. This indicates that the amount of adsorbed lactoferrin increased with increasing protein concentration. The emulsion droplets stabilized with lactoferrin had a positive ζ-potential value, which gradually became

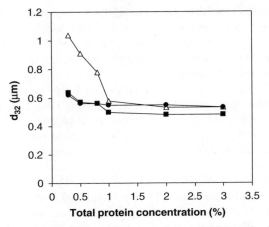

Figure 1 *Changes in average particle diameter d_{32} of emulsions made at pH 7.0 with 30 wt.% soya oil and lactoferrin (●), β-lactoglobulin (■) or a mixture of lactoferrin + β-lactoglobulin (1:1 by weight) (△) as a function of protein concentration. Each data point is the average of determinations on two separate emulsions*

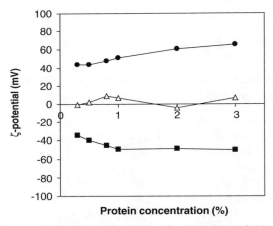

Figure 2 *The ζ-potential of emulsion droplets made at pH 7.0 with 30 wt.% soya oil and lactoferrin (●), β-lactoglobulin (■) or a mixture of lactoferrin + β-lactoglobulin (1:1 by weight) (△) as a function of protein concentration*

more positive with an increase in the protein concentration. The ζ-potential of the emulsion droplets made with a mixture of lactoferrin + β-lactoglobulin was close to zero at all protein concentrations.

At pH 7.0, the lactoferrin and β-lactoglobulin molecules carry opposite net charges, which may result in the formation of a complex between the two proteins *via* electrostatic interactions.[4] The overall net charge of both proteins would decrease with the gradual attachment of other protein molecules. Adsorption of such a complex onto the droplet surface during emulsion formation would result in emulsion droplets with little overall charge density. The ζ-potential data confirm that emulsion droplets stabilized by lactoferrin + β-lactoglobulin have a very low overall net charge.

It is noteworthy that the emulsion droplets formed by a mixture of lactoferrin + β-lactoglobulin at total protein concentrations above 1 wt.% were stable, as shown by the emulsion droplet size data and creaming measurements (not shown), even though their charge density was very low or close to zero. This suggests that the interfacial layers formed by these binary protein complexes provide good steric stabilization.

At low protein concentration (<1 wt.%), some droplet flocculation was evident in the emulsions formed with a mixture of lactoferrin + β-lactoglobulin (Figure 1). This was probably because of the formation of a complex between these proteins in aqueous solution, prior to emulsion formation, reduced the number concentration of protein molecules available for adsorption, and also increased the molecular weight of the protein complexes. Larger droplet sizes were probably formed, mostly as a result of bridging flocculation of the emulsion droplets, whereby one protein complex is adsorbed on to more than one droplet. Some re-coalescence may have also occurred during emulsion formation under those conditions.

11.3.2 Addition of β-Lactoglobulin to Lactoferrin-Stabilized Emulsions

Figure 3 shows ζ-potential values of emulsions formed with lactoferrin (1 wt.% protein, pH 7.0) and then diluted with aqueous phase containing a range of β-lactoglobulin concentrations.

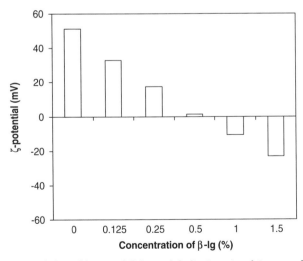

Figure 3 *Influence of the addition of β-lactoglobulin into emulsions made with 1 wt.% lactoferrin (30 wt.% soya oil, pH 7.0) on the ζ-potential of the emulsion droplets*

In the absence of β-lactoglobulin, the ζ-potential of the emulsion droplets was around +50 mV, because the lactoferrin used to stabilize the droplets has a net positive charge at pH 7.0, as shown above. The ζ-potential became less positive, and eventually changed from positive to negative, as the β-lactoglobulin concentration in the emulsion was increased. This change suggests that β-lactoglobulin, which has a net negative charge at pH 7.0, adsorbed to the surface of the lactoferrin-coated emulsion droplets. The ζ-potential was very close to zero (*i.e.*, no net charge on the droplet surface) at 0.42 wt.% addition of β-lactoglobulin. Further increases in the β-lactoglobulin concentration resulted in negative values of the ζ-potential. The reversal of charge upon the adsorption of charged polymers on the surface of oppositely charged colloidal particles has been shown by many workers.[10,11] It presumably occurs because only a fraction of the charged groups on the polymer is required to neutralize the oppositely charged groups on the particle surface, and the remaining charged groups remain on the particle surface or may protrude into the aqueous phase.

The changes in ζ-potential were also reflected in measurements of the surface protein concentration, which showed an increase in adsorbed protein up to 0.42 wt.% added β-lactoglobulin, but no further change above this concentration (Figure 4). Further evidence of β-lactoglobulin adsorption onto the lactoferrin-stabilized droplets was obtained by SDS-PAGE analysis of the emulsion droplets; it was shown that the intensity of the β-lactoglobulin band at the droplet surface increased with increasing β-lactoglobulin concentration in the emulsion, whereas the intensity of the lactoferrin band remained essentially constant (see inset to Figure 4). The adsorption of β-lactoglobulin

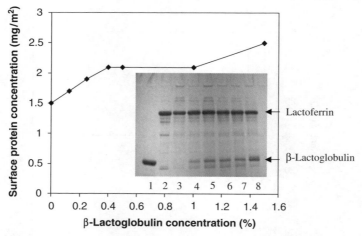

Figure 4 *Changes in protein surface coverage of emulsion droplets made with 30 wt.% soya oil and 1 wt.% lactoferrin (pH 7.0) as a function of β-lactoglobulin added into the emulsion. The inset shows SDS-PAGE patterns of adsorbed protein; lane 1 represents lactoferrin alone, and lanes 2–7 show increasing additions of β-lactoglobulin*

appeared to increase considerably with increase in β-lactoglobulin concentration up to 0.42 wt.%, with very little change above this concentration.

There were no significant changes in d_{32} values or creaming stability with an increase in the β-lactoglobulin concentration in the emulsions (data not shown), suggesting that these emulsions remained very stable despite major changes in the net charges of the droplets. There was no evidence of droplet aggregation, even at 0.42 wt.% β-lactoglobulin addition, which corresponded to the compositions where there was almost complete charge neutralization. This suggests that steric repulsion may play a more important role than electrostatic interactions in the stabilization of these emulsions.

11.3.3. Addition of Lactoferrin to β-Lactoglobulin-Stabilized Emulsions

In the absence of lactoferrin, the ζ-potential of the β-lactoglobulin-stabilized droplets (1 wt.% protein, pH 7.0) was about -50 mV (Figure 5), indicating that these emulsion droplets had a relatively high negative charge. The ζ-potential became increasingly less negative and eventually changed from a negative value to a positive value as the lactoferrin concentration in the emulsion was increased. A slightly positive value was obtained when the lactoferrin concentration was about 0.83 wt.%; the ζ-potential continued to become more positive with further increase in lactoferrin concentration.

The surface protein coverage of the emulsion droplets was found to increase almost linearly from ∼1.5 to ∼3.5 mg m^{-2} with increase in the lactoferrin concentration up to 1.2 wt.%, and to reach a plateau thereafter (Figure 6). The

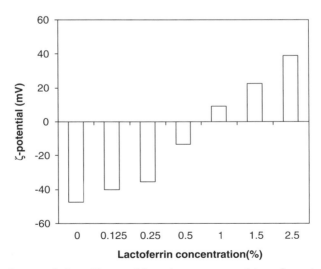

Figure 5 *Influence of the addition of lactoferrin into emulsions formed with 1 wt.% β-lactoglobulin (30 wt.% soya oil, pH 7.0) on the ζ-potential of the emulsion droplets*

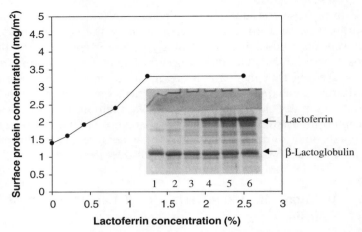

Figure 6 *Changes in protein surface coverage of emulsion droplets made with 30 wt.% soya oil and 1 wt.% β-lactoglobulin (pH 7.0) as a function of lactoferrin added into the emulsion. The inset shows SDS-PAGE patterns of adsorbed protein: lane 1 represents β-lactoglobulin alone, and lanes 2–7 show increasing additions of lactoferrin*

SDS-PAGE analysis of the emulsion droplets indicated clearly that, with increasing lactoferrin concentration in the emulsion, the intensity of the β-lactoglobulin band remained essentially constant, whereas the lactoferrin band became more intense with increasing lactoferrin concentration up to 1.25 wt.%, which is consistent with the surface protein coverage results. The particle size and creaming stability measurements show that all emulsions were stable to droplet aggregation, as there appeared to be no significant changes in the d_{32} values or the rate of creaming with increasing lactoferrin concentration in the emulsion (data not shown).

These combined results show that multilayered emulsions can be produced by interactions of oppositely charged proteins at neutral pH. A primary emulsion, containing either anionic droplets coated with β-lactoglobulin or cationic droplets coated with lactoferrin, can be produced. A secondary emulsion can then be made by mixing either β-lactoglobulin solution or lactoferrin solution with the primary emulsion. The amount of protein required to saturate the droplet surface in these emulsion systems can be estimated, as described by McClements:[11]

$$C_{\text{sat}} = 3\phi \Gamma_{\text{sat}}/r_{32}. \quad (1)$$

Here, ϕ is the volume fraction of the oil, r_{32} the volume–surface mean radius of the droplets, Γ_{sat} the surface protein coverage at saturation and C_{sat} the minimum concentration of protein in the whole system that is required to saturate the surfaces. Equation (1) is based on the assumption that, below the saturation concentration ($C < C_{\text{sat}}$), all of the protein added to the system is adsorbed to the droplet surfaces. Based on this assumption, a minimum concentration of $C_{sat} \approx 8.8$ kg m^{-3} of β-lactoglobulin would be required to saturate the surfaces of emulsion droplets coated with lactoferrin. Similarly, a minimum concentration of $C_{sat} \approx 14$ kg m^{-3} of lactoferrin would be needed to saturate the surfaces of emulsion droplets formed with β-lactoglobulin.

Oil-in-water emulsions containing oil droplets surrounded by multilayered interfacial layers consisting of low-molecular-weight emulsifiers (SDS or lecithin) and biopolymers (gelatin, chitosan, pectin) have been studied extensively.[12–14] These systems tend to show extensive bridging flocculation upon charge neutralization when biopolymers with an opposite charge from that of the droplets are added to the emulsion. In contrast, the multilayered emulsion droplets formed here using β-lactoglobulin + lactoferrin appear to be stable to droplet aggregation even at very low net electrical charges, presumably because of strong steric repulsion associated with the protein interfacial layers. Further work is underway to determine the behaviour of these emulsions under different environmental conditions, such as pH, temperature and ionic strength.

Acknowledgements

We gratefully acknowledge financial support from the New Zealand Foundation for Research, Science and Technology. We thank Dandan Chen, Jeniene Gilliland and Michelle Tamehana for assistance with the experimental work.

References

1. E.N. Baker and H.M. Baker, *Cell. Mol. Life Sci.*, 2005, **62**, 2531.
2. E.N. Baker, H.M. Baker and R.D. Kidd, *Biochem. Cell Biol.*, 2002, **80**, 27.
3. S.A. Moore, B.F. Anderson, C.R. Groom, M. Haridas and E.N. Baker, *J. Mol. Biol.*, 1997, **274**, 222.
4. F. Lampreave, A. Piniero, J.H. Brock, H. Castiplo, L. Sanchez and M. Calvo, *Int. J. Biol. Macromol.*, 1990, **12**, 2.
5. M.C. Wahlgren, T. Arnebrant and M.A. Paulsson, *J. Colloid Interface Sci.*, 1993, **158**, 46.
6. M. Laurence and H.J. Griesser, *Colloids Surf. B*, 2002, **23**, 125.
7. J.R. Lu, P. Shiamalee and X. Zhao, *Langmuir*, 2005, **21**, 3354.
8. A. Ye and H. Singh, *J. Colloid Interface Sci.*, 2006, **295**, 249.
9. M. Srinivasan, H. Singh and P.A. Munro, *Int. Dairy J.*, 1999, **9**, 337.
10. P. Chodanowski and S. Stoll, *J. Chem. Phys.*, 2001, **115**, 4951.
11. D.J. McClements, *Langmuir*, 2005, **21**, 9777.
12. S. Ogawa, E.A. Decker and D.J. McClements, *J. Agric. Food Chem.*, 2003, **51**, 2806.
13. S. Ogawa, E.A. Decker and D.J. McClements, *J. Agric. Food Chem.*, 2004, **52**, 3595.
14. Y.S. Gu, E.A. Decker and D.J. McClements, *Food Hydrocoll.*, 2005, **19**, 83.

Chapter 12
β-Lactoglobulin Aggregates from Heating with Charged Cosolutes: Formation, Characterization and Foaming

Gerlinde Unterhaslberger,[1] Christophe Schmitt,[1] Sabrina Shojaei-Rami[1] and Christian Sanchez[2]

[1] NESTLÉ RESEARCH CENTER, P.O. BOX 44, CH-1000 LAUSANNE 26, SWITZERLAND
[2] LABORATOIRE DE PHYSICO-CHIMIE ET GÉNIE ALIMENTAIRES, INPL-ENSAIA, 2 AVENUE DE LA FORÊT DE HAYE, BP 172, F-54505, VANDOEUVRE-LÈS NANCY CEDEX, FRANCE

12.1 Introduction

Foamed products are gaining more importance in the food industry because of their smooth and light texture. Food foams can be stabilized by emulsifiers and fats, but proteins are the most commonly used foaming agents. Globular proteins (*e.g.*, whey proteins) exhibit good foaming properties, but in order to achieve sufficient shelf-life stability of foamed food products and to maintain consumer acceptability, the addition of food stabilizers (polysaccharides, sugars, gelatin) is often needed.

Foam stability is governed by very complex (de)stabilization mechanisms.[1] Surface activity, molecular properties (protein size, flexibility, secondary structure and surface charge) and environmental conditions like pH and ionic strength all play an important role in determining protein adsorption and protein packing density at the interface.[2,3] It is known that proteins exhibit the highest foam stability and the lowest drainage rates around their isoelectric point and/or in the presence of ions. These influencing factors reduce repulsive forces due to charge neutralization between the protein molecules allowing dense packing at the air–water interface and leading to stable films.[4,5] In addition, globular proteins are able to form cross-links (by SH-group activation, and hydrophobic and electrostatic interactions) and stable interfacial networks that can prevent interbubble gas diffusion.[6–8]

Recently, the use of silica or latex particles with different surface properties and degree of hydrophobicity was demonstrated to stabilize interfacial films against disproportionation.[6,9–11] Proteins have been also demonstrated to exhibit a barrier against interbubble gas diffusion. However, gas diffusion stability of protein films has been found to be less effective, with values of 40–80 min for bubbles prepared from 0.05 wt.% protein solutions, as compared to lifetimes up to 300 min or even longer for bubbles stabilized by silica particles (0.08 wt.% solution) under similar conditions.[6,12]

It has been further demonstrated[13–15] that foam stability can be significantly enhanced upon formation of gel-like protein layers and incorporation of protein aggregates or coacervates in the interfacial film. It was also found in the past[13,16,17] that one possible way to improve foam functionality is by thermal treatment. The pH and ionic strength play an important role in controlling the formation of soluble and insoluble protein aggregates upon heat treatment. During thermal treatment, free SH groups are activated, surface charges are screened and protein aggregation and protein hydrophobicity increases.[18–21] Specifically tailored protein particles might therefore be used to prevent gas diffusion between bubbles leading to disproportionation, film thinning, bubble coalescence and film rupture. The aim of this study was to attempt to establish a correlation between the surface properties of soluble aggregates formed during heat treatment of β-lactoglobulin (β-LG) in the presence of charged cosolutes and their ability to improve foam stability.

12.2 Materials and Methods

12.2.1 Materials

β-LG-enriched whey protein isolate (BioPure β-LG, lot JE002-8-922, 13-12-2000) was obtained from Davisco Foods (Le Sueur, MN, USA). The composition of the powder (wet basis) was 89.7% protein, 8.85% moisture, fat ($<$ 1%), lactose ($<$1%) and 1.36% ash (major ionic components were 0.079% Ca^{2+}, 0.013% Mg^{2+}, 0.097% K^+, 0.576% Na^+, 0.05% Cl^-). The following protein profile was determined by RP-HPLC for major fractions: 12.9% α-lactalbumin (α-LA), 35% β-LG A, 32.4% β-LG B, and 1.2% bovine serum albumin (BSA). The amount of non-native protein in the powder was about 4% as measured at 20°C from the ratio between total protein at pH 4.6 and 7.[17] Arginine HCl (BioChemika, lot 431391/1) and GdnHCl (BioChemika, lot 450290/1) were obtained from Fluka Chemie (Buchs, Switzerland). All other reagents were of analytical or HPLC grade (Merck Darmstadt, Germany). Owing to the high content of β-LG (>67%) in the powder, the β-LG-enriched whey protein isolate is mainly referred to here as β-LG or just 'protein', unless stated otherwise.

12.2.2 Sample Preparation

The protein dispersion was prepared by addition of β-LG powder to Millipore water, stirring at 20°C for 2 h, and storing for 12 h at 4°C to allow complete hydration. Afterwards the dispersion (natural pH 7.0 ± 0.1) was centrifuged at 1.08 × 10^4 g (Rotor SLA-3000, Sorvall Instruments RC5C, Kendro, Switzerland) for 60 min to remove insoluble matter. The protein solution was then filtered on paper filter (Schleicher & Schuell filter $597\frac{1}{2}$, Germany). The protein concentration was determined by UV/visible spectroscopy ($\varepsilon_{278} = 1.013$ cm^{-1} g^{-1} L) using an Uvikon 810 spectrophotometer (Kontron, Flowspec, Switzerland) and adjusted to 40 g L^{-1}. The cosolute solutions (arginine HCl, NaCl and GdnHCl; 0–800 mM) were prepared in Millipore water. Mixtures of protein and cosolute solutions were prepared by mixing equal volumes of protein and cosolute to a final protein concentration of 10 g L^{-1} and 0–400 mM of cosolute and were adjusted to pH 7.0 by addition of 1 M HCl or NaOH. The amount of acid and base (additional Na$^+$ and Cl$^-$ ions) needed to adjust the pH of the solutions was less than 2% of the final molarity for pH 7. These amounts were neglected in the final cosolute molarity calculation.

For the sake of clarity, the following nomenclature is used for the sample coding: A, Arginine HCl; N, NaCl; G, GdnHCl; these letters are followed by a numerical code, with '0' for β-LG alone, '10' for a cosolute concentration of 10 mM and so forth.

12.2.3 Determination of Denaturation Kinetics by RP-HPLC

To study the heating kinetics, protein + cosolute solutions were heated at 80°C for time intervals between 0 and 30 min. The time to reach the set temperature was ~2.5 min. Following the heat treatment, the samples were cooled in ice water to 4°C. The denaturation degree of the heated β-LG solutions was determined by RP-HPLC.

To measure the content of soluble protein after heat treatment, the solutions were adjusted to pH 4.6 and centrifuged at 2.69 × 10^4 g (Sorvall Instruments RC5C, rotor Sorvall SS34) for 15 min.[17] After centrifugation, the samples were filtered for injection through 0.45 μm filters (Orange Scientific, Waterloo, Belgium) and adjusted to a protein concentration of ~0.1–0.3 g L^{-1}. The injector system was composed of a Hewlett Packard series 1050 apparatus equipped with an autosampler and an UV detector. The stationary phase was the hydrophobic interaction column PLRP-S3 (150 × 4.6 mm) of Polymer Laboratoires (Ercatech AG, Bern, Switzerland). Elution solvent A was composed of 0.1% TFA (Pierce Rockford, IL, USA) in water. Eluent B contained 0.1% TFA in 80% acetonitrile and 20% Lichrosolv water (Merck, Darmstadt, Germany). The chromatography was carried out at 50°C and 205 nm, with a nonlinear elution gradient (43% B at $t = 0$ min to 100% B at $t = 28$ min) with a flow rate of 1 mL min^{-1}, and an injection volume of 20 μL.[22]

The kinetic parameters were calculated using the general rate equation

$$-\frac{dC_t}{dt} = k_n C^n, \qquad (1)$$

where C_t (g L^{-1}) represents the concentration of protein at a given reaction time t, the parameter n the reaction order and k_n ((g L^{-1})$^{1-n}$ s^{-1}) the reaction rate constant. Using the integrated form

$$\left(\frac{C_t}{C_0}\right)^{1-n} = 1 + (n-1)\,kt \qquad (2)$$

the reaction rate k can be derived for a given reaction order n. The quantity C_t/C_0 is the protein concentration at a reaction time t divided by the initial protein concentration.[23,24]

12.2.4 Determination of Stability Ratio

The aggregation mechanism of Brownian particles can be depicted as belonging to two limiting regimes.[25] Diffusion-limited cluster aggregation (DLCA) is defined by the power-law dependence

$$D_\mathrm{h}(t) = D_\mathrm{h}(0)(1+\alpha c_\mathrm{s} t)^{z/d_\mathrm{f}}, \quad c_\mathrm{s} = \frac{4k_\mathrm{B}T}{3\eta} N_0 \qquad (3)$$

In Equation (3), $D_\mathrm{h}(0)$ is described as the cluster or particle diameter extrapolated to time zero, α is the sticking probability, and d_f the fractal dimension of the particles. The dynamic exponent z denotes the reactivity between the particles and is close to unity for a pure DLCA mechanism. The quantity c_s is the Smoluchowski rate constant with $k_\mathrm{B}T$ as the thermal energy, η the viscosity of the solvent and N_0 the number of monomeric particles at $t=0$.

Several authors have suggested that the aggregation kinetics of colloids can be described using a so-called stability ratio W.[26] It represents the ratio of the rate constant for rapid coagulation kinetics, where every collision is effective (DLCA), and slow aggregation kinetics, where not every collision is effective (RLCA). The parameter W is linked to the sticking probability α by

$$W = \frac{1}{\alpha}. \qquad (4)$$

Thus, the DLCA aggregation regime is reached for W values close to unity, *i.e.*, where the sticking probability is highest. As the calculation of the stability ratio W relates to the disappearance of monomers at early reaction times, we choose to calculate it from the available reaction rates determined by RP-HPLC.

12.2.5 Determination of Protein Aggregate Molecular Weight and Second Virial Coefficient

The weight-averaged molecular weight M_w and the second virial coefficient A_2 of the aggregates formed during heat treatment were determined using the Nanosizer ZS (Malvern Instruments, UK). The measurements were done using static light scattering (SLS), which measures the time-averaged intensity of the scattered light. The samples before heat treatment were filtered in a two-step process using 0.1 μm and 0.025 μm filters (NC 03 Membrane Filters, Schleicher & Schuell, Germany) to eliminate dust particles and non-native protein. The samples were placed in a square quartz cell (Hellma, pathlength 1 cm) and heated at 80°C for 10 min in the Nanosizer. After heat treatment, the solutions were diluted to different concentrations with water and measured against water and toluene as a SLS standard. Application of the so-called Debye plot leads to the determination of M_w and A_2:

$$\frac{KC}{R_\theta} = \left(\frac{1}{M_w} + 2A_2C\right). \tag{5}$$

In Equation (5) C is the concentration of protein particles, K the optical constant (depending on the apparatus), and R_θ the Rayleigh factor at a given scattering angle θ (173°) using toluene as the reference scattering medium. The second virial coefficient is a measure of the affinity between the protein and the solvent. A value of $A_2 > 0$ indicates high protein–solvent affinity, $A_2 = 0$ shows equivalent protein–solvent and protein–protein affinities and $A_2 < 0$ indicates low protein–solvent affinity. A negative A_2 value can be considered to be a sign of protein aggregation due to strong protein–protein interactions.[27]

12.2.6 Determination of Protein Aggregate Size and Electrophoretic Mobility

Aggregate size was determined in the samples after heat treatment using the Nanosizer ZS (Malvern Instruments, UK). The apparatus is equipped with a laser emitting at 633 nm and with 4.0 mW power source. The instrument was used in the backscattering configuration where detection is done at a scattering angle of 173°. This allows considerable reduction of the multiple scattering signals in turbid samples. The heated protein + cosolute solutions were filtered using 0.22 μm filters (Millipore) to eliminate dust and other insoluble particles. The filtered solutions were placed in square plastic cuvettes. Depending on the sample turbidity (attenuation) the pathlength of the light was set automatically by the apparatus. The autocorrelation function was calculated from the fluctuation of the scattered intensity. From the fit of the correlation function using the cumulant technique,[28] the particle diffusion coefficient D_{app} was calculated. The hydrodynamic diameter D_h was calculated from the Stokes–Einstein equation:[29]

$$D_{app} = \frac{k_B T}{3\pi\eta D_h}. \tag{6}$$

The reported hydrodynamic diameter D_h was averaged from two runs each of 20 measurements.

For determination of the electrophoretic mobility, protein + cosolute solutions were filled in 22-mL glass vials (with a screw top and solid cap with PTFE Liner) (Supelco USA, Milian, Switzerland) and heated in a water-bath while stirring at 80°C for 10 min. After the heat treatment, the samples were cooled in ice water to 4°C. The solutions were filtered using 0.22 μm filters (Millipore) to eliminate dust and other insoluble particles. The filtered solutions were placed in the cell and analysed using the Nanosizer ZS (Malvern Instruments, UK). The effective voltage in the measurement cell was between 50 and 150 V depending on the ionic strength of the samples. The electrophoretic mobility experiments were performed at room temperature and recorded values are averages of two runs of 20 measurements.

12.2.7 Determination of Protein Aggregate Interfacial Properties

Dynamic surface tension measurements of the heated β-LG + cosolute solutions at pH 7.0 and protein concentration 10 g L^{-1} were carried out using a dynamic pendant-drop Tracker tensiometer (IT Concept, Longessaigne, France). The filtered β-LG + cosolute solutions (0.45 μm filter) were placed in a square optical quartz cuvette (volume 7 mL). The tip of the aluminium needle (0.9 mm o.d., 0.55 mm i.d., Popper, Milian, Switzerland) of a syringe (0.25 mL, Micro Syringe, Milian Switzerland) was put in contact with the liquid in the cuvette. An axisymmetric air bubble was formed in the solution with a constant area of 16 mm^2. The image of the bubble was recorded by a CCD camera and the interfacial tension γ was calculated by application of the Laplace equation to the profile of the bubble image. Once equilibrium of the surface tension was reached, oscillations at frequency 0.01–0.1 Hz and amplitude 10% were performed. From these measurements the dynamic elastic modulus E was calculated:[30,31]

$$|E| = A(\Delta\gamma/\Delta A) = E' + iE''. \quad (7)$$

Experiments were performed at 25 ± 1°C. In the presence of too many insoluble aggregates or too strong aggregation, the formation of a stable bubble was not possible due to film rupture or lack of sufficient surface-active protein due to particle sedimentation.

12.2.8 Determination of Protein Aggregate Foaming and Foam-Stabilizing Properties

Foaming properties of β-LG + cosolute solutions (pH 7.0, protein concentration 10 g L^{-1}, cosolute concentration 0–200 mM) have been determined by the method of Guillerme and co-workers.[32] The principle of the method is to foam a defined quantity of protein dispersion by sparging gas through a glass frit with controlled porosity and gas flow. The generated foam rises along a glass column, where the foam volume evolution is followed by image analysis using a CCD camera. A second CCD camera equipped with a 'macro' objective was used to image the variation of the air bubble size at a foam height of about 10 cm (corresponding to half of the foam height). Image analysis was performed on a foam area of 0.38 cm^2 using software developed in-house based on mathematical analysis as described elsewhere.[33] A modified version of the Foamscan apparatus (ITConcept, Longessaigne, France) has been used to perform the experiments. A square glass column was used to minimize optical artifacts associated with air bubble imaging. The temperature within the column and the cuvette was set to $27 \pm 2°C$. Foaming of the protein + cosolute solutions was measured by sparging with N_2 at 80 mL min^{-1} in 50 mL of the heated protein + cosolute solution. This flow-rate was found to allow efficient foam formation prior to strong gravitational drainage. The porosity of the glass frit allowed the formation of air bubbles with diameters between 10 and 16 μm. Sparging with N_2 was stopped after a foam volume of 180 cm^3 was reached. At the end of bubbling, the mean air-bubble diameter was followed with time. In addition, evolution of the liquid fraction in the foam was followed at two different positions – at the top and bottom of the foam – by means of electrical conductimetry.

12.3 Results and Discussion

12.3.1 Protein Denaturation and Aggregation in Presence of Cosolutes

The picture of the sample tubes in Figure 1 shows the appearance of the β-LG dispersions at pH 7.0 after heating at 80°C for 10 min in presence of arginine HCl. With increasing cosolute concentration, turbidity increased gradually because of the formation of protein aggregates, but the samples remained homogeneous with no sign of sedimentation. At a critical cosolute concentration (C_{cs}^*), which was different for the three cosolutes (results for NaCl and GdnHCl are not shown), precipitation occurred and large cloudy aggregates were formed, leading to a macroscopic phase separation. Overall, the

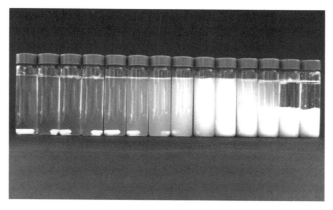

Figure 1 *Appearance of glass tubes containing solutions of β-LG (10 g L^{-1}) + cosolute after heat treatment (80°C, 10 min, pH 7.0). The arginine HCl concentrations increase from left to right: 0, 5, 10, 20, 30, 40, 50, 60, 70, 80, 90, 100, 150 and 200 mM*

aggregation pattern were similar in presence of the three cosolutes, but C_{cs}^* was significantly lower for arginine HCl (70 mM) and GdnHCl (50 mM) than for NaCl (150 mM).

The kinetics of denaturation and aggregation of the β-LG + cosolute mixtures has been investigated within 30 min at 80°C by determination of the remaining native soluble protein at pH 4.6 using RP-HPLC. Figure 2 shows the calculation of the kinetic parameters for heat denaturation and aggregation of the β-LG variants A and B in presence of arginine HCl. The application of the general rate equation (1) demonstrates that β-LG aggregation follows a reaction order of $n = 1.8$–1.9 for variant A and $n = 2$ for variant B. With increased cosolute concentration the reaction order decreases to $n = 1.5$ for both variants. This is in agreement with results by Verheul and co-workers,[19] who explained the decrease of the reaction order in terms of increased protein stability at higher salt concentrations, and therefore reduced protein unfolding leading to rate limitation. With further increase in cosolute concentration (100 mM arginine HCl for both β-LG variants, 200 mM NaCl and 50 mM GdnHCl for variant B), the reaction order increases slightly again to values of $n \approx 1.8$. Only for 400 mM of arginine HCl, the value of n increases to 2.2 and 2.0 for β-LG variants A and B, respectively. Possible explanations for this latter behaviour are more complex reactions due to arginine-induced protective effects on the protein structure and/or suppressed aggregation. Overall, it was found that the data for most of the conditions with the three cosolutes fitted well with a reaction order of 1.5 ($R^2 > 0.99$) with a few exceptions for β-LG variant B ($R^2 = 0.9736$) and for cosolute concentrations higher than 200 mM ($R^2 \sim 0.97$–0.98). The reaction rate k was calculated from the HPLC heating kinetics, with a reaction order of 1.5 for β-LG A and B (Figure 2). The reaction rate was found to increase with increasing cosolute concentration up

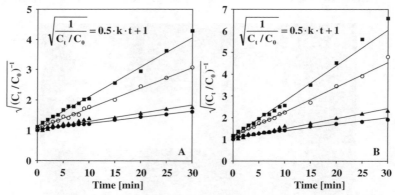

Figure 2 *Representation of the kinetics of heat denaturation and aggregation of (A) β-LG A and (B) β-LG B in presence of arginine HCl (pH 7.0, 80°C, 0–30 min, protein concentration 10 g L^{-1}, cosolute concentration 0–400 mM):* ●, *protein alone;* ○, *A20;* ■, *A70;* ▲, *A400. The lines indicate the decrease in protein concentration for a given reaction order n, from which the reaction rate constant k was derived*

to a maximum (k_{max}) that corresponds to C_{cs}^* where the protein becomes insolubilized. At higher concentrations than C_{cs}^*, k decreases significantly (with exception of GdnHCl for β-LG A and B). This behaviour was found to be similar for all three cosolutes. Normalization of all reaction rates by the fastest one allowed calculation of the stability ratio W.

Looking closer at the stability ratio data reveals that, upon increased arginine HCl concentration, W decreases and the sticking probability α increases significantly, as shown in Figure 3; similar results were obtained for NaCl and GdnHCl. This indicates that the proteins aggregate in the presence of cosolutes, the likely reason being the surface charge neutralization of the denatured proteins.[34] To understand the formation of soluble aggregates in presence of cosolutes and the changes in the protein + solvent behaviour, the second virial coefficient A_2 of the β-LG + cosolute mixtures was determined. Even if overall positive values were found for A_2, increasing cosolute concentration led to a three to fourfold decrease in A_2 before precipitation occurred (Figure 4). This indicates a lowered affinity of the protein for the solvent compared to β-LG alone that might be explained by an increased surface hydrophobicity, protein unfolding and aggregation.[35] In the tested conditions, up to C_{cs}^*, the value of A_2 remained still positive after heating in presence of the cosolute, indicating that the protein tends to aggregate but remains soluble ($A_2 > 0$). Therefore sufficient repulsive forces are still present between the protein molecules due to incomplete surface charge neutralization. At the starting point of precipitation (C_{cs}^* at A70, N150 and G50), the stability ratio reaches values close to unity, implying that the highest sticking probability for

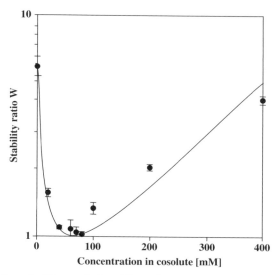

Figure 3 *Stability ratio W, as calculated from the reaction rate constant k, determined by RP-HPLC (pH 7.0, 80°C, 0–30 min, protein concentration 10 g L^{-1}, arginine HCl concentrations 0–400 mM). Vertical error bars represent the standard deviation*

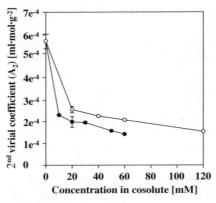

Figure 4 Second virial coefficient A_2 of β-LG solutions heated in presence of cosolutes (pH 7.0, 80°C, 10 min): ●, arginine HCl; ○, NaCl. Vertical error bars represent the standard deviation

the denatured proteins was reached (Figure 3). Particle surface charges are sufficiently neutralized and heteroaggregation of oppositely charged particles is favoured. Under these conditions, aggregation is purely diffusive (DLCA) and it depends only on solvent viscosity, which explains why the aggregation proceeds very fast ($k = k_{max}$). It is important to note that, under these reduced electrostatic repulsion conditions between denatured protein monomers, chemical aggregation is also driven by covalent interaction (an increase in the activation of thiol groups of more than 35% was measured compared to β-LG alone) beside physical interaction (attractive forces and hydrophobic interaction).[18] At higher cosolute concentrations, the value of W increases again for all three cosolutes, leading to sticking probability values of $\alpha = 0.74$ for A100, $\alpha = 0.48$ for A200 or $\alpha = 0.69$ for N200 (Figure 3). This behaviour can be explained by an 'overcharging effect', resulting in slower protein aggregation due to repulsive forces and the necessity for more protein/particle contacts until aggregation occurs (RLCA). Several studies on colloidal aggregation have reported similar behaviour[25,26] with α shown to depend on the surface charges and on overcharging effects at higher electrolyte concentrations.

The apparent surface properties of the aggregates were investigated by means of electrophoretic mobility measurements. Figure 5 shows the averaged electrophoretic mobility of aggregate populations formed during heat treatment of β-LG at 80°C for 10 min at pH 7.0 in the presence of arginine HCl, NaCl and GdnHCl. The mobility was found to decrease with increasing cosolute concentration and to level off at concentrations higher than 60 mM of cosolute. The change in mobility distribution with increasing cosolute concentration (results not shown) demonstrates overall charge neutralization and a significant broadening of the particle distributions. The presence of various populations with different surface charges is consistent with the hypothesis of heteroaggregation between particles.

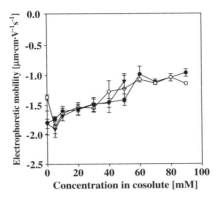

Figure 5 *Electrophoretic mobility of β-LG aggregates heated in presence of cosolutes (protein concentration 10 g L^{-1}, cosolute concentration 0–100 mM, pH 7.0, heated at 80°C for 10 min:* ●, *arginine HCl;* ○, *NaCl;* ▼, *GdnHCl. Vertical error bars represent the standard deviation*

Based on this set of results, a tentative mechanism of β-LG denaturation and aggregation at pH 7.0 in presence of cosolute can be proposed. Considering that β-LG unfolds at a temperature ($T_d \sim 75°C$) below the actual heating temperature of 80°C, the unfolding process should not be rate-limiting in those conditions. However, due to an overall negative surface charge on the β-LG molecules, repulsive forces certainly restrict immediate protein aggregation upon cluster collision. Results in Figures 2–5 indicate that the overall negative net charge of the protein is neutralized by interaction with positively charged cations. In the presence of low concentrations of cosolute, negative protein charges are partially screened by arginine, sodium and guanidinium cations. Therefore the repulsive forces decrease, leading to an increased sticking probability (Figure 3) and fast particle aggregation (Figure 2). Consequently, diffusion-limited aggregation gains an importance compared to rate-limited aggregation, although DLCA and RLCA do coexist.[36] The higher the cosolute concentration, the more particles are neutralized until a certain degree of overall charge screening is reached, characterized by very heterogeneously charged populations. In this case, the probability of interaction and the aggregation rate increase. The reactivity between the particles becomes very high, resulting in fast aggregation ($W \sim 1$, see Figure 3). With increased cosolute concentration the overall electrophoretic mobility of the particles decreases for all three cosolutes to the same extent, until it levels off. This indicates a certain degree of saturation with counter-ions (cosolutes) on the particle surface, supporting the overcharging hypothesis.[37] At a certain degree of surface charge neutralization, partial overcharging leads to higher repulsive forces meaning that more particle contacts are necessary to induce aggregation. We note that the effect of the guanidinium group on charge neutralization and protein aggregation seems to be stronger than that of Na^+. However, concerning the overcharging effect, arginine HCl and NaCl lead to significantly higher protein overcharging than GdnHCl. This might be explained by a higher accessibility and diffusion rate of Na$^+$, but also by additional charges on arginine HCl due to the amino acid group.

12.3.2 Physicochemical Properties of Protein Aggregates in Relation to Interfacial and Foaming Properties

Figure 6A shows the evolution of the hydrodynamic diameter D_h of the soluble aggregates formed during heat treatment of β-LG + arginine HCl or NaCl. It was found that β-LG alone heated at 80°C for 10 min gives a higher D_h value (65 ± 1.8 nm) than at low cosolute concentrations (5–30 mM for arginine HCl or GdnHCl, and up to 80 mM for NaCl). The effect of the low concentration of cosolute could be related to a salting-in effect, as described by Arakawa and Timasheff.[38] However, it is more likely that a few larger aggregates dominate the scattered signal in a mainly monomeric and dimeric protein solution (β-LG alone and mixtures with low cosolute concentration). With increasing particle size and denaturation degree this effect gets less important as higher quantities of larger particles dominate the signal. With increasing cosolute concentration, D_h increases up to a maximum of ∼130 nm for A70 and N150 corresponding to the critical cosolute concentration where insolubility of the proteins occurred. At concentrations higher than C_{cs}^*, the soluble aggregates still present in the sample decrease in size until D_h reaches values of 8–14 nm. Figure 6B shows the molecular weight M_w of the soluble aggregates formed in presence of arginine HCl and NaCl as a function of the hydrodynamic radius of the aggregates. Interestingly, a master curve is obtained, even though the cosolute concentration to obtain aggregates with similar D_h is different for the two cosolutes (arginine HCl or NaCl). In other words, it is possible to produce

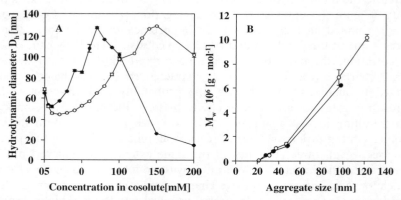

Figure 6 *Dynamic light scattering of heated β-LG solutions (pH 7.0, protein concentration 10 g L^{-1}, 0–200 mM cosolute, heated at 80°C for 10 min):* ●, *arginine HCl;* ○, *NaCl. (A) Aggregate size D_h plotted as a function of the cosolute concentration. (B) Molecular weight M_w plotted as a function of the aggregate size. Vertical error bars represent the standard deviation*

β-LG soluble aggregates having a similar condensation state using various cosolutes providing that charge neutralization of the denatured monomers is achieved. This statement is corroborated by the superimposed values of the apparent electrophoretic mobility of these aggregates for the three cosolutes (Figure 5).

The air–water interfacial properties of the soluble aggregates have been investigated by dynamic pendent drop tensiometry, allowing determination of the surface viscoelastic parameters. Figure 7 shows a comparison of the frequency behaviour of the surface elasticity E' and the phase angle δ for β-LG heated alone or in the presence of cosolutes close to the critical concentration C_{cs}^* (A50 and N140). The heated β-LG + cosolute dispersions exhibited a considerable increase in E' of more than 100% over the whole frequency range. The surface viscosity $\eta_d = E''/\omega$ (at 0.01 Hz) exhibited very similar behaviour, increasing from 160 mN s m^{-1} for β-LG heated alone to 280 mN s m^{-1} when heated with cosolutes (results not shown). The frequency dependence of η_d can be explained in terms of conformational rearrangements at the interfacial layer upon deformation. The gel-like character of the interface composed of soluble aggregates generated in presence of cosolutes is emphasized by the change in phase angle from 22° to 9° for β-LG alone and with cosolutes, respectively (Figure 7).[31] Here, it is important to note that the viscoelastic patterns for A50 and N140 (conditions with > 80% of soluble aggregates) could be superimposed, indicating similar interfacial properties for aggregates having similar molecular weight, hydrodynamic diameter and apparent surface charge.

The foaming and foam-stabilizing properties of these soluble aggregates have been investigated using the Foamscan^{TM} apparatus. The foamability of aggregates generated with or without cosolute presence was determined by measuring the average number of bubbles initially present in a given volume of foam after N$_2$ sparging. It has been found that 2–3 times more bubbles were present

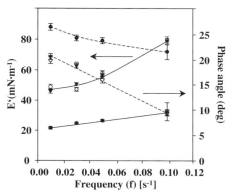

Figure 7 *Frequency dependence of surface elasticity E' (—) and phase angle (- - - -) at air–water interface: ●, heated β-LG; ○, A50 (β-LG + 50 mM arginine HCl); ▲, N140 (β-LG + 140 mM NaCl). Vertical error bars represent the standard deviation*

in the foam when β-LG aggregates were generated in presence of cosolutes (data not shown). This initial number of bubbles levelled off at C_{cs}^*, and it was identical whatever cosolute used. This result can be interpreted in terms of similar interfacial properties of the soluble aggregates that were present. The foams stabilized with aggregates generated with cosolutes exhibited significantly lower drainage rates as compared to β-LG alone. Figure 8 shows the variation of the normalized liquid fraction for the two foam heights of 6.5 and 11.5 cm. It can be seen that, in presence of cosolutes, aggregates are able to retain much more liquid within the thin films and Plateau borders, as the drainage rate exponent is two–threefold lower than that of foams stabilized by aggregates of β-LG heated alone. This behaviour can be directly linked to the increased surface viscosity that was reported previously[39] in combination with the high surface elasticity. Hence it has been shown that, in the case of the interface-dominated drainage regime, because of the high surface elasticity compared to bulk, the liquid fraction scaling exponent is close to -1, whereas it reaches much higher values (-3 to -4) if the drainage is limited by the bulk.[40]

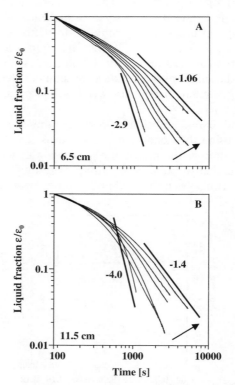

Figure 8 *Normalized liquid fraction in the foam versus time for solutions heated at pH 7.0 and 80°C for 10 min (protein concentration 10 g L^{-1}, arginine HCl concentration 0–70 mM). The quantity ε_0 is defined as the liquid fraction at the end of N_2 sparging at two height positions in the foam: A, 6.5 cm; B, 11.5 cm. The arrow indicates increasing cosolute concentration up to C_{cs}^**

The ability of the β-LG aggregates to prevent coarsening of the foam has been evaluated by following the evolution of the normalized mean bubble diameter over time. It should scale as t^α with $\alpha = 1$ for purely diffusive foam coarsening (Figure 9).[41] Heated β-LG solutions (10 or 18 g L^{-1}, the latter corresponding to an elevated content of dry matter due to presence of cosolutes, e.g., N140) showed low foam stability ($t \ll$ 3 ks) and high gas diffusion rates. Calculation of the coarsening exponent resulted in values of 2.4–2.5. Disproportionation in the foams decreased significantly with increasing cosolute concentration to α values from 1.7 to 2.0. In the presence of soluble and insoluble aggregates, and no matter what type of cosolute was present, the foams exhibited significantly higher stability against gas diffusion with α values from 1.3 to 1.7. However, the presence of insoluble particles was in some cases (e.g., A70) detrimental to the gas diffusion properties, whereas several samples (G50, N140, N150 and N200) showed very high resistance against gas diffusion (not all the results are shown in Figure 9). This might be explained[9,10,12] by specific particle and surface properties like size, smoothness, shape, surface hydrophobicity and surface charge. In general, for lower cosolute concentrations (and the protein controls), the coarsening exponent reflects the fact that, besides disproportionation, additional destabilization mechanisms also participate in bubble growth. This might be due to the formation of insufficiently stable interfacial networks, leading to film thinning, film rupture and bubble coalescence or collapse, which is in agreement with observations made for foam volume stability under these same conditions (results not shown). In conditions where soluble aggregates of size of > 100 to 130 nm were present in the heated protein + cosolute mixture, and in some cases also insoluble particles, the foams demonstrated significantly higher stability and lower gas diffusion rates. This might be explained[6,8] by the formation of highly viscoelastic interfacial films and gel-like interfacial layers, which entrap air bubbles and control drainage.

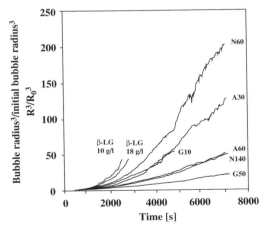

Figure 9 *Time evolution of normalized cube of average air bubble radius as calculated from image analysis of foams stabilized by β-LG aggregates at pH 7.0*

12.4 Conclusions

This study has demonstrated that the use of soluble and/or insoluble aggregates formed during heat treatment of β-LG in the presence of cosolutes increases the foam stability parameters significantly. The presence of charged cosolutes initiates electrostatic charge screening, leading to protein aggregation during heat treatment, and also to reduced repulsive forces of the proteins on adsorption at the air–water interface. It has been further demonstrated that the nature of the cosolute has no direct effect on foam stabilization, but only on the extent of formation of soluble and insoluble aggregates. As a result the foam volume stability was increased, the foam drainage was reduced (controlled by interfacial viscoelasticity) and the rate of bubble disproportionation considerably lowered. This was explained by the formation of stable interfacial networks, comparable with gel-like networks entrapping the air bubbles. Owing to low repulsive forces between the proteins adsorbed at the air–water interface, the interfacial films are stabilized by increased intermolecular cross-linking, hydrophobic and electrostatic interactions between the proteins/particles. Interfacial elasticity and viscosity were found to reach maximum values under conditions prior to aggregate precipitation (C_{cs}^*).

The presence of insoluble aggregates (1–10 μm) has been found significant to improve foam stability, presumably through contributions to disjoining pressure forces from steric and electrostatic repulsion. The presence of large amounts of soluble aggregates (>80%), and the incorporation of insoluble protein aggregates in a stable interfacial network, thereby entrapping the air bubbles, was found to reduce foam coarsening significantly. However, when aggregation was too strong, foam destabilization occurred due to film rupture, presumably induced by bridging of the thin films by the particles. The formation of a stable gas diffusion barrier against disproportionation and film rupture depends, therefore, not only on the presence of soluble and insoluble protein aggregates with specific sizes and molecular characteristics, but also on the ability to cross-link at the interface *via* electrostatic, hydrophobic and covalent SH–SS exchange.

References

1. S. Damodaran, *J. Food Sci.*, 2005, **70**, 54.
2. A.H. Martin, K. Grolle, M.A. Bos, M.A. Cohen Stuart and T. van Vliet, *J. Colloid Interface Sci.*, 2002, **254**, 175.
3. P. Suttiprasit, V. Krisdhasima and J. McGuire, *J. Colloid Interface Sci.*, 1992, **154**, 316.
4. D.C. Clark, A.R. Mackie, L.J. Smith and D.R. Wilson, in *Food Colloids*, R.D. Bee, P. Richmond and J. Mingins (eds), Royal Society of Chemistry, Cambridge, 1989, p. 97.
5. V. Petkova, C. Sultanem, M. Nedyalkov, J.-J. Benattar, M.E. Leser and C. Schmitt, *Langmuir*, 2003, **19**, 6942.

6. E. Dickinson, R. Ettelaie, B.S. Murray and Z. Du, *J. Colloid Interface Sci.*, 2002, **252**, 202.
7. R. Ettelaie, E. Dickinson, Z. Du and B. S. Murray, *J. Colloid Interface Sci.*, 2003, **263**, 47.
8. W. Kloek, T. van Vliet and M. Meinders, *J. Colloid Interface Sci.*, 2001, **237**, 158.
9. Z. Du, M.P. Bilbao-Montoya, B.P. Binks, E. Dickinson, R. Ettelaie and B.S. Murray, *Langmuir*, 2003, **19**, 3108.
10. G. Kaptay, *Colloids Surf. A*, 2004, **230**, 67.
11. K.P. Velikov and O.D. Velev, in *Emulsions, Foams and Thin Films*, K.L. Mittal and P. Kumar (eds), Marcel Dekker, New York, 2000, p. 279.
12. E. Dickinson, R. Ettelaie, T. Kostakis and B.S. Murray, *Langmuir*, 2004, **20**, 8517.
13. A. Bals and U. Kulozik, *Int. Dairy J.*, 2003, **13**, 903.
14. J.T. Petkov and T.D. Gurkov, *Langmuir*, 2000, **16**, 3703.
15. C. Schmitt, E. Kolodziejczyk and M.E. Leser, in *Food Colloids: Interactions, Microstructure and Processing*, E. Dickinson (ed), Royal Society of Chemistry, Cambridge, 2004, p. 284.
16. J.P. Davis, E.A. Foegeding and F.K. Hansen, *Colloids Surf. B*, 2004, **34**, 13.
17. H. Zhu and S. Damodaran, *J. Agric. Food Chem.*, 1994, **42**, 846.
18. J.G.S. Ho, A.P.J. Middelberg, P. Ramage and H.P. Kocher, *Protein Sci.*, 2003, **12**, 708.
19. M. Verheul, S.P.F.M. Roefs and C.G. de Kruif, *J. Agric. Food Chem.*, 1998, **46**, 896.
20. K. Watanabe and H. Klostermeyer, *J. Dairy Res.*, 1976, **43**, 411.
21. Y.L. Xiong, K.A. Dawson and L. Wan, *J. Dairy Sci.*, 1993, **76**, 70.
22. P. Resmini, L. Pellegrino, J.A. Hogenboom and R. Andreini, *Ital. J. Food Sci.*, 1989, **1**, 51.
23. F. Dannenberg and H.-G. Kessler, *J. Food Sci.*, 1988, **53**, 258.
24. S.P.F.M. Roefs and C.G. de Kruif, *Eur. J. Biochem.*, 1994, **226**, 883.
25. Y. Georgalis, P. Umbach, J. Raptis and W. Saenger, *Acta Cryst.*, 1997, **D53**, 691.
26. J.A. Molina-Bolivar, F. Galisteo-González and R. Hidalgo-Álvarez, *J. Colloid Interface Sci.*, 1998, **208**, 445.
27. P. Aymard, D. Durand and T. Nicolai, *Int. J. Biol. Macromol.*, 1996, **19**, 213.
28. P. Štepánek, in *Dynamic Light Scattering: The Methods and Some Applications*, W. Brown (ed), Clarendon Press, London, 1993, p. 177.
29. U.M. Elofsson, P. Dejmek and M.A. Paulsson, *Int. Dairy J.*, 1996, **6**, 343.
30. J. Lucassen and M. Van den Tempel, *Chem. Eng. Sci.*, 1972, **27**, 1283.
31. J.M. Rodriguez Patino, M.R. Rodriguez Nino and C. Carrera Sanchez, *J. Agric. Food Chem.*, 1999, **47**, 3640.
32. C. Guillerme, W. Loisel, D. Bertrand and Y. Popineau, *J. Texture Stud.*, 1993, **24**, 287.
33. C. Schmitt, T. Palma da Silva, C. Bovay, S. Rami-Shojaei, P. Frossard, E. Kolodziejczyk and M.E. Leser, *Langmuir*, 2005, **21**, 7786.

34. G. Unterhaslberger, C. Schmitt, C. Sanchez, C. Appolonia-Nouzille and A. Raemy, *Food Hydrocoll.*, 2006, **20**, 1006.
35. E.Y. Chi, S. Krishnan, B.S. Kendrick, B.S. Chang, J.F. Carpenter and T.W. Randolph, *Protein Sci.*, 2003, **12**, 903.
36. R. Vreeker, L.L. Hoekstra, D.C. den Boer and W.G.M. Agterof, *Food Hydrocoll.*, 1992, **6**, 423.
37. S. Ravindran and J. Wu, *Langmuir*, 2004, **20**, 7333.
38. T. Arakawa and S.N. Timasheff, *Biochemistry*, 1987, **26**, 5147.
39. S.A. Koehler, S. Hilgenfeldt, E.R. Weeks and H.A. Stone, *Phys. Rev. E*, 2002, **6**, 1.
40. A. Saint-Jalmes and D. Langevin, *J. Phys. Condens. Matter*, 2002, **14**, 9397.
41. D. Weaire and S. Hutzler, *The Physics of Foams*, Clarendon Press, Oxford, 1999.

Chapter 13

Manipulation of Adsorption Behaviour at Liquid Interfaces by Changing Protein–Polysaccharide Electrostatic Interactions

Renate A. Ganzevles,[1,2,3] Ton van Vliet,[1,3] Martien A. Cohen Stuart[2] and Harmen H.J. de Jongh[1,4]

[1] WAGENINGEN CENTRE FOR FOOD SCIENCES, P.O. BOX 557, 6700 AN WAGENINGEN, THE NETHERLANDS
[2] LABORATORY OF PHYSICAL CHEMISTRY AND COLLOID SCIENCE, WAGENINGEN UNIVERSITY, P.O. BOX 8038, 6700 EK WAGENINGEN, THE NETHERLANDS
[3] DEPARTMENT OF AGROTECHNOLOGY AND FOOD SCIENCES, WAGENINGEN UNIVERSITY, P.O. BOX 8129, 6700 EV WAGENINGEN, THE NETHERLANDS
[4] TNO QUALITY OF LIFE, P.O. BOX 360, 3700 AJ ZEIST, THE NETHERLANDS

13.1 Introduction

Protein–polysaccharide interactions have been investigated in a diversity of contexts such as heparin and blood coagulation,[1] protection of enzymes against high pressure or high temperature, enzyme–substrate binding, and recovery and fractionation of milk proteins. A well-known example from the food industry is the use of pectin to stabilize casein micelles in acidified milk drinks. Due to electrostatic interaction the negatively charged pectin molecules adsorb on the casein micelles preventing them from acid-induced aggregation *via* electrostatic and steric repulsion.[2–4]

On mixing an anionic polysaccharide with a globular protein, four different interaction regimes can be distinguished, depending on pH, ionic strength and mixing ratio, as described for mixtures of whey protein + arabic gum by Weinbreck *et al.*:[5]

(i) At neutral pH, both protein and polysaccharide are net negatively charged, and therefore co-soluble; but there can be some attractive interaction between the positively charged groups on the protein and the negatively charged polysaccharide.

(ii) On lowering the pH close to the isoelectric pH of the protein (or below), soluble protein–polysaccharide complexes are formed at low ionic strength (below ~100 mM, depending on the binding affinity).
(iii) A further decrease in pH leads to aggregation of the soluble complexes and subsequently to complex coacervation.
(iv) At pH values below pH ≈ 2.5, the complexation can be suppressed by protonation of the acidic groups on the polysaccharide.[5]

The adsorption kinetics of proteins has been extensively studied in relation to foam and emulsion formation. Several steps in protein adsorption have been identified:[6–8] transport of the molecules to the interface by diffusion/convection; adsorption at the interface; and possibly conformational changes once adsorbed at the interface. Large variations have been found between different proteins – for example, β-lactoglobulin (β-Lg) is known to increase surface pressure rapidly,[9] whereas for lysozyme the process is much slower.[10] According to Wierenga and co-workers,[11,12] a kinetic barrier for protein adsorption can explain these differences: when a protein molecule approaches the air–water interface, a balance between its hydrophobic exposure and its net charge determines the probability of adsorption. Obviously all this holds for pure protein solutions; but the presence of anionic polysaccharide molecules that interact with the protein molecules also affects the adsorption kinetics to air–water interfaces.[13–15] Because protein adsorption kinetics at air–water and oil–water interfaces can be different,[16] the impact of protein–polysaccharide interactions on adsorption kinetics could also differ for the two types of interfaces.

Based on previous observations at the air–water interface, a mechanistic model for mixed protein + polysaccharide adsorption has been proposed (Figure 1).[17] The protein–polysaccharide-binding affinity determines how much protein is uncomplexed in solution and how much is bound to the polysaccharide (Figure 1, equilibrium A). Protein molecules can diffuse freely to the air–water interface when not bound to a polysaccharide (route D). Once protein approaches the interface, it might encounter a kinetic barrier for adsorption (depicted by E). Protein–polysaccharide complexes can also diffuse to the interface (route B), but their diffusion coefficient is much lower than that for the free protein due to the larger hydrodynamic radius.

One may wonder how a protein–polysaccharide complex adsorbs at the air–water interface. Presumably it is the protein molecule(s) in the complex – and not so much the polysaccharide one(s) – that are responsible for the increase in surface pressure. Complexes that are highly negatively charged may have a low density. If a polysaccharide can trap protein molecules in such a diffuse complex, it may prevent efficient packing of protein at the interface. In this case the mobility of the protein molecules within the complex determines how fast a compact protein layer can be formed. This is represented by factor C in Figure 1. This factor most likely depends on the protein–polysaccharide-binding affinity. Understanding the mechanism of mixed adsorption should allow better control of adsorption kinetics to the air–water interface (as monitored, for example, by drop tensiometry) and therefore the processes

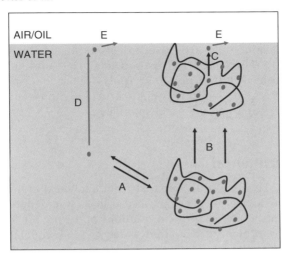

Figure 1 *Schematic model for polysaccharide-controlled protein adsorption at the air–water or oil–water interface: A, partitioning of free protein and protein bound to polysaccharide; B, diffusion of protein–polysaccharide complex in the bulk; C, availability of complexed protein for the interface; D, diffusion of free protein in the bulk; E, kinetic barrier to protein adsorption*

of foam and emulsion formation. Parameters that can be used to manipulate protein–polysaccharide interactions in the bulk are the protein–polysaccharide-binding affinity (through the pH, ionic strength and charge density/distribution on individual ingredients) and the mixing ratio *via* paths A and C, and the molecular weight of the polysaccharide *via* path B.

Anionic polysaccharides can be used in combination with proteins to prevent aggregation and creaming of emulsion droplets.[18–23] However, depending on the relative concentrations, the added polysaccharides can also decrease emulsion stability due to bridging flocculation.[13] Through layer-by-layer deposition on emulsion droplets of oppositely charged proteins, polysaccharides and surfactants,[24,25] one may control the net charge on the droplets and enhance the emulsion stability by electrostatic repulsion. Through their electrostatic interaction with proteins, anionic polysaccharides that are not surface-active are able to affect surface rheological properties by coadsorption.[14,26] Ducel and co-authors[27] have reported that oil–water interfacial rheological properties of plant protein–arabic gum coacervates are related to the interfacial rheological properties of the protein used. Schmitt *et al.*[15] have described an effect of ageing time of β-Lg–acacia gum complexes on air–water surface rheology in terms of reorganization of the complexes at the interface. The elastic behaviour of adsorbed protein layers at air–water or oil–water interfaces may affect foam and emulsion stabilizing ability.[28] A mechanistic understanding is still lacking, however, of how the surface rheological behaviour depends on parameters like charge density of the polysaccharides, protein/polysaccharide mixing ratio, ionic strength and the order of adsorption to the interface – simultaneous or sequential.

The aim of our work is to demonstrate how protein–polysaccharide interactions can be used to control both the adsorption kinetics and the surface rheological behaviour of mixed adsorbed layers at air–water and oil–water interfaces, and to determine which parameters are mainly involved. Here, a general mechanism of mixed protein + polysaccharide adsorption to liquid interfaces is presented, while focusing particularly on the comparison of behaviour at air–water and oil–water interfaces.

13.2 Experimental

Acetate buffer (pH = 4.5) was prepared from analytical grade chemicals and deionized water. Bovine β-Lg was purified using a non-denaturing method as described previously.[29] Stock protein solutions of 0.2 mg mL^{-1} were prepared by dissolving the protein in deionized water and subsequently diluting with concentrated acetate buffer solution to a final acetate concentration of 5 mM with an ionic strength of 2 mM. The protein solutions were kept at $-40°$C until further use. Two pectin samples with different degree of methyl esterification (DM) were used: low-methoxyl pectin (LMP, DM = 30, meaning that 30% of the galacturonic acid subunits are methyl esterified) and high-methoxyl pectin (HMP, DM = 70) supplied by CP Kelko (Copenhagen, Denmark). Only the non-methylated galacturonic acid sub-units have a free carboxyl group (weak acid, pK_a ~4.5). Polysaccharide solutions were prepared by first wetting the powder with ethanol, then dissolving in buffer, and subsequently heating at 70°C for 30 min. After overnight storage at 4°C, the samples were centrifuged at 6000 g at room temperature for 10 min. Sunflower oil (Reddy, Vandermoortele, Roosendaal, the Netherlands) was purified by stirring it under vacuum with silica gel 60 (70–230 mesh, Merck, Germany) as described before.[30] After performing this procedure twice, the oil was stored at $-20°$C until use.

Second-cumulant diffusion coefficients of β-Lg–pectin complexes were determined at 23 ± 0.3°C by dynamic light scattering[31] using an ALV light-scattering instrument equipped with a 400 mW argon laser tuned at a wavelength of 514.5 nm.[32] Hydrodynamic radii were calculated according to the Stokes–Einstein relation, assuming that the pectin molecules and the complexes adopt spherical conformations. The protein concentration was 0.1 g L^{-1} for all the samples, and the pectin concentrations were varied over the protein/pectin mixing ratio range 0–15 (weight basis).

The surface tension at the air–water or oil–water interface was measured as a function of time for pure protein and mixed solutions of protein + polysaccharide using an automated drop tensiometer (ITCONCEPT, Longessaigne, France). The arrangement has been described in detail elsewhere.[33] Protein + polysaccharide mixtures were freshly prepared for each experiment and equilibrated for 30 min to allow formation of protein–polysaccharide complexes. Mixing ratios mentioned in the text are on a weight basis. Unless mentioned otherwise, the protein concentration was always 0.1 g L^{-1}. Each experiment started with a clean interface of a newly formed air bubble (7 µL) or oil droplet (21 µL) in a cuvette containing the sample solution. Temperature was controlled at 22 ± 1°C. The surface tension was determined by axisymmetric bubble shape analysis and the results are presented in terms of surface pressure, $\Pi = \gamma_0 - \gamma$, where γ_0 is the air–water tension (72 mN m^{-1}) or oil–water tension (30 mN m^{-1}) and γ is the measured solution tension. Surface rheology measurements were performed with the same apparatus. By subsequent expansion and compression of the interfacial area A (amplitude of area oscillation = 4%, frequency = 0.1 Hz), and recording the resulting change in surface tension γ, the dilatational modulus $\varepsilon = d\gamma/d \ln A$ was followed in time. Periods of five oscillations were alternated with equally long resting periods. Average dilatational moduli were measured using the last three oscillations of each period. All experiments were performed in duplicate.

13.3 Adsorption Kinetics

In our previous work,[17] the β-Lg–pectin interaction in bulk solution was characterized as a function of mixing ratio by dynamic light scattering. Figure 2 shows that, with a small addition of β-Lg to LMP, the scattered light intensity is found to increase, indicating formation of soluble protein–polysaccharide complexes as suggested previously.[5] On further increasing the mixing ratio, the scattered light intensity slightly increases while the hydrodynamic radius remains fairly constant, indicating that the soluble complexes are gradually filled with protein. From the point where the positive charge on the protein compensates the negative charge on the polysaccharides and the net charge of the complexes approaches zero,[34] the soluble complexes aggregate and phase separate. This is shown in Figure 2 by the strongly increasing hydrodynamic radius at a mixing

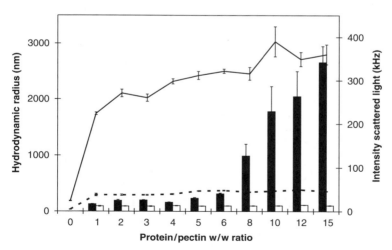

Figure 2 *Dynamic light scattering at various β-Lg/pectin mixing ratios for solutions (pH = 4.5, ionic strength = 2 mM) with protein concentration kept constant at 0.1 g L^{-1}. The plot shows hydrodynamic radius (■) and intensity-scattered light (—) for mixtures with LMP, and hydrodynamic radius (□) and intensity-scattered light (⋯) for mixtures with HMP*

ratio of 8 and higher. In the case of HMP, an increase in scattered light intensity is observed as well. Based on the lower charge density of HMP as compared with LMP, aggregation could have been expected at a lower mixing ratio, but no aggregation was actually observed with this pectin sample.

To study the impact of the presence of polysaccharide on the adsorption kinetics of the protein, the surface pressure was measured as a function of time for β-Lg solutions and β-Lg + pectin solutions, both at the air–water interface [Figure 3(a)] and at the oil–water interface [Figure 3(b)]. The protein concentration in all the samples was 0.1 g L^{-1}, and the protein/polysaccharide mixing ratio was 0.5. Under these conditions, the majority of the protein is present in soluble complexes with the polysaccharide because the charge on the polysaccharide is in excess. Both LMP and HMP were used, in order to compare the effect of their different protein-binding affinities due to their different charge densities. Compared to proteins, the polysaccharides on their own do not give a significant increase in surface pressure at the concentrations used (<2 mN m^{-1} in 2×10^4 s for a concentration of 0.05 g L^{-1}, data not shown). The pure β-Lg solution demonstrated a fast increase in surface pressure at both types of interfaces. The presence of LMP led to a 'lag time' before the surface pressure increased at the air–water interface. The lag time is defined as the time from the beginning of the experiment – with a clean interface – until the surface pressure starts increasing, or, more precisely, the time of the cross-over point between the initial nearly horizontal slope and the steepest slope in the surface pressure *versus* time curve. The presence of HMP causes a lag time at the air–water interface as well, but it is much shorter compared to the one with LMP. This can be explained by the lower charge density and thus the lower binding affinity for β-Lg. At the oil–water interface this lag time is absent, but the increase in surface pressure is clearly delayed with LMP, and to a lesser extent with HMP.

One could imagine that the difference in shape of the curves for the air–water and oil–water interfaces is caused by a difference in the effect of pectin at both interfaces. However, Figure 4 shows that the surface pressure *versus* time curve at the oil–water interface for a pure β-Lg solution of lower concentration also

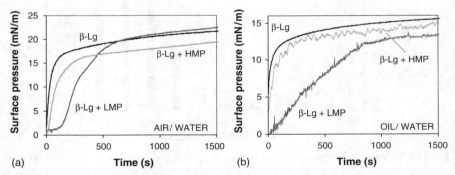

Figure 3 *Surface pressure as a function of time for β-Lg and mixed β-Lg + pectin solutions (pH = 4.5, ionic strength = 2 mM) with mixing ratio = 0.5: (a) air–water interface and (b) oil–water interface*

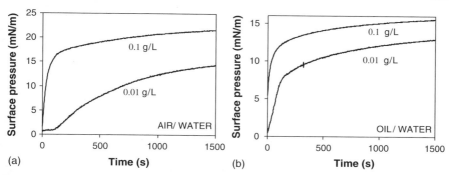

Figure 4 *Surface pressure as a function of time for 0.01 and 0.1 g L^{-1} β-Lg solutions (pH = 4.5, ionic strength = 2 mM): (a) air–water interface, and (b) oil–water interface*

differs from the curve at the air–water interface. That is, at the air–water interface a lag time is observed, whereas at the oil–water interface this lag time is absent and only the initial slope differs from that at the higher concentration. It is known[16,35] that both the adsorbed amount of protein as a function of time and the relation between surface pressure and adsorbed amount can differ at the oil–water interface from those at the air–water interface.

Since the delay in surface pressure development observed for the mixtures (Figure 3) is caused by electrostatic interaction of protein with polysaccharide, it should reduce with increasing salt concentration. It has been demonstrated[17] that the lag time at the air–water interface is indeed gradually diminished on increasing the salt concentration until it completely disappears at an ionic strength of 80 mM. At the oil–water interface the effect of pectin is manifested not as a lag time but as a lower initial slope of the surface pressure *versus* time curve. The inverse of this initial slope is plotted in Figure 5 as a function of ionic strength for a 2:1 β-Lg + LMP mixture and for pure β-Lg. It shows that at the oil–water interface the effect of LMP on protein adsorption also diminishes with increasing ionic strength, and eventually disappears at an ionic strength between 50 and 100 mM.

To get a semi-quantitative insight into the extent to which the increase in surface pressure is delayed by the polysaccharide, an apparent diffusion constant was calculated from the data at the air–water interface. Because the presence of the polysaccharide could alter the surface activity of the protein, one should consider the adsorbed amount instead of the surface pressure in order to assess adsorption kinetics. In the absence of well-defined values for the adsorbed amount of the mixed systems, only a rough calculation is made here based on the surface pressure and the relationship of surface pressure to the adsorbed amount of pure β-Lg at the air–water interface. The adsorbed amount of β-Lg at the air–water interface from the point where surface pressure starts to increase (Γ^*) is ~1 mg m^{-2} (data not shown). The value of Γ^* for the β-Lg + LMP mixture is of the same order of magnitude as indicated by ellipsometry measurements (data not shown). Using the time t^* at which

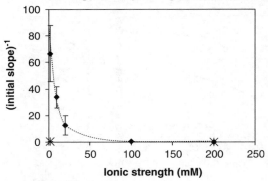

Figure 5 *Inverse of initial slope of surface pressure versus time curves at the oil–water interface as a function of ionic strength for solutions (β-Lg concentration = 0.1 g L^{-1}, pH = 4.5, ionic strength = 2 mM) of 2:1 β-Lg + LMP mixtures (♦) and pure β-Lg (✻)*

surface pressure starts increasing (defined earlier as the 'lag time'), we can roughly estimate an effective diffusion coefficient D_{eff} using the equation of Ward and Tordai,[36]

$$\Gamma^* = 2c\left[\left(\frac{D_{eff}t^*}{\pi}\right)\right]^{1/2}, \tag{1}$$

where c is the total protein bulk concentration. The effective diffusion coefficient for pure β-Lg was obtained by this method using the protein concentration range 0.002–0.02 g L^{-1}, where the possible contribution of convection during droplet formation is relatively low. The calculated value is 1×10^{-10} m^2 s^{-1}, which corresponds well with the value reported by Le Bon et al.[37] From this it can be inferred that adsorption of β-Lg is diffusion-controlled and not significantly hindered by a kinetic barrier for adsorption (as depicted by route E in Figure 1).

We consider that the diffusion rate to the oil–water interface is not significantly faster than to the air–water interface, and that presumably the difference in the surface pressure *versus* time curves for β-Lg at air–water and oil–water interfaces is caused by a difference in the relation between surface pressure and adsorbed amount. A difference in this Π–Γ relation was previously suggested by Benjamins[35] for BSA at air–water and *n*-tetradecane–water interfaces. The effective diffusion coefficient obtained for a mixture of β-Lg (0.1 g L^{-1}) + LMP (mixing ratio = 2) is 3×10^{-13} m^2 s^{-1}. The diffusion coefficient obtained from dynamic light scattering of the complexes is $D_{DLS} = 1 \times 10^{-12}$ m^2 s^{-1}. If all the β-Lg is complexed to LMP, the experimentally obtained effective diffusion coefficient should be equal to the one obtained by dynamic light scattering, since all the protein diffuses as complexes in this case. However, D_{eff} appears to be about three times smaller than D_{DLS}. Since dynamic light scattering is biased to the larger components in a polydisperse sample,[38] because it determines the weight-averaged diffusion coefficient, it is likely that there are many smaller complexes with a much larger diffusion coefficient, leading to faster overall adsorption kinetics. Although diffusion rate seems to play an important role, it is unlikely that retarded diffusion is the only reason for a delayed increase in surface pressure. In absence of a lag time for the oil–water interface, we observe that the surface pressure for 0.1 g L^{-1} β-Lg at the oil–water interface has increased to 5 mN m^{-1} within the first second of the experiment, while for the 2:1 β-Lg + LMP mixture it takes ~250 s before a surface pressure of 5 mN m^{-1} is reached (data not shown). Therefore the delay in surface pressure increase at the oil–water interface caused by the presence of LMP is of the same order of magnitude (~300) as found for the air–water interface (assuming that possible differences in the Π–Γ relation in the presence and absence of pectin are small compared with this number). For HMP the delay is of the order of a factor of 10 for both types of interfaces. Although these calculations are only rough estimations, they show that the effect of the presence of pectin on protein adsorption kinetics at the air–water and oil–water interfaces may be comparable. When using a protein like chicken-egg ovalbumin,[12] that has a kinetic barrier for adsorption, the effect could be different for both types of interfaces. If the affinity of the protein differs for the two interfaces, the balance of affinity of protein for the interface and for the polysaccharide may shift and different effects might come into play.

In conclusion, a retarded diffusion as a result of the larger hydrodynamic radius of the protein–polysaccharide complexes makes a large contribution to the slower increase in surface pressure at both air–water and oil–water interfaces, as observed in the presence of polysaccharides. However, the availability of protein molecules in the complexes at the interface (Figure 1, route C) also presumably plays an important role. This availability is expected to depend on the protein density (number of protein molecules per unit volume) in the complex and the mobility of the protein molecules through the complex.

13.4 Surface Rheology and Structure of Adsorbed Layers

It is supposed that if a polysaccharide involved in protein–polysaccharide complex adsorption can affect the protein density in the interface, it should be detectable by surface rheology. Presumably the protein density in the complexes in the bulk depends on the protein–polysaccharide mixing ratio, and consequently on the net charge of the complexes. For β-Lg + LMP complexes of mixing ratio = 0.5, a ζ-potential of -32 ± 1 mV was measured; and for mixing ratios of 2, 4 and 8, the values were -27, -21 and -9 ± 1 mV, respectively.[34] The ζ-potential determined for the protein molecules was 0 ± 1 mV.

The surface dilatational modulus at the oil–water interface of these complexes and of pure β-Lg was followed in time and plotted as a function of surface pressure (Figure 6). In this way the different adsorbed layers could be compared independently of adsorption kinetics. The shape of the resulting modulus *versus* surface pressure curves is very similar to that found before for

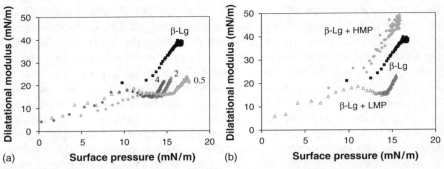

Figure 6 *Dilatational modulus as a function of surface pressure at the oil–water interface as a function of time for (a) pure β-Lg and β-Lg + LMP mixtures at different mixing ratios (as indicated) and (b) pure β-Lg and β-Lg + LMP or HMP mixtures with mixing ratio = 2 (β-Lg concentration = 0.1 g L^{-1}, pH = 4.5, ionic strength = 2 mM)*

the air–water interface,[34] where a minimum in dilatational modulus, depending on mixing ratio, was observed between surface pressures of 16 and 20 mN m^{-1} (data not shown). Comparison of the dilatational modulus for the different adsorbed layers at the air–water interface at a surface pressure of 20 mN m^{-1} shows an increase with mixing ratio from ∼25 mN m^{-1} for a mixing ratio of 2 to ∼62 mN m^{-1} for a mixing ratio of 8. The pure protein layer at the same surface pressure gave a dilatational modulus of 72 mN m^{-1}.

Based on these results a schematic picture of the different layers is proposed (Figure 7). Layer I represents a pure protein monolayer, and layer II represents a layer built up from negatively charged complexes, e.g., for mixing ratio = 0.5. Due to electrostatic repulsion within and between the complexes, the layer could be rather diffuse, and the polysaccharide could prevent the formation of a dense protein layer at the interface. The fact that such a diffuse layer would be easily compressible can account for the lower dilatational modulus. When the mixing ratio is so high that the net charge on the complexes is close to neutral, electrostatic repulsion is minimized and a compact layer can be formed (layer III). On injection of NaCl into the aqueous phase up to an ionic strength of 100 mM, the dilatational modulus of what we assume to be a diffuse layer, built up from highly negatively charged complexes (layer II), is rapidly increased from 25 to 65 mN m^{-1}. The presence of salt reduces the strong electrostatic repulsion and the layer can become denser (like layer III). The data for the oil–water interface show qualitatively the same trend as observed for the air–water interface; when comparing the dilatational modulus at the oil–water interface for the different mixing ratios at a surface pressure of 15 mN m^{-1}, it increases with increasing mixing ratio. (See Figure 6(a), where the measurements were terminated before the dilatational modulus reached a steady-state value). The curve for the highest mixing ratio of 4, for which the complex has the lowest net charge, comes closest to the curve for the pure protein layer.

Figure 6(b) compares the effects of the two pectins with different charge densities on the dilatational modulus at a mixing ratio of 2. Surprisingly, whereas the presence of LMP decreases the dilatational modulus, in the presence of HMP the dilatational modulus is higher than that for the pure protein layer. The same trend was observed at the air–water interface. A possible explanation is that, since HMP has a lower charge density than LMP, and therefore the binding of protein molecules to HMP is weaker, it is unable to prevent the protein molecules from forming a dense protein layer at the interface. An adsorbed complex structure with HMP, as in layer II of Figure 7, could therefore quickly transform into a structure like layer IV, *i.e.*, a dense protein layer, reinforced by coadsorption of HMP, and possibly even containing more β-Lg than the pure protein layer. This hypothesis is supported by the observation that, when a protein layer was formed at the air–water interface from a pure protein solution, and subsequently LMP was injected into the bulk solution, the dilatational modulus increased to a value higher than that of the pure protein layer. The resulting mixed β-Lg + LMP adsorbed structure could be depicted as layers IV or V in Figure 7, depending on the protein/polysaccharide mixing ratio in the bulk solution. Apparently, when LMP is present from the beginning of adsorption, it can prevent the formation of a dense protein layer at the air–water interface (and presumably also at the oil–water interface). Conversely, it can reinforce a protein layer formed prior to LMP addition.

The effect of different mixing ratios with LMP, as well as the effect of the sequence of adsorption, has been shown to influence the surface shear rheology at the air–water interface even stronger than dilatational rheology. The build-up

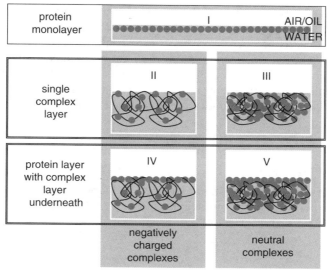

Figure 7 *Schematic representation of various kinds of adsorbed layers at the air–water or oil–water interface: I, protein monolayer; II and III, mixed layers from simultaneous adsorption, where layer II is from negatively charged complexes and layer III is from neutral complexes; IV and V, complexes adsorbed at a previously formed protein layer, where layer IV is for negatively charged complexes and layer V is for neutral complexes*

of the mixed layers seems to be at least qualitatively the same at air–water and oil–water interfaces, with respect to both adsorption kinetics and surface dilatational rheology. Also the surface shear rheology at the oil–water interface would be expected to depend strongly on mixing ratio and adsorption sequence.

13.5 Concluding Remarks

The carrier-mediated protein adsorption mechanism, as depicted in Figure 1, provides an opportunity to control adsorption kinetics at air–water and oil–water interfaces *via* protein–polysaccharide interactions in the bulk. This is a new approach offering a variety of parameters to manipulate adsorption kinetics to liquid interfaces, and presumably therefore to manipulate foam and emulsion formation. Small changes in ionic strength or pH can change protein adsorption kinetics by changing the relative amounts of complexed and free protein or the binding strength. In addition to system properties like ionic strength and pH, the mechanism can be controlled by molecular parameters like the charge density of the protein or polysaccharide. The charge distribution on the protein surface and along the polysaccharide chain is also expected to play a role, but this needs to be further explored. Moreover, since diffusion of the complexes plays such a dominant role, the hydrodynamic radius of the polysaccharide molecules can also affect adsorption kinetics.

Besides adsorption kinetics, also the rheological behaviour of the adsorbed layers is affected by the mixing ratio, by the protein–polysaccharide-binding affinity, and consequently by the charge density on the protein and polysaccharide, the pH, and the ionic strength. Furthermore, the sequence of adsorption of the individual components influences the layer structure and surface rheological properties. Understanding the mechanism of mixed and sequential protein + polysaccharide adsorption should be of great value in manipulating surface rheological properties at both air–water and oil–water interfaces. This could then help to improve understanding and control of foam and emulsion stability.

Acknowledgements

The authors thank Jan Benjamins for many valuable discussions and Frouke Duijzer for performing the surface tension and surface rheological measurements at the oil–water interface.

References

1. Y.C. Lee and R.T. Lee, *Acc. Chem. Res.*, 1995, **28**, 321.
2. R. Pereyra, K.A. Schmidt and L. Wicker, *J. Agric. Food Chem.*, 1997, **45**, 3448.
3. A. Syrbe, W.J. Bauer and H. Klostermeyer, *Int. Dairy J.*, 1998, **8**, 179.

4. A. Maroziene and C.G. de Kruif, *Food Hydrocoll.*, 2000, **14**, 391.
5. F. Weinbreck, R. de Vries, P. Schrooyen and C.G. de Kruif, *Biomacromolecules*, 2003, **4**, 293.
6. F. MacRitchie and A.E. Alexander, *J. Colloid Sci.*, 1963, **18**, 453.
7. F. MacRitchie and A.E. Alexander, *J. Colloid Sci.*, 1963, **18**, 458.
8. D.E. Graham and M.C. Phillips, *J. Colloid Interface Sci.*, 1979, **70**, 403.
9. M. Paulsson and P. Dejmek, *J. Colloid Interface Sci.*, 1992, **150**, 394.
10. B.C. Tripp, J.J. Magda and J.D. Andrade, *J. Colloid Interface Sci.*, 1995, **173**, 16.
11. P.A. Wierenga, M.B.J. Meinders, M.R. Egmond, A.G.J. Voragen and H.H.J. de Jongh, *J. Phys. Chem. B*, 2005, **109**, 16946.
12. P.A. Wierenga, M.B.J. Meinders, M.R. Egmond, F.A.G.J. Voragen and H.H.J. de Jongh, *Langmuir*, 2003, **19**, 8964.
13. E. Dickinson and K. Pawlowsky, *J. Agric. Food Chem.*, 1997, **45**, 3799.
14. R. Baeza, C.C. Sanchez, A.M.R. Pilosof and J.M. Rodriguez Patino, *Food Hydrocoll.*, 2005, **19**, 239.
15. C. Schmitt, T.P. da Silva, C. Bovay, S. Rami-Shojaei, P. Frossard, E. Kolodziejczyk and M.E. Leser, *Langmuir*, 2005, **21**, 7786.
16. T. Sengupta and S. Damodaran, *Langmuir*, 1998, **14**, 6457.
17. R.A. Ganzevles, M.A. Cohen Stuart, T. van Vliet and H.H.J. de Jongh, *Food Hydrocoll.*, 2006, **20**, 872.
18. E. Dickinson, *Food Hydrocoll.*, 2003, **17**, 25.
19. A. Benichou, A. Aserin and N. Garti, *J Dispersion Sci. Technol.*, 2002, **23**, 93.
20. L. Moreau, H.J. Kim, E.A. Decker and D.J. McClements, *J. Agric. Food Chem.*, 2003, **51**, 6612.
21. S. Laplante, S.L. Turgeon and P. Paquin, *Food Hydrocoll.*, 2005, **19**, 721.
22. U. Einhorn-Stoll, *Nahrung*, 1998, **42**, 248.
23. E.S. Tokaev, A.N. Gurov, I.A. Rogov and V.B. Tolstoguzov, *Nahrung*, 1987, **31**, 825.
24. S. Ogawa, E.A. Decker and D.J. McClements, *J. Agric. Food Chem.*, 2004, **52**, 3595.
25. D. Guzey, H.J. Kim and D.J. McClements, *Food Hydrocoll.*, 2004, **18**, 967.
26. E. Dickinson, M.G. Semenova, A.S. Antipova and E.G. Pelan, *Food Hydrocoll.*, 1998, **12**, 425.
27. V. Ducel, J. Richard, Y. Popineau and F. Boury, *Biomacromolecules*, 2005, **6**, 790.
28. B.S. Murray, *Curr. Opin. Colloid Interface Sci.*, 2002, **7**, 426.
29. H.H.J. de Jongh, T. Gröneveld and J. de Groot, *J. Dairy Sci.*, 2001, **84**, 562.
30. P.E.A. Smulders, Formation and stability of emulsions made with proteins and peptides, Ph.D. Thesis, Wageningen University, Wageningen, the Netherlands, 2000.
31. B.J. Berne and R. Pecora, *Dynamic Light Scattering with Applications to Chemistry, Biology, and Physics*, Wiley, New York, 1976.

32. S. van der Burgh, A. de Keizer and M.A. Cohen Stuart, *Langmuir*, 2004, **20**, 1073.
33. J. Benjamins, A. Cagna and E.H. Lucassen-Reynders, *Colloids Surf. A*, 1996, **114**, 245.
34. R.A. Ganzevles, K. Zinoviadou, T. van Vliet, M.A. Cohen Stuart and H.H.J. de Jongh, *Langmuir*, 2006, **22**, 10089.
35. J. Benjamins, Static and dynamic properties of proteins adsorbed at liquid interfaces, Ph.D. Thesis, Wageningen University, Wageningen, the Netherlands, 2000.
36. A.F.H. Ward and L. Tordai, *J. Chem. Phys.*, 1946, **14**, 453.
37. C. Le Bon, T. Nicolai, M.E. Kuil and J.G. Hollander, *J. Phys. Chem. B*, 1999, **103**, 10294.
38. M.A.V. Axelos, J. Lefebvre and J.F. Thibault, *Food Hydrocoll.*, 1987, **1**, 569.

Chapter 14
Adsorption Experiments from Mixed Protein + Surfactant Solutions

Veneta S. Alahverdjieva,[1] Dmitry O. Grigoriev,[1] James K. Ferri,[2] Valentin B. Fainerman,[3] Eugene V. Aksenenko,[4] Martin E. Leser,[5] Martin Michel[5] and Reinhard Miller[1]

[1] MAX-PLANCK-INSTITUT FÜR KOLLOID- UND GRENZFLÄCHEN FORSCHUNG, AM MÜHLENBERG 1, 14424 POTSDAM, GERMANY
[2] DEPARTMENT OF CHEMICAL ENGINEERING, LAFAYETTE COLLEGE, EASTON, PA 18042, USA
[3] MEDICAL PHYSICOCHEMICAL CENTRE, DONETSK MEDICAL UNIVERSITY, 16 ILYCH AVENUE, 83003 DONETSK, UKRAINE
[4] INSTITUTE OF COLLOID CHEMISTRY AND CHEMISTRY OF WATER, 42 VERNADSKY AVENUE, 03680 KIEV, UKRAINE
[5] NESTLÉ RESEARCH CENTER, VERS-CHEZ-LES-BLANC, CH-1000 LAUSANNE 26, SWITZERLAND

14.1 Introduction

There is great practical importance in the study of the adsorption of proteins with or without surfactants at fluid interfaces for a wide range of areas, such as food foams and emulsions,[1] cosmetics, coating processes,[2] pharmacy and biotechnology. This has led to the development of various theoretical models which describe the equilibrium and dynamic behaviour of protein adsorption layers. The properties of protein adsorption layers differ in many aspects from those characteristic of simple surfactants.[3] First, for a protein, surface denaturation may occur, leading to the unfolding of the protein molecule within the surface layer, at least at low surface pressure. In contrast to surfactants, the partial molar surface area for proteins is large and it can vary. This feature, and also the fact that the number of configurations for an adsorbed protein molecule exceeds significantly that of the protein molecules in the bulk, leads to a significant increase in the non-ideality of the surface layer entropy. This also makes it impossible to apply the simplest models (*i.e.*, Henry and Langmuir isotherms) to the description of protein adsorption layers.

When surfactant is added, the adsorption mechanism may be changed, as well as the rheological characteristics of the surface layer.[3-11] The protein–surfactant interactions depend mostly on two factors – the nature of the surfactant (non-ionic, ionic, and zwitter-ionic) and its concentration in the solution bulk phase. For protein + non-ionic surfactant systems, the adsorption is competitive,[9,12-24] whereas for protein + ionic surfactant systems the mechanism is totally different. Here several concentration ranges have to be distinguished in which different kinds of interactions dominate. So, Coulombic interactions are observed at low surfactant concentration, whereas hydrophobic interactions become more important with increasing surfactant concentration.

In the present work, the adsorption behaviour of the globular protein lysozyme with and without surfactant will be described. First, we consider the dynamic surface tension for lysozyme, presented as surface tension *versus* time curves at the air–water interface and compared with the theoretical model. Secondly, we discuss mixed systems of lysozyme at a constant concentration (7×10^{-7} mol L^{-1}) with varying concentrations of the anionic surfactant SDS. The equilibrium surface tension *versus* concentration isotherms for lysozyme alone and with SDS present will be analysed and compared with the behaviour of a theoretical model.

14.2 Theoretical Approach

14.2.1 Thermodynamic Model

The theory of protein adsorption with changing partial molar area has been described in detail elsewhere.[25] It is based on the idea of Joos and Serrien,[26] according to which the protein component adsorbs at the interface with two different molar areas. This model has been generalized to allow protein molecules to exist in multiple states in the interfacial layer.

Our model assumes that protein molecules can exist in a number of states of different molar area, varying from the maximum value (ω_{max}) at very low surface coverage (low surface pressure) to the minimum value (ω_{min}) at high surface coverage (high surface pressure). The molar areas of two 'neighbouring' conformations differ from each other in terms of the molar area increment ω_0, chosen to be equal to the molar area of the solvent molecules. If the total number of possible states of an adsorbed protein molecule is n, the molar area in the ith state is $\omega_i = \omega_1 + (i-1)\omega_0$ and the maximum area is $\omega_{max} = \omega_1 + (n-1)\omega_0$, where $\omega_1 = \omega_{min} \gg \omega_0$.

The thermodynamic model is based on Butler's equation of the chemical potentials of the ith state of the protein molecule within the interfacial layer and in the solution bulk.[27,28] The equation of state is based on a first-order model, accounting for non-ideal entropy and enthalpy of mixing for the surface layer:[25]

$$-\frac{\Pi\omega_0}{RT} = \ln(1 - \Theta_P) + \Theta_P\left(1 - \frac{\omega_0}{\omega}\right) + a_P\Theta_P^2. \quad (1)$$

Here, Π is the surface pressure, R the gas constant, T the temperature, and a_P the intermolecular interaction parameter.

When a protein molecule with m ionized groups is interacting with an ionic surfactant, Coulombic interactions occur and complexes will be formed. The formation of such complexes is determined by the average activity of ions participating in the reaction, $(c_P{}^m c_S)^{1/(1+m)}$, assuming the average activity coefficient to be equal to 1. Other complexes, such as between protein and buffer ions, can also be formed, but these are less surface-active and therefore their contribution to the surface pressure can be neglected. The respective equation of state of the surface layer, consisting of proteins and ionic surfactants, is

$$-\frac{\Pi\omega_0}{RT} = \ln(1 - \Theta_{PS} - \Theta_S) + \Theta_{PS}\left(1 - \frac{\omega_0}{\omega}\right) + a_{PS}\Theta_{PS}^2 + a_S\Theta_S^2 + 2a_{SPS}\Theta_{PS}\Theta_S. \quad (2)$$

The subscript PS refers to the protein–surfactant complex and the subscript S refers to the surfactant. The adsorption isotherms for the protein–surfactant complex in state 1 and the surfactant not bound to the protein are

$$b_{PS}\left(c_P^m c_S\right)^{1/(1+m)} = b_{PS} c_P^{m/(1+m)} c_S^{1/(1+m)}$$
$$= \frac{\omega\Gamma_1}{(1 - \Theta_{PS} - \Theta_S)^{\omega_1/\omega}} \exp\left[-2a_{PS}\left(\frac{\omega_1}{\omega}\right)\Theta_{PS} - 2a_{SPS}\Theta_S\right], \quad (3)$$

$$b_S(c_S c_C)^{1/2} = \frac{\Theta_S}{(1 - \Theta_{PS} - \Theta_S)} \exp[-2a_S\Theta_S - 2a_{SPS}\Theta_{PS}], \quad (4)$$

where we have $\omega = (\sum_{i=1}^{n} \Gamma_i)/\Gamma$ (the subscript PS is here omitted for the sake of simplicity), and $\Theta_{PS} = \omega\Gamma$. The quantity c_C is the surfactant counter-ion concentration, so that we have $(c_S c_C)^{\frac{1}{2}} = c_S$ in the absence of inorganic salt. The quantity a_{SPS} is the intermolecular interaction parameter describing the interaction of free surfactant with the protein–surfactant complex.

The distribution of protein adsorption over the states is given by an expression which follows from Equation (3) when written for any arbitrary jth state of the complex:

$$\Gamma_j = \Gamma \frac{(1 - \Theta_{PS} - \Theta_S)^{\frac{\omega_j - \omega_1}{\omega}} \exp\left[2a_{PS}\Theta_{PS}\frac{(\omega_j - \omega_1)}{\omega}\right]}{\sum_{i=1}^{n}(1 - \Theta_{PS} - \Theta_S)^{\frac{\omega_i - \omega_1}{\omega}} \exp\left[2a_{PS}\Theta_{PS}\frac{(\omega_i - \omega_1)}{\omega}\right]}. \quad (5)$$

The above set of equations is sufficient to describe the adsorption behaviour of mixed protein + surfactant solutions. Therefore, the theoretical description of such a mixture can be formulated in the following way. Once the values of T, ω_0, ω_{min}, ω_{max}, a_S, a_P, a_{PS}, a_{SPS}, m, b_P, b_{PS}, c_P, and b_S are known, the dependencies of ω, Θ_P, Θ_S, Θ_{PS}, and Π as a function of the surfactant concentration c_S can be calculated. Assuming that the approximations $a_{PS} = a_P$, $a_{SPS} = 0$ (or $a_{SPS} = (a_S + a_P)/2$)[29] and $b_{PS} = b_P$ are all valid, it is possible to calculate the adsorption behaviour of mixtures using only the characteristics of the individual components.

Among all the parameters necessary for the calculations, there are five parameters corresponding to the individual protein solution. The values of ω_{min} and ω_{max} can be determined from the geometrical dimensions of the protein molecule in the completely folded and unfolded states, respectively. The value of ω_0 is almost the same for all proteins, being in the range $2 - 3 \times 10^5$ m^2 mol^{-1}. All the rest of the parameters can be estimated from a fitting procedure, using some of the experimental dependencies of Π *versus* c_P, Π *versus* Γ, and Π *versus* c_P, or all simultaneously.[25] The remaining three parameters related to the individual surfactant solution (ω_S, a_S, and b_S) can be determined by fitting the experimental dependency Π *versus* c_S. The parameter m is either known from independent experimental data, or it can be estimated by fitting the model to the surface tension data of the mixture. We note that this would be the only additional parameter obtained by the fitting, with all others coming from the pure compounds.

The assumption $a_{SPS} = 0$ is more realistic than $a_{SPS} = (a_S + a_P)/2$ when the surface layer is strongly inhomogeneous, *i.e.*, when protein and surfactant molecules do not mix in the surface layer but form domains containing essentially one of the components.[19–20,22–23] The assumption $b_{PS} = b_P$ is based on the fact that m is small (10–100 times lower than the number of amino-acid groups in the protein molecule), and therefore the adsorption activity of the protein–surfactant complex only changes slightly.

14.2.2 Adsorption Kinetics

The analysis of adsorption kinetics and dynamic surface tension under quasi-equilibrium conditions is based on the equation of state for the surface layer and the adsorption isotherm. The most general relation between dynamic adsorption $\Gamma(t)$ and sub-surface concentration $c(0,t)$ is described by the Ward and Tordai equation.[30] For a freshly formed non-deformed surface this equation has the following form:

$$\Gamma(t) = 2\sqrt{\frac{D}{\pi}}\left[c_0\sqrt{t} - \int_0^{\sqrt{t}} c(0, t - t')\mathrm{d}\left(\sqrt{t'}\right)\right], \qquad (6)$$

where c_0 is the protein bulk concentration, D the diffusion coefficient, t time, and t' the integration variable. The dependence of $\Gamma(t)$ on the lifetime of the surface can be derived by combining Equation (6) with the adsorption isotherm for the protein molecule (Equation (3)), or the respective expression for the distribution of protein molecules over states (Equation (5)), which in the case of diffusion-controlled adsorption serve as the boundary conditions for Equation (6).

When the adsorption takes place on a spherical surface (drop or bubble), the interfacial curvature can be approximately taken into account for sub-surface concentrations $c(0,t)$ far from equilibrium. This is done by introducing an additional term into Equation (6), allowing for radial diffusion to the surface:

$$\Gamma(t) = 2\sqrt{\frac{D}{\pi}} \left[c_0 \sqrt{t} - \int_0^{\sqrt{t}} c(0, t - t') \mathrm{d}\left(\sqrt{t'}\right) \right] \pm \frac{c_0 D}{r} t. \qquad (7)$$

Here, r is the radius of curvature, and the $+$ and $-$ signs refer to the diffusion in the drop or in the solution around the bubble, respectively.

The corresponding equations for the separate contributions to the time-evolution behaviour of the adsorption of protein and surfactant in a mixture have the following forms:

$$\Gamma_P(t) = 2\sqrt{\frac{D_P}{\pi}} \left[c_{0P} \sqrt{t} - \int_0^{\sqrt{t}} c_P(0, t - t') \mathrm{d}\left(\sqrt{t'}\right) \right] \pm \frac{c_{0P} D_P}{r} t, \qquad (8)$$

$$\Gamma_S(t) = 2\sqrt{\frac{D_S}{\pi}} \left[c_{0S} \sqrt{t} - \int_0^{\sqrt{t}} c_S(0, t - t') \mathrm{d}\left(\sqrt{t'}\right) \right] \pm \frac{c_{0S} D_S}{r} t, \qquad (9)$$

where D_P and D_S are the diffusion coefficients of protein and surfactant, respectively, and c_{0P} and c_{0S} are the corresponding bulk concentrations.

14.3 Materials and Methods

Lysozyme from chicken egg-white (Sigma, L-6876, 14.3 kDa, pI ~ 11) was used without further purification. All protein solutions were prepared in 10 mM sodium phosphate buffer (pH = 7.0) from appropriate stock solutions of Na_2HPO_4 and NaH_2PO_4. The solutions were prepared with Millipore water. The surface tension of the buffer solution at 21°C was 72.7 mN m^{-1}. Mixed protein + surfactant solutions were prepared keeping the concentration of the protein constant and varying the surfactant concentration. The surfactant was sodium dodecyl sulfate (SDS) with a critical micelle concentration of 8×10^{-3} mol L^{-1}.

The dynamic surface tension of pure and mixed protein solutions at short adsorption times was measured using the maximum bubble pressure tensiometer BPA-1P (Sinterface Technologies, Germany). This technique is based on the measurement of the maximum pressure in a bubble growing at the tip of a steel capillary (0.25 mm i.d.) immersed into the liquid under study. The apparatus allows measurements down to a few milliseconds of surface age.

To measure the dynamic surface tension at long adsorption times from seconds to hours (or days) the profile analysis tensiometer PAT1 (Sinterface Technologies, Germany) was used. The principle of this method is to determine the surface tension of the studied solution from the shape of a pendant drop or a buoyant bubble.[31] By means of an active control loop, the instrument allows long-time experiments with a constant drop (or bubble) volume or surface area. The equipment is also suitable for transient and harmonic relaxation studies, yielding the dilatational elasticity and viscosity of the interfacial layer.

In order to measure the adsorption layer thickness and the adsorbed amount, ellipsometry measurements have been performed with a Multiskope apparatus (Optrel, Germany). The operation of this equipment and the procedure to calculate the layer thickness and adsorbed amount have been described in detail elsewhere.[32] Briefly, the Multiskope set-up consists of a conventional polarizer/compensator/sample/analyser (PCSA) null-ellipsometer. A low-capacity laser (532 nm) serves as the light source of beam diameter 0.5–1 mm. The light beam passes through a first quarter-wave plate to produce circularly polarized light. Then the light is linearly polarized by a Glan–Thompson prism mounted in a rotatable divided circle, which can be read to very high precision. A second quarter-wave plate and the analyser (a second Glan–Thompson prism) are mounted in a similar manner. The angle of incidence of the light is 50°. A photodiode serves as detector. Both incidence and reflection arms are motorized and computer controlled. The highly precise computer-controlled motors of polarizer and analyser allow exact determination for both optical elements of the null positions, *i.e.*, positions at which a minimum intensity of transmitted light is registered. From these nul positions of the polarizer and analyser are obtained the two ellipsometric angles, Δ and Ψ. These angles characterize the polarization state of the reflected light, which can be attributed to the properties of the reflecting surface.[33]

14.4 Results and Discussion

The dynamic surface tension of lysozyme measured with the maximum bubble pressure method is shown in Figure 1. We can see that it does not change significantly during the experimental time period of 30 s, which suggests a long induction period. There is much speculation in the literature on protein adsorption as to what might be the origin of the induction period. For lysozyme, Xu and Damodaran[34] proposed that there is a negative surface excess of protein during the induction period and that the surface pressure increases only after conformational rearrangements have occurred. However, many authors have assumed that during the initial stages of adsorption the surface concentration is not zero. Graham and Phillips[35] reported a non-zero surface concentration for lysozyme from radio-labelling studies, and similar

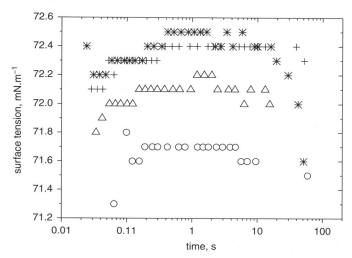

Figure 1 *Dynamic surface tension of lysozyme from maximum bubble pressure method at different bulk concentrations:* ✻, 10^{-6} mol L^{-1}; +, 5×10^{-6} mol L^{-1}; △, 10^{-5} mol L^{-1}; ○, 9×10^{-5} mol L^{-1}

results have been reported for BSA.[36] On the basis of fluorescent images, Stebe et al.[37] have concluded that the induction time of lysozyme is caused by a first-order phase transition from gaseous to liquid-expanded state.

Figure 2 presents dynamic surface tension data of lysozyme determined from the profile of a buoyant bubble recorded over 24 h. The decrease in surface tension for various lysozyme concentrations as measured from pendant drops is shown in Figure 3. The adsorption was followed over 12–15 h. It can be seen that no equilibrium has been reached and the dynamic surface tension continues to decrease over a long period of time. In contrast to behaviour at the low lysozyme concentrations measured with a buoyant bubble, no intermediate increase in surface tension was observed.

The dynamic surface tension of mixtures of lysozyme (7×10^{-7} M) + SDS at varying concentrations has been measured with buoyant bubbles. At pH = 7 the protein has a positive net charge (ζ-potential ~ 5.7 mV) which favours complexation with the anionic SDS. Figure 4 presents the decrease of surface tension of lysozyme (7×10^{-7} M) mixed with various concentrations of SDS.

The pure lysozyme system and the lysozyme + SDS systems have been studied ellipsometrically. The refractive index, the adsorption layer thickness and the adsorbed amount (mg m^{-2}) have all been determined. The changes in adsorbed layer thickness for lysozyme alone and of SDS in the presence of lysozyme (7×10^{-7} M) are shown in Figure 5. For the single component system at very low concentrations, the thickness of the adsorption layer seems to begin from *ca.* 225 Å, but apparently this is due to the very low difference between refractive index of water (1.3334) and that of the adsorbed layer, which makes

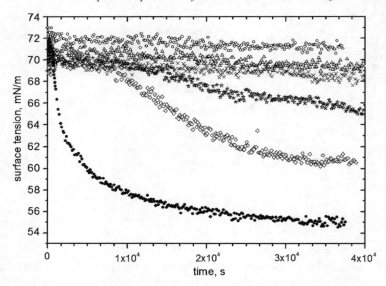

Figure 2 *Dynamic surface tension of lysozyme from buoyant bubbles at different bulk concentrations: ○, 10^{-10} mol L^{-1}; △, 2×10^{-10} mol L^{-1}; ×, 2×10^{-9}; ☆, 2×10^{-8} mol L^{-1}; ◇, 5×10^{-8} mol L^{-1}; ●, 1×10^{-7} mol L^{-1}*

the apparent probing depth deeper. So, instead of the thickness of the actual layer with adsorbed molecules, one obtains the thickness of this surface layer and the adjacent sub-layer. With increasing lysozyme concentration, the difference between the refractive indices of water and the adsorbed layer becomes larger, and so the adsorbed layer thickness can be evaluated more accurately. The layer thickness up to a concentration of 2×10^{-5} M can be considered more or less constant, after which it starts increasing. Figure 6 shows that the adsorbed amount increases continuously throughout the whole concentration range, although one can see that above 7×10^{-5} M the surface concentration increases sharply. This is in agreement with the findings of Lu et al.[38,39] that, at a concentration of 0.1 wt% (7×10^{-5} M), lysozyme rearranges its conformation from a side-on state to an end-on state. When SDS is added to a lysozyme solution, the thickness of the adsorbed layer is not significantly changed, but the refractive index and adsorbed amount increase slightly. This increase in protein adsorption and in the overall protein + ionic surfactant adsorption is confirmed by theory, although the theoretical model predicts an increase of the overall adsorption only up to a concentration of 10^{-5} M.

The plateau which is observed for the equilibrium surface pressure isotherm for lysozyme (Figure 7) and for the adsorption from ellipsometric measurements (Figure 6) begins at 10^{-7} up to 10^{-6} M. In this concentration range the adsorbed lysozyme molecules are still in a side-on conformation,[38,39] and at 7×10^{-5} M they rearrange to an end-on conformation. Of course, this could

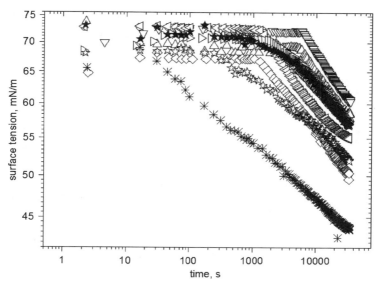

Figure 3 *Dynamic surface tension of lysozyme from pendant drop at different bulk concentrations:* $\triangledown$, 1×10^{-6} mol L^{-1}; $\triangle$, 2×10^{-6} mol L^{-1}; $\triangleright$, 5×10^{-6} mol L^{-1}; $\bigstar$, 7×10^{-6} mol L^{-1}; $\triangleleft$, 1×10^{-5} mol L^{-1}; $\diamondsuit$, 2×10^{-5} mol L^{-1}; -, 5×10^{-5} mol L^{-1}; $\star$, 7×10^{-5} mol L^{-1}; $\ast$, 1×10^{-4} mol L^{-1}

explain the subsequent increase in surface pressure and film thickness. In their radio-labelling studies, Graham and Phillips[40] observed the same behaviour for lysozyme adsorption and its surface pressure isotherms, although their plateau in the surface pressure isotherm was between $\sim 7 \times 10^{-8}$ and $\sim 5.6 \times 10^{-6}$ M. The surface pressure isotherm for lysozyme has been calculated from the model and the theoretical parameters summarized in Table 1. The theory describes the experimental data reasonably well with the listed set of parameters down to a concentration of 10^{-7} M. Nevertheless, in the region of low bulk concentrations, and hence low surface coverage, the theoretical predictions deviate somewhat from the experimental points. In order to improve the agreement, another set of parameters for the low concentrations would be required. The difference between the sets involves the area parameters ω_0 and ω_{max}. This result is not surprising. When the concentration is very low the molecules adsorb in their completely unfolded state, and so the value of ω_{max} is higher. The area parameter ω_{min} (ω_1) does not play a role in this low protein concentration region, as it accounts for the completely folded state, and hence the lowest surface area state of the adsorbed molecules, which is the case at high concentrations. In order to fit the experimental data better, the value for the other area parameter ω_0 (the area of one segment of an adsorbed protein molecule) should be increased as well. Finally, the parameters a and b are taken as fitting parameters, and best fit values are reported.

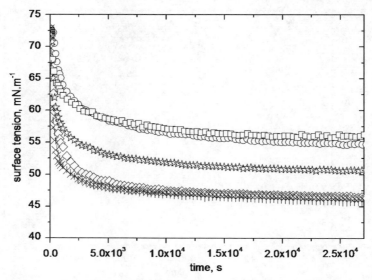

Figure 4 *Dynamic surface tension of 7×10^{-7} mol L^{-1} lysozyme in mixtures with different concentrations of SDS:* ○, 10^{-6} mol L^{-1}; □, 2×10^{-6} mol L^{-1}; ☆, 10^{-5} mol L^{-1}; ★, 5×10^{-5} mol L^{-1}; ◇, 7×10^{-5} mol L^{-1}; ×, 10^{-4} mol L^{-1}

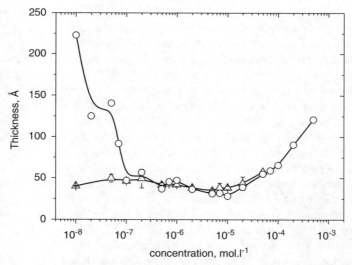

Figure 5 *Thickness of adsorbed lysozyme layer at different concentrations:* ○, *pure protein;* △, 7×10^{-7} mol L^{-1} *lysozyme + SDS at different concentrations*

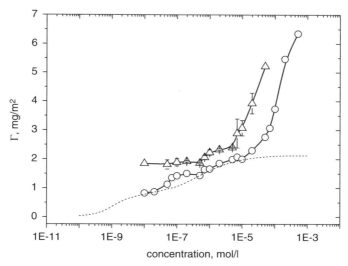

Figure 6 *Adsorbed amount Γ of lysozyme at different concentrations:* ○, *pure protein;* △, 7×10^{-7} *mol* L^{-1} *lysozyme + SDS at different concentrations;* - - -, *theory*

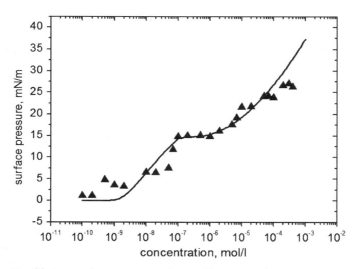

Figure 7 *Equilibrium surface tension isotherm of lysozyme;* ▲ – *experimental data, solid line – theoretical calculations*

Table 1 *Theoretical parameters for the equilibrium surface pressure isotherm of lysozyme: a = Frumkin intermolecular interaction parameter; α = coefficient; ω_0 = molar area of one segment of the protein molecule; ω_1 = minimal molar area of the protein molecule; ω_{max} = maximal molar area of the protein molecule; b = adsorption equilibrium constant; π^* = critical surface pressure; ε = additional parameter for the critical regime. The symbol * refers to parameters obtained from fitting the surface pressure data at low concentrations*

a	0.69 (0.2*)
α	2 (2.82*)
ω_0, m² mol⁻¹ (×10⁵)	4.95 (9.95*)
ω_1, m² mol⁻¹ (×10⁶)	6.72
ω_{max}, m² mol⁻¹ (×10⁷)	2.54 (5.08*)
ε	0.032
π^*, mN m⁻¹	14.5
b, m³ mol⁻¹	2×10³

The equilibrium surface pressure of lysozyme having been analysed in terms of the theoretical model, the dynamic surface tension can be considered. The fitting of each of the experimental dynamic surface tension curves was performed on the basis of the parameters listed in Table 2. As is well known from the literature,[35,41,42] the adsorption of proteins at fluid interfaces follows three regimes of surface tension lowering. First, proteins adsorb at the interface, but the surface tension remains unchanged, or it changes only slightly (the induction period). As time passes, the adsorption proceeds, the interfacial concentration increases, and the surface tension decreases. At longer times, adsorption and conformational rearrangements continue; multilayers may develop; and interfacial aggregation and gelation might occur. Figures 8–10 present the dynamic surface tension curves for 10^{-7}, 5×10^{-7} and 7×10^{-7} M lysozyme as representative examples, and the respective model parameters are summarized in Table 2. We can see that the correlation between the experimental curves and the theory is very good, with the exception of the last part of the curves at very long adsorption times, where the surface tension passes through a minimum as described above. Regarding the parameters of the thermodynamic model, they are practically the same as in the fitting of the equilibrium isotherm with only one additional parameter – the diffusion coefficient D. In order to match the experimental data for each protein concentration, and to keep the value of the diffusion coefficient constant, the parameters ω_0 and ω_{max} were varied too. As discussed already, this is required by the fact that, at low concentrations, the proteins adsorb in an unfolded state at the interface, and their molar surface areas are larger in comparison to those at higher concentrations. Nevertheless, a detailed quantitative interpretation of the dynamic surface tension of lysozyme remains to be provided.

Table 2 Theoretical parameters for the dynamic surface tension fitted at various bulk concentrations of lysozyme: a = Frumkin intermolecular interaction parameter; ω_0 = molar area of one segment of the protein molecule; ω_1 = minimal molar area of the protein molecule; ω_{max} = maximal molar area of the protein molecule; b = adsorption equilibrium constant; D = diffusion coefficient

Concentration, $mol\,L^{-1}$	a	$\omega_0, m^2\,mol^{-1}$ $(\times 10^5)$	$\omega_1, m^2\,mol^{-1}$ $(\times 10^6)$	$\omega_{max}, m^2\,mol^{-1}$ $(\times 10^7)$	$b, m^3\,mol^{-1}$ $(\times 10^3)$	$D, m^2 s^{-1}$ $(\times 10^{-10})$
1×10^{-8}	0.2	9.95	6.72	5.08	2	3
1×10^{-8}	0.2	9.95	6.72	5.08	2	1
1×10^{-8}	0.8	6.05	6.72	2.94	3	1
1×10^{-8}	0.69	4.95	6.72	2.54	3	1
1×10^{-8}	0.69	4.95	6.72	2.54	3	1
1×10^{-8}	0.7	3.08	6.72	1.02	2	1
1×10^{-8}	0.69	4.95	6.72	2.54	2	0.03
1×10^{-8}	0.69	4.95	6.72	2.54	1	0.10
1×10^{-8}	0.69	4.95	6.72	2.54	1	0.02

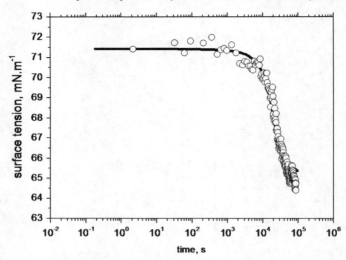

Figure 8 *Dynamic surface tension of 10^{-7} mol L^{-1} lysozyme: ○, experiment; —, theory*

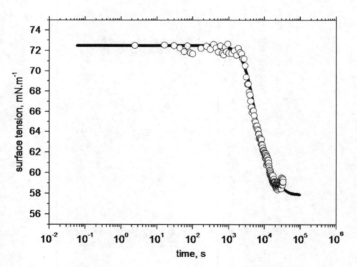

Figure 9 *Dynamic surface tension of 5×10^{-7} mol L^{-1} lysozyme: ○, experiment; —, theory*

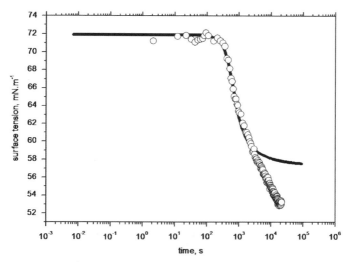

Figure 10 *Dynamic surface tension of 7×10^{-7} mol L^{-1} lysozyme: ○, experiment; —, theory*

14.5 Conclusions

The adsorption kinetics of lysozyme has been studied by measuring the dynamic surface tension for different bulk concentrations and applying a thermodynamic model assuming adsorption with different molar areas for each subsequent adsorption step. The theoretical analysis of the kinetic curves shows satisfactory agreement between the experimental data and the model, and the obtained diffusion coefficients suggest diffusion-controlled adsorption for low lysozyme concentration and a mixed mechanism for higher concentrations. The equilibrium surface tension isotherm is described well by the theory, but better agreement is found if the fitting procedure adopts a higher value for the maximum molar area, ω_{max}. This is a parameter which is highly significant at low surface pressures, where the conformation of the protein is most unfolded.

Acknowledgement

The work was financially supported by a project of the European Space Agency (FASES MAP AO-99-052).

References

1. E. Dickinson and J.M. Rodriguez Patino (eds), *Food Emulsions and Foams: Interfaces, Interactions and Stability*, Royal Society of Chemistry, Cambridge, UK, 1999.

2. R. Wüstneck and J. Krägel, in *Proteins at Liquid Interfaces*, Vol 7, D. Möbius and R. Miller (eds), Studies in Interface Science Series, Elsevier, Amsterdam, 1998, p. 433.
3. R. Miller, V.B. Fainerman, A.V. Makievski, J. Krägel, D.O. Grigoriev, V.N. Kazakov and O.V. Sinyachenko, *Adv. Colloid Interface Sci.*, 2000, **86**, 39.
4. A. Dussaud, G.B. Han, L. Ter Minassian-Saraga and M. Vignes-Adler, *J. Colloid Interface Sci.*, 1994, **167**, 247.
5. J. Krägel, R. Wüstneck, D. Clark, P.J. Wilde and R. Miller, *Colloids Surf. A*, 1995, **98**, 127.
6. N.J. Turro, X.-G. Lei, K.P. Ananthapadmanabhan and M. Aronson, *Langmuir*, 1995, **11**, 2525.
7. R. Wüstneck, J. Krägel, R. Miller, P.J. Wilde and D.C. Clark, *Colloids Surf. A*, 1996, **114**, 255.
8. J. Krägel, R. Wüstneck, F. Husband, P.J. Wilde, A.V. Makievski, D.O. Grigoriev and J.B. Li, *Colloids Surf. B*, 1999, **12**, 399.
9. E. Dickinson, *Colloids Surf. B*, 1999, **15**, 161.
10. R. Miller, V.B. Fainerman, A.V. Makievski, J. Krägel and R. Wüstneck, *Colloids Surf. A*, 2000, **161**, 151.
11. J. Oakes, *J. Chem. Soc. Faraday Trans. 1*, 1974, **70**, 2200.
12. E.H. Lucassen-Reynders, *Colloids Surf. A*, 1994, **91**, 79.
13. M.A. Cohen Stuart, G.J. Fleer, J. Lyklema, W. Norde and J.M.H.M. Scheutjens, *Adv. Colloid Interface Sci.*, 1991, **34**, 477.
14. P. Joos, *Dynamic Surface Phenomena*, VSP, Utrecht, 1999.
15. P. Joos and G.J. Serrien, *J. Colloid Interface Sci.*, 1991, **145**, 291.
16. M.A. Bos and T. van Vliet, *Adv. Colloid Interface Sci.*, 2001, **91**, 437.
17. J. Chen and E. Dickinson, *Food Hydrocoll.*, 1995, **9**, 35.
18. B.S. Murray, *Colloids Surf. A*, 1997, **125**, 73.
19. A.R. Mackie, A.P. Gunning, P.J. Wilde and V.J. Morris, *Langmuir*, 2000, **16**, 8176.
20. A.R. Mackie, A.P. Gunning, P.J. Wilde and V.J. Morris, *J. Colloid Interface Sci.*, 1999, **210**, 157.
21. C.M. Wijmans and E. Dickinson, *Langmuir*, 1999, **15**, 8344.
22. A.R. Mackie, A.P. Gunning, M.J. Ridout, P.J. Wilde and V.J. Morris, *Langmuir*, 2001, **17**, 6593.
23. A.R. Mackie, A.P. Gunning, M.J. Ridout, P.J. Wilde and J.M. Rodriguez Patino, *Biomacromolecules*, 2001, **2**, 1001.
24. V.B. Fainerman, S.A. Zholob, M. Leser, M. Michel and R. Miller, *J. Colloid Interface Sci.*, 2004, **274**, 496.
25. V.B. Fainerman, E.H. Lucassen-Reynders and R. Miller, *Adv. Colloid Interface Sci.*, 2003, **106**, 237.
26. P. Joos and G.J. Serrien, *J. Colloid Interface Sci.*, 1991, **145**, 291.
27. V.B. Fainerman, R. Miller and R. Wüstneck, *J. Colloid Interface Sci.*, 1996, **183**, 25.
28. V.B. Fainerman, E.H. Lucassen-Reynders and R. Miller, *Colloids Surf. A*, 1998, **143**, 141.

29. V.B. Fainerman and E.H. Lucassen-Reynders, *Adv. Colloid Interface Sci.*, 2002, **96**, 295.
30. A.F.H. Ward and L. Tordai, *J. Phys. Chem.*, 1946, **14**, 453.
31. G. Loglio, P. Pandolfini, R. Miller, A.V. Makievski, F. Ravera, M. Ferrari and L. Liggieri, in *Novel Methods to Study Interfacial Layers*, Vol 11, D. Möbius and R. Miller (eds), Studies in Interface Science Series, Elsevier, Amsterdam, 2001, p. 439.
32. R. Miller, V.B. Fainerman, A.V. Makievski, J. Krägel, D.O. Grigoriev, F. Ravera, L. Liggieri, D.Y. Kwok and A.W. Neumann, in *Encyclopaedic Handbook of Emulsion Technology*, J. Sjöblom (ed), Marcel Dekker, New York, 2001, p. 1.
33. D.O. Grigoriev, V.B. Fainerman, A.V. Makievski, J. Krägel, R. Wüstneck and R. Miller, *J. Colloid Interface Sci.*, 2002, **253**, 257.
34. S. Xu and S. Damodaran, *Langmuir*, 1992, **8**, 2021.
35. D.E. Graham and M.C. Phillips, *J. Colloid Interface Sci.*, 1979, **70**, 403.
36. D. Cho, G. Narsimhan and E.I. Franses, *J. Colloid Interface Sci.*, 1997, **191**, 312.
37. J.S. Erickson, S. Sundaram and K.J. Stebe, *Langmuir*, 2000, **16**, 5072.
38. J.R. Lu, T.S. Su, R.K. Thomas, J. Penfold and J. Webster, *J. Chem. Soc. Faraday Trans.*, 1998, **94**, 3279.
39. J.R. Lu, T.J. Su and B.J. Howlin, *J. Phys. Chem. B.*, 1999, **103**, 5903.
40. D.E. Graham and M.C. Phillips, *J. Colloid Interface Sci.*, 1979, **70**, 415.
41. C.J. Beverung, C.J. Radke and H.W. Blanch, *Biophys. Chem.*, 1999, **81**, 59.
42. M.E. Freer, K.S. Yim, G.G. Fuller and C.J. Radke, *J. Phys. Chem. B*, 2004, **108**, 3835.

Chapter 15

Role of Electrostatic Interactions on Molecular Self-Assembly of Protein + Phospholipid Films at the Air–Water Interface

Ana Lucero Caro,[1] Alan R. Mackie,[2] A. Patrick Gunning,[2] Peter J. Wilde,[2] Victor J. Morris,[2] Mª. Rosario Rodríguez Niño[1] and Juan M. Rodríguez Patino[1]

[1] DEPARTAMENTO DE INGENIERÍA QUÍMICA, FACULTAD DE QUÍMICA, UNIVERSIDAD DE SEVILLE, C/PROF. GARCÍA GONZÁLEZ, 1, 41012, SEVILLE, SPAIN
[2] INSTITUTE OF FOOD RESEARCH, NORWICH RESEARCH PARK, COLNEY, NORWICH NR4 7UA, UK

15.1 Introduction

Many food formulations are emulsions or foams. The constituent droplets or bubbles are microstructural entities stabilized by the formation of an interfacial emulsifier layer around the particles.[1] The properties of the interfacial layer are governed by the composition and structure of the adsorbed material.[2-4] Biopolymers (proteins, polypeptides, and polysaccharides) and low molecular weight surfactants have been used for many decades as emulsifiers in the production of food colloids.[1,5] On the molecular level, these ingredients and their chemical properties, including specific modifications, can be used to add new functionalities to these emulsifiers.

The way emulsifiers interact with each other at the interface influences the formation and stability of individual emulsion droplets or foam bubbles and the interaction between groups of droplets or bubbles.[6,7] This means that the bulk rheological or textural properties can be improved by controlling nanostructure formation at the interface.[8] The key component of this nanoscopic approach is molecular self-assembly. Thus, a better understanding of supra-molecular structuring principles will reveal new phenomena and lead to new manufacturing processes for high-added-value food products and emulsifiers. A prerequisite for the success of this approach is a proper understanding of how the fundamental principles of non-covalent interactions lead to supramolecular structures, and so influence the formation and stabilization of higher hierarchical structures (droplets or bubbles) present in most food colloids.[8] For investigating such molecularly assembled materials, there has been rapid development of advanced equipment,[9–11] such as atomic force microscopy (AFM), Brewster angle microscopy (BAM), imaging ellipsometry, reflection-absorption spectrometry, transmission fluorescence spectrometry, 2-D X-ray reflectometry, surface plasmon resonance, *etc*. Furthermore, traditional apparatus such as the Langmuir trough has been equipped with new features to produce constant/dynamic shearing and/or dilatation, *in situ* observations, molecular deposition, and so on.

In this study, we analyse the role of electrostatic interactions on the structure, topography, and film thickness of dipalmitoylphosphatidyl choline (DPPC), β-casein, and their mixtures in spread monolayers at the air–water interface. This involves surface pressure *versus* area isotherms coupled with BAM and AFM. Phospholipids and proteins are often used as emulsifiers in emulsions and foams due to their amphiphilic character.[12] Thus, the capacity of proteins and phospholipids to form supramolecular structures that self-assemble at a fluid interface is of practical importance for the manufacture of food formulations. Since phospholipids and proteins can be present at interfaces with different net charges, depending on the pH, an analysis of the effect of the pH on structure formation will give insight into the importance of electrostatic interactions on self-assembly in the pure and mixed monolayers.

15.2 Experimental

DPPC was supplied by Sigma (>99%). The β-casein (99% pure) was supplied and purified from bulk milk from the Hannah Research Institute (Ayr, Scotland). To form the surface film, DPPC was spread in the form of a solution, using chloroform/ethanol (4:1 by volume) as a spreading solvent. Analytical grade chloroform (Sigma, 99%) and ethanol (Merck, >99.8%) were used without further purification. Samples for interfacial characteristics of protein and DPPC films were prepared using Milli-Q ultrapure water at pH 7. The water used as sub-phase was purified by means of a Millipore filtration device (Millipore, Milli Q™). To adjust the sub-phase pH, buffer solutions were used. Acetic acid + sodium acetate aqueous solution (CH_3COOH/CH_3COONa) was used to achieve pH 5, and a commercial buffer solution called Trizma (($CH_2OH)_3CNH_2$/$(CH_2OH)_3CNH_3Cl$) for pH 7 and 9. All these products were supplied by Sigma (>99.5%). The ionic strength was 0.05 M in all the experiments.

Measurements of surface pressure (π) *versus* average area per molecule (A) were performed on fully automated Langmuir-type film balance, as described elsewhere for pure[13,14] and mixed[15] monolayers. The π–A isotherm was measured at least three times. The reproducibility of the results was better than ± 0.5 mN m^{-1} for surface pressure and ± 0.05 m^2 mg^{-1} for area. The sub-phase temperature was controlled at $20 \pm 0.3°C$ by water circulation from a thermostat.

For microscopic observation of the monolayer structure, the Brewster angle microscope BAM 2 plus (NFT, Germany) was used as described elsewhere.[13,14] The BAM was positioned over the film balance. Measurements of surface pressure, area, and grey level during monolayer compression were carried out simultaneously. To measure the reflectivity of the monolayer from grey level measurements a previous camera calibration was necessary.[13,14] The imaging conditions were adjusted to optimize both image quality and quantitative measurement of reflectivity; thus, as surface pressure increased, the shutter speed was also increased.

AFM images of β-casein + DPPC monolayers were taken of Langmuir–Blodgett films deposited onto hydrophilic mica substrates in the manner described elsewhere.[9] Images were obtained in the constant force contact mode in air or under redistilled *n*-butanol (Sigma) using an AFM manufactured by East Coast Scientific Limited (Cambridge, UK). The imaging under *n*-butanol yields reversible force–distance curves with no adhesive component under retraction of the tip. Moreover, DPPC is soluble in *n*-butanol, a fact that can be used advantageously in the analysis of images of DPPC + β-casein mixed films.

15.3 Results and Discussion

15.3.1 Effect of pH on Electrostatic Interactions

Figure 1 illustrates how pH affects the ionization of DPPC and β-casein in monolayers at the air–water interface. At pH 5, some DPPC molecules will be positively charged because the pH is close to the pK_a value (3.8–4.0) corresponding to dissociation of the first acid group.[16] Weak hydrogen bonds between adjacent positively charged head-groups could lead to the formation of a stable monolayer. At pH 7, the DPPC is zwitterionic which reduces the formation of intermolecular hydrogen bonds. At pH 9, the DPPC monolayer becomes completely ionized since the second acidic group dissociates (pK_a = 8.0–8.5).[16] At this pH the incorporation of counter-ions into charged phosphatidylcholine groups causes monolayer expansion by decreasing the formation of intermolecular hydrogen bonds and increasing electrostatic repulsion (see Figure 1A).[17] The β-casein molecule, a flexible linear polyelectrolyte, has a nonuniform distribution of hydrophilic and hydrophobic amino acid residues. At pH 5, it will be close to having a net neutral charge, as this pH is close to its p*I* (see Figure 1B). At pH 7, due to the amino acid content, β-casein still has only a small net negative charge, which becomes much greater at pH 9. However, the charge is located on the 40–50 terminal residues, with the rest of the molecule remaining highly hydrophobic. Thus, although the force fields that the protein molecules generate are predominantly electrostatic in nature, the role of hydrophobic interactions must be significant for β-casein at the interface,[18,19] these interactions being the main driving force for reversible self-assembly in dilute solution.[20] The charged phosphoserine residues in β-casein are considered to be repelled from the interface by electrostatic forces, leaving the remainder of the hydrophobic groups strongly anchored to the interface.[21]

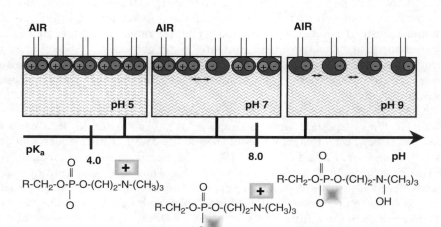

Figure 1 *Cartoon illustrating the molecular charge distribution in (A) DPPC and (B) β-casein at the air–water interface as a function of the pH of the aqueous phase. The values of pK_a (4.0 and 8.0) for DPPC and pI (4.6) for β-casein are indicated by vertical lines*

By changing the pH, therefore, we can alter the nature of the electrostatic interactions between adsorbed β-casein and DPPC. At pH 5, weak electrostatic interactions are possible due to near neutral net charges exhibited by each molecule. At pH 7, these interactions become weaker, as the net charge on β-casein increases. At pH 9, there are strong repulsive electrostatic interactions because both molecules are negatively charged. Thus, the ionization of the adsorbed molecules hinders the formation of a compact monolayer structure by increasing electrostatic repulsion, and thus reducing intermolecular hydrogen bond formation.

In summary, the importance of electrostatic interactions on molecular self-assembly of DPPC, β-casein, and their mixtures in monolayers at the air–water interface is expected to depend strongly on the aqueous phase pH. Nevertheless, the free energy of the system is associated not only with the long-range electrostatic interactions, but also with the liberation of structured water molecules around hydrophobic regions of the emulsifier molecules and the increased short-range van der Waals forces between the hydrophobic regions (mainly the hydrocarbon chains in DPPC).

15.3.2 DPPC Monolayers

Figure 2A shows the π–A isotherm of DPPC on aqueous sub-phases for the three pH values in the range 5–9, and Figure 2B shows the elasticity modulus deduced from the slope of each π–A isotherm ($E = -A(\partial\pi/\partial A)_T$). A minimum in the E–π curve suggests the presence of a first-order phase transition in the monolayer structure. The π–A isotherm shows three distinct regions: liquid expanded (LE), liquid condensed (LC), and solid (S) phases. The transition region between LE and LC phases at pressure $\pi(t,\text{DPPC})$ is denoted by the region of intermediate slope prior to the full transition to the LC phase (Table 1). At higher surface pressure ($\pi = 28$ mN m^{-1}) the transition to the solid phase occurs, before the monolayer collapse and at a surface pressure higher than the equilibrium spreading pressure $\pi(e,\text{DPPC})$ (Table 1). The BAM and AFM images[22] corroborate the structural polymorphism deduced from the

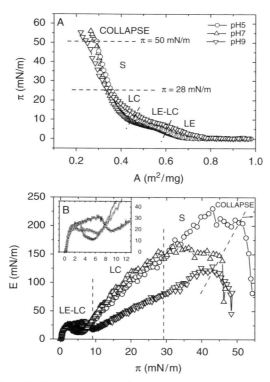

Figure 2 *The effect of pH on (A) the surface pressure versus area isotherm and (B) the elasticity modulus for DPPC monolayers spread at the air–water interface at 20°C and ionic strength 0.05 M; O, pH 5; Δ, pH 7; ∇, pH 9*

Table 1 *The effect of pH on the equilibrium spreading pressure for DPPC, $\pi(e,DPPC)$, and for β-casein, $\pi(e,\beta\text{-}cas)$, and the critical surface pressure for the main transition between LE and LC structures for spread monolayers of DPPC, $\pi(t,DPPC)$, and structures 1 and 2 for β-casein, $\pi(t,\beta\text{-}cas)$, at the air–water interface at 20°C and ionic strength 0.05 M*

Surface pressure (mN m^{-1})	pH 5	pH 7	pH 9
$\pi(e,\text{DPPC})$	49.0	47.0	46.0
$\pi(e,\beta\text{-cas})$	21.8	20.9	17.1
$\pi(t,\text{DPPC})$	4.3–9.8	5.5–9.4	5.5–11.7
$\pi(t,\beta\text{-cas})$	15.6	10.6	—

π–A isotherms. The pH of the aqueous phase has an effect on the structural characteristics of DPPC (Figure 2A). In fact, the value of π at the beginning of the first-order phase transition between the LE and LC phases was highest at pH 9 (see Table 1, and inset to Figure 2B), suggesting a more expanded monolayer structure. The lower values of E (Figure 2B) and $\pi(e,\text{DPPC})$ at pH 9 are consistent with weaker molecular interactions due to electrostatic repulsion between negatively charged head-groups in the DPPC molecules. In addition, the mass per unit area at the beginning of the monolayer formation (A_o) (i.e., the area A at which $\pi > 0$) is somewhat similar at pH 5 and 7 ($A_o = 0.69$ and 0.68 m^2 mg^{-1}, respectively), but it is higher at pH 9 ($A_o = 0.78$ m^2 mg^{-1}). The value of A_o represents the maximum distance at which monolayer molecules begin to interact. Thus the long-range electrostatic forces between the DPPC head-groups are clearly important within the LE phase. The limiting molecular area, A_{lim}, is determined by extrapolating from the steepest part of the π–A isotherm to the A axis. This describes the isotherm shape in the more condensed state and is usually interpreted as the cross-sectional area of the molecule in the monolayer. Surprisingly, A_{lim} does not depend on the aqueous phase pH. The A_{lim} values are 0.41, 0.39, and 0.40 m^2 mg^{-1} for pH 5, 7, and 9, respectively. Such modest effects of sub-phase pH on A_{lim} are unexpected. It suggests that, in the more condensed state, repulsive electrostatic interactions between head-groups may be overcome by attractive van der Waals interactions between hydrocarbon chains, which do not depend upon the aqueous phase pH. Thus, whereas in bulk solution the intermolecular interactions determine the formation of supramolecular structures, there is in contrast a complex counterbalance between intermolecular and molecular sub-phase interactions controlling the two-dimensional self-assembly at interfaces. From the different effects of pH on A_o and A_{lim}, it can be inferred that repulsive electrostatic interactions between charged head-groups of DPPC at pH 9 are predominant over attractive van der Waals interactions between hydrocarbon chains at long range (in the more expanded state of the monolayer); but the opposite appears to be the case at short range (in the more condensed state of the monolayer).

From the reflectivity *versus* π curve, it is possible to identify the structural polymorphism implied by the π–A isotherm as a function of pH. In Figure 3,

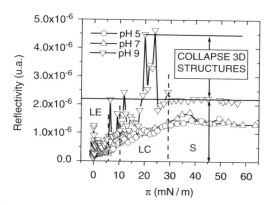

Figure 3 *The effect of pH on the reflectivity of BAM images (shutter time = 1/250 s) as a function of the surface pressure for DPPC monolayers spread at the air–water interface at 20°C and ionic strength 0.05 M; O, pH 5; △, pH 7; ▽, pH 9*

from the LE to the LC region, the thickness increases with increasing π because the hydrocarbon chains acquire a greater inclination with respect to the interfacial plane. At higher π the thickness becomes constant, coinciding with the formation of the solid structure within the monolayer. The increase in the monolayer reflectivity with pH in Figure 3 could be due to the increased affinity between water molecules and the DPPC polar head-group ionized in the more basic pH aqueous phase. That is, the higher reflectivity of DPPC monolayers at pH 9 may be due to compression against the greater electrostatic repulsion. This will force some head-groups to be displaced further into the aqueous phase to maintain head-group spacing as required by the electrostatic repulsion forces. This will give the impression that $A_{\lim}$ remains independent of pH, but only in the two-dimensional plane parallel to the interface.

The BAM and AFM images for DPPC monolayers as a function of pH show similar structures when viewed at equivalent magnifications. However, the AFM data reveal a previously unexpected heterogeneity within the domains and at the surfaces of the domain structures. Figure 4 shows that the pH of the aqueous phase has a significant effect on the topography of the DPPC monolayers. The topographic images suggest the presence of numerous small holes in the interior of the domains (possibly LE regions) and ragged surface protrusions at the domain boundaries. The main geometrical characteristics and topographical parameters deduced from the BAM images for the LE and LC domains are included in Table 2. Electrostatic interactions clearly play a role in the self-assembly of domains within the DPPC layers as observed in the BAM images. At pH 5, the DPPC domains are small and compact in shape. As the pH increases, so the domains increase in size. In addition, the shape of the domains changes: they become clearly dendrite like in appearance. This is presumably because increasing the boundary length minimizes the effects of repulsive interactions between DPPC molecules within the domain. The greatest thickness of DPPC domains is observed at pH 9 because these domains are

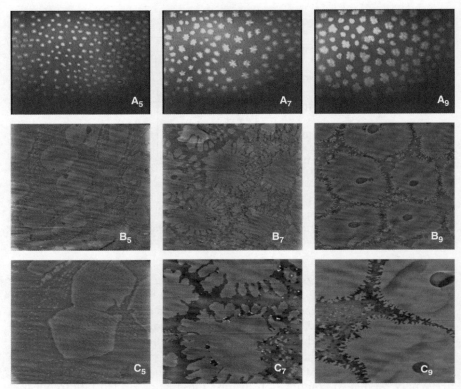

Figure 4 *Topography by BAM (A_5, A_7, and A_9; image size $630 \times 470\,\mu m$) and AFM at 40 μm (B_5, B_7, and B_9) and at 20 μm (C_5, C_7, and C_9) for DPPC spread at the air–water interface at the transition between LC and LE phases (at 8 mN m^{-1} for BAM images and 6.7 mN m^{-1} for AFM images) at 20°C and ionic strength 0.05 M; A_5, B_5, C_5, pH 5; A_7, B_7, C_7, pH 7; A_9, B_9, C_9, pH 9*

partially submerged into the aqueous phase sub-layer to reduce the repulsion between the DPPC polar groups.

We have observed recently[22] that surface shear rheology is clearly sensitive to both the monolayer structure and the nature of the electrostatic interactions between adsorbed DPPC molecules. It was observed that both the elastic and viscous components of the surface shear modulus (G', η_s) increase with increasing surface density. In addition, they both also increase with decreasing pH. It has been suggested above that the repulsive interactions between DPPC polar groups at pH 9 produces a less dense packing of the monolayer and larger domains, as observed by BAM. It seems reasonable to suppose that these structural features are responsible for the reduced values of G' and η_s at pH 9.

15.3.3 β-Casein Monolayers

The results from the π–A isotherms in Figure 5A confirm[14] that β-casein monolayers have a LE-like structure under these experimental conditions. The

Table 2 *The effect of pH on characteristic BAM geometrical and topographical parameters (from image morphology and reflectivity) for LE and LC domains of DPPC monolayers spread at the air–water interface, at the transition between LC and LE phases (at $\pi = 6.7$ mN m^{-1}) at 20°C and ionic strength 0.05 M*

pH	5	7	9
Domain size			
Radius (μm)	6.7	10.2	14.8
Area (μm^2)	143.5	397.0	816.6
Domain shape			
Roundness	1.1	1.5	1.3
Border	Circular	Star shape	Ellipsoidal
Thickness (a.u. × 10^7)			
LE	2.9	2.9	5.5
LC	6.3	6.3	8.5
LE–LC transition	3.4	3.4	3

sub-phase pH exerts a significant influence on the self-assembly of β-casein monolayers at the air–water interface. The π–A isotherm is displaced towards the surface pressure axis, and the monolayer structure is more condensed, in the acidic subphase. From the π–A isotherm itself (Figure 5A) and the elasticity modulus deduced from its slope (Figure 5B), it appears that there is a critical surface pressure $\pi(t,\beta\text{-cas})$ (see Table 1) at which the monolayer properties change significantly. With β-casein films we distinguish two different structures (Structures 1 and 2) around the monolayer collapse at the highest surface pressure, close to its equilibrium spreading pressure value, $\pi(e,\beta\text{-cas})$ (see Table 1). According to Graham and Phillips,[23] a tail–train structure for β-casein occurs at $\pi < \pi(t,\beta\text{-cas})$ (Structure 1). At higher surface pressures, and up to $\pi = \pi(e,\beta\text{-cas})$, amino acid segments are extended into the underlying aqueous solution and adopt the form of loops and tails (Structure 2). The $\pi(e,\beta\text{-cas})$ value decreases monotonically as the pH increases, a phenomenon which is consistent with a reduction in the interactions between amino acid residues at the air–water interface with the basic sub-phase. The acidic sub-phase caused an increase in the $\pi(t)$ value. This transition was observed at $\pi(t,\beta\text{-cas}) = 15.6$ and 10.6 mN m^{-1} at pH 5 and 7, respectively (Table 1). However, at pH 9 this transition was not observed. The presence of a minimum E plateau value, which is characteristic of a second-order phase transition, corroborates the existence of two structures for β-casein monolayers at pH 5 and 7. However, such a minimum E plateau was not observed at pH 9. Clearly, the electrostatic repulsions between amino acid residues at pH 9 do not favour Structure 2 with the formation of loops and tails. At $\pi < \pi(t,\beta\text{-cas})$, the E values are higher at pH 5 and 7 than at pH 9, but the opposite was observed at $\pi > \pi(t,\beta\text{-cas})$.

The BAM images confirm that only a homogeneous phase is present during the compression of β-casein up to the monolayer collapse (Figure 6A). That is,

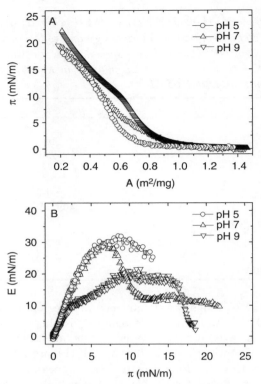

Figure 5 *The effect of pH on (A) the surface pressure versus area isotherm and (B) the elasticity modulus for β-casein monolayers spread at the air–water interface at 20°C and ionic strength 0.05 M; O, pH 5; Δ, pH 7; ∇, pH 9*

from the BAM images it is impossible to distinguish between the different structures that β-casein adopts at the air–water interface. The images suggest that the β-casein residues spread at the air–water interface have the same isotropy in the plane vertical to the molecular chain, no matter what is the structure adopted by β-casein residues at the interface. This isotropy is also confirmed by the evolution during the monolayer compression of the reflectivity of the BAM images. Figure 6 shows that the reflectivity increases as the monolayer is compressed and goes to a maximum at the collapse point, which means that during the monolayer compression a denser film is formed. The surface pressure dependence on reflectivity (Figure 6) and film thickness (Table 3) for β-casein spread monolayers is essentially the same when viewed at higher levels of magnification by AFM.[24] The differences in surface coverage and packing are clearly seen. The monolayer thickness is 1.1 and 1.6 nm at π values below and above π(t), respectively. A value of 1.1 nm is consistent with a flat two-dimensional network composed of individual β-casein molecules. The increase in film thickness at π > π(t) is consistent with a deformable protein structure influenced by surface packing.[24] The arrows in Figure 6 indicate the

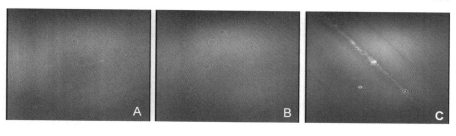

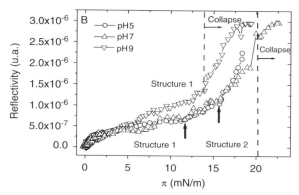

Figure 6 *BAM images for β-casein monolayers spread at the air–water interface at pH 9, 20°C, and ionic strength 0.05 M: A, 5 mN m^{-1}; shutter time = 1/50 s; B, 13.2 mN m^{-1}, 1/125 s; C, 20.5 mN m^{-1}, 1/250 s. Image sizes are 470 × 600 μm. Also shown is the effect of pH on the evolution of reflectivity with surface pressure for β-casein monolayers spread at the air–water interface at 20°C and ionic strength 0.05 M: O, pH 5; Δ, pH 7; ∇, pH 9*

Table 3 *The effect of pH on the maximum thickness deduced from the reflectivity of BAM images for β-casein spread monolayers at the air–water interface at 20°C and ionic strength 0.05 M*

pH	Thickness (nm) (Structure 1)	Thickness (nm) (Structure 2)	Thickness (nm) (collapse)
5	1.0	1.4	(1.5)^a
7	0.9	1.4	1.7
9	1.2	—	1.7

^{a} Just before the collapse point.

changes in the reflectivity *versus* π evolution at pH 5 and 7, but the reflectivity increases monotonically with π up the collapse point at pH 9. These data are also in agreement with those of a density profile of adsorbed β-casein perpendicular to the interface as inferred from neutron reflectance experiments at neutral and acidic aqueous sub-phases.[25] The higher reflectivity (and thickness) in β-casein monolayers was observed at pH 9 because amino acid residues are partially submerged into the aqueous phase sub-layer to reduce the repulsion between negatively charged residues.

15.3.4 DPPC + β-Casein Monolayers

The role of electrostatic interactions in the molecular self-assembly of DPPC and β-casein molecules in mixed monolayers at the air–water interface is of practical importance because real foods contain mixtures of both classes of emulsifiers with the potential to stabilize fluid interfaces.[2,6,12] However, the analysis of electrostatic interactions between components in mixed monolayers is more complex than for the pure components discussed in the previous sections. In a previous contribution we concluded[26] that these interactions have a very complex dependency on the pH, surface pressure, and monolayer composition. In fact, at $\pi < \pi(e,\beta\text{-cas})$, DPPC and β-casein form a mixed monolayer at the air–water interface at pH 5 and 7, but with weak interactions. However, significant repulsive interactions exist between adsorbed DPPC and β-casein at pH 9, where the DPPC molecules are negatively charged. The collapsed β-casein is displaced from the interface by DPPC at $\pi > \pi(e,\beta\text{-cas})$. This displacement is further facilitated by increased electrostatic repulsion at pH 9 and by higher DPPC concentrations.

The results derived from BAM and AFM confirm the importance of electrostatic repulsion on both the displacement of β-casein by phospholipids and the promotion of phase separation at the air–water interface. At the lowest surface pressure (at $\pi \approx 0$, at the beginning of the compression or at the end of the expansion), the DPPC + β-casein films form two-dimensional foams (Figure 7) at all pH values, similar to pure DPPC monolayers. Thus, at low surface coverage the influence of attractive hydrophobic and/or electrostatic interactions at pH 5 or 7, or repulsive electrostatic interactions at pH 9, does not appear to have a significant effect. The influence of electrostatic interactions on molecular self-assembly at higher π is more complex, depending on both the surface pressure and the protein/phospholipid ratio expressed as the mass fraction of DPPC in the mixture, X_{DPPC}. It is convenient to discuss three separate cases, depending on the surface pressure.

For $\pi < \pi(e,\beta\text{-cas})$ (Figure 8), domains of both β-casein and DPPC are present at the interface, but with different morphology depending on the pH of the aqueous phase. At pH 5 (image A) and 7 (image B), the modest interactions allowed more mixing of DPPC domains within regions dominated by β-casein, with some evidence of multilayer formation. But, at pH 9 (image C), homogeneous β-casein phases containing no DPPC were observed, suggesting increased levels of phase separation driven by the increased electrostatic repulsion at pH 9. From the π–A isotherm of the mixed monolayers (data not shown), we have deduced the surface pressure, $\pi(t,\text{DPPC}/\beta\text{-cas})$, at the transition between Structures 1 and 2 of β-casein in the mixtures (see Table 4). A noteworthy observation is that the presence of DPPC increases $\pi(t,\text{DPPC}/\beta\text{-cas})$ at pH 5,

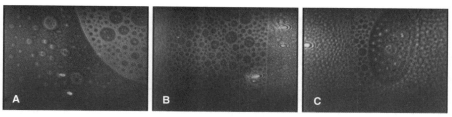

Figure 7 BAM images for DPPC + β-casein mixed monolayers spread at the air–water interface (20°C, ionic strength = 0.05 M, shutter time = 1/50 s), just after the spreading of the monolayer or at the end of the expansion of the previously compressed monolayer at $\pi \approx 0$ mN m^{-1}: A, pH 5; B, pH 7; C, pH 9. Mass fraction of DPPC in the mixture is X_{DPPC} 60%. Image sizes are 470 × 600 μm

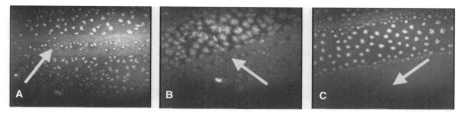

Figure 8 BAM images for DPPC + β-casein mixed monolayers spread at the air–water interface (20°C, ionic strength = 0.05 M, shutter time = 1/50 s) at $\pi > \pi(e,\beta\text{-}cas)$: A, pH 5; B, pH 7; C, pH 9. Mass fraction of DPPC in the mixture is X_{DPPC} 60%. Image sizes are 470 × 600 μm. The arrows indicate regions with moderate mixing of DPPC + β-casein domains (images A and B) and a segregated homogeneous β-casein phase (image C)

whereas it decreases it at pH 7. A similar effect has been observed[27] when comparing the displacement of protein by ionic and nonionic surfactants. This suggests that greater electrostatic attraction leads to greater mixing and requires higher surface pressures to cause competitive displacement or interfacial transitions. Therefore, as pH increases, the greater electrostatic repulsion leads to more well-defined phase separation.[16] This increases the pressure on the β-casein-rich regions of the film, forcing structural transitions to occur at lower surface pressures.

For $\pi \approx \pi(e,\beta\text{-}cas)$ (Figure 9), the DPPC begins to displace the β-casein collapsed domains from the interface into the sub-phase. At pH 5 (image A) and 7 (image B), white fringes of thick, collapsed β-casein domains are observed at the interface. The area occupied by these fringes increases with the content of β-casein in the mixture. At this pH attractive hydrophobic and/or electrostatic interactions may exist and the protein displaced to the sub-phase interacts with DPPC forming β-casein–DPPC complexes with high reflectivity (thickness). However, at pH 9 (image C) no fringes of collapsed β-casein domains are observed in the mixed film. Thus, the morphology of the monolayer is dominated by the presence of the DPPC domains. At this pH

Table 4 *The effect of pH and mass fraction of DPPC in the mixture, X_{DPPC}, on the critical surface pressure for the transition between the Structures 1 and 2 for β-casein in DPPC–β-casein mixed monolayers, $\pi(t,DPPC/β\text{-}cas)$ spread at the air–water interface at 20°C, and ionic strength 0.05 M*

X_{DPPC} (%)	$\pi(t,DPPC/β\text{-}cas)$ $(mN\ m^{-1})$ at pH 5	$\pi(t,DPPC/β\text{-}cas)$ $(mN\ m^{-1})$ at pH 7
0	15.6	10.6
20	17.5	11.5
40	16.0	9.4
60	15.8	9.0
80	17.4	7.2

Figure 9 *BAM images for DPPC + β-casein mixed monolayers spread at the air–water interface (20°C, ionic strength 0.05 M, shutter time 1/125 s) at $\pi \approx \pi(e,β\text{-}cas)$: A, pH 5; B, pH 7; C, pH 9. Mass fraction of DPPC in the mixture is $X_{DPPC} = 60\%$. Image sizes are 470 × 600 μm*

the β-casein–DPPC complexes are not formed as a consequence of the repulsive electrostatic interactions between the components.

For $\pi > \pi(e,β\text{-}cas)$ (Figure 10), the β-casein is displaced from the interface by DPPC. At pH 5 (image A), the bright domains of collapsed β-casein aggregates bind to the phospholipid from the sub-phase and at the interface. The presence of β-casein is evident even at the highest surface pressures (at the collapse point of the mixed monolayer). At pH 7 (image B), the densities of the bright domains of collapsed β-casein are reduced and the domains are distributed disorderly over the interface. At pH 9 (image C), repulsive electrostatic interactions between the components inhibit complex formation, leading to the early and complete displacement of β-casein from the interface, with no evidence of β-casein aggregates associating with the DPPC monolayer.

While the AFM observations can obviously yield higher resolution, the phenomenological changes observed as a function of pH by BAM and AFM in the DPPC + β-casein mixed monolayers are essentially the same.[26,28] The topographic AFM images do, however, allow quantitative data analysis of thickness. This will be used in future studies to gain further insight into the complex behaviour of protein + phospholipids mixed films at the air–water interface as a function of electrostatic interactions.

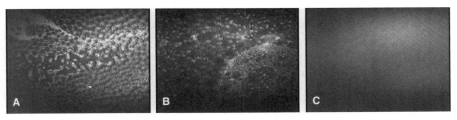

Figure 10 *BAM images for DPPC + β-casein mixed monolayers spread at the air–water interface (20°C, ionic strength 0.05 M, shutter time = 1/125 s) at $\pi > \pi(e,\beta\text{-}cas)$: A, pH 5; B, pH 7; C, pH 9. Mass fraction of DPPC in the mixture is $X_{DPPC} = 60\%$. Image sizes are 470×600 μm*

15.4 Conclusions

The structure, morphology, and film thickness of DPPC, β-casein, and their mixtures at the air–water interface has been determined using a combination of π–A isotherms, BAM, and AFM. Pure DPPC monolayers show a structural polymorphism dependent on π and pH, displaying film anisotropy, and heterogeneous domain structures. In contrast, pure β-casein monolayers display a homogeneous morphology, with a flat two-dimensional network composed of individual β-casein molecules at low π value and protein structure influenced by surface packing at high π value. This behaviour was observed at all surface pressures and at pH 5 and 7.

For DPPC + β-casein mixed films, the role of attractive interactions (hydrophobic and/or electrostatic) or repulsive interactions (electrostatic) is more complex, being dependent on the surface pressure and the DPPC/β-casein ratio. At microscopic and nanoscopic levels, respectively, BAM and AFM have corroborated the structural polymorphism deduced from the π–A isotherms and the role of electrostatic interactions on mixed film structure. Increasing the repulsive electrostatic interactions between DPPC and β-casein by raising the pH results in a greater degree of phase separation, and an earlier onset of structural transitions of the β-casein component. Nonetheless, in addition to the purely electrostatic forces, other types of interactions – hydrophobic forces, the restructuring of solvent and ionic distribution around different parts of the molecules, and the loss of mobility of the hydrophobic chains – must also be considered in the self-assembly of mixed protein + lipid films at fluid interfaces.

Acknowledgements

The authors acknowledge the support of CICYT through grants AGL2001-3843-C02-01 and AGL2004-01306/ALI. Research at IFR is funded through the core BBSRC grant to the Institute.

References

1. E. Dickinson, *An Introduction to Food Colloids*, Oxford University Press, Oxford, 1992.
2. J.M. Rodríguez Patino, M.R. Rodríguez Niño and C. Carrera, *Curr. Opin. Colloid Interface Sci.*, 2003, **8**, 387.
3. D.S. Horne and J.M. Rodríguez Patino, in *Biopolymers at Interfaces*, 2nd edn, M. Malmsten (ed), Marcel Dekker, New York, 2003, p. 857.
4. M.R. Rodríguez Niño, J.M. Rodríguez Patino, C. Carrera, M. Cejudo and J.M. Navarro, *Chem. Eng. Commun.*, 2003, **190**, 15.
5. S.E. Friberg, K. Larsson and J. Sjöblom (eds), *Food Emulsions*, 4th edn, Marcel Dekker, New York, 2004.
6. P.J. Wilde, *Curr. Opin. Colloid Interface Sci.*, 2000, **5**, 176.
7. E. Dickinson, *Colloids Surf. B*, 2001, **20**, 197.
8. M.E. Lesser, M. Michael and H.J. Watzke, in *Food Colloids: Biopolymers and Materials*, E. Dickinson and T. van Vliet (eds), Royal Society of Chemistry, Cambridge, 2003, p. 3.
9. A.R. Mackie, A.P. Gunning, P.J. Wilde and V.J. Morris, *J. Colloid Interface Sci.*, 1999, **210**, 157.
10. V.J. Morris, A.R. Kirby and A.P., Gunning, *Atomic Force Microscopy for Biologists*, Imperial College Press, London, 1999.
11. C.W. Hollars and R.C. Dunn, *Biophys. J.*, 1998, **75**, 342.
12. M. Bos, T. Nylander, T. Arnebrant and D.C. Clark, in *Food Emulsions and their Applications*, G.L. Hasenhuette and R.W. Hartel (eds), Chapman & Hall, New York, 1997, p. 95.
13. J.M. Rodríguez Patino, C. Carrera and M.R. Rodríguez Niño, *Langmuir*, 1999, **15**, 2484.
14. J.M. Rodríguez Patino, C. Carrera and M.R. Rodríguez Niño, *Food Hydrocoll.*, 1999, **13**, 401.
15. J.M. Rodríguez Patino, C. Carrera and M.R. Rodríguez Niño, *J. Agric. Food Chem.*, 1999, **47**, 4998.
16. G. Cevec, *Phospholipids Handbook*, Marcel Dekker, New York, 1993.
17. J. Miñones, Jr., J.M. Rodríguez Patino, J. Miñones, P. Dynarowicz-Latka and C. Carrera, *J. Colloid Interface Sci.*, 2002, **249**, 388.
18. D.G. Dalgleish, *Food Res. Int.*, 1996, **29**, 541.
19. E. Dickinson, *Int. Dairy J.*, 1999, **9**, 305.
20. E. Leclerc and P. Calmettes, *Phys. Rev. Lett.*, 1997, **78**, 150.
21. D.S. Horne and J. Leaver, *Food Hydrocoll.*, 1995, **9**, 91.
22. J.M. Rodríguez Patino, A. Lucero, M.R. Rodríguez Niño, A.R. Mackie, A.P. Gunning and V.J. Morris, *Food Chem.*, in press.
23. D.E. Graham and M.C. Phillips, *J. Colloid Interface Sci.*, 1979, **70**, 427.
24. A.P. Gunning, P.J. Wilde, D.C. Clark, V.J. Morris, M.L. Parker and P.A. Gunning, *J. Colloid Interface Sci.*, 1996, **183**, 600.
25. F.A.M. Leermakers, P.J. Atkinson, E. Dickinson and D.S. Horne, *J. Colloid Interface Sci.*, 1996, **178**, 681.

26. A. Lucero, M.R. Rodríguez Niño, C. Carrera, P.A. Gunning, A.R. Mackie and J.M. Rodríguez Patino, in *Food Colloids: Interactions, Microstructure and Processing*, E. Dickinson (ed), Royal Society of Chemistry, Cambridge, 2005, p. 160.
27. P.A. Gunning, A.R. Mackie, A.P. Gunning, N.C. Woodward, P.J. Wilde and V.J. Morris, *Biomacromolecules*, 2004, **5**, 984.
28. C. Lucero, Ph.D. Thesis, University of Seville, 2005.

Chapter 16

Theoretical Study of Phase Transition Behaviour in Mixed Biopolymer + Surfactant Interfacial Layers Using the Self-Consistent-Field Approach

Rammile Ettelaie,[1] Eric Dickinson,[1] Lei Cao[1] and Luis A. Pugnaloni[2]

[1] PROCTER DEPARTMENT OF FOOD SCIENCE, UNIVERSITY OF LEEDS, LEEDS LS2 9JT, UK
[2] INSTITUTE OF PHYSICS OF LIQUIDS AND BIOLOGICAL SYSTEMS, CC565, CALLE 59, NO 789, 1900, LA PLATA, ARGENTINA

16.1 Introduction

The simultaneous presence of large amphiphilic biopolymers and low-molecular-weight surfactant-like molecules is a common feature of many food emulsions and foams. While both groups of molecules deliver their required functionality through their strong tendency for adsorption at interfaces, they play quite different roles in the formation and stability of emulsion drops or bubbles.[1]

Low-molecular-weight surfactants can greatly reduce the interfacial tension, thus promoting the creation of new interfaces and therefore aiding the emulsification process. They also have fast adsorption kinetics, with bulk–interface equilibration times rarely above a few seconds.[2] Thus, they can quickly adsorb at newly created interfaces to provide short-term stability through mechanisms such as the Marangoni effect.[3] In contrast, adsorption/desorption times associated with proteins and other amphiphilic biopolymers are far longer – they can be of the order of hours or even days.[4] Furthermore, globular proteins undergo unfolding and substantial structural changes during adsorption. This often leads to the formation of intermolecular bonds as the reactive segments of the chains become exposed. The resulting viscoelastic interfacial films are crucial in providing much of the required long-term stability in many food emulsions and foam systems.[3] A different stabilization mechanism, more closely associated with disordered proteins like β-casein, involves the ability of such biopolymers to act as effective steric stabilizers, preventing close approach of droplets through the provision of long-range repulsive forces.[1]

The different roles played by protein stabilizers and low-molecular-weight emulsifiers necessitates that both type of molecules be present in many food colloid formulations. This gives rise to some more complex types of behaviour, the study of which has attracted much interest in recent years.[5-10] The association and formation of complexes between surfactant molecules and biopolymers can have important implications for the surface activity, the net charge, and the subsequent functionality of both kinds of species. For example, it has been reported[9] that sodium caseinate forms complexes with certain surfactants at moderately low pH values (~ 5.5), and that this increases the foam stabilizing properties of the protein. While some surfactants interact favourably with protein, there are others, mostly non-ionic water-soluble surfactants, which do not show any tendency for association[11] and may even exhibit a degree of incompatibility in bulk solution. Such surfactants compete with the amphiphilic biopolymers for coverage of the interface. At typical concentrations of practical interest in foods, the competitive adsorption leads to displacement of protein from the surface, which is sometimes followed by undesirable destabilization and break up of the emulsion,[6,12] particularly if the system is sheared.

Due to the relatively large size of protein molecules and their high adsorption energies, their displacement from the interface by surfactant species is a rather slow process. It is therefore reasonable, at any stage during the displacement process, to view the interfacial layer as one with a fixed degree of coverage by the biopolymer. In contrast, the surfactant molecules can quickly exchange between the bulk solution and the interface. Therefore the amount of adsorbed surfactant is determined by the equilibrium conditions. In turn, these conditions are dictated by the bulk concentration of surfactant and the amount of polymer still remaining on the film at any given time.

We have recently considered[13] the phase separation behaviour in binary interfacial layers in which either one or both components maintains equilibrium with the bulk solution. It was concluded that, for the latter case, no matter how unfavourable are the interactions between the species, no phase separation can occur and the composition of the surface will remain homogeneous. However, in the former situation, where the surface coverage of one of the components remains fixed, a certain degree of incompatibility between the two species is sufficient to induce surface phase separation.[13] These predictions were based on a simple lattice model which took no account of the internal configurational degrees of freedom of the molecules. For biopolymers, however, the configurational entropy is expected to make an important contribution to the phase properties of the mixed layer. A number of theoretical studies have attempted to overcome this problem, by assuming a small number of finite states, each with a different degree of surface area coverage, that the chains may adopt.[14,15] In these studies the interactions between surfactant and biopolymers have generally been assumed to be net attractive.

In what follows we describe a theoretical study to investigate the phase behaviour of surfactant + biopolymer layers based on the well-known self-consistent-field (SCF) approach.[16] This method is chosen here as it accounts for all the possible configurations that the polymer chains can take at the interface. Also, unlike most previous theoretical studies, the current work focuses on situations where the mixed surfactants and biopolymers exhibit some degree of unfavourable unlike interactions. The origins of such interactions are briefly discussed below.

16.2 Methodology and Model

16.2.1 Self-Consistent-Field Calculations

In SCF theories, the quantities of interest are a set of density profiles, $\phi_i^\alpha(z)$, representing the variation of the concentration of each kind of monomer α, belonging to chains of type i, with distance z away from a planar interface. The interface is assumed to be placed at $z = 0$ and for the purpose of the current study this is taken to be a solid–liquid interface. The calculations implement an iterative procedure, which results in the set of concentration profiles that minimize the free energy of the system. It is this set of profiles which is assumed to dominate the thermodynamic behaviour of the surface layer. The approximation is well justified for dense interfacial layers, such as the ones in this study, where the fluctuations about the most probable profile are generally small.[16]

The calculations begin by first choosing a rough starting estimate of the most probable density profiles. Using these initial concentrations, a set of auxiliary fields, $\psi^\alpha(z)$, is calculated. These fields represent the sum of interactions that each kind of monomer feels due to the presence of the neighbouring segments and the solvent molecules. For a segment of type α, at a distance z from the interface, the field is related to the density profiles by

$$\psi^\alpha(z) = \psi_{hd}(z) + \sum_i \sum_\beta \chi_{\alpha\beta} \left(\phi_i^\beta(z) - \Phi_i^\beta \right). \tag{1}$$

In Equation (1), the bulk concentration of monomers of type β, belonging to species i, is given by Φ_i^β. For any species only present at the interface, of course, the corresponding value of Φ_i^β is set to zero. The set of parameters $\chi_{\alpha\beta}$ defines the strength of interaction between different monomer kinds (α and β). From Equation (1), it is noticed that all the interactions are assumed to be short range and local. Fields $\psi^\alpha(z)$ also contain a hard-core potential term, $\psi_{hd}(z)$, that enforces the incompressibility condition

$$\sum_i \sum_\alpha \phi_i^\alpha(z) = \sum_i \sum_\alpha \Phi_i^\alpha, \tag{2}$$

which needs to be satisfied at each point z. The value of $\psi_{hd}(z)$ is the same for all segments, irrespective of their type.[16] For monomers also in contact with the solid surface, i.e., at position $z = 0$, there are additional interaction parameters,

$\chi_{\alpha s}$, not explicitly shown in Equation (1), specifying the adsorption energies for each type of segment.

The next step in the calculation involves the evaluation of the probability of each possible configuration that can be taken up by different polymer species under the influence of the fields $\psi^{\alpha}(z)$. This in turn provides us with a new set of density profiles, generally different from the starting solution. Due to the very large number of configurations available to a macromolecule, at first such a numeration seems like an unfeasible task. However, by using the Scheutjens–Fleer scheme,[16] such a calculation can be carried out very efficiently. The newly obtained density profiles are then fed back into Equation (1), and the whole process is repeated until convergence is obtained. The resulting self-consistent solutions can be shown to be a local minimum of the free energy, and they are stable against any small concentration fluctuations about the calculated profiles.[16] They may not, however, correspond to the global minimum. That is, the method does not preclude the calculation of metastable solutions. In fact, it is precisely this feature of the procedure that the current work seeks to exploit.

A more detailed account of the method and its implementation can be found in a number of studies[17–22] where SCF calculations have been applied successfully to a variety of problems involving food biopolymers.

16.2.2 The Mixture Model

In the model adopted here, both biopolymer and surfactant are represented as chains consisting of hydrophobic and hydrophilic monomers of equal size. The presence of both types of monomer is essential to capture the amphiphilic nature of the molecules. The biopolymer is a diblock copolymer consisting of 100 hydrophobic and 100 hydrophilic segments. To ensure a fixed amount of polymer at the interface we assume that the last hydrophobic segment of each chain is anchored onto the solid surface. The much smaller surfactant molecules consist of two hydrophobic and two hydrophilic monomers; as discussed before, they are free to adsorb or desorb from the surface.

The degree of hydrophobicity of the monomers is specified through their interactions with the solvent molecules. For simplicity, and to keep the number of parameters in the model small, we shall assume that the hydrophobic segments of both biopolymer and surfactant molecules have the same unfavourable interaction with the solvent. This is specified by setting the values of the appropriate Flory–Huggins χ parameters to 1 (in units of kT). As for hydrophilic segments, the solvent is assumed to be athermal (*i.e.*, $\chi = 0$). Both sets of hydrophobic monomers have a strong affinity for the surface, with adsorption energies of $-1\ kT$ and $-3\ kT$ per monomer, for the polymer and surfactant molecules, respectively.

An important feature of the model studied is the strong degree of incompatibility between the macromolecules and the low-molecular-weight surfactants. Thus, the competition to occupy the interface is influenced by factors beyond simple geometric considerations. Note that, while the actual number of biopolymer chains at the interface is constant, the segments of the polymers in contact with the solid surface can vary greatly as chains adopt different configurations. The χ parameter for the unfavourable interactions between hydrophobic monomers of the polymer and surfactant species is set to 4 kT here. It should be mentioned that the presence of such a strong interaction, between hydrophobic groups of two different species, is unlikely to arise directly in food systems. However, in our recent Brownian dynamics simulation study, we showed[23] that the effect of the formation of strong (yet reversible) bonds between some segments on one set of molecules, but not the other, is effectively similar to the presence of a strong repulsion between the two species. We have attributed[23] the observed surface phase separation in mixed biopolymer layers, as reported[24] for some combinations of adsorbed protein molecules, to the presence of such an effective repulsion between the different species.

16.3 Results and Discussion

16.3.1 Adsorption Isotherms

We begin the discussion of the results by presenting our calculated adsorption curves for the surfactant molecules, as they exchange between the bulk and an interface containing a fixed amount of biopolymer. As an example, we take a case where the amount of biopolymer in the entire interfacial layer is set to 0.005 chains per unit monomer area (a_0^2). Figure 1 shows the variation in the volume fraction of segments belonging to the surfactant molecules, Γ_c^s, in the first layer in direct contact with the solid surface. This is plotted against the bulk concentration of surfactant c_s. At low values of the surfactant bulk concentration ($c_s < 0.0265$), the number density of surfactant segments in contact with the surface is seen to increase smoothly (the solid line), following a typical adsorption isotherm. However, as the bulk concentration reaches the value of 0.0265, we find that the amount of the adsorbed surfactant shows a sudden jump, with the value of Γ_c^s changing abruptly from a relatively low value of 0.125 to a much higher value of 0.614. This discontinuous change is indicated by the arrow in Figure 1. For $c_s > 0.0265$, the curve is once again smooth, increasing monotonically with surfactant concentration. Associated with this change in Γ_c^s, we also observe a similar corresponding abrupt drop in the amount of biopolymer in contact with the solid surface. This drop occurs at precisely the same surfactant bulk concentration of 0.0265. This is shown in Figure 2, where the volume fraction of the polymer chains, Γ_c^p, in the layer in direct contact with the wall, is plotted against c_s. The value of Γ_c^p at $c_s = 0.0265$ falls from 0.452 down to a much lower value of 0.0096. This is interpreted as signalling a sudden change in the configuration of the chains, from one with chains originally laying flat on the surface to one with chains extending into the bulk. These results are not unique to the particular case considered here. We have found that they can also be reproduced for a variety of other similar systems, with somewhat different interaction strengths, adsorption energies,

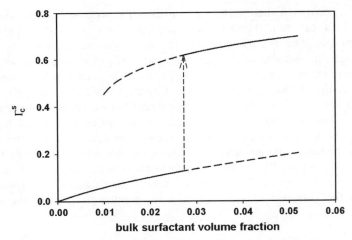

Figure 1 *The variation of the volume fraction of segments belonging to surfactant molecules in the layer in direct contact with the solid surface, Γ_c^s, plotted as a function of the volume fraction of the surfactant in the bulk solution. The solid and the dashed lines indicate the stable and metastable values of Γ_c^s, respectively. The number of biopolymer chains at the interfacial region is kept fixed at a value of 0.005 chains per unit monomer area (a_0^2)*

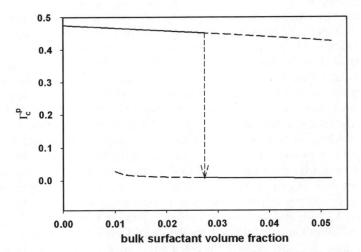

Figure 2 *As Figure 1, but now showing the variation of the volume fraction of the segments belonging to the biopolymer chains, Γ_c^p, in contact with the surface*

chain sizes, and other system-dependent parameters. The important feature required is the presence of a reasonably strong degree of incompatibility between polymer and surfactant species. Of course, the exact threshold values for the occurrence of the abrupt change, as seen in Figures 1 and 2, does depend on the precise values of the chosen parameters.

The sudden change in the amount of adsorbed surfactant with increasing surfactant bulk concentration is reminiscence of a first-order phase transition. The existence of a genuine first-order phase transition in any system implies the presence of metastable states. As mentioned in the previous section, our numerical scheme should be able to detect such metastable solutions. We have investigated this proposition, by feeding into our SCF calculations a variety of starting guesses for our initial profiles. Indeed, at and close to the threshold value for c_s, we have found that the SCF calculations do converge onto a second solution, different from the one shown by the solid lines in Figures 1 and 2. As one might expect for a metastable state, for each surfactant bulk concentration the value of the free energy associated with this secondary solution is greater than the corresponding original solution. The volume fractions of surfactant and polymer segments in direct contact with the solid surface, for the second solution, are indicated by the dashed lines in Figures 1 and 2. It is clear that these metastable solutions form the continuation of the stable parts of the curves, for the low- and high-coverage surfactant phases, below and above the threshold value of $c_s = 0.0265$, respectively.

The existence of a first-order phase transition can further be corroborated by studying the way in which the difference between the free energy of the high- and the low-coverage surfactant phase varies with bulk surfactant concentration. This is illustrated in Figure 3, where we present the result of the free energy difference, calculated once again from SCF theory. The graph clearly shows that, at surfactant bulk concentrations below the transition point, the high-coverage surfactant phase has a higher free energy and therefore constitutes the metastable state. At the transition concentration, the graph passes through zero, indicating that free energies of the two solutions are

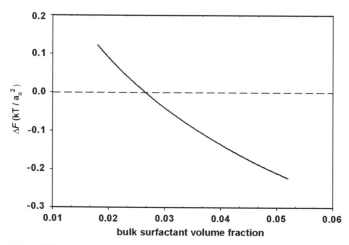

Figure 3 *The difference in the free energy (ΔF) per interfacial area between the surfactant phases of high- and low-surface coverage. At the transition point, the two phases have identical free energy per unit area ($\Delta F = 0$)*

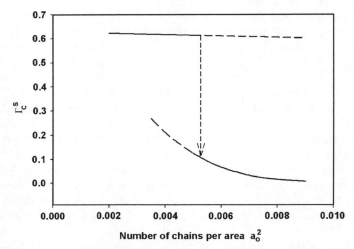

Figure 4 *As Figure 1, but now the bulk volume fraction of surfactant molecules is kept fixed at 0.0265, with the number of biopolymer chains at the interfacial region being made to vary*

identical. As the surfactant concentration is further increased above 0.0265, however, it is now the high-coverage surfactant phase that has the lower free energy and is therefore the stable phase. These results are precisely those one would expect arising from the existence of a first-order phase transition in the system, with the volume fraction of surfactant in the first layer in contact with the solid surface acting as a measure of the order parameter for this transition.

During the gradual displacement of protein by surfactant in food systems, it is the amount of biopolymer at the interface that slowly decreases. It is therefore useful to consider a slightly different situation in which the bulk concentration of the surfactant is kept constant and the number of chains at the interface is altered. The results of such an exercise, for the same biopolymer and surfactant system as the one above, are displayed in Figure 4. The bulk concentration of surfactant is now fixed at 0.0265. Starting from a high surface coverage of protein molecules, as the amount of protein on the surface decreases, the number of surfactant segments in contact with the wall increases smoothly at first. However, as the number of chains per unit monomer area (a_0^2) reaches the transition value of 0.005, once again there is a rapid and abrupt jump in the value of Γ_c^s from 0.13 to 0.61. The above result indicates that the displacement of protein molecules from the interface may be accompanied by configurational and structural phase transitions in the mixed monolayer, involving the sudden large uptake of surfactant molecules. Phase transition behaviour of this type also has further consequences for the emergence of phase-separated surface domains, resulting in a heterogeneous film if the protein molecules possess a reasonable degree of lateral mobility.

16.3.2 Density Profiles

In adsorbed layers involving macromolecules, the interfacial region can be rather thick. The variation of the density profile of the biopolymers in this interfacial region can provide useful information on possible configurations adopted by the adsorbed chains.[22] In particular, changes in the structure of the mixed films, at the transition point, can be obtained by studying the concentration profiles of the biopolymers, for both the low- and the high-coverage surfactant phases. Using the SCF calculations, we have determined the variation of the concentration profile of the protein precisely at the transition points; this is plotted as a function of distance away from the solid surface in Figure 5. The profile for the high-coverage surfactant phase is represented by the solid line and that for the low-coverage surfactant phase by the dashed line. Although the free energies associated with each profile are exactly identical at the transition point, the variations in the biopolymer concentration profiles, with the distance away from the solid surface, are manifestly different. For the low-coverage surfactant phase, a high number of predominantly hydrophobic polymer segments are in contact with the wall, with the hydrophilic parts of the chains dangling into the solution. For the high-coverage surfactant phase, on the other hand, the number of contacts between the chains and the solid surface is much reduced. In the latter phase, even the hydrophobic sections of these molecules form extended loops protruding into the bulk solution. This gives rise to a distinct peak in the shape of the profile, clearly visible at a distance of around five monomer units away from the surface. The interfacial layer, not surprisingly, is also found to be somewhat more extended for the high-coverage surfactant phase.

In Figure 6, we have plotted the average distance away from the surface, $<R_i>$, for each polymer segment i. The segments are numbered consecutively

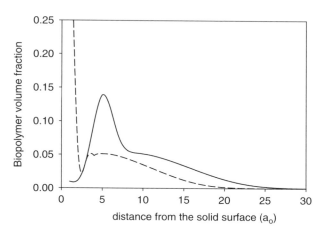

Figure 5 *Variation in the biopolymer density profile for the high-coverage (solid line) and low-coverage (dashed line) surfactant phases, plotted as a function of distance (in monomer unit size a_0) from the solid interface. The results are obtained at the transition point*

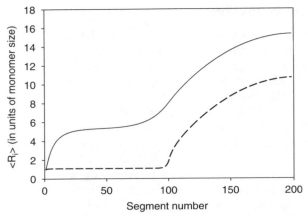

Figure 6 *The average distance of biopolymer segments from the surface at the transition point ($<R_i>$). The first monomer of the hydrophobic end of the diblock copolymer chain is numbered $i = 1$. The dashed line represents the results for the low-coverage surfactant phase, and the solid line that for the high-coverage surfactant phases*

along the chain backbone. The first monomer starting from the hydrophobic end of the diblock copolymer is numbered as 1. Once again the solid and the dashed lines represent the high- and low-coverage surfactant phases, respectively. The results support a similar conclusion to that of Figure 5. In the low-coverage phase, the hydrophobic part of the polymer lies completely flat on the surface. But this is seen to be largely displaced by the surfactant, for the high surface coverage phase at the transition point. A large proportion of hydrophobic monomers is now situated at an average distance of 5–6 monomer units away from the surface, providing an explanation for the peak seen in Figure 5 at the same distance. The more extended nature of the interfacial layer is also evident for the latter phase. At the transition, the average distance away from the surface for the last hydrophilic segment on the chain makes a jump from a value of $<R_i> = 10.7$ to $<R_i> = 15.3$ (in units of monomer size).

The current study has focused on a particular type of biopolymer structure, roughly representative of a disordered amphiphilic protein such as β-casein. However, the same phenomenon is also predicted for other types of polymer architecture. For example, results not presented here show the occurrence of the phase transition in mixed layers containing chains with a large number of alternating hydrophobic and hydrophilic short blocks. These structures may be rather closer representatives of globular-type proteins – at least in their denatured state.

16.4 Conclusions

We have used SCF calculations to explore the behaviour of mixed surfactant + biopolymer interfacial layers for the case where there is some degree of incompatibility between these two species. The study predicts the existence of complex behaviour in these layers, involving abrupt changes in the configuration of the macromolecules and rapid adsorption and desorption of surfactant from the interface, as the bulk concentration of surfactant or the amount of biopolymer at the interface is varied. We have shown that such behaviour is associated with the existence of a first-order phase transition in such systems.

The presence of such sudden changes in the composition of the interfacial films can have important implications for the processing and formulation of food products, making these more difficult and unpredictable close to the transition point. It will be of interest in future studies to consider the implication that the behaviour of such mixed layers might have on the colloidal interaction that these mediate between emulsion droplets. It is expected that a similar phase transition may occur at a critical separation distance between the droplet surfaces. Indeed, for systems involving pure ionic surfactants, such a phenomenon has already been predicted,[25,26] although the driving force for the predicted transition is somewhat different to the one arising here in our system.

Future experiments involving ellipsometry or Brewster angle microscopy may provide further useful evidence for the existence of the phenomenon predicted here.

References

1. E. Dickinson, *An Introduction to Food Colloids*, Oxford University Press, Oxford, 1992.
2. C.-H. Chang and E.I. Frances, *Colloids Surf. A*, 1995, **100**, 1.
3. P.J. Wilde, *Curr. Opin. Colloid Interface Sci.*, 2000, **5**, 176.
4. R. Miller, D.O. Grigoriev, J. Krägel, A.V. Makievski, J. Maaldonado-Valderrama, M. Leser, M. Michel and V.B. Fainerman, *Food Hydrocoll.*, 2005, **19**, 479.
5. J.M. Rodriguez Patino, M.R. Rodriguez Niño and C.C. Sanchez, *Curr. Opin. Colloid Interface Sci.*, 2003, **8**, 387.
6. A.R. Mackie, A.P. Gunning, P.J. Wilde and V.J. Morris, *Langmuir*, 2000, **16**, 2242.
7. L.A. Pugnaloni, E. Dickinson, R. Ettelaie, A.R. Mackie and P.J. Wilde, *Adv. Colloid Interface Sci.*, 2004, **107**, 27.
8. B. Lindman, A. Carlsson, S. Gerdes, G. Karlström, L. Picullel, K. Thalberg and K. Zhang, in *Food Colloids and Polymers: Stability and Mechanical Properties*, E. Dickinson and P. Walstra (eds), Royal Society of Chemistry, Cambridge, 1993, p. 113.
9. M.M. Il'in, M.S. Anokhina, M.G. Semenova, L.E. Belyakova and Y.N. Polikarpov, *Food Hydrocoll.*, 2005, **19**, 455.
10. A.S. Antipova, M.S. Semenova, L.E. Belyakova and M.M. Il'in, *Colloids Surf. B*, 2001, **21**, 217.
11. P. Walstra, *Physical Chemistry of Foods*, Marcel Dekker, New York, 2003.
12. E. Dickinson, *Colloids Surf. B*, 2001, **20**, 197.

13. L.A. Pugnaloni, R. Ettelaie and E. Dickinson, *J. Chem. Phys.*, 2004, **121**, 3775.
14. V.B. Fainerman, S.A. Zholob, E.H. Lucassen-Reynders and R. Miller, *J. Colloid Interface Sci.*, 2003, **261**, 180.
15. R. Miller, V.B. Fainerman, M.E. Leser and M. Michel, *Colloids Surf. A*, 2004, **233**, 39.
16. G.J. Fleer, M.A. Cohen Stuart, J.M.H.M. Scheutjens, T. Cosgrove and B. Vincent, *Polymers at Interfaces*, Chapman & Hall, London, 1993.
17. F.A.M. Leermakers, P.J. Atkinson, E. Dickinson and D.S. Horne , *J. Colloid Interface Sci.*, 1996, **178**, 681.
18. E. Dickinson, D.S. Horne, V.J. Pinfield and F.A.M. Leermakers, *J. Chem. Soc., Faraday Trans.*, 1997, **93**, 425.
19. R. Ettelaie, B.S. Murray and E.L. James, *Colloids Surf. B*, 2003, **31**, 195.
20. E.B. Zhulina, O.V. Borisov, J. van Male and F.A.M. Leermakers, *Langmuir*, 2001, **17**, 1277.
21. R. Ettelaie, E. Dickinson and B.S. Murray, in *Food Colloids: Interactions, Microstructure and Processing*, E. Dickinson (ed), Royal Society of Chemistry, Cambridge, 2005, p. 74.
22. E.L. Parkinson, R. Ettelaie and E. Dickinson, *Biomacromolecules*, 2005, **6**, 3018.
23. L.A. Pugnaloni, R. Ettelaie and E. Dickinson, *Langmuir*, 2003, **19**, 1923.
24. T. Sengupta and S. Damodaran, *J. Colloid Interface Sci.*, 2000, **229**, 21.
25. F.A.M. Leermakers, L.K. Koopal, W.J. Lokar and W.D. Ducker, *Langmuir*, 2005, **21**, 10089.
26. L.K. Koopal, F.A.M. Leermakers, W.J. Lokar and W.A. Ducker, *Langmuir*, 2005, **21**, 11534.

Chapter 17

Interactions during the Acidification of Native and Heated Milks Studied by Diffusing Wave Spectroscopy

Marcela Alexander, Laurence Donato and Douglas G. Dalgleish

DEPARTMENT OF FOOD SCIENCE, UNIVERSITY OF GUELPH, GUELPH, ONTARIO, CANADA, N1G 2W1

17.1 Introduction

The acid gelation of milk has often been thought of as a classic case of the destabilization of a colloidal system to form a particulate gel. Skim milk is a dispersion of particles (casein micelles) that are essentially composed of proteins and protected by a surface 'hairy' layer of κ-casein.[1] This surface layer carries some charge (with a reasonably large negative ζ-potential of around −20 mV),[2] but the layer is believed to exert its stabilizing effect mainly because it protrudes from the surface to provide steric stabilization of the particles.[3] Many studies suggest that the hairy layer has a hydrodynamic thickness between 5 and 10 nm, and that its removal by chymosin treatment at neutral pH destabilizes the casein micelle.[4,5] The chymosin treatment diminishes both the surface charge and the steric stabilization at the same time.

Conversely, during the acidification of milk, the hairy layer remains on the micellar surface. But it is believed that, as the pH decreases, the surface charges are titrated and the hairy layer collapses, so that steric and electrostatic stabilization are diminished, and at pH ≈ 5.0 the casein micelles coagulate to form a gel.[6] This gel is weak,[7] presumably because the hairy layer, although collapsed, prevents touching of the (putatively hydrophobic) micellar 'cores', so that the forces holding the particles together are not as strong as for the case when the hairy layer is removed. This is evident when the strengths of acid gels made from partially renneted milks are compared with those from unrenneted milk.[8] Thus far, therefore, it has been possible to regard acid milk as simply a destabilizing hard-sphere colloid system, although some reasons why this may not be so have also been proposed.[9]

The acidification of milk becomes much more difficult to describe if the milk is heated before the acidification step. Here, the acid gelation is known to occur at a higher pH than in unheated milk and the gel strength is greatly increased.[7] These effects have been attributed to the incorporation of denatured whey proteins (WP) into the acid gel, but the increase in strength is much greater than would be accounted for by simply incorporating about 20% more protein. The heating of milk at temperatures in excess of about 75°C causes the denaturation of the WP, which then form complexes with the κ-casein (κ-cas) of the casein micelles.[10,11] These WP–κ-cas complexes are distributed between the serum of the milk and the surfaces of the casein micelles, and the distribution depends on the pH at which the milk is heated. At the normal pH of milk, the particles are partly on the micellar surface and mainly free in the serum; the free particles have diameters in the region of 50 nm,[12] and so they are much smaller than the approximately 200 nm diameter of the casein micelles. The result of these changes in the particulate structure of the milk is that the apparently well-behaved system (in colloid chemistry terms) of unheated milk becomes considerably more complicated. The casein micelles now have modified surfaces, partly because they have lost some of their κ-casein to form the soluble WP–κ-cas particles, and partly because there are some regions of denatured WP on their surfaces. In addition, the serum contains a new population of colloidal particles – the WP–κ-cas complexes. The detailed behaviour and interactions of these two kinds of particles during acidification are essentially unknown, although it has been demonstrated that the extent of solubilization of the complexes affects the gelation.[13–15] A simplistic view is that the small complexes begin to aggregate into chains at pH ~ 5.3 because they have a higher isoelectric point than the caseins. These aggregates then interact with the modified casein micelles to give the acid gel, and they may be imagined as providing stronger inter-micellar links than those formed in unheated milk.

The research described in this paper was designed to try to elucidate some of the changes that occur in unheated and heated milks during acidification. For this study, we have used mainly the technique of diffusing wave spectroscopy (DWS), which is capable of measuring the scattering of light in turbid suspensions such as milk. This technique was combined with measures of the bulk rheological behaviour of the different milks.

17.2 Diffusing Wave Spectroscopy

The DWS technique has been extensively described previously,[16] and only basic details will be given here. When light is passed into a turbid suspension, which is sufficiently concentrated that every photon is extensively multiply scattered, the paths of the photons may be considered to be random walks, and it is possible to measure a correlation function from the scattered transmitted light.

This correlation function has the form[16]

$$g_{(1)}(t) \approx \frac{\left(\frac{L}{l^*}+\frac{4}{3}\right)\sqrt{\frac{6t}{\tau}}}{\left(1+\frac{8t}{3\tau}\right)\sinh\left[\frac{L}{l^*}\sqrt{\frac{6t}{\tau}}\right]+\frac{4}{3}\sqrt{\frac{6t}{\tau}}\cosh\left[\frac{L}{l^*}\sqrt{\frac{6t}{\tau}}\right]} \quad (1)$$

for $t \ll \tau$, where $\tau = (Dk^2)^{-1}$, and D is the particle diffusion coefficient, $k = 2\pi n/\lambda$ the wave vector of the light, n the refractive index of the medium, and L the thickness of the sample being measured. This form of correlation function is only valid when $L \gg l^*$ and $t \ll \tau$, which is true for all of our experiments. The parameter l^* is the photon transport mean free path, defined as the length scale over which the direction of the scattered light has been completely randomized. Practically, it is essentially a turbidity parameter and is directly related to the total scattered light intensity of the system. The parameter l^* is not measured as part of the measurement of the correlation function, but it can be determined from the long-time average of the intensity of the scattered light.[17] If l^* is known, it can be seen that the only unknown quantity in Equation (1) is the relaxation time τ. So the quantity τ can be derived from the measurement of the correlation function once l^* has been measured, and from τ the diffusion coefficient D can be calculated. If the particles in the suspension can be considered as freely diffusing identical spheres, their radius R can be determined from the Stokes–Einstein equation,

$$D = \frac{kT}{6\pi\eta R}, \quad (2)$$

where k is Boltmann's constant, T the temperature, and η the viscosity of the medium. The factor l^* in turn is defined in terms of the scattering properties of the particles in the suspension, as described by the form factor $F(q)$ and the structure factor $S(q)$:[16]

$$l^* \propto \left(\int F(q)S(q)q^3 dq\right)^{-1} \quad (3)$$

The form factor $F(q)$ is related to the size, shape and refractive index contrast of the scatterers as used in the scattering of dilute suspensions. The structure factor $S(q)$ describes the effects of the positional correlations between the scatterers on the scattering of light. In highly diluted systems, we have $S(q) = 1$ since the particles are far apart and their positions are completely uncorrelated, and there is no multiple scattering. However, in milk or food emulsions, where there is maybe a substantial volume fraction ($\phi \geq 0.1$) of dispersed material, the particles are relatively close to one another. For instance, the average distance between the surfaces of the casein micelles in milk is about 0.75 of the particle diameter. Thus, spatial correlations between the scattering particles become important in the light scattering, and these in turn depend on the interactions between the particles. If the volume fraction and refractive index contrast between the particles and the continuous phase are constant throughout a process, a change in the value of l^* will indicate some degree of increase in the interactions between particles *via* the change in $S(q)$. Therefore, changes in this parameter offer the possibility of interpreting changing organization within the suspension arising from the interactions between the particles.

A further parameter is available from the analysis of the correlation functions. As the motion of particles in a suspension begins to be restricted, the motion of the particles is no longer freely diffusive, and Equation (1) is no longer completely descriptive of the system. We must distinguish here between aggregates that diffuse freely (although more slowly than the original particles) and aggregates that are part of, or contained within, a gel and have restricted motion inside the gel framework. In the latter case, the correlation function can be related to the mean-squared displacement (MSD) of the particles, $<\Delta r^2(t)>$, and the correlation function (Equation (1)) can be analysed to determine the MSD at any time by replacing t/τ by $k_0^2<\Delta r^2(t)>$. The MSD is an indication of whether the movements of the particles are being constrained. Once a gel starts to form, the use of the Stokes–Einstein equation is invalid because the particles are probably non-spherical and are certainly not free-diffusing. It has been shown that, for a fractal type of gel, the slope of MSD against time decreases from a value of 1 for freely diffusing particles to 0.7 for a gelled material.[18]

This study of the gelation of milk will make use of all of these parameters derived from the DWS experiment, namely the apparent particle size, the slope of the MSD against time, and the quantity $1/l^*$.

17.3 Materials and Methods

Fresh milk, obtained from the University of Guelph dairy farm, was skimmed by centrifugation in the laboratory, followed by filtration through glass-fibre filters. Heated milks were prepared by placing 10 mL tubes of milk in a water bath at 85°C and leaving for 10 min before removing and cooling rapidly to room temperature in an ice bath. The milk was acidified by the addition of glucono-δ-lactone (GDL, 1 and 1.5% w/w). The GDL was added to the milk at a temperature of 30°C, and it was agitated briefly to dissolve the acidulant. The part of the sample used for DWS or rheological measurement was placed in the measuring equipment. The rest of the sample was used for monitoring the pH value, which was followed continuously until it had decreased to pH $= 4.8$.

The DWS equipment has been described elsewhere.[20] Light from a laser (532 nm, 100 mW) was passed through 5 mm path length cuvettes filled with milk or mixtures maintained at a temperature of 30°C. The transmitted light was detected and analysed by photomultipliers and a correlator to give the measured intensity correlation function. The value of l^* was measured from the total intensity of the scattered light, and by comparison with the light scattered by a latex of known size and scattering properties. The correlation function was corrected using the values of l^* and analysed to give the value of τ. In appropriate cases, the whole correlation function was analysed in detail to give the MSD information.

Time-sweep oscillatory measurements were performed at a frequency of 1 Hz and a stress of 0.02 Pa using a controlled-stress rheometer (AR 1000, TA Instruments, Newcastle, DE, USA) equipped with a Peltier temperature controller and a Couette device consisting of two concentric cylinders of diameters 28 and 30 mm. The GDL was added to the milk which was then transferred to the rheometer. Paraffin oil was put on top of the sample to prevent evaporation. Experiments were carried for 12 h at 30°C, by which time the sample had reached pH = 4.8.

17.4 Results

17.4.1 Unheated Milk

The behaviour of the parameters measured by DWS and rheology as a function of pH during acidification of unheated milk is shown in Figure 1. Several different types of changes are apparent in the parameters derived from the DWS experiment. The casein micelles are the only scattering particles present in the unheated skim milk, and so it is evident that there are two early changes in the properties of the particles. The apparent radius decreases as the pH decreases, to a minimum at a pH of about 5.5, after which it increases again, but relatively slowly, until a point is reached at which there is a rapid increase in apparent particle size, which presumably arises from the formation of

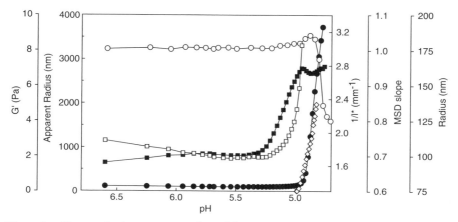

Figure 1 *Changes in the parameters derived from DWS and rheology during acidification of unheated milk with 1.5% GDL. Various quantities are plotted against pH:* ●, *apparent radius from value of τ calculated from correlation function, and converted using the Stokes–Einstein relation (left scale);* □, *apparent radius measured during early part of the reaction (far right scale);* ○, *slope of MSD against time, calculated from correlation functions (middle right scale);* ■, $1/l^*$ *calculated from turbidity (right scale);* ◊, G' *from bulk rheometry (far left scale)*

structures in the milk. It is noticeable that the beginning of the increase in particle size occurs considerably earlier than the 'structure formation' point, obtained by extrapolation to zero of the linear portion of the apparent radius *versus* pH curve. At the same time as the apparent radius of the particles decreases, the value of $1/l^*$ increases slightly, and it comes to a plateau at about the point where the radius has flattened off.

These results can be generally explained by changes in the casein micellar structure and properties. The decrease in apparent radius is assumed to result from the collapse of the hairy layer on the casein micellar surface; this has been described from both theory and experiment.[19,20] The extent of the decrease in radius (12 nm) is, however, somewhat larger than might be expected from other estimates of the thickness of the hairy layer. We believe that the concomitant small increase in the value of $1/l^*$ is a result of the collapse of the hairy layer on to the particle surface. This slightly increases the refractive index of the micellar core and changes the value of $F(q)$. It can be calculated that only a very small change in refractive index is needed to change the value of l^* by the observed amount. Unfortunately, more detailed calculation of the optical properties of the micelles is not possible without knowing some details of the structure of the hairy layer. Removal of the hairy layer by the action of chymosin does not cause a change in the value of $1/l^*$,[21] and so it is evident that the core of the casein micelle does the bulk of the scattering of light, and the hairy layer, being diffuse, does not play a significant part in the particle scattering, although it is important in the hydrodynamic behaviour of the casein micelles.

During the next stage of the reaction as the pH is lowered below 5.3, the apparent radius of the casein micelles increases slowly (compared with the later structure-forming stage). At the same time the value of $1/l^*$ increases considerably. It is possible to interpret the change in apparent radius in two ways: either there is genuine aggregation of particles, but to a low level, or the changes in $1/l^*$ and apparent radius arise from a progressive change in the relative positions of the casein micelles as the inter-particle steric/electrostatic repulsive forces diminish, attractive forces begin to become important, and the motions of the particles and their orientations with respect to one another change (*i.e.*, the $S(q)$ factor in Equation (3) begins to increase). At the end of the increase in $1/l^*$, the apparent radius of the particles is approximately doubled. During this period, the slope of the MSD remains close to unity, indicating that the particles are freely diffusing. It is possible that this stage in the reaction may represent the establishment of equilibrium between aggregated and non-aggregated micelles.[6]

At a pH between 5.0 and 4.9, the apparent radius of the particles begins to increase very rapidly, because the particles become highly aggregated. At this stage the Stokes–Einstein relation (Equation (2)) cannot be regarded as valid; the change in apparent radius simply demonstrates the very rapid change in the aggregation state of the milk. The slope of the MSD at around this time begins to decrease, although it seems that aggregation can proceed for some time before the change in MSD is apparent. However, at pH < 4.9, the slope of the MSD decreases rapidly. These DWS measurements suggest that the milk is forming a gel, and this is confirmed by the rheological measurements. At pH = 4.9 we have the gel point as indicated by the crossover of G' and G''. The elastic modulus increases also almost exactly in synchronization with the change in apparent radius (or more correctly, the average τ) in the system. At the beginning of this gelation process, the value of $1/l^*$ appears to stabilize. If we interpret the change in l^* as being defined by $S(q)$, this must imply that, at the moment of gelation, this function has stopped changing because the particles have reached their final positions in the gel, and that subsequently no change in their relative positions occurs.

Thus, for the unheated milk, we are seeing agreement between the two experimental methods used, showing that a single network is formed. The situation is more complex in milk that has been heated before acidification.

17.4.2 Heated Milk

Observation of the heated milk before acidification shows that the particle size is very similar to that of unheated milk (Figure 2). This has generally been found to be the case in measurements made with milk that has been heated at around its natural pH.[22] However, the initial value of $1/l^*$ is larger than it is in unheated milk. This also has been commonly observed in the experiments that we have described previously. Since the overall concentration of protein in the milk does not change during heating, the increase in $1/l^*$ must arise from changes in the optical properties of the particles, and this may in turn result from the binding of denatured WP to the casein micelles, and the formation of

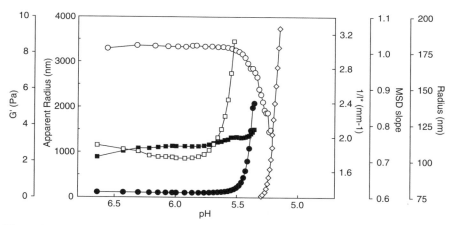

Figure 2 *Changes in the parameters derived from DWS and rheology during acidification of heated milk with 1.5% GDL. Various quantities are plotted against pH: ●, apparent radius from value of τ calculated from correlation function, and converted using the Stokes–Einstein relation (left scale); □, apparent radius measured during early part of the reaction (far right scale); ○, slope of MSD against time, calculated from correlation functions (middle right scale); ■, $1/l^*$ calculated from turbidity (right scale); ◇, G' from bulk rheometry (far left scale)*

soluble WP–κ-cas particles in the milk. The overall effect of this is to increase the scattering power of the milk (because new particles are formed), and this leads to an increase in the value of $1/l^*$. After the GDL is added, we see a decrease in the particle radius with pH that is exactly the same as that seen in unheated milk. This appears to suggest that, although the casein micelles in heated milk have lost quantities of their κ-cas, there is still enough left to form a hairy layer that collapses as the milk is acidified. Similarly, the value of $1/l^*$ increases slightly during this period, as was seen for unheated casein micelles. These changes exactly parallel those seen in unheated milk.

At pH ≈ 5.9, the apparent radius of the particles begins to increase, as does the value of $1/l^*$. However, compared to the behaviour in unheated milk, the increase in $1/l^*$ is very small. From the plateau value at pH ≈ 6.0 to the value at pH ≈ 5.5, where rapid increase in the apparent radius starts in the milk, the increase in $1/l^*$ is ~ 0.1 mm^{-1} compared with 1.0 mm^{-1} in unheated milk between the plateau and the start of the rapid change in radius. In the heated milk, the value of $1/l^*$ is at a plateau at pH ≈ 5.5, although there is an indication of an increase for pH < 5.4. Unfortunately, because of experimental noise, reliable results are difficult to obtain beyond this point. Over this pH range, the apparent radius of the particles increases to about double the original value. This is essentially similar to the behaviour of unheated milk, apart from the much smaller change in $1/l^*$ and the higher pH at which the changes occur. At pH ≈ 5.6 the MSD slope begins to decrease as the particles are inhibited from diffusing freely, and the very rapid increase in particle size begins. It should be noted, however, that, in contrast to the apparently very rapid transition between slow and rapid aggregation that is seen in unheated milk, there is a somewhat longer transition stage in the heated milk. These phenomena are taking place at pH values around 0.5 units higher than those in unheated milk.

A remarkable difference between the unheated and heated milks is that, for the latter systems, the changes in the rheological parameters G' and G'' do not occur at the same pH as the changes in the DWS parameters. The rheological gel point is found to be at pH ≈ 5.3, as has been shown by other studies.[13,23] The G' measured for heated milk is much higher than for unheated milk, and the final gel obtained is about 20 times more elastic than for unheated milk; the data shown in Figure 2 refer to the earliest stages of the gelation only. Whereas in unheated milk the G' value becomes higher than G'', and increases in parallel with the apparent particle radius and the change in MSD slope, in heated milk the light scattering changes are well established before there are detectable changes in the rheological parameters. We have consistently observed in DWS experiments a major increase in the apparent particle radius and changes in the MSD slope at higher pH values than the point where gelation is observed in the rheometer.[24] This seems to present us with the paradox that, at higher pH values, there is extensive aggregation of the particles in the system, and restricted diffusion, even though there is no evidence of gel formation registered by the rheometer.

17.5 Discussion

These two sets of contrasting results raise questions about the interpretation of DWS in terms of defining gelation. Clearly, the 'gelation' that is seen in unheated milk using all of the criteria of DWS agrees with the rheology, by any of the common definitions of a gel. The definition of a gel is difficult to make in practice, because although we know that a gel is a space-filling network, it is not clear what criteria can be used to establish its existence. In terms of rheological measurements, the gel point has been identified in different ways by different authors. The point at which G' and G'' cross (where $\tan \delta = 1$) has been used,[25] or where G' attains a value of 1 Pa,[26] or where the approximately linear plot of G' versus pH extrapolates back to the baseline.[27]

In unheated milk, the variation of the defined parameters from DWS can be used as indicators to define the gel point. Gelation can be imagined to occur in the region where abrupt changes are seen in the MSD slope and rapid increase in apparent radius (or more properly, the average relaxation time τ). In contrast, the behaviour of the heated milk suggests that the mobility of the particles is severely restricted well in advance of the changes in the elastic and loss moduli in rheology. So for this suspension of particles there can be no unique definition of the gel point.

If the behaviour of $1/l^*$, in the region where the apparent radius begins to increase, is dominated by the effects of $S(q)$, then we may take the changes in this parameter (or lack of them) to indicate that the relative positions of the casein micelles do not change very much in heated milk, whereas they change considerably in unheated milk, perhaps as a result of the establishment of an aggregation equilibrium. This may be taken as an indication that the casein micelles in heated milk cannot (or do not) approach one another as closely as they do in unheated milk. This in turn argues that, in heated milk, in the pH region around 5.6, we are not 'seeing' direct interactions between the casein micelles, but rather the formation of a weak network that restricts the motion of the particles. It should be remembered that the concentration of WP in the milk is much less than is necessary to produce a self-supporting gel. The diffusion of the casein micelles is not only slowed (as it would be if there were simple aggregation reactions of freely diffusing particles), but it is also restricted, which is typical of the formation of a network. We may then ask of what species is this network composed: it is clear that the only candidates can be the WP–κ-cas complexes. The casein micelles are not the primary structure forming units at this stage because the lack of change in $1/l^*$ suggests that they do not approach closely. It is arguable that if the WP–κ-cas complexes form links between the casein micelles, by forming chains between the large particles, the diffusion of the micelles could be considerably affected without there being a strong enough gel network formed to be detectable by rheology. It should be remembered that the WP–κ-cas complexes are small and of a low volume fraction relative to the casein micelles, so that their contribution to the overall light scattering is small; that is, we are not directly seeing the behaviour of the complexes themselves, only their effect. In this view, the WP–κ-cas complexes serve at the same time to hold the micelles together and also to force them to remain sufficiently far apart so that there is no direct interaction between the micelles leading to gelation, and so $S(q)$ will change only a little. None of this can happen if the milk is not heat-treated, because in unheated milk there are no intermediary particles to form an initial gel network.

As the pH of the heated milk continues to fall below the point where this proposed network is formed, the charges on the casein micelles decrease to a point where there begins aggregation of the micelles themselves and with it the formation of a more rigid gel network. It is at this stage that the gelation process begins to be detected by the rheometer. This implies that the original weak gel would be kept in its state not simply by the interactions between caseins and WP particles, but possibly by some steric or charge repulsions on the micelles that diminish as pH is lowered. That is, the chains of WP may no longer be able to keep the casein micelles apart, and therefore there occurs more general aggregation, and the formation of a network of casein micelles, aided by the WP complexes on their surfaces. Basically, the original 'gel' is presumably formed between specific sites on the casein micellar surface (where there are bound WP–κ-cas complexes) and the WP–κ-cas complexes in the serum. The subsequently formed gel contains these sites, but in addition it has a general interaction between the surfaces of the casein micelles. That is, the final strong gel is produced by the partial collapse of the first weak one.

References

1. C. Holt and D.S. Horne, *Neth. Milk Dairy J.*, 1996, **50**, 85.
2. D.G. Dalgleish, *J. Dairy Res.*, 1984, **51**, 425.
3. C.G. de Kruif and E.B. Zhulina, *Colloids Surf. A*, 1996, **117**, 151.
4. D.S. Horne, *Biopolymers*, 1984, **23**, 989.
5. C. Holt and D.G. Dalgleish, *J. Colloid Interface Sci.*, 1986, **114**, 513.
6. C.G. de Kruif, *J. Colloid Interface Sci.*, 1997, **185**, 19.
7. J.A. Lucey and H. Singh, *Food Res. Int.*, 1998, **30**, 529.
8. J.A. Lucey, M. Tamehana, H. Singh and P.A. Munro, *J. Dairy Res.*, 2000, **67**, 415.
9. D.S. Horne, *Colloids Surf. A*, 2003, **213**, 255.
10. S.G. Anema and H. Klostermeyer, *J. Agric. Food Chem.*, 1997, **45**, 1108.
11. A.J. Vasbinder and C.G. de Kruif, *Int. Dairy J.*, 2003, **13**, 669.
12. C. Rodriguez and D.G. Dalgleish, *Food Res. Int.*, 2005, **39**, 472.
13. F. Guyomarc'h, C. Queguiner, A.J.R. Law, D.S. Horne and D.G. Dalgleish, *J. Agric. Food Chem.*, 2003, **51**, 7743.
14. S.G. Anema, E.K. Lowe and S.K. Lee, *Lebensm. Wiss. Technol.*, 2004, **37**, 779.
15. A.J. Vasbinder, A.C. Alting and C.G. de Kruif, *Colloids Surf. B.*, 2003, **31**, 115.

16. D.A. Weitz and D.J. Pine, in *Dynamic Light Scattering*, W. Brown (ed), Oxford University Press, Oxford, 1993, 652.
17. M. Alexander, L.F. Rojas-Ochoa, M. Leser and P. Schurtenberger, *J. Colloid Interface Sci.*, 2002, **253**, 34.
18. S. Romer, F. Scheffold and P. Schurtenberger, *Phys. Rev. Lett.*, 2000, **85**, 4980.
19. C. G. de Kruif, *Int. Dairy J.*, 1999, **9**, 183.
20. M. Alexander and D. G. Dalgleish, *Colloids Surf. B*, 2004, **38**, 83.
21. D.G. Dalgleish, M. Alexander and M. Corredig, *Food Hydrocoll.*, 2004, **18**, 747.
22. S.G. Anema and Y. Li, *J. Dairy Res.*, 2003, **70**, 73.
23. J.A. Lucey, *J. Dairy Sci.*, 2002, **85**, 281.
24. M. Alexander and D.G. Dalgleish, *Langmuir*, 2005, **21**, 11380.
25. S.B. Ross-Murphy, *J. Texture Stud.*, 1995, **26**, 391.
26. J.A. Lucey, P.A. Munro and H. Singh, *Int. Dairy J.*, 1999, **9**, 275.
27. D.S. Horne, *Int Dairy J.*, 1999, **9**, 261.

Chapter 18

Computer Simulation of the Pre-Heating, Gelation, and Rheology of Acid Skim Milk Systems

Joost H. J. van Opheusden

MATHEMATICAL AND STATISTICAL METHODS GROUP, BIOMETRIS, WAGENINGEN UNIVERSITY, BORNSESTEEG 47, 6708 PD WAGENINGEN, THE NETHERLANDS

18.1 Introduction

Milk is, and has been for maybe half a billion years, an essential consumption product. Breast-feeding of sucklings is natural behaviour among mammals, and in human culture animal milk is considered a valuable element of the diet for people of all ages. Whether intended for direct consumption or further processing, milk is routinely pasteurized to reduce microbial contamination. Different heat-treatment procedures are used, depending on the final consumption product intended.[1] As well as enhancing the safety of the product, the heat treatment also changes milk's properties, such as the taste. But even more important for the production of secondary milk products, such as cheese, yoghurt, and butter, are the heat-induced microscopic changes.

Milk is a very stable colloidal system of fat globules and proteins dispersed in water. A substantial part of the protein content (~ 3 wt%) is present in the form of casein micelles of size ~ 100 nm; the casein micelles also contain substantial amounts of calcium phosphate. The smaller whey proteins (3 nm) constitute a minor fraction (~ 0.5 wt%). Skim milk is prepared by removing the large fat globules (~ 1 μm) from the suspension.

In this modelling study, we first concentrate on the description of skim milk samples during heat treatment at slightly different levels of acidity. Small differences in acidity of milk occur mainly due to seasonal variation. For different temperatures and different pH values, there are changes in the properties of the casein micelles and the whey proteins, as well as in their mutual interaction. At high temperatures the denaturation of whey proteins leads to destabilization, but the casein micelles themselves are not so much affected by heat treatment.[2] Denatured whey proteins can bind with each other and the casein micelles. Depending on the physical circumstances, different complexes of whey proteins and casein micelles are formed, influencing the properties of the products after further processing.[3,4] We consider here the subsequent gelation of the pre-heated samples, and also the properties of the resulting binary gels, specifically their break-up under shear.

18.2 Simulation Model

We model the colloidal skim milk solution as a bidisperse distribution of soft spherical particles in a homogeneous solvent. The larger particles correspond to the casein micelles, and the smaller ones to the whey proteins. The particles move through the solute by Brownian motion. There are two distinct mutual particle interactions: a medium-range reversible attraction or repulsion and an irreversible short-range bonding interaction. During the shearing stage, we allow this bond to break beyond a certain maximal stretching length, in order to study rupture. But during the gelation stage, the bonds once formed are permanent.

The Brownian dynamics (BD) simulation model is based on the Langevin equation, the dynamical equation of motion for a system of diffusing particles. The total force is the sum of the net force of interaction between particles, the random fluctuating Brownian force, and the hydrodynamic interactions. The solvent is regarded as continuous and the Brownian force mimics thermal collisions between the solvent and the dispersed particles. The total force on particle i is given by

$$F = m\frac{d^2 r_i}{dt^2} = S_i + H_i + \sum_{\text{neighbours}} P_{ij} + \sum_{\text{bonds}} B_{ij}, \qquad (1)$$

where S_i is the stochastic (Brownian) force, H_i the force modelling hydrodynamic interactions, r_i the position of particle i, and t the time. We approximate the hydrodynamic force H_i by simple Stokesian friction, neglecting hydrodynamic interactions between particles. The liquid drag force on a single particle, H_i, is proportional to the particle velocity, i.e.,

$$H_i = -\gamma_\alpha \frac{dr_i}{dt}, \qquad (2)$$

where $\gamma_\alpha = 6\pi\eta R_\alpha$ is the Stokes drag, η the solvent shear viscosity, and R_α the particle radius with the index α referring to the large (l) or small (s) particles. The physical interactions P_{ij} between the neighbouring particles are a constant repulsive or attractive force when two particles are within a cut-off distance, but not overlapping. When the soft elastic particles overlap there is a Hookean restoring force between small and/or large particles. The chemical or binding interaction is governed by a binding probability, a binding distance, and a binding strength. There can be only one direct bond between any two particles. Once a bond is formed it produces a force

$$B_{ij} = k_{\alpha\beta}(b_j - b_i), \qquad (3)$$

with b_i being the point at the particle surface to which the Hookean spring of strength $k_{\alpha\beta}$ is attached. Due to their non-central nature the bonds also induce particle rotation. If a bond is stretched during deformation of the sample beyond a maximal length, it is removed; the energy stored in the stretched bond is immediately dissipated.

The size of the cubic or rectangular simulation box determines the volume fraction of the particles. Periodic boundary conditions are used to avoid edge effects. All parameters corresponding to sizes or distances are normalized to the radius of the small particles ($R_s = 1$), and all quantities corresponding to energies are normalized to units of $k_B T$ ($k_B T = 1$).

By choosing the time step much larger than the relaxation time of the particles, we can neglect the second-order term in Equation (1). So we have

$$\frac{dr_i}{dt} = -\frac{1}{6\pi\eta R_\alpha}\left[S_i + \sum_{\text{neighbours}} P_{ij} + \sum_{\text{bonds}} B_{ij}\right]. \quad (4)$$

The differential equations are solved numerically using an Euler forward method:

$$r_i(t+\Delta t) = r_i(t) + \frac{\Delta t}{6\pi\eta R_\alpha}\left[S_i(t) + \sum_{\text{neighbours}} P_{ij}(t) + \sum_{\text{bonds}} B_{ij}(t)\right]. \quad (5)$$

The stochastic force S_i gives the random translational displacements that, on average, obey Einstein's law for an isolated particle. The dimensionless root-mean-square displacement in the absence of interactions is $(2D_T \Delta t)^{1/2}$ in each direction. The translational diffusion coefficient, $D_T = 1/6\pi\eta R_s$ (taking $k_B T = 1$), for the small particles is normalized to unity. The large particles ($R_l = 8$) move substantially slower than the small ones. As well as the translational motion, each individual particle also undergoes rotational diffusion. The rotational motion is governed by the diffusion coefficient D_R:

$$D_R = \frac{3D_T}{4R_\alpha^2}. \quad (6)$$

The implementation of rotational diffusion in the model is similar to that of translational diffusion. Rotational diffusion of clusters results from the combined translations and rotations of individual particles. Rheological properties are studied by affine shear deformation of the image box at a constant low shear-rate and applying the Lees–Edwards boundary conditions.

18.3 The Pre-Heating Phase

18.3.1 Methodology

During the heating of milk, the denaturated whey proteins can bind to each other and to casein micelles. As the casein micelles themselves do not bind directly to each other, the solution still remains as a stable suspension, at least over the time period of the process encountered in practice. For the model this simplifies matters considerably. Because we do not have complexes involving several casein micelles, we can restrict the simulated system to a single casein micelle. Moreover, as the rate at which the large casein micelles diffuse is substantially less than that of the small whey proteins, we can reasonably ignore their motion altogether.

In the first set of pre-heating simulations, there is one single large particle of radius $R_l = 8$ fixed at the origin, surrounded by $N_s = 200$ small particles of size $R_s = 1$ in a cubic box of dimensions $30 \times 30 \times 30$ units. This corresponds to volume fractions of 7.9% and 3.1% for the casein and whey protein components, respectively. This relative size difference is quite a bit less than that which occurs in reality, and also the relative volume fraction is different. The amount of whey protein in real milk is only sufficient to cover about 40% of the surface,[5] and so by using these values the same applies in the simulated system. A realistic size ratio of $R_l/R_s \approx 30$ would result in about 60 times as many of the small particles, and the consequent increase in computational effort in the later gel stage would make that simulation unfeasible.

It was not the purpose of this study to give quantitative predictions of the properties of the real systems. Rather the aim was to provide some qualitative trends in the possible behaviour of a binary system with a rather large size ratio, not only in relation to the complexation, but also for the subsequent gelation and large deformation rheology. A quantitative study of a model imitating all relevant aspects of the real system would require an extremely long simulation. But, in any case, there are not sufficient experimental observations available to validate or calibrate the parameters within such a refined model. Hence, we have chosen the convenient set of parameter values mentioned above.

The effect of pH was translated into different values of the interaction parameter for the physical interaction strength between the small particles and the surface of the large particle (A_{sl}) and the bonding probability for binding of the whey proteins to the casein micelle (p_{sl}). The purpose of this preliminary numerical model was to reproduce the salient features of the real system during heat treatment by changing as few parameter values as possible. Hence, the bonding distance (0.3), the small–small bonding probability ($p_{ss} = 0.01$), the small–small physical interaction strength and distance ($A_{ss} = -5$ and $C_{ss} = 1$, respectively, with $5k_BT$ repulsion), the small–large interaction distance ($C_{sl} = 1$), and the spring constants ($k_{\alpha\beta} = 250$) were all kept unchanged. The time-step used was $\Delta t = 0.00025$. In reality the pH values during the heat treatment process vary only over a relatively small range, from 6.35 to 6.9. Samples are typically heated to 80°C over a period of 10 min. The temperature plays no explicit role in our model, other than it being responsible for controlling the rate of denaturation. This particular set of parameter values was chosen so that small whey aggregates would form, because of the irreversible binding, but the rate would be slowed down due to the $5k_BT$ barrier that has to be overcome before binding can take place. All simulations were run over the same time period (50,000 steps).

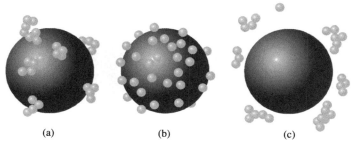

Figure 1 *Illustration of different kinds of complexes of casein micelles and whey proteins formed by heat treatment at different pH values: (a) patchy coating with small clusters at low pH (6.35), (b) homogeneous coating at intermediate pH (6.55), and (c) small free clusters in solution at high pH (6.9)*

With these calculations we can try to compare with the experimental results of Vasbinder et al.,[6] who hypothesized on the basis of both acidic and renneted gelation that (i) at low pH (6.35) the whey proteins produce a patchy coating of the casein micelles, (ii) at intermediate pH (6.55) a homogeneous coating is formed, and (iii) at high pH (6.9) no coating is formed. These three cases are illustrated in Figure 1.

18.3.2 Results and Discussion

To keep track of the aggregation dynamics we monitor four descriptive parameters: the fraction of free monomers, the average cluster size, the fraction of small particles having a bond with the large particle, and the fraction of small particles in clusters bound to the large. At the beginning of the heat treatment procedure some monomers ($\sim 20\%$) happen to be within the binding range, and so they immediately form dimers. After this rapid initial drop in free monomers (Figure 2), there is a slow but steady decrease of free monomers. At the end still $\sim 30\%$ are free in solution.

Four curves are shown in Figure 2. The low pH curve corresponds to parameter values $A_{sl} = 25$ and $p_{sl} = 10^{-5}$. Small particles are trapped within a potential well of $25k_BT$ when they come close to the surface, but only have a small probability of actually forming a bond with that surface. The medium pH curve has parameter values $A_{sl} = 25$ and $p_{sl} = 10^{-3}$, and the high pH system has $A_{sl} = -5$ and $p_{sl} = 0.1$. Also a reference sample is considered with $A_{sl} = 0$ and $p_{sl} = 0$, i.e., with only the repulsive core interaction. These are just the data of a single set of simulation runs, but they are typical with respect to the slow decrease of the monomer fraction. There is little significant discrimination between the pH curves, the differences being within error of reproducibility. Note that the pH level indicated is actually just a name given to the system conditions. It remains still to be shown that for the given parameter values the expected behaviour is obtained.

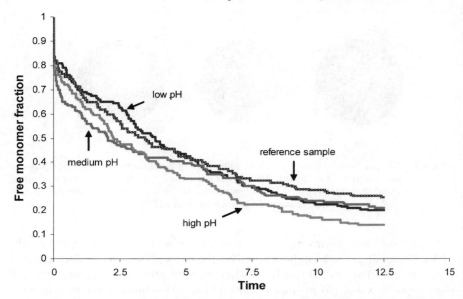

Figure 2 *Fraction of free monomers as a function of time during the heating process. Due to sample preparation some dimers are very rapidly formed. The irreversible binding process results in a slow but steady decrease in number of the repulsive monomers. Curves for different parameter values are identical within reproducibility*

Over the duration of the simulated heat-treatment procedure, the denaturated whey proteins by themselves form small (free) clusters, as can be seen from the reference sample in Figure 3. Mainly dimers are formed, and a few larger clusters of five or six particles. The average cluster size in the high pH sample develops similarly, as can be expected. The repulsive surface prevents bonds from occurring, and the whey proteins behave as if the casein micelle is not interacting. At medium pH they do come close to the casein surface, and when they do they have a high probability of binding. Once they bind, they are immobilized and cannot form further bonds with other whey proteins at the surface. This can explain the lower average cluster size found. At low pH the whey proteins get trapped at the casein surface, but do not bind to it. Instead they diffuse over the surface, where they have a higher probability of meeting with other whey proteins than in solution. Hence larger clusters form, mainly at the surface. At the surface the diffusion process is more efficient in bringing monomers into proximity than in solution; this is an example of a surface-enhanced reaction.

Figure 4 shows the number of bonds as a function of time. At medium pH a sizeable fraction of the whey protein binds to the casein micelle surface. If also particles not bound directly are taken into account (Figure 5), the fraction associated with the casein micelle is even slightly larger, but most particles bind as monomers. The coverage at high pH values is somewhat lower than that at medium pH, but the difference is quantitative rather than qualitative. Many

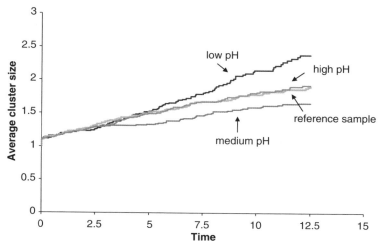

Figure 3 *Average cluster size of the whey proteins during the heating process under different conditions. In all cases mainly dimers are formed, and some larger clusters. The high pH system is similar to the reference one; at medium pH the clusters are somewhat smaller; at low pH they are somewhat larger*

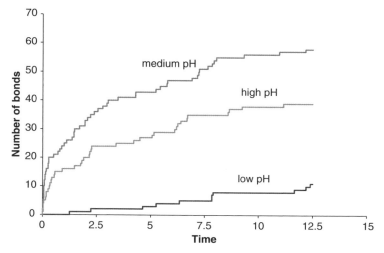

Figure 4 *Number of bonds formed between the small particles and the surface of the large particle. In the medium pH and high pH systems, a coating is formed. The surface area covered is significantly larger only in the medium pH system*

whey particles manage to overcome the relatively low potential barrier, and stay long enough in proximity of the casein micelle to form a bond. Note that the binding probability, once they have crossed the barrier, is quite high. Moreover, the binding possibility is checked at each time step rather than for each collision event. In diffusive motion the concept of a single collision is a

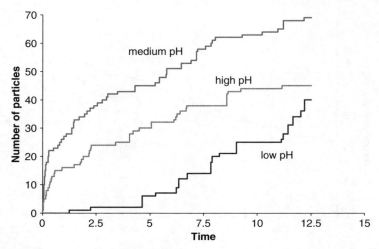

Figure 5 *Number of small particles bound in clusters to the large particle surface. Comparison with Figure 4 shows that, at high pH and medium pH, most particles bind as monomers. There are few bonds at low pH, but many particles are part of clusters bound to the surface with one or two links*

somewhat vague notion, whereas in the simulation it can easily be checked whether particles are within the binding distance. At medium pH the binding probability is two orders of magnitude smaller, while still a higher surface coverage is observed. The low pH system shows qualitatively different behaviour. The rate at which bonds are formed is very much lower than that for either of the other cases, but the rate at which particles collect near the surface is more or less the same, as governed by the diffusion towards the casein micelle surface. Once they are trapped – now because of the deep potential well rather than the binding – a depletion zone develops and the rate slows down.[7] After a while the particles trapped near the surface start forming clusters, which show up in the low pH curve of Figure 5. Eventually for this system the surface coverage obtained is similar, but the small particles are mainly in clusters of average size ~ 4. The coverage is therefore much less homogeneous. In all cases the major fraction of the whey protein is not part of this coating layer, but is free in solution, mainly in the form of dimers. Figure 6 shows a snapshot of the simulated low pH system at the end of the heat-treatment phase.

We have performed some simulations for sets of systems with parameter values in the ranges $-5 \leq A_{sl} \leq 25$ and $0 \leq p_{sl} \leq 1$, and we have determined what type of complex was established at the end of the heat-treatment process. The results are collected in Table 1. For a binding probability of zero, obviously no whey proteins can bind to the casein micelle surface. An almost bare surface, with free whey protein clusters in solution, is also found for a slightly repulsive or neutral surface and not too high binding probability. For high binding probabilities many small particles are bound to the surface, most of them directly, forming a more-or-less homogeneous coating of the large particle. The

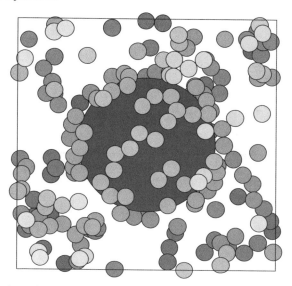

Figure 6 *Snapshot of a system configuration at the end of the simulated heat treatment. About 20% of the whey proteins form a patchy covering of the casein micelle surface. Light small particles are towards the front of the box, and dark ones towards the back*

actual surface coverage depends on the parameters, but even for the most extreme case considered only 25% is covered, with 65% of the small particles forming free clusters or monomers in solution. When a very low binding probability is combined with a moderate or high surface attraction, small particles form aggregates which are bound to the surface by only a few bonds. They form patches at the micelle surface; and if they would not be held close to that surface by the reversible attraction, only a very small fraction of the surface would be covered. We conclude that, by changing two different parameters describing the complexation process, corresponding to the physical and chemical binding (in an opposite way with pH), it is possible to have the same type of complexing as hypothesized on the basis of experimental observations.[6] Table 1 shows that indeed both the parameters are needed, but that just two can suffice. We note that a small change in pH seems to be related to a large change in the parameter values, a situation which is not unexpected. In reality, of course, many different kinds of interactions within the system will change with acidity to varying extents.

In the full simulations discussed below, there are 27 large particles of radius $R_l = 8$ and 5400 small particles of size $R_s = 1$, the same size ratio as for the pre-processing. Rather than just making 27 identical copies of the small system, a single large system is initialized. The only difference is that the large particles may diffuse, but in practice they move only a little. The system is contained in a cubic box of size $90 \times 90 \times 90$ units with periodic boundary conditions. This corresponds to volume fractions of 7.9% and 3.1% for the casein and whey

Table 1 *Characterization of the casein micelle coverage as a function of the interaction parameters: F ≡ free, C ≡ coating, P ≡ patches. Free (F) implies there is no whey protein particle (or very few) bound to the surface. This occurs for a repulsive surface (A < 0), or one of low binding probability (p ≪ 1). For a high binding probability, and a not too repulsive surface, a more-or-less homogeneous coating (C) is formed. A special situation arises when a low binding probability is combined with high attraction, for which the whey proteins form a patchy cover (P) of the casein micelle*

	Parameter p_{sl}						
A_{sl}	0	10^{-5}	10^{-4}	10^{-3}	10^{-2}	10^{-1}	1
−5	F	F	F	F	F	F	C
0	F	F	C	C	C	C	C
5	F	P	C	C	C	C	C
10	F	P	C	C	C	C	C
25	F	P	C	C	C	C	C

protein components, respectively. For these large-scale simulations we use slightly different parameter values from those indicated above. High pH corresponds to $A_{sl} = -5$ and $p_{sl} = 10^{-3}$ (the lower binding probability gives less coating); medium pH corresponds to $A_{sl} = 5$ and $p_{sl} = 10^{-4}$; and low pH corresponds to $A_{sl} = 25$ and $p_{sl} = 10^{-5}$, the same as above. In Table 1 these parameter values can be seen to lead to free micelles, coated micelles, and patchy coatings, respectively. The other parameters are the same. In particular, the large particles have only a repulsive Hookean overlap interaction.

The difference between the three sets of conditions in the pre-heating phase is most obvious if we consider the number of bonds between the large particles and the small ones (Figure 7). As a pair of particles can only form a single bond, the plotted number is also the number of whey protein particles directly bound to the casein surface. For the high pH system the number at the end of the simulation is of the order of 100, implying that only about 2% of the whey proteins are bound to the casein. In the medium pH system the number is substantially larger, but still only ∼20% are actually bound to the casein micelles, and so the average surface area covered is ∼15%. The patchy coating at low pH has only 7% of the whey proteins actually bound.

18.4 The Gelation Phase

The second stage is the gelation phase. In the heat-treatment phase, we chose to distinguish three sets of values of the interaction parameters, corresponding to high, medium, and low pH. For comparison, we added a fourth case in which the heat-treatment phase is omitted, *i.e.*, with no complexation, but with reactive whey proteins. During the gelation phase, the interaction parameters are assumed to be the same for all four cases. Other differences from the

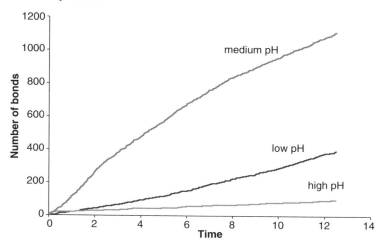

Figure 7 *Number of whey proteins bound to casein micelles as a function of time during the heating process at different pH values. Of the 5400 small particles, only 2% get bound at high pH, 20% at medium pH, and 7% at low pH*

pre-heating phase are in unit binding probabilities for all the particles ($p_{sl} = p_{ss} = p_{ll} = 1$) and all the physical interactions being set to zero ($A_{sl} = A_{ss} = A_{ll} = 0$).

When irreversible bonds are formed readily, the clusters start growing. Because of the high value of the bonding probability, the aggregation process is largely diffusion dominated, leading to very open structures, with low fractal dimensionalities. In the case of a binary gel, however, such a structural parameter is of much less interest than in a monodisperse system. Locally the structure of the whey protein gel strands may be of importance, since the gel in all cases is formed by such strands forming connections between the casein micelles. In that sense we have a hybrid gel, with the whey proteins forming percolating clusters in the gaps between the large casein micelles, which act essentially as spacer particles. In terms of the mass, the casein micelles dominate the system, and hence also the mass autocorrelation function, which is the key ingredient from which the fractal dimensionality is derived.

The monitored structural parameters include the number of bonds between the particles, the number of particles in the largest cluster, and the average number of particles per cluster. The number of bonds between the small and large particles is plotted in Figure 8. There appears to be a distinct difference between the systems at low and medium pH on the one hand, and the high pH and untreated systems on the other. For the first two the average number of whey proteins directly bound to a casein micelle is around 60, whereas for the last two it is around 40; these values represent surface coverages of ~20% and ~12%, respectively.

The situation for the number of bonds involving only whey proteins is just the opposite: the average number of bonds per whey protein shows the same quantitative dynamics for all four systems. Whey proteins bound to micelles are immobilized and do not form bonds with other particles unless they are

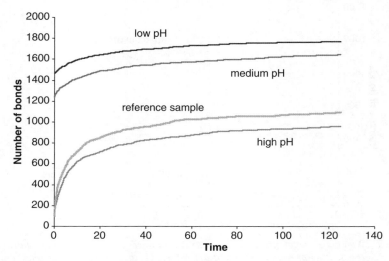

Figure 8 *Number of whey proteins bound to casein micelles as a function of time during the gelation process for pre-heated samples at different pH. The low pH system shows a rapid binding in the first few time steps. Eventually in both low pH and medium pH systems about 30% of the whey particles are directly bound to the casein micelles, as against 15% in the high pH system and the untreated reference sample*

approached by another whey protein. Though initially the high pH and untreated systems have very few bonds, they rapidly catch up, and even slightly overtake the low and medium pH systems. This is due to the effect of the whey proteins attached to the surface of the casein micelles, which can form fewer bonds with other whey proteins. The differences in the end are, however, very small. We note that a bond between two particles is only counted once, so that a value of 2.5 in Figure 9 implies that, on average, a whey particle has five bonds with neighbouring particles. The networks in all the systems are heavily branched at the particle level, but they still form open structures, as could be concluded from the fractal dimensionalities. In Figure 10 we have plotted the average number of neighbouring particles within a sphere of radius r of a central particle for the whey proteins only. All the systems at the end of the gelation phase as calculated show a linear regime on a double logarithmic scale with a slope of about 1.7, the fractal dimension D. On extrapolating the linear regime between $r = 4$ and $r = 10$ down to $r = 2$, we find an effective coordination number of *ca.* 5, in accordance with the number of bonds (Figure 9). Extrapolating the fractal regime upwards leads to a crossover to the homogeneous regime at a correlation length of $r \approx 12$. That is rather larger than the size assumed for the casein micelles in this model, and considerably less than the average distance (~ 30) between them at the given volume fraction.

The other two parameters indicative of the gelation behaviour of the system are the largest and the average cluster sizes. These are plotted in Figures 11

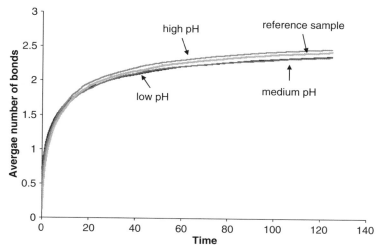

Figure 9 *Average number of bonds per whey protein particle during the gelation phase. All systems show more or less the same behaviour. The high pH and untreated systems even have slightly more bonds, as less are bonded to the large particles. The final structure has about 2.5 pair bonds per particle, implying that each whey particle is bound to about 5 neighbours*

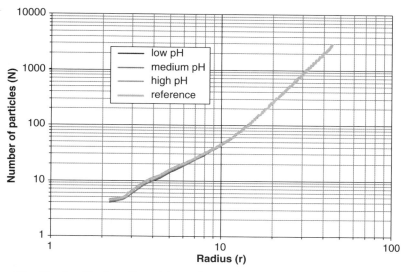

Figure 10 *Number N of small particles within a sphere of radius r around a central small particle as a function of that radius. The fractal dimension D is given by the scaling relation $N \sim r^D$. A value of about $D = 1.7$ is found for all systems. The number of neighbours in the first shell at $r = 2$ is around 5, in accordance with the number of bonds (from Figure 9.)*

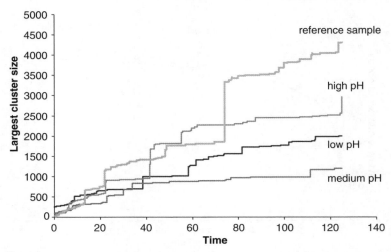

Figure 11 *Number of particles (small and large) in the largest cluster of the gelling sample as a function of time in the simulations. Only the untreated sample actually shows the characteristic S-shaped curve of the cluster–cluster aggregation process. The low pH and (especially) the medium pH sample may not have fully formed a multiply percolating network in the image box*

and 12, respectively. Monodisperse samples exhibiting diffusion-limited aggregation have an S-shaped curve that gives a clear indication of the transition between the initial stage, in which cluster–cluster aggregation is the dominant phenomenon, and the final stage, in which smaller clusters attach to the gel backbone. We can identify the gelation time with the symmetry point of the curve, when about 50% of the particles are part of the largest cluster. In reality the percolation point will be reached slightly earlier, but that time is more difficult to assess, and for comparison between the different systems it probably does not matter very much which is taken. In order to have a sample with gel-like behaviour, a single percolating strand does not suffice. Actually, Figure 11 shows that only the untreated sample gives more or less that distinctive shape, indicating that the pre-heated simulations may not actually have formed a percolating cluster. The largest cluster in the high pH system does contain over 50% of the particles. Upon visual inspection of the configurations in the simulations with the largest cluster drawn in a different colour, it could be observed to stretch across the image box – as do the largest clusters in low and medium pH systems, although they do not actually form a continuous connection across the box.

The average cluster size in the percolating samples is higher than in the non-percolating ones, as can be expected; but the difference is much less clear than for the largest cluster. Most of the particles in all the simulated systems studied here are part of rather large clusters. Figure 12 shows that the average cluster size for the low pH system is somewhat lower than that for the medium pH system, even though the largest cluster is larger (see Figure 11). This apparent

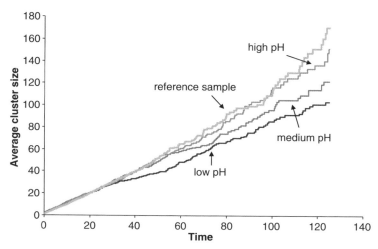

Figure 12 *Average number of particles (small and large) per cluster during the gelation phase as a function of time. All systems consist of rather large clusters, even if the largest one is not percolating. The low pH system has smaller clusters than the medium pH one, even though the largest cluster is larger. These differences are probably within the statistical scatter of the simulations*

contradiction relates to the fact that the size of the largest cluster cannot be expected to be a very reproducible sample parameter, even though the overall shape of the curves in Figure 11 is indicative of a gel transition.

18.5 The Deformation Phase

All interaction parameters were set to zero in the deformation phase, apart from the bond and overlap potentials. Bonds could be broken when stretched to beyond a length of 0.3. It was decided to diminish the Brownian motion of the particles by lowering the diffusion coefficients by a factor of 10. In practice, particle gels of this type are mainly enthalpic in character, in contrast to polymer gels that have a large entropic character. Lowering the temperature greatly reduces the stress fluctuations in the system and avoids bonds being broken by the fluctuations. Again the same values are used for all the four systems. Parameters are kept constant during each phase; and between the gelation phase and the deformation phase, a short equilibration period (1000 steps) is included during which no bonds are formed and all interactions but the bonds and overlap potentials are set to zero. The samples are strained at a low constant rate, allowing individual bonds to relax relatively rapidly. Relaxation of all bonds within the larger clusters is a much slower process, and so we may expect to see stresses build up in all samples. If the stresses become large, the clusters will break up. In the percolating strands the stresses can never relax without the breaking of bonds.

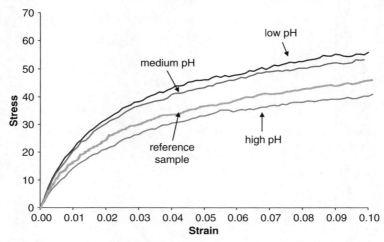

Figure 13 *Small-deformation stress–strain behaviour of the simulated systems under slow steady shearing deformation. The low pH and medium pH systems have a somewhat higher modulus than the high pH system and the untreated sample. The linear regime is very short, with weakening apparent above a strain of 0.005*

Figure 13 indicates that small-deformation behaviour of all the systems is more-or-less the same. The linear deformation regime is very short, and at a strain of 0.005 all the systems show weakening. The explanation for the weakening is that, even at the low temperatures and following the equilibration procedure, some bonds already break at the very small overall deformation, and the bonds keep breaking due to the deformation, be it rather slowly. The low and medium pH systems appear to have a somewhat higher modulus than the high pH and reference systems, and they remain stiffer in the non-linear regime.

The stress reaches a maximum at a strain of about 0.25 (Figure 14). Whether rupture indeed occurs and actual yielding takes place is not observed directly. Only two of the systems are percolating, and so in principle could show a yielding transition. No qualitative difference between the samples is observed; the quantitative differences are small but significant. It appears that the number of whey particles bonded to the casein micelles correlates positively with the observed stiffness of the samples. There is no clear immediate explanation for that. Maybe the affine deformation we use results in added stresses in the coating of the casein micelles, which does not relax as quickly as that in clusters of whey proteins because of the large size of the micelle. If so, that would be a simulation artefact. In a real system the stress generated by external deformation forces is transmitted to the sample through either the particle network or the interstitial fluid. Neither would be expected to result in added stresses at the surfaces of the casein micelles. In our earlier studies of large-deformation shear rheology,[8] we investigated the difference between methods with affine deformation, which is mainly a simulation technique, and deformation through stresses in the network, which is what is expected in these type of particle gels at low shear-rate. There were no large differences found for rates similar to the

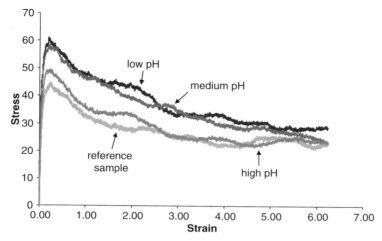

Figure 14 *Large-deformation stress–strain behaviour of the samples under slow steady shearing deformation. At a strain of ∼0.25 all the systems start to show rupture of the structure, whether it is percolating or not. Also large clusters do not relax, but break up into smaller ones. At large strains the stresses are similar for all the systems*

present one, but maybe the previous conclusions cannot be extended to these bidisperse systems. For all the simulations considered here, at the end only ∼20% of all bonds have been broken, with the small particles having around four bonds with neighbours, and still rather large clusters remaining.

18.6 Comparison with Experiment

Heat treatment is one of the main processing factors influencing the texture of acid-induced milk gels,[9,10] and it leads to much stiffer gels than without the treatment.[11] We cannot observe this distinction directly with the present simulation. This is because the reference sample we use is not a system in which the aggregation of casein micelles is compared with that of a whey protein + casein mixture, but rather a system in which no complexation whatever takes place between the whey proteins and the casein. So, in our model, the whey proteins are still denaturated, and a whey protein gel will form.

The denaturation of the β-lactoglobulin during pre-heating is the most important aspect of the pre-heating influencing the gel formation.[5,12] Artificial mixtures of casein + β-lactoglobulin behave like skim milk.[13] The amount of whey protein forming complexes with casein during pre-heating decreases with increasing pH during the pre-heating. Heating above pH = 6.6 leads to partial coverage. Below pH = 6.6 all the denaturated whey protein is attached to the casein. At low pH the coating is inhomogeneous.[6,14] We have been able to reproduce this generic effect in our study by changing only two different parameters describing the interaction between the whey protein and the casein.

The amount of free whey protein in our simulation, however, is quite high: real samples typically have 30% free aggregates and 70% coating.[15] To obtain such figures, longer complexation runs would be needed. More experimental measurements could directly investigate whether indeed the interaction between the whey proteins and the casein is consistent with our parameter choice.

Higher pH leads to a shorter gelation time, a higher gelation pH, and stiffer gels.[14] The shorter gelation time can be explained by the denaturation. Even if the probability of complexation is low compared with the binding probability of the caseins, the rapid diffusion will lead to faster gelation. The denatured β-lactoglobulin in solution is mainly responsible for this.[14] We have found the same effect for all the simulated systems, as they all had about the same amount of whey protein in solution. The complexation appears less relevant, as denatured β-lactoglobulin added to casein has the same effect as pre-heating.[15,16] That effect was also found in our reference sample. Disulfide bonds, formed at ambient temperatures and under acidic conditions, play an important role during gel formation.[17] For the rheology the complexation did not appear to have much of an effect in this simulation study, the effects on the gelation kinetics being stronger. Systems corresponding to low pH were found to gel more rapidly. The gelation kinetics for these bidisperse systems is quite different from cluster aggregation in monodisperse models[7] due to the slow diffusion rate of the casein micelles, which act as spacers in the final gel structure.

Acid gelation of pre-heated milk has been suggested[18] to be a two-step process: first the whey proteins form a gel, and later maybe the caseins coagulate. From our results this explanation seems unlikely, as the coated casein micelles in the simulation are trapped in the whey protein network, in agreement with observations from confocal scanning laser microscopy.[15] Only substantial rearrangement of that network would result in casein micelle gelation. Because of the timescales involved such rearrangements cannot be investigated using the present model, though rearrangements at the level of the whey proteins are included. We have used a similar model[19,20] with flexible bonds in earlier studies of rennet-induced casein gels, containing only the large particles. Simplified models would have to be developed in order to study ageing in bidisperse models with a large size difference.

Acknowledgement

The work was financially sponsored by Unilever Research and Development, Vlaardingen, the Netherlands.

References

1. P. Walstra and R. Jenness, *Dairy Chemistry and Physics*, Wiley, New York, 1984.
2. A.J.R. Law, D.S. Horne, J.M. Banks and J. Leaver, *Milchwissenschaft*, 1984, **49**, 125.

3. M. Corredig and D.G. Dalgleish, *Food. Res. Int.*, 1996, **55**, 529.
4. D.J. Oldfield, H. Singh, M.W. Taylor and K.N. Pearce, *Int. Dairy J.*, 2000, **10**, 509.
5. A.J. Vasbinder, P.J.J.M. van Mill, A. Bot and C.G. de Kruif, *Colloids Surf. B.*, 2001, **21**, 245.
6. A.J. Vasbinder and C.G. de Kruif, *Int. Dairy J.*, 2003, **13**, 669.
7. A.A. Rzepiela, J.H.J. van Opheusden and T. van Vliet, *J. Colloid Interface Sci.*, 2001, **244**, 43.
8. A.A. Rzepiela, J.H.J. van Opheusden and T. van Vliet, *J. Rheol.*, 2004, **48**, 863.
9. T. van Vliet, C.M.M. Lakemond and R.W. Visschers, *Curr. Opin. Colloid Interface Sci.*, 2004, **9**, 298.
10. R.B. Pereira, H. Singh, P.A. Munro and M.S. Luckman, *Int. Dairy J.*, 2003, **13**, 655.
11. J.A. Lucey, *J. Dairy Sci.*, 2002, **85**, 281.
12. J.F. Graveland-Bikker and S.G. Anema, *Int. Dairy J.*, 2003, **13**, 401.
13. M.-H. Famelart, J. Tomazewski, M. Plot and S. Pezennec, *Int. Dairy J.*, 2003, **13**, 123.
14. S.G. Anema, S.K. Lee, E.K. Lowe and H. Klostermeyer, *J. Agric. Food Chem.*, 2004, **52**, 1640.
15. A.J. Vasbinder, F. van de Velde and C.G. de Kruif, *J. Dairy Sci.*, 2003, **87**, 1167.
16. C. Schorsch, D.K. Wilkins, M.G. Jones and I.T. Norton, *J. Dairy Res.*, 2001, **68**, 471.
17. A.J. Vasbinder, A.C. Alting, R.W. Visschers and C.G. de Kruif, *Int. Dairy J.*, 2003, **13**, 29.
18. D.G. Dalgleish, M. Alexander and M. Corredig, *Food Hydrocoll.*, 2004, **18**, 747.
19. M. Mellema, J.W.M. Heesakkers, J.H.J. van Opheusden and T. van Vliet, *Langmuir*, 2000, **16**, 6847.
20. M. Mellema, P. Walstra, J.H.J. van Opheusden and T. van Vliet, *Adv. Colloid Interface Sci.*, 2002, **98**, 25.

Chapter 19

Xanthan Gum in Skim Milk: Designing Structure into Acid Milk Gels

Pierre-Anton Aichinger,[1] Marie-Lise Dillmann,[1] Sabrina Rami-Shojaei,[1] Alistair Paterson,[2] Martin Michel[1] and David S. Horne[3]

[1] NESTLÉ RESEARCH CENTER, VERS-CHEZ-LES-BLANC, CH-1000 LAUSANNE 26, SWITZERLAND
[2] DEPARTMENT OF BIOSCIENCE, UNIVERSITY OF STRATHCLYDE, 204 GEORGE STREET, GLASGOW G1 1XW, SCOTLAND, UK
[3] CHARIS FOOD RESEARCH, HANNAH RESEARCH PARK, AYR KA6 5HL, SCOTLAND, UK

19.1 Introduction

Polysaccharides such as xanthan gum (XG) are widely used to enhance and maintain textural and visual properties in food products. However, the underlying mechanisms are not sufficiently well understood. The interactions of polysaccharides with other food ingredients are often ignored, although playing an important role in structure formation during food processing. In dairy products, various phenomena may occur due to the milk protein interacting with the added polysaccharide.[1] It has been demonstrated[2,3] that the addition of XG as 'thickener/stabilizer' to skim milk results in phase separation at the natural pH of the mixtures. Furthermore, it has been shown[4] that XG incorporation has a great impact in drastically modifying the microstructure of acid-induced gels made from pre-heated skim milk; this was also attributed[4] to phase separation phenomena. These combined observations point to the importance of the interactions of XG and milk protein in contributing to the microstructure of materials made from their mixtures.

The microstructure of a food material determines its physical properties, and is therefore crucial to product performance and acceptance by the consumer. Hence, a profound understanding of structure formation is a prerequisite for the improvement of existing food products and the development of novel food materials with desirable properties. In the framework of the present study we have investigated the impact of XG on acid-induced gel formation in skim milk, acidified through addition of the acidulant glucono-δ-lactone (GDL). Our aim is to understand the underlying mechanisms and to demonstrate how the microstructure of a food material can be designed by controlling the colloidal interactions between the casein particles in this model skim milk system.

19.2 Materials and Methods

19.2.1 Solution Preparation

The low-heat skim milk powder (SMP) (Caledonian Cheese Company, UK) contained 35.6 wt.% protein and had a whey protein nitrogen index of 8 mg of (undenatured) whey protein nitrogen per gram. Therefore, with respect to whey protein denaturation, the reconstituted unheated skim milk could be regarded as a satisfactory model for the behaviour of raw skim milk. The number-average radius R_n, the weight-average radius R_w and the molecular weights, M_n and M_w, of the xanthan sample (Keltrol SF/NF; CPKelco, UK) were: $R_n = 208$ nm, $R_w = 238$ nm, $M_n = 4.65 \times 10^6$ g mol^{-1} and $M_w = 5.77 \times 10^6$ g mol^{-1}. These values were determined by high-performance size-exclusion chromatography (SEC) coupled with multi-angle laser light scattering (MALLS) and differential refractometry.

Dispersions of low-heat SMP and XG were separately prepared in distilled water under magnetic stirring at 50°C (SMP) or 80°C (XG). Sodium azide was added to protect against microbial growth in the samples. After storage overnight in a refrigerator (~ 5°C), distilled water was added to adjust the final concentration of the dispersions (17.5 wt.% SMP, 0.125 wt.% XG, 0.1 wt.% XG and dilutions thereof). For experiments performed with pre-heated skim milk, the SMP dispersions (pH 6.54) were heated to 90°C (within 4 min), held at 90°C for 10 min, and then rapidly cooled to room temperature in a batch-wise process in a double-walled reactor under constant stirring. The SMP and XG dispersions were separately equilibrated in a water-bath at 40°C for at least 1 h, and then blended to obtain mixtures containing 14 wt.% SMP (5 wt.% milk protein) + 0.005–0.025 wt.% XG. Blends of SMP dispersions with distilled water served as reference (14 wt.% SMP + 0 wt.% XG). The subsequent experiments were all carried out at 40°C.

19.2.2 Phase Separation

The blends were produced in culture tubes (Pyrex, Bibby Sterilin, UK), kept in a water-bath for at least 10 min, and then mixed for 1 min with a Vortex Genie 2 mixer (Scientific Industries, USA). The mixtures were then either loaded into the temperature-controlled sample cell of the microscope or transferred to an incubator. Phase separation was observed at the natural pH of the mixtures.

Initial microscopic phase separation was investigated by confocal laser scanning microscopy (CLSM) with an upright Zeiss Axioplan 2 microscope equipped with the LSM 510 confocal unit (Carl Zeiss, Germany), operated in reflection mode at 488 nm with a 40× objective LD Achroplan. The focus position was set to 100 μm below the interface of the coverslide in contact with the liquid sample. The time course of phase separation was followed by taking a series of images of frame size 512 × 512 pixels (*i.e.*, 230 × 230 μm) with a scan time of 6.2 s at intervals of 30 s.

Photographs of the tubes were taken with a digital camera after two days of storage in the incubator (Material testing cabinet 2391, Köttermann, Germany). For all samples in which macroscopic phase separation into two layers could be observed, the height of the top layer (H_{top}) and the total height (H_{col}) of the liquid were determined from the photographs using the millimetre scale included in the photos, with analysis in Microsoft Photo Editor.

19.2.3 Gel Formation

Samples of the blends (100 g) were stirred in a beaker at 40°C for 10 min using a three-blade propeller-mixer operated at high speed. After addition of 3 g of GDL (Sigma Chemical, USA) the mixtures were stirred for another 2 min at 40°C and then loaded by means of a syringe (BD Plastipak, Drogheda, Ireland) into test tubes for pH measurement, and into the sample cell of the rheometer (19 mL) or into the tubes in which the samples for later microscopic analysis were produced (11.5 mL). For all the experiments, pH curves were recorded in parallel to monitor the evolution in sample pH during acidification. The pH and gelation profiles were corrected for the initial time delay between GDL addition (time = 0) and the start of the measurement.

The pH was measured with a combination glass electrode with integrated temperature probe (InLab 410, Mettler Toledo, Switzerland) connected to a pH meter (MP 230, Mettler Toledo, Switzerland). The decline in pH due to GDL hydrolysis was monitored by continuous acquisition of pH data using a Laptop (Latitude C510, Dell, USA), connected to the pH meter via RS-232 link, and software written by NRC-IT (NRC Lausanne) using Visual Basic 6.0 software (Microsoft, USA).

Gel formation was monitored using a Paar Physica MCR 500 rheometer equipped with a concentric cylinder geometry (CC 27) with thermal unit TEZ 180-C connected to water-bath Viscotherm VT2 (Anton Paar, Graz, Austria) and equipped with solvent trap system. The rheometer was operated in continuous oscillation mode at a frequency of 0.1 Hz and 0.04% strain. Measurements were recorded as a function of time at 30 s intervals (240 data points). Under applied strain, the rheometer software calculated the stress and the phase angle, from which the elastic modulus G' was calculated.

The gel samples for microscopic analysis were prepared in 14-mL round-bottom polypropylene tubes (Falcon No. 352059, Becton Dickinson Labware, USA). The tubes (closed with their caps) were kept in a water-bath to restore sample temperature rapidly to 40°C, and transferred to the incubator 10 min after GDL addition. The samples were taken out of the incubator 60 min after GDL addition, when the samples had reached pH 4.6, and sample preparation for microscopy was started immediately at room temperature. Encapsulation in agar gel tubes, and fixation and embedding, were performed as described earlier.[5] Semi-thin sections of 1 μm thickness, obtained using an Omu2 Ultramicrotome (Reichert, Austria), were placed between slide and coverslide using immersion oil (518c, Zeiss, Germany) as mounting agent. The sections were then examined by phase-contrast light microscopy (PCLM) using a Zeiss Axioplan 2 light microscope (Carl Zeiss, Germany) in phase contrast mode with 40× objective Ph 2 Plan-Neofluar. For each sample, 20–30 micrographs were taken using a ProgRes 3008 camera (Kontron, Germany). Image analysis was performed using a mathematical morphology technique for shape characterization and segmentation,[6,7] implemented by means of the SDC morphology toolbox in the Matlab environment (Version 6.5, MathWorks, USA). The ratio of the image surface area taken by the aggregates and the total image surface area, $A_{\text{aggregates}}/A_{\text{image}}$, was calculated from the segmented images.

19.3 Manipulation of Particle Interactions

The structural elements present in unheated skim milk are the roughly spherical colloidal casein particles of average radius $R \sim 100$ nm as determined by dynamic light scattering, and the globular whey proteins (a few nanometres in size), both dispersed in milk serum containing mainly lactose and salts.

Xanthan is an anionic extracellular heteropolysaccharide of remarkably high chain stiffness, produced by the bacterium *Xanthomonas campestris*.[8] The radius of gyration R_g of the XG sample used was estimated to be of the order of 200 nm, based on the values of the radii determined by SEC-MALLS.

The incorporation of the polymer XG, the continuous acidification, and the skim milk pre-heating are presented in what follows as effective means to manipulate the casein particle interactions in skim milk, to tailor the gel microstructure.

19.3.1 Depletion-Induced Phase Separation

In unheated skim milk, the inclusion of ≥ 0.015 wt.% XG resulted in phase separation at 40°C at the natural pH of the mixtures (pH 6.5). The process of initial microscopic phase separation, investigated by CLSM after mixing, is illustrated in Figure 1 for an unheated skim milk sample containing 0.02 wt.% polysaccharide. At this XG level, phase separation was already visible in the first image taken 3.5 min after the end of mixing, where the appearance of slightly dark zones indicated the formation of areas depleted of protein. Ongoing phase separation resulted then in the formation of discrete growing domains deficient in protein (rich in polysaccharide), as illustrated by the series of CLSM images taken with time intervals of 5 min.

Phase separation in skim milk containing XG can be explained by a depletion interaction mechanism, depicted in Figure 2 at the level of a pair of spherical colloidal particles in a common solvent with non-adsorbing polymers. The centres of mass of the polymer molecules, depicted here as random coils, are excluded from a region away from the surface of the particles – the depletion layer (shown as dashed lines). Owing to the lower polymer segment concentration, the osmotic pressure due to the polymer is smaller in the depletion layer

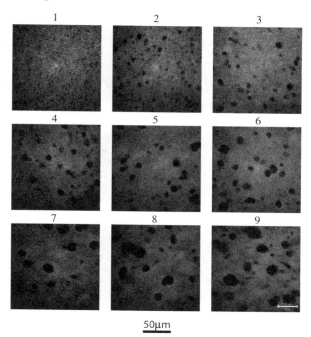

Figure 1 *Phase separation observed by CLSM at 40°C in unheated skim milk (14 wt.% SMP) containing 0.02 wt.% XG (pH 6.5). Image 1 was recorded at 3.5 min after the end of mixing. The set of images 2–9 depicts the evolution of the microstructure over intervals of 5 min. Dark spots represent areas rich in polysaccharide and deficient in protein.*
(Copyright Nestec Ltd (2006))

than in the bulk, resulting in an osmotic pressure gradient. For a single sphere this pressure is isotropic. As the particles approach due to Brownian motion, the depletion layers overlap, with the result that there is a volume of solution between the particles from which polymer molecules are excluded. As a consequence, the osmotic pressure becomes anisotropic, resulting in a net osmotic force pushing the particles together, as indicated by the arrows in Figure 2. Therefore, addition of a non-adsorbing polymer to a dispersion of (spherical) colloidal particles induces an effective attraction between the particles, resulting in phase separation at sufficiently high polymer concentration.[9–11]

At later stages of the phase separation process, macroscopic demixing into two liquid layers became visible. After incubation for two days, we observed a slightly turbid top layer (rich in XG, deficient in protein) and a very turbid bottom layer (rich in protein, deficient in XG), with a sharp boundary between them.

19.3.2 Acid-Induced Aggregation and Gel Formation

Slow acidification of skim milk is known to result in gel formation,[12,13] involving aggregation of the destabilized casein micelles into a three-dimensional network

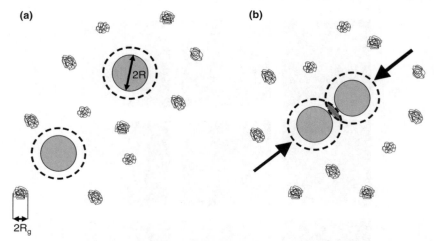

Figure 2 *Schematic picture of two spherical colloidal particles of radius R in a solution containing non-adsorbing polymers of radius of gyration R_g, with polymer coils excluded from a depletion zone (dashed line) near the particle surface: (a) no overlap of depletion layers (osmotic pressure on the spheres is isotropic); (b) overlapping depletion layers (net attractive force between particles from unbalanced osmotic pressure). The excess osmotic pressure is indicated by the arrows (After Ref. 10.)*

entrapping the serum phase. Thus, a prerequisite for gel formation to occur is acid-induced destabilization of the casein particles.

In unheated milk at physiological pH (≈ 6.7), the stability of casein micelles against aggregation is due to the presence of a 'hairy' layer of κ-casein molecules protruding from the casein micelle surface, providing primarily steric repulsion upon collision of two casein micelles.[14,15] Considering this 'hairy' κ-casein layer as a salted brush of weak polyelectrolyte molecules, de Kruif and Zhulina applied[16] the scaling relation derived for the sharp conformational transition from a highly extended to a strongly collapsed state of the brush. Within the emerging casein brush model,[17,18] acidification is seen to cause brush collapse below pH ≈ 5, resulting in the loss of polymeric stabilization of the casein micelles, which finally aggregate.

The development of elasticity during gelation is illustrated in Figure 3 for acid-induced gel formation in unheated skim milk. Figure 3(a) shows the acidification kinetics common to all the gelation experiments performed. When pH had dropped to a critical value (4.95), the storage modulus G' increased from its baseline value and continued to develop in the manner shown in Figure 3(b).

19.3.3 Effects of Pre-Heating of Skim Milk

Heat treatment of milk at temperatures above *ca.* 70°C induces denaturation of the whey proteins, which subsequently aggregate and also interact with the casein micelles and more specifically with κ-casein.[19] This results in the

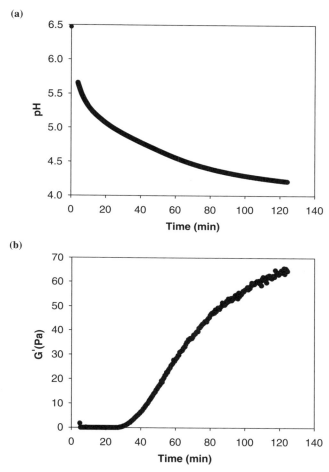

Figure 3 *Acid-induced gel formation in unheated skim milk (14 wt.% SMP) at 40°C after addition of 3 wt.% GDL (at time = 0): (a) sample pH; (b) storage modulus G' as determined by oscillatory rheometry (frequency = 0.1 Hz, strain = 0.04%)*

formation of aggregates of heat-denatured whey proteins, either remaining in the serum phase (soluble aggregates) or associated with casein micelles (micelle-bound aggregates). It has been established[19,20] that both the soluble and the micelle-bound denatured whey proteins, generally known to become insoluble on lowering the pH (*e.g.*, to 4.6), play an active role in acid-induced gel formation, while the native whey proteins present in unheated skim milk remain soluble during acidification and therefore play no such role.

Pre-heating of skim milk results in a shift of the onset of gelation to significantly higher pH, as shown in Figure 4. This phenomenon, demonstrated previously using diffusing-wave spectroscopy,[21] strongly suggests a significant change in particle interactions due to skim milk pre-heating. Interestingly, skim

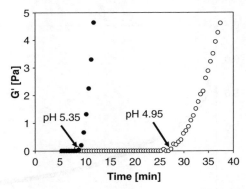

Figure 4 *Onset of gelation in unheated skim milk (○) and in pre-heated skim milk (●) based on rise of storage modulus G' from its baseline*

milk pre-heating was also found to have a great impact on the phase separation of skim milk + XG mixtures. In pre-heated skim milk, macroscopic phase separation into two liquid layers first became visible at lower polysaccharide concentration (0.005 wt.%) than in unheated skim milk (0.015 wt.% XG). Moreover, in the polysaccharide content range investigated, the XG-rich upper layer was found to be thicker in pre-heated skim milk than in unheated skim milk. To describe the latter observation in a more quantitative way, the volume fraction of the XG-rich phase (Φ_{PS}) was estimated (see Figure 5).

The effects of skim milk pre-heating on the phase separation behaviour can be attributed to the change in particle interactions and to the increased volume fraction of colloidal protein particles. The latter is due to both the increase in casein micelle size on coating with denatured whey protein[22] and the presence of soluble whey protein aggregates as new particle species of colloidal dimensions.[23,24] The diameter of the soluble whey protein aggregates formed during heat treatment of skim milk has been estimated to vary between > 10 nm[23] and ~ 60 nm,[24] *i.e.*, of the order of the size of the smaller casein micelles.

19.4 Tailoring Gels from Particle Dispersions

Acid-induced gels were prepared with systematic XG level variation in unheated and pre-heated skim milk, acidified in the same manner at 40°C with 3 wt.% GDL. To compare the gel strength developed, the values of the storage modulus G' were determined from the gelation profiles at the point where the samples had reached pH 4.60. The data are presented in Figure 6. Some PCLM images of gels selected according to the rheological behaviour observed in unheated skim milk samples are shown in Figure 7; the images are for the reference sample (no added XG), the strongest gel (0.015 wt.% XG) and the weakest gel (0.025 wt.% XG). Images of the corresponding gel samples made from pre-heated skim milk, formed at the same XG levels, are presented for comparison.

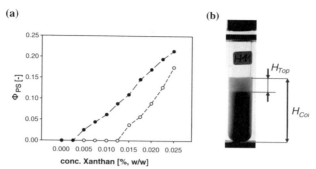

Figure 5 *Macroscopic phase separation observed in the mixtures after incubation at 40°C for two days at pH 6.5. (a) Estimated volume fraction (Φ_{PS}) of the XG-rich phase in unheated skim milk (○) and in pre-heated skim milk (●) containing XG at various levels. The value of Φ_{PS} is calculated as the ratio of the height of the top layer (H_{top}) to the total height of the liquid (H_{col}), both determined from the photographs of the tubes as illustrated in (b) for phase separation in pre-heated skim milk containing 0.025 wt.% XG (no account is taken of concave tube bottom)*
(Copyright Nestec Ltd (2006))

In unheated skim milk, an increase in XG level resulted in an increase in G' up to a maximum observed in this series for 0.015 wt.% XG incorporated (Figure 6(a)). Further increase in XG content (0.02 and 0.025 wt.%) then caused G' to drop even below its value for the pure skim milk gel. The protein network microstructure of the gel made from unheated skim milk without added polysaccharide appeared fine and regular (see Figure 7), but more coarse and irregular in the presence of 0.015 wt.% XG. The maximum gel strength observed at the latter polysaccharide concentration can be attributed to the crowding of protein particles in protein-rich domains resulting in more massive network-constituting entities, with still high connectivity between them. The impact of polysaccharide inclusion on network microstructure was much more pronounced at the highest XG level investigated (0.025 wt.%), where the protein was concentrated in domains appearing as large, rather isolated regions. Here the poor apparent connectivity between the massive protein aggregates is the likely reason for the low gel strength observed, since only material integrated well into the stress-bearing network can be expected to contribute to gel rigidity.

Pre-heating of skim milk affects substantially both the microstructure and the storage modulus of gels formed in the absence of polysaccharide addition. The increase in G' by almost one order of magnitude (Figure 6) is in agreement with the work of Lucey *et al.*[25] When XG was present, the modulus increased almost linearly with increasing polysaccharide concentration, similar to the behaviour of unheated skim milk, but the maximum in G' was shifted to a higher XG level and decreased then only moderately upon further addition of XG to 0.025 wt.% (see Figure 6(b)). The difference in network microstructure due to skim milk pre-heating was already visible in the reference gels made from

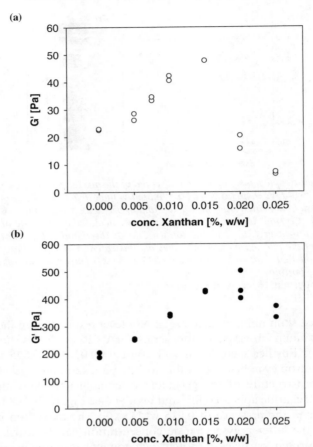

Figure 6 *Effect of XG content on storage modulus G' of gels at pH 4.60 determined by oscillatory rheometry for gel formation: (a) unheated skim milk and (b) pre-heated skim milk*

skim milk alone (Figure 7), which is in agreement with observations of Kalab et al.;[26] but it was much more pronounced in gels formed with 0.015 or 0.025 wt.% XG. The aggregates were shaped like islands when formed in unheated skim milk, but appeared like a web of causeways in gels made from pre-heated skim milk. This morphological description implies a difference in connectivity between the network-constituting entities. Connectivity between protein aggregates appeared higher in the gels made from pre-heated skim milk. Moreover, for an increase in XG concentration from 0.015 to 0.025 wt.%, the connectivity remained quite high in the pre-heated skim milk, explaining the only moderate decrease in G', while the connectivity and modulus decreased dramatically in the unheated skim milk.

The above results demonstrate that the inclusion of XG is an effective means to tailor protein network microstructure. As can be seen in both the image

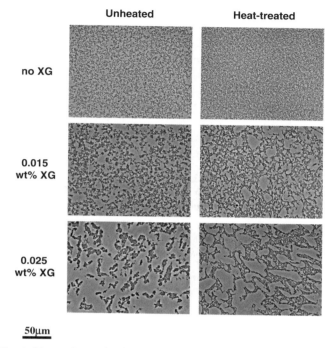

Figure 7 *Semi-thin sections of gels investigated by PCLM after chemical fixation and embedding. Gels were prepared by acidification of unheated and pre-heated skim milk containing XG at three concentrations (0, 0.015 and 0.025 wt%). Protein network microstructure appears as dark grey in the cross-sections of the gels.* (Copyright Nestec Ltd (2006))

series shown in Figure 7, the presence of XG causes a concentration of the protein particles into protein-rich domains, resulting in more clearly delimited and massive network-constituting entities following acid-induced aggregation. The extent to which this occurs depends strongly on the polysaccharide content. To describe this in a more quantitative manner, image analysis was performed on the PCLM pictures. The proportion of the two-dimensional image surface area covered by the aggregates, $A_{\text{aggregates}}/A_{\text{image}}$, was determined to estimate the three-dimensional solution volume fraction effectively occupied by the particle aggregates constituting the protein network. This surface area ratio, ultimately representing a measure of the extent of local particle crowding, was found to decrease with increasing XG concentration, as shown in Figure 8. Furthermore, this decrease reflects well the phase separation observed at the natural pH of the mixtures, quantitatively described by the estimated volume fraction of the XG-rich phase (Φ_{PS}) in Figure 5. In unheated skim milk, the inclusion of 0.015 wt.% XG resulted in a slight decrease in surface area ratio, indicating rather weak phase separation (low Φ_{PS}), whereas incorporation of 0.025 wt.% XG resulted in a pronounced decrease in surface area ratio, reflecting strong phase separation (high Φ_{PS}). In pre-heated skim milk the

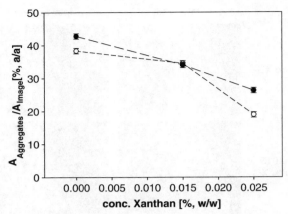

Figure 8 Ratio of the image surface area taken by the aggregates ($A_{aggregates}$) and the total image surface area (A_{image}) determined from PCLM micrographs of gels prepared from unheated skim milk (○) and pre-heated skim milk (●) containing XG at various levels (0, 0.015 and 0.025 wt.%). Points represent mean values calculated from the 20 to 30 images that were analysed for each gel sample. Error bars represent the least significant difference (LSD)

surface area ratio was found to decrease almost linearly with increasing XG level, with phase separation (and a linear increase in Φ_{PS}) observed over almost the entire range of polysaccharide contents investigated.

The observation that, at 0.025 wt.% XG, phase separation appears stronger in pre-heated skim milk (higher Φ_{PS}, Figure 5), while the extent of crowding of protein particles is lower (higher $A_{aggregates}/A_{image}$, Figure 8) and the connectivity between the regions of gel is higher, seems contradictory at first. However, the change in particle interactions due to skim milk pre-heating makes the onset of gelation occur at much higher pH, trapping the evolving microstructure at an earlier stage of the phase separation process. For the given acidification conditions (40°C, 3 wt.% GDL), the onset of gelation in the unheated skim milk took place at pH 4.94–5.06, while it occurred significantly earlier in the pre-heated skim milk, namely when the pH had just dropped into the range 5.33–5.38. Hence, although phase separation at a given polysaccharide content is stronger in pre-heated skim milk, the acid-induced gel formation results in fixation of the phase-separating mixture at a stage at which the resulting gel strands are still well connected to each other and thereby integrated into the force-transmitting network through linkage to their neighbours.

19.5 Conclusions

We have demonstrated the importance of phase separation to the process of gel formation in acidified dairy products with XG added as a supposed 'thickener/stabilizer'. The gel microstructure can be designed by triggered arrest of the phase-separating structure through acid-induced particle aggregation and network formation. Incorporation of polysaccharide has been found to be an effective means to influence gel strength even at levels at which no phase separation was observed in the initial mixtures at their natural pH, as demonstrated in unheated skim milk.

Overall, a broad range of systems with defined microstructure and mechanical properties has been created from skim milk by manipulating the colloidal interactions through skim milk pre-heating, incorporation of XG, and controlled acidification. To understand fully the mechanisms of acid gel formation in skim milk + XG mixtures, future work should be aimed at the investigation of phase separation kinetics and particle dynamics as a function of the acidification kinetics imposed. Ultimately, the study of the materials prepared should be extended to the large-scale deformation behaviour and be complemented by sensory analysis, exploiting the inherent opportunity to improve our understanding of the relationships between microstructure, mechanical properties and sensory texture perception.

Acknowledgement

We thank Dr Christophe Schmitt for assistance with the SEC-MALLS experiments and Dr Eric Kolodziejczyk for help with CLSM.

References

1. A. Syrbe, W.J. Bauer and H. Klostermeyer, *Int. Dairy J.*, 1998, **8**, 179.
2. Y. Hemar, M. Tamehana, P.A. Munro and H. Singh, *Food Hydrocoll.*, 2001, **15**, 565.
3. S. Thaiudom and H.D. Goff, *Int. Dairy J.*, 2003, **13**, 763.
4. C. Sanchez, R. Zuniga-Lopez, C. Schmitt, S. Despond and J. Hardy, *Int. Dairy J.*, 2000, **10**, 199.
5. P.A. Aichinger, M. Michel, C. Servais, M.L. Dillmann, M. Rouvet, N. D'Amico, R. Zink, H. Klostermeyer and D.S. Horne, *Colloids Surf. B*, 2003, **31**, 243.
6. S. Beucher, in *Handbook of Pattern Recognition and Computer Vision*, C.H. Chen, L.F. Pau and P.S.P. Wang (eds), World Scientific, River Edge, 1993, p. 443.
7. C. Vachier, Ph.D. Thesis, Ecole des Mines de Paris, 1995.
8. B.T. Stokke, B.E. Christensen and O. Smidsrod, in *Polysaccharides: Structural Diversity and Functional Versatility*, S. Dumitriu (ed), Marcel Dekker, New York, 1998, p. 433.
9. R.A.L. Jones, *Soft Condensed Matter*, Oxford University Press, New York, 2002, p. 61.
10. R. Tuinier, J. Rieger and C.G. de Kruif, *Adv. Colloid Interface Sci.*, 2003, **103**, 1.
11. P. Walstra, *Physical Chemistry of Foods*, Marcel Dekker, New York, 2003, p. 466.

12. J.A. Lucey and H. Singh, in *Advanced Dairy Chemistry, Volume 1 – Proteins*, Part A, P.F. Fox and P.L.H. McSweeney (eds), Kluwer Academic/Plenum, New York, 2003, p. 1001.
13. J.A. Lucey and H. Singh, *Food Res. Int.*, 1998, **30**, 529.
14. C. Holt and D.S. Horne, *Neth. Milk Dairy J.*, 1996, **50**, 85.
15. P. Walstra, *J. Dairy Sci.*, 1990, **73**, 1965.
16. C.G. de Kruif and E.B. Zhulina, *Colloids Surf. A*, 1996, **117**, 151.
17. C.G. de Kruif, *Int. Dairy J.*, 1999, **9**, 183.
18. C.G. de Kruif and C. Holt, in *Advanced Dairy Chemistry, Volume 1 – Proteins*, Part A, P.F. Fox and P.L.H. McSweeney (eds), Kluwer Academic/Plenum, New York, 2003, p. 233.
19. Y.D. Livney, M. Corredig and D.G. Dalgleish, *Curr. Opin. Colloid Interface Sci.*, 2003, **8**, 359.
20. T. van Vliet, C.M.M. Lakemond and R.W. Visschers, *Curr. Opin. Colloid Interface Sci.*, 2004, **9**, 298.
21. D.S. Horne and C.M. Davidson, in *Protein and Fat Globule Modifications*, IDF Seminar (25–28 August 1993, Munich, Germany), IDF Special Issue 9303, International Dairy Federation, Brussels, 1993, p. 267.
22. S.G. Anema and Y.M. Li, *J. Dairy Res.*, 2003, **70**, 73.
23. F. Guyomarc'h, A.J.R. Law and D.G. Dalgleish, *J. Agric. Food Chem.*, 2003, **51**, 4652.
24. A.J. Vasbinder, A.C. Alting and C.G. de Kruif, *Colloids Surf. B*, 2003 **31**, 115.
25. J.A. Lucey, C.T. Teo, P.A. Munro and H. Singh, *J. Dairy Res.*, 1997 **64**, 591.
26. M. Kalab, P. Allan-Wojtas and B.E. Phipps-Todd, *Food Microstruct.*, 1983, **2**, 51.

PART III
Particles, Droplets and Bubbles

Chapter 20

Particle Tracking as a Probe of Microrheology in Food Colloids

Eric Dickinson, Brent S. Murray and Thomas Moschakis

PROCTER DEPARTMENT OF FOOD SCIENCE, UNIVERSITY OF LEEDS, LEEDS LS2 9JT, UK

20.1 Introduction

Food colloids are typically heterogeneous, and the degree of heterogeneity is susceptible to change during processing and long-term storage. For a thorough understanding of the relationship between bulk properties and colloidal interactions, we require detailed information about structure and mechanical properties on various length scales. Confocal microscopy is emerging as a powerful technique for monitoring and quantifying the evolution of structural change on the microscopic scale as a function of variables such as time, temperature and shear deformation. In addition, the dynamical analysis of sequences of images from confocal microscopy has the potential to provide new insight into local rheological properties within heterogeneous systems.

Microrheology is the study of viscous and viscoelastic properties of regions within a material *via* the tracking of the motion of microscopic tracer particles.[1,2] There are two classes of microrheology techniques: those involving *active* manipulation of probe particles within the sample, and those involving *passive* observation of thermal fluctuations of probe particles. The active manipulation of micrometre-sized magnetic particles by magnetic fields was pioneered more than 80 years ago in connection with the gelation behaviour of gelatin.[3] Since then, magnetic particles have been used extensively for investigating the microrheology of biological systems, including living cells.[4] Nevertheless, it is the technique of passive particle tracking in confocal microscopy which probably has the most obvious potential in the field of food colloids. As an example of this potential, we report here on the passive tracking of inert colloidal particles to probe the *in situ* viscoelasticity of a model food emulsion.

The fundamental assumption of tracer microrheology is that the Brownian motion of the diffusing particles is controlled by the mechanical properties of the surrounding medium. Typically the experimental information consists of the mean-square displacement (MSD) of particles as a function of lag time τ.[5] For instance, the two-dimensional time-averaged MSD is defined by

$$\langle \Delta r^2(\tau)\rangle = \langle [x(t+\tau) - x(t)]^2 + [y(t+\tau) - y(t)]^2 \rangle, \quad (1)$$

where $x(t)$ and $y(t)$ are the time-dependent coordinates of the centres of the particles and the angular brackets indicate an average over many starting times for the ensemble of particles in the field of view. For spherical monodisperse particles of radius a in a Newtonian fluid, the viscosity η of the surrounding medium can be determined from the measured diffusion coefficient D using the standard Stokes–Einstein relation,

$$D = kT/6\pi\eta a, \quad (2)$$

where k is Boltzmann's constant and T the absolute temperature. For particles tracked in viscoelastic media, the MSD may be analysed using a generalized Stokes–Einstein equation to give frequency-dependent viscous and elastic moduli.[6-8] For a macroscopically uniform system, the analysis allows direct comparison of the viscosity η or the complex modulus $G^*(\omega)$ with the equivalent property determined by conventional bulk rheometry.

For particles undergoing free diffusion, the ensemble-averaged MSD is a linear function of time, *i.e.*, in two dimensions, we have

$$\text{MSD} = \langle \Delta r^2(\tau)\rangle = 4D\tau. \quad (3)$$

When there is a steady drift or flow in the system (*e.g.*, due to convection), the plot of MSD *versus* τ is no longer linear, but has *positive* curvature. For a constant drift velocity V, the MSD is given by[9]

$$\text{MSD} = 4D\tau + (V\tau)^2. \quad (4)$$

Even if the drift is only slow, the contribution from the quadratic term $(V\tau)^2$ becomes predominant at longer times. It is as if the particle diffusive motion becomes more vigorous the further the particle moves. In contrast, for diffusion in a finite region (*e.g.*, a pore or cage), the plot of MSD *versus* τ has *negative* curvature. The observed behaviour is most simply interpreted in terms of interactions between diffusing particles and other mobile or immobile structures.[9]

Different scenarios for the way in which colloidal probe particles may be embedded in a biopolymer network have been discussed recently by Weitz and co-workers.[10] One important factor is the particle size in relation to the characteristic structural length-scale(s) of the material. The ensemble-averaged MSD can be reliably related to $G^*(\omega)$ of the bulk biopolymer network for the case of probe particles that are large compared to macromolecular network length scales (Figure 1A). In contrast, the bulk viscoelastic response does not influence so much the dynamics of probe particles that are smaller than the structural length-scales (Figure 1B). In this case, individual particle movements are determined by the solvent viscosity, the effects of macromolecular crowding and interactions with the network (electrostatic, steric, hydrodynamic). For probing microenvironments of a heterogeneous material with particles smaller

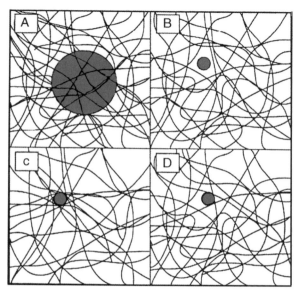

Figure 1 Schematic representation of a spherical probe particle in a biopolymer network. (A) For a particle of radius a much greater than the mesh size ξ, the MSD is directly related to the bulk linear rheology. (B) For a small non-interacting particle $(a < \xi)$, the MSD is related to the solvent viscosity and local repulsive interactions with the network. (C) For a highly sticky particle $(a < \xi)$, the particle diffusion is greatly restricted, but the MSD differs from the bulk linear rheology because strong particle–protein interactions modify the local biopolymer network structure. (D) For a slightly sticky particle $(a < \xi)$, its Brownian motion in the solvent environment is hindered by competing attractive interactions with the biopolymer network (Taken from Ref. 10, and reproduced with permission of the Biophysical Society.)

than the structural length-scales $(a < \xi)$, it has been noted[10] that complications from particle–protein interactions can be very significant. With interactions that are strong, small particles can become completely immobilized within the entangled network (Figure 1C). And the presence of even fairly weak protein adsorption can cause particles to associate preferentially with network strands or cavity walls, thereby hindering their diffusion and limiting their exploration of smaller pores (Figure 1D).

In this paper we report specifically on the use of particle tracking to probe the microrheology in the phase-separating regions of a heterogeneous protein-stabilized emulsion exhibiting depletion flocculation. The heterogeneity is induced by the presence of a very small concentration of a non-adsorbing polysaccharide (xanthan gum). On the macroscopic scale, this type of emulsion exhibits substantial serum separation[11–14] and strongly shear-thinning rheological behaviour.[12,15] Our objective is to monitor the local dynamics of incorporated probe particles to establish which regions of the evolving heterogeneous system are predominantly responsible for determining the bulk rheology, and hence the physical shelf life, of the emulsion with respect to gravity-induced creaming.

20.2 Experimental

The probe particles were carboxylate-modified polystyrene (COOH–PS) microspheres of three different nominal mean diameters (0.21, 0.5 and 0.89 μm). Spray-dried sodium caseinate (> 82 wt.% dry protein, < 6 wt.% moisture, < 6 wt.% fat and ash, 0.05 wt.% calcium) was obtained from DeMelkindustrie (Veghel, Netherlands). Food-grade xanthan gum powder (Kelzan 'S' F850414) was obtained from Kelco (San Diego, CA). Some of this xanthan had been labelled covalently with fluorescein-5-isothiocyanate (FITC).[16] 1-Bromohexadecane (density 0.99 g ml^{-1}) was used as the oil phase in order to minimize oil droplet buoyancy effects in the cell of the confocal microscope.[14] The emulsion aqueous phase was a 20 mM imidazole buffer solution, adjusted to pH 6.8 with HCl, with sodium azide (0.01% w/v) included as antimicrobiological agent. Further details of the materials used can be found elsewhere.[17]

Oil-in-water emulsions were made with a single-pass laboratory-scale jet homogenizer[18] operating at a pressure of 300 bar. Aqueous solutions of sodium caseinate (2 wt.%) and xanthan were prepared by adding protein powder and xanthan gum powder to the buffer solution (imidazole) and then gently stirring overnight at room temperature to ensure complete dispersion. In the final emulsion, the protein concentration was 1.4 wt.% and the amount of dispersed oil phase was always adjusted to 30 vol.%. This protein/oil ratio was roughly the minimum required to emulsify all the bromohexadecane without bridging flocculation, while avoiding any significant amount of unadsorbed protein which might cause depletion flocculation by caseinate nanoparticles.[19,20] Droplet-size distributions of the sodium caseinate-stabilized emulsions with xanthan added prior to emulsification, as determined using a Malvern Mastersizer 2000, were monomodal with ∼95% of droplets in the size range from 0.1 to ∼3 μm. The presence of xanthan in the range 0–0.15 wt.% did not affect the droplet-size distribution: the average diameters remained constant at $d_{32} = 0.35 \pm 0.03$ μm and $d_{43} = 0.95 \pm 0.05$ μm.

Steady-state shear viscosities of emulsions and xanthan solutions (0.02–0.07 wt.%) were measured at 20°C using a Bohlin CVO-R rheometer with a double-gap geometry (DG 24/27). Samples were covered with a thin layer of silicone oil and a solvent trap was employed. Freshly prepared emulsion samples were initially subjected to a high shear rate (150 s^{-1}) to break up any existing aggregates or phase-separated regions; half an hour after this treatment, the zero shear rate viscosity was determined for each sample.

Confocal microscopy was carried out with a Leica TCS SP2 confocal laser scanning microscope (CLSM) operated in fluorescence mode with a 40 times oil-immersion objective. The oil phase was stained with Nile Red dye, and fluorescence from the sample was excited with the 488 nm Ar laser line. A 5 ml sample of the freshly made emulsion was transferred to a small beaker. A 20 μl aliquot of Nile Red solution (0.01 wt.%) was immediately added to give a final concentration of 0.4 µg of dye per 1 ml of the solution. The solution was thoroughly mixed at a shear rate of 1 rad s^{-1} for 5 min at 20°C. The stained emulsion was then immediately placed into a laboratory-made welled slide, filling it completely. A coverslip was placed on top of the well, and it was ensured that there were no bubbles trapped between sample and coverslip. The edge of the coverslip was sealed to prevent moisture loss and exposure to the atmosphere.

Trajectories of fluorescently labelled COOH–PS microspheres were recorded at 20°C in real time using the CLSM. The probe particles were added at a concentration of 0.2–0.4 vol.%. The position of each fluorescent microsphere was identified by finding the brightness-averaged centroid position with a sub-pixel accuracy of approximately 20 nm spatial resolution.[21] While the microscope was capable of tracking hundreds of particles simultaneously, the number of tracked microspheres per field of view was actually kept relatively low (between 10 and 30) to eliminate complications due to particle–particle interactions. A voxel size of 183 × 183 × 668 nm was chosen for most of the experiments (image dimensions 93.75 × 23.44 nm), but the pinhole size was sometimes increased to record thicker slices and facilitate more extended tracking of microspheres in the focal plane. The acquisition time for each pixel slice in the horizontal $x-y$ plane was typically 0.134 s. Images were scanned approximately 10–20 µm below the level of the coverslip to minimize hydrodynamic (and other) interactions with the coverslip.

Movies recording the changing positions of the diffusing microspheres in the $x-y$ plane were analysed according to the procedures of Crocker and Grier.[21] A tracking macro-routine was applied between successive frames. The trajectories were analysed with a slightly modified version of the IDL tracking software series (Research Systems, Boulder, CO) comprising a series of macros developed originally by Crocker and Weeks.[22] Microsphere positions were matched frame by frame, using a routine incorporated into the IDL software, identifying each individual particle and then generating its trajectory. Another macro was employed to diagnose and eliminate any background drift due to unwanted convective flow or movement of the microscope stage. Systematic drift was found by examining the average values of dx and dy for all the visible particles from one frame to the next, and then subtracting the computed background drift component from the raw data for each individual particle. Further experimental details can be found elsewhere.[17]

20.3 Results and Discussion

We describe first the behaviour of COOH–PS microspheres of diameter 0.5 µm dispersed in aqueous solutions of 60 wt.% glycerol (0.010 Pa s) and 88 wt.% glycerol (0.147 Pa s). It can be seen from Figure 2 that the ensemble-averaged MSD of the microspheres is a linear function of lag time τ for these simple Newtonian liquids. The two straight lines in Figure 2 correspond to theoretical

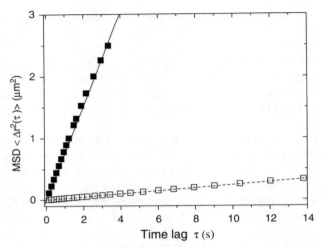

Figure 2 *Ensemble-averaged MSD of 0.5 μm particles as a function of time τ in Newtonian liquids: ■, 60 wt.% aqueous glycerol solution; □, 88 wt.% aqueous glycerol solution. The lines are theoretical values calculated from Equation (2) using the known viscosities of glycerol solutions (0.0107 and 0.147 Pa s)*

plots of MSD *versus* τ calculated from the Stokes–Einstein equation (2) and the known values of the viscosity of the solutions.[23] This excellent agreement with the theory demonstrates the validity of the particle tracking technique for determining the shear viscosity under these ideal conditions.

For the case of non-Newtonian media, it is convenient also to plot the MSD as a function of time on a double logarithmic plot. The fitted slope α, which corresponds to the exponent in the relationship

$$<\Delta r^2(\tau)> \sim \tau^\alpha, \tag{5}$$

lies somewhere between the purely viscous limit ($\alpha = 1$) and the elastic limit ($\alpha = 0$). A value of $\alpha < 1$ indicates that the motion is sub-diffusive, or hindered, whereas a slope of $\alpha \sim 1$ indicates that the motion can be regarded as entirely diffusive (as in the glycerol solutions). Even for a purely Newtonian liquid, the fitted value of α does not always work out to be exactly unity because of various factors: image resolution errors (spatial and temporal), statistical fluctuations and small uncorrected drifts. We make the practical assumption here that a system that can be fitted with a value of $\alpha > 0.9$ is effectively viscous on the timescale of the experimental observation.

Before moving on to consider the microrheology of the emulsion systems, we present data on their bulk rheological properties. Figure 3 shows apparent viscosity as a function of shear rate for the caseinate-stabilized emulsions (30 vol.% oil, 1.4 wt.% protein) with different xanthan concentrations in the aqueous phase (0.03–0.07 wt.%). In agreement with previous work on emulsions of this type,[15,24] the curves show pronounced shear-thinning behaviour

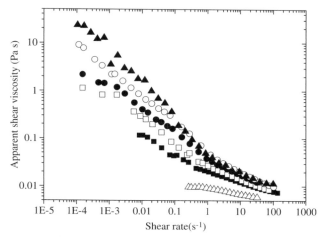

Figure 3 *Apparent viscosity against shear rate at 20°C for oil-in-water emulsions (30 vol.% oil, 1.4 wt.% sodium caseinate, pH 6.8) containing different concentrations of xanthan in the aqueous phase: △, none; ■, 0.03 wt.%; □, 0.04 wt.%; ●, 0.05 wt.%; ○, 0.06 wt.%; ▲, 0.07 wt.%*

and increasingly high viscosities at low shear rates for increasing polysaccharide concentrations. Figure 4 shows the limiting zero shear rate viscosity of the emulsions (left ordinate axis) plotted against the aqueous phase xanthan concentration. Also plotted on a different scale in Figure 4 (right ordinate axis) are corresponding data for 0.03–0.07 wt.% aqueous xanthan solutions. We note the enormous difference in absolute viscosity values for the emulsions and the solutions. That is, the viscosities of the xanthan-containing emulsions are of the order of a thousand times larger than those of the equivalent xanthan-containing solutions.

Figure 5 shows particle tracking data for the caseinate-stabilized emulsion (30 vol.% oil, 1.4 wt.% protein) without any added xanthan. The two sets of data refer to two sizes of microspheres (0.5 and 0.89 μm). To allow direct comparison of results from the two kinds of probe particles, the quantity $a\langle\Delta r^2(\tau)\rangle$ (*i.e.* particle radius × MSD) is plotted against lag time τ on a log–log scale. In each case the slope of the best linear fit to the data is close to unity indicating that the particle motion is predominantly diffusive ($\alpha > 0.9$). The combined data can be fitted by the Stokes–Einstein relation [Equation (2)] with an effective viscosity $\eta = (6.5 \pm 1.5)$ mPa s. This value is in satisfactory agreement with the value $\eta = (9.5 \pm 2)$ Pa s measured at low shear rate in the rheometer (see Figure 3).

As has been reported previously,[14] the emulsions containing > 0.03 wt.% xanthan appear heterogeneous and structurally time dependent under the confocal microscope. In the polysaccharide concentration range 0.03–0.05 wt.%, we observed time-dependent microscopic separation into oil-rich and oil-depleted regions. These regions exhibited relaxation, coarsening and coalescence phenomena. In particular, after stirring stopped, the interfacial tension

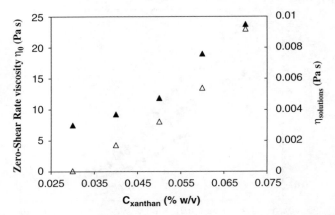

Figure 4 *Effect of xanthan concentration on zero shear rate viscosity of emulsions (30 vol.% oil, 1.4 wt.% sodium caseinate) (left ordinate scale, △) and aqueous solutions (right ordinate scale, ▲)*

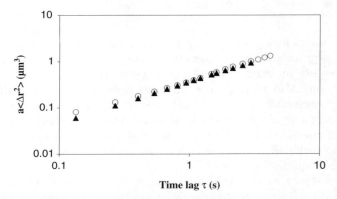

Figure 5 *Normalized MSD of two kinds of probe particles in emulsions (30 vol.% oil, 1.4 wt.% sodium caseinate, pH 6.8) at 20°C without added xanthan: ▲, 0.5 μm microspheres; ○, 0.89 μm microspheres. The quantity $a\langle \Delta r^2(\tau)\rangle$ is plotted as a function of time τ, where a is the particle radius*

drove the elongated shapes of the phase-separating xanthan-rich domains into more circular shapes.[14] With increasing xanthan concentration, the rate of this relaxation process was increasingly retarded; and for ≥ 0.06 wt.% xanthan there was complete inhibition of the microstructural evolution.

When incorporated into the xanthan-containing emulsions, the microspheres were seen to be well dispersed within both the oil-droplet-rich and the xanthan-rich regions. Figure 6 shows two images for a system with 0.05 wt.% xanthan containing 0.21 μm diameter microspheres. The oil phase was stained with Nile Red. Image (a) was recorded soon after the stirring had stopped: clearly visible are the deformed and elongated xanthan-rich blobs (depleted of oil droplets).

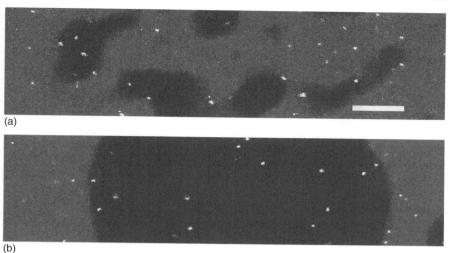

Figure 6 *Confocal images of 0.21 μm probe particles in phase-separating emulsion (30 vol.% oil, 1.4 wt.% sodium caseinate, pH 6.8) with 0.05 wt.% xanthan in the aqueous phase: (a) shortly after stopping stirring and (b) after xanthan-rich blobs had relaxed to circular (spherical) scale bar = 10 μm*

Image (b) was recorded 1–2 h later: following coarsening, a large phase-separated xanthan-rich blob has relaxed into a circular (spherical) shape.

The probe microspheres were quite evenly distributed within the phase-separated emulsion regions. Nevertheless there appeared to be a gradual migration of particles from the oil-droplet-rich microphase to the xanthan-rich microphase. That is, there seemed to be a thermodynamic preference for probe particles to be located in the xanthan-rich (water-rich) phase. Presumably this can be attributed to unfavourable repulsion (due to excluded volume interactions) between the charged microspheres and the protein-coated emulsion droplets in the highly concentrated oil-droplet-rich phase. For instance, at the composition illustrated in Figure 6, there was initially observed to be an excess particle number density of $\sim 9\%$ in oil-droplet-rich regions (compared with the random distribution), whereas after around 3 h there was a slight excess particle density ($\sim 4\%$) in the xanthan-rich regions. Migration of probe particles from the oil-droplet-rich phase to the xanthan-rich phase occurred much more slowly than the timescale of individual particle Brownian movement. So, fortunately, it did not affect the interpretation of the diffusion coefficients derived from the particle tracking experiments.

Some representative particle trajectories in the emulsion with 0.05 wt.% xanthan are shown in Figure 7. We can clearly see that, during the same time interval (10 s), a probe particle located in the xanthan-rich microphase moves much further (trajectory A) than one located in the oil-droplet-rich microphase (trajectory B). Also illustrated in Figure 7 is the path of a particle that starts in the xanthan-rich phase but ends up in the oil-droplet phase (trajectory C). While

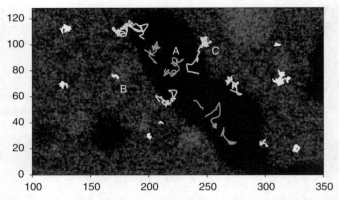

Figure 7 Some trajectories of 0.21 µm probe particles in phase-separating emulsion (30 vol.% oil, 1.4 wt.% sodium caseinate, pH 6.8) with 0.05 wt.% xanthan in the aqueous phase: (A) trajectory in xanthan-rich region; (B) trajectory in oil-droplet-rich region; (C) trajectory passing across boundary between microphases. Axis dimensions are in pixels (1 pixel = 0.183 µm)

some care was taken to ensure that such trajectories were omitted as much as possible from our statistical analysis, their existence conveniently demonstrates the ability of probe particles to move easily across the interface between phase-separating regions of contrasting composition and microstructure.

Figure 8 shows a log–log plot of MSD *versus* lag time τ for microspheres of diameter 0.21 µm in the xanthan-rich regions of the emulsions containing different concentrations of xanthan. Slopes of the straight line fits were close to unity ($\alpha > 0.9$) indicating a predominantly viscous response. The inferred viscosities in the xanthan-rich regions are quite low (< 5 mPa), being slightly higher than those inferred from particle tracking in the corresponding simple xanthan solutions,[14] and fairly consistent with the sort of values measured by bulk viscometry (Figure 4).

Figure 9 shows a log–log plot of MSD *versus* lag time τ from the tracking of microspheres of diameter 0.5 µm in the oil-droplet-rich regions of emulsions of different xanthan contents. Predominantly diffusive motion (slope > 0.9) is obtained only for emulsions containing 0.03 or 0.04 wt.% xanthan; at higher concentrations the microsphere mobility appears substantially hindered. Nevertheless, at these higher polysaccharide contents, the MSD is still reasonably linear in τ for $\tau > 2$ s, and so for these relatively long-time processes in the viscous regime, effective viscosities could be approximately estimated.

Figure 10 shows that the inferred viscosity of the flocculated oil droplet region of the emulsions increases dramatically with xanthan concentration c. Also indicated on the same graph are the apparent viscosities in the xanthan-rich regions of the same emulsions, as determined from the particle tracking data in Figure 8. A large contrast in absolute values can be seen: the inferred viscosities in the oil-droplet-rich regions are some 10^2–10^3 times larger than those in the xanthan-rich regions. Hence, while the viscosity of the xanthan-containing phase does increase with increasing biopolymer concentration, the main

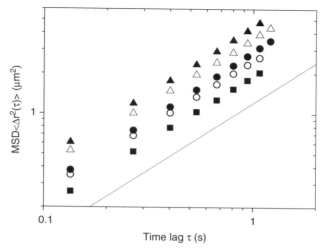

Figure 8 *Ensemble-averaged MSD of 0.21 μm particles as a function of time τ in xanthan-rich regions of emulsions (30 vol.% oil, 1.4 wt.% sodium caseinate, pH 6.8) with different concentrations of xanthan in the aqueous phase:* ▲, *0.03 wt.%;* △, *0.04 wt.%;* ●, *0.05 wt.%;* ○, *0.06 wt.%;* ■, *0.07 wt.%. The straight line has a slope of unity*

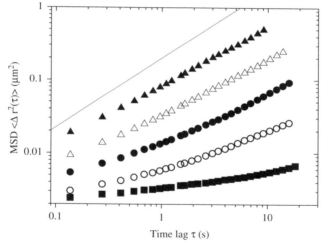

Figure 9 *Ensemble-averaged MSD of 0.5 μm particles as a function of time τ in oil-droplet-rich regions of emulsions (30 vol.% oil, 1.4 wt.% sodium caseinate, pH 6.8) with different concentrations of xanthan in the aqueous phase:* ▲, *0.03 wt.%;* △, *0.04 wt.%;* ●, *0.05 wt.%;* ○, *0.06 wt.%;* ■, *0.07 wt.%. The straight line has a slope of unity*

influence of the added biopolymer on the bulk emulsion rheology (see Figure 3) and stability is *via* its effect on the oil droplet network. This is because it is the microrheology of the interconnected flocculated droplet network which is the determining factor in relation to the susceptibility of the emulsion to separation

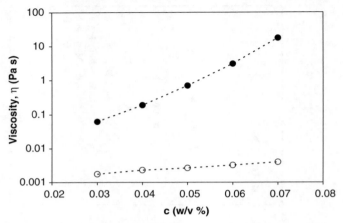

Figure 10 *Comparison of viscosities determined by particle tracking in the phase-separated regions of emulsions (30 vol.% oil, 1.4 wt.% sodium caseinate, pH 6.8) as a function of the concentration c of xanthan in the aqueous phase:* ○, *xanthan-rich aqueous regions;* ●, *flocculated oil-droplet-rich regions*

under gravity – and not the viscoelasticity of the xanthan-containing microregions (with or without dispersed oil droplets).

20.4 Concluding Remarks

The use of particle tracking combined with confocal microscopy appears to offer new opportunities for understanding the texture and stability of food colloids and for systematically controlling the processing and shelf life of food products. Here we have illustrated the application of the technique for the case of a protein-stabilized emulsion with added hydrocolloid thickening agent. From qualitative visual observations of the particle diffusive motions, and from quantitative analysis of trajectories in the viscous regime, we have demonstrated unequivocally that the viscosity in the oil-droplet-rich regions is very much greater than in the xanthan-rich phase. Hence we assert that the kinetics of phase separation in this type of emulsion system is predominantly determined by the rheological behaviour of the interconnected oil droplet regions.

These findings complement our confocal microscopy study[14] of the evolving morphology of the xanthan-containing blobs. The shape relaxation time determined by image analysis can now be unambiguously related to the dominant viscoelasticity of the surrounding regions of the concentrated protein-coated oil droplets, and not to the viscosity of the xanthan-rich blobs themselves. Owing to the flexibility of the weak interdroplet bonds, the flocculated droplet network undergoes reorganization and rearrangement under the combined influence of Brownian motion and gravity. Eventually, the restructuring and reorganization of the flocculated droplets leads to a loss of connectivity in the aggregated microstructure and the transient network collapses. The time of onset of network collapse depends on the effective yield stress of the oil-droplet network and on the relative densities of the microphases. Such a system typically exhibits a characteristic delay time before the emulsion gel finally collapses completely.[11,25] While the stabilizing emulsion gel structure is inherently metastable, it can be very long-lasting for a sufficiently high xanthan concentration.

Hydrocolloid thickeners like xanthan gum are commonly added to pourable salad dressings and other food emulsions to enhance the shelf life during extended storage. The traditional explanation for the stabilizing effect of the hydrocolloid is *via* control of the rheology of the aqueous continuous phase. This explanation appears quite sound at low oil volume fractions, where individual droplets can be separately immobilized in an entangled biopolymer network, and the small buoyancy force acting on each droplet is considered insufficient to overcome the effective yield stress of the surrounding weak gel-like biopolymer matrix. But, a different explanation is required for concentrated emulsions because the evolving microstructure is highly heterogeneous,[14] and it has long been recognized[15,26,27] that the emulsion (in)stability correlates more with the rheology of the emulsion than with that of the hydrocolloid aqueous phase.

On the basis of experimental evidence presented here, we therefore can confirm the stabilization mechanism proposed by Parker and co-workers.[25] That is, the concentrated emulsion is prevented from creaming by the visco-elastic character of interconnected regions of droplets that have become flocculated (and locally further concentrated) under the influence of attractive depletion interactions induced by the non-adsorbed polysaccharide. In essence, the incipiently unstable salad-dressing-like emulsion is transformed, through changes in microrheology, into a stable 'thick' mayonnaise-like emulsion containing embedded blobs of low-viscosity biopolymer-structured water.

Acknowledgements

This research was supported by an EPSRC Studentship with ICI plc (UK). TM thanks the State Scholarships Foundation of Greece for financial support.

References

1. F.C. MacKintosh and C.F. Schmidt, *Curr. Opin. Colloid Interface Sci.*, 1999, **4**, 300.
2. A. Mukhopadhyay and S. Granick, *Curr. Opin. Colloid Interface Sci.*, 2001, **6**, 423.
3. H. Freundlich and W. Seifriz, *Z. Phys. Chem.*, 1922, **104**, 233.
4. A.R. Bauch, W. Moller and E. Sackmann, *Biophys. J.*, 1999, **76**, 573.
5. T. Gisler and D.A. Weitz, *Curr. Opin. Colloid Interface Sci.*, 1998, **3**, 586.
6. T.G. Mason and D.A. Weitz, *Phys. Rev. Lett.*, 1995, **74**, 1250.

7. F. Gittes, B. Schnurr, P.D. Olmsted, F.C. MacKintosh and C.F. Schmidt, *Phys. Rev. Lett.*, 1997, **79**, 3286.
8. B.R. Dasgupta, S.-Y. Tee, J.C. Crocker, B.J. Frisken and D.A. Weitz, *Phys. Rev. E*, 2001, **65**, 051505.
9. H. Qian, M.P. Sheetz and E.L. Elson, *Biophys. J.*, 1991, **60**, 910.
10. M.T. Valentine, Z.E. Perlman, M.L. Gardel, J.H. Shin, P. Matsudaira, T.J. Mitchison and D.A. Weitz, *Biophys. J.*, 2004, **86**, 4004.
11. Y. Cao, E. Dickinson and D.J. Wedlock, *Food Hydrocoll.*, 1990, **4**, 185.
12. Y. Cao, E. Dickinson and D.J. Wedlock, *Food Hydrocoll.*, 1991, **5**, 443.
13. E. Dickinson, J.G. Ma and M.J.W. Povey, *Food Hydrocoll.*, 1994, **8**, 481.
14. T. Moschakis, B.S. Murray and E. Dickinson, *J. Colloid Interface Sci.*, 2005, **284**, 714.
15. E. Dickinson, M.I. Goller and D.J. Wedlock, *Colloids Surf. A*, 1993, **75**, 195.
16. R.H. Tromp, F. van de Velde, J. van Riel and M. Paques, *Food Res. Int.*, 2001, **34**, 931.
17. T. Moschakis, B.S. Murray and E. Dickinson, *Langmuir*, 2006, **22**, 4710.
18. I. Burgaud, E. Dickinson and P.V. Nelson, *Int. J. Food Sci. Technol.*, 1990, **25**, 39.
19. E. Dickinson, M. Golding and M.J.W. Povey, *J. Colloid Interface Sci.*, 1997, **185**, 515.
20. S.J. Radford and E. Dickinson, *Colloids Surf. A*, 2004, **238**, 71.
21. J.C. Crocker and D.G. Grier, *J. Colloid Interface Sci.*, 1996, **179**, 298.
22. J.C. Crocker and E.R. Weeks, Software package for particle tracking, available from: http://www.physics.emory.edu/~weeks/idl/tracking.html.
23. *CRC Handbook of Chemistry and Physics*, 85th edn, CRC Press, D.R. Lide (ed), Boca Raton, FL, 2004–2005.
24. E. Dickinson, M.I. Goller and D.J. Wedlock, *J. Colloid Interface Sci.*, 1995, **172**, 192.
25. A. Parker, P.A. Gunning, K. Ng and M.M. Robins, *Food Hydrocoll.*, 1995, **9**, 333.
26. E. Dickinson, *Food Hydrocoll.*, 2003, **17**, 25.
27. E. Dickinson, in *Gums and Stabilisers for the Food Industry*, Vol. 12, P.A. Williams and G.O. Phillips (eds), Royal Society of Chemistry, Cambridge, 2004, p. 394.

Chapter 21

Optical Microrheology of Gelling Biopolymer Solutions Based on Diffusing Wave Spectroscopy

Frédéric Cardinaux, Hugo Bissig, Peter Schurtenberger and Frank Scheffold

DEPARTMENT OF PHYSICS, UNIVERSITY OF FRIBOURG, CH-1700 FRIBOURG, SWITZERLAND

21.1 Introduction

Optical microrheology has emerged as one of the most important new techniques to study and characterize rheological properties of complex fluids. The main principle of passive microrheology is to study the thermal response of small (colloidal) particles embedded in the system under study.[1-4] The particles are either artificially introduced, which is then called 'tracer microrheology', or they can be part of the system itself, *e.g.*, as in the case of yoghurt. By analysing the thermal motion of the particles, it is possible to obtain quantitative information about the storage and loss moduli, $G'(\omega)$ and $G''(\omega)$, over an extended range of frequency ω.

One of the most popular techniques to study the thermal motion of particles is diffusing wave spectroscopy (DWS), which is an extension of the standard technique of photon correlation spectroscopy to turbid media.[5,6] DWS-based microrheology is perfectly suited to a fast and non-invasive determination of the key rheological properties as given by the frequency dependence of $G'(\omega)$ and $G''(\omega)$. In this article we demonstrate how optical microrheology based on DWS can be used to characterize the critical gelation behaviour in physical gels that rapidly evolve in time.

21.2 Critical Gelation Microrheology

The sol–gel transition of a polymeric system is characterized by the instant at which the connectivity of the network extends over the entire sample; this characteristic gel point is considered to be a material parameter and so its value should not be influenced by the method of determination. A commonly employed technique to follow the gelation of colloidal polymer solutions is small-amplitude oscillatory rheometry. The time at which the G' and G'' curves cross each other has been proposed as a simple method for locating the gel point, but in practice it obviously depends on the frequency of the oscillatory measurement.[7] A better way to detect the gel point is the method of Winter and Chambon,[8,9] where the criterion is that the ratio $G''(\omega)/G'(\omega)$ becomes independent of frequency over a wide range of frequencies at the gel point. From the slopes of G' and G'' versus ω at the gel point, a relaxation exponent n may be calculated. Its value has been found to range between 0 and 1 depending on the composition and on the molecular structure of the material.[10,11] The validity of the Winter–Chambon criterion has been demonstrated in various studies including chemically and physically cross-linked gels.[11–13]

A complication with classical rheometry is that the investigation of gelation of weak physical gels may be hampered by the fact that the physical bonds are shear sensitive. Weak physical gels show a significant decrease of the critical strain,[12] which limits the linear viscoelastic region close to the gel point. Moreover the determination of the entire relaxation spectrum in a single run is difficult and time consuming, and in many cases it is not even feasible for such a time-evolving sample. In contrast, DWS provides non-invasive access to the linear viscoelastic properties of the gelling sample at any time during the gelation process. In a recent article[14] we demonstrated that an accurate determination of the gel point of starch dispersions is possible using DWS. For a different system here, we demonstrate that the short time of DWS data acquisition – 1 min or less – turns out to be a highly advantageous factor. Time-dependent changes in the mechanical properties on a minute scale can be resolved while still covering a frequency range of up to seven decades (ω from $\sim 10^{-1}$ to $\sim 10^6$ rad s^{-1}). Furthermore the technique always works in thermal equilibrium; hence no structural changes or distortions of the systems arise due to the application of external shear forces.

21.3 Sample Preparation

A dispersion of sulfated polystyrene latex spheres (diameter 720 nm) (Interfacial Dynamics, Portland, USA) at a concentration of 8.2 g per 100 g of solution was added to a gelatin solution to give a final concentration of 0.5 g of latex particles + 2 g gelatin in 100 g of solution. The polystyrene spheres act as tracer particles to probe the local viscoelastic properties *via* Brownian motion. The particle size is chosen such that the typical relevant length scale ξ (several nanometres) of the biopolymer solutions is much smaller than the particle size ($\xi \ll 720$ nm).

21.4 Diffusing Wave Spectroscopy Based on Optical Microrheology

DWS is the extension of dynamic light scattering (DLS) to soft materials exhibiting strong multiple scattering.[6] It has been shown that DWS allows monitoring of the displacement of micrometre-sized particles with sub-nanometre precision. In recent years significant progress has been made in the development of the DWS approach,[15,16] and it has been successfully applied to the study of fluid and solid media, *e.g.*, colloidal suspensions, gels, and biocolloids (yoghurt and cheese), and also ceramic slurries and green bodies.[17–24]

In a DWS experiment, coherent laser light impinges on one side of a turbid sample and the intensity fluctuations of the light propagated through the sample are analysed either in transmission or in a back-scattering geometry. A diffusion model is used to describe the propagation of photons across the sample, where the diffusion approximation allows determination of the distribution of scattering paths and calculation of the temporal autocorrelation of the intensity fluctuations. DWS provides quantitative information on the average mean-square displacement of the scattering particles from the measured intensity autocorrelation function (ICF) over a very broad range of timescales. Analogous to DLS, for the case of noninteracting particles, we can express the measured ICF,

$$g_2(\tau) - 1 = \langle I(t)I(t+\tau)\rangle/\langle I\rangle^2 - 1, \qquad (1)$$

in terms of the mean-square displacement of the scattering particles:

$$g_2(\tau) - 1 = \left[\int_0^\infty ds\, P(s) \exp\left(-(s/l^*)k^2\langle \Delta r^2(\tau)\rangle\right)\right]^2. \qquad (2)$$

The quantity $k=2\pi n/\lambda$ is the wave number of light in a medium of refractive index n. The function $P(s)$ is the distribution of photon trajectories of length s in the sample, which can be calculated within the diffusion model by taking the experimental geometry into account. The transport mean free path l^* characterizes the typical step length of the photon random walk, as given by the individual particle scattering properties and the particle concentration. The value of l^* can be determined independently, and it enters the analysis as a constant parameter. From Equation (2) it is possible to calculate the particle mean-square displacement ($\langle \Delta r^2(t)\rangle$) numerically from the measured autocorrelation function $g_2(\tau)$.

From the Laplace transform of the particle mean-squared displacement, $\langle \Delta \tilde{r}^2(s)\rangle$, the complex modulus of the sample can be determined *via* a generalized Stokes–Einstein relation using $s = i\omega$:

$$G^*(\omega) = \frac{k_B T}{\pi a\, i\omega \langle \Delta \tilde{r}^2(i\omega)\rangle} = G'(\omega) + iG''(\omega). \qquad (2)$$

The details of this procedure can be found elsewhere.[1,25]

The measurements were performed with a DWS Rheolab instrument from LS Instruments (Switzerland). This DWS instrument allows us to record the intensity correlation function over a wide range of τ (i.e., 10^{-6}–10 s) within approximately a minute. This superior time resolution can be achieved by a recently introduced two-cell echo principle.[15] It allows efficient ensemble averaging of the signal even in the case of slow or arrested dynamics. This is achieved by putting a fast rotating diffuser in the optical path between laser and sample as seen in Figure 1. A 7 mW He–Ne laser (JDS Uniphase) operating at $\lambda = 632.8$ nm is used to illuminate a circular ground glass, which is mounted on a stepper motor. Through scattering and dephasing, the ground glass creates a speckle with a nearly Gaussian optical field. The transmitted light from the ground glass is collected by a lens and focused onto the sample with a resulting spot-size diameter of roughly 5 mm. The scattered light is then collected with a monomode fibre and analysed by a photodetector and a digital correlator (Correlator.com, NJ, USA). The recorded (multi-speckle) correlation echoes provide an ensemble-averaged signal that does not require any additional time averaging. Absolute values of the correlation function were obtained by calibrating the instrument with a solid block of teflon (for which $g_2(t) - 1 \equiv 1$). Access to the short-time dynamics was obtained with the motor at rest. The time-averaged correlation function was corrected for non-ergodicity using the method of Pusey and van Megen.[26] The whole procedure was controlled by the instruments' software package, including l^* calibration, motor control, data handling, and microrheological analysis. A full intensity correlation function could be recorded within 1 min.

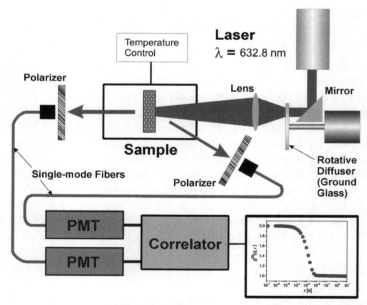

Figure 1 *The DWS experimental arrangement*

21.5 Results and Discussion

The gelatin solution was filled into a 5-mm path length cuvette (Hellma, Germany). Before the measurement the sample was heated to ∼60°C, *i.e.*, well above the melting point, and subsequently kept at (20 ± 0.1)°C during the actual experiment. Rapid quenching was ensured by immersing the sample in a temperature-controlled water bath for 60 s before transferring it to the sample cell holder (kept in air). We define this instant as time zero.

Figure 2 shows the time evolution of the ICF in transmission geometry. Just after the quench the ICF fully decays to zero, as expected for freely diffusing particles in a purely viscous fluid. As gelation proceeds the recorded ICF shows an increasing slowing down of its characteristic decay time reflecting the growing viscoelastic response felt by the tracers. Once the gel is formed, tracers are trapped in the elastic cage formed by the surrounding biopolymer matrix

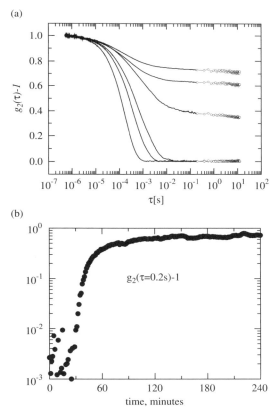

Figure 2 *Time evolution of the recorded intensity correlation function $g_2(\tau)$. Data sets were recorded for 60 s at a rate of one experiment every 75 s. (a) Correlation functions at waiting times of 0, 12, and 24 min and 1, 2, and 4 h. (b) Correlation function for a lag time of $\tau = 0.2$ s. The noise level was found to be well below 1%, despite the short measurement time of only 1 min*

and the ICF develops a plateau whose height can directly be interpreted in terms of the elastic modulus of the system. Figure 2(b) shows the plateau height recorded with the echo technique at a lag time of $\tau = 0.2$ s (corresponding to the time evolution of the elastic modulus G'). In our measurement, the onset of a detectable non-zero plateau value is visible after 40–50 min (Figure 3). The measurement at a single relaxation time, however, does not clearly indicate the location of the gel point. As mentioned above, a better way to detect the gel point is the method of Winter and Chambon,[8,9] where the criterion is that the ratio $G'(\omega)/G''(\omega)$ becomes constant and independent of frequency over a wide range of frequencies at the gel point. Taking advantage of the large window of frequencies accessible with DWS, we have plotted this ratio in Figure 4 for several waiting times. We find that the data set measured at a time of 24 min

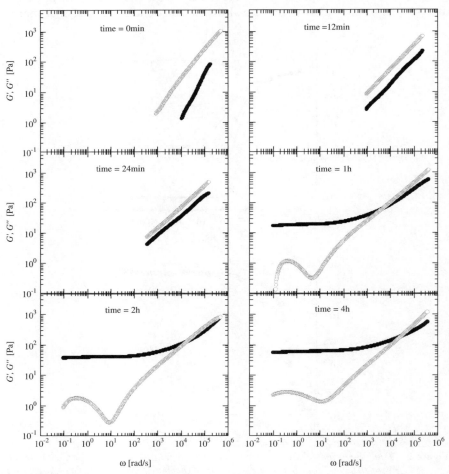

Figure 3 *Frequency dependence of the storage modulus G' (filled circles) and the loss modulus G'' (open circles) of a 2 wt.% gelatin solution kept at 20°C for various times as determined by optical microrheology*

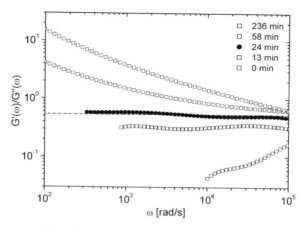

Figure 4 *Frequency dependence of the ratio of the storage and loss moduli, $G'(\omega)/G''(\omega)$, for different waiting times. The ratio is found to be constant over the full range of accessible frequencies at time 24 min (●), thereby defining the gel point*

gives the best fit the Winter–Chambon criterion. At this waiting time the slope is $n = 0.8$ over the full range of accessible frequencies.

21.6 Conclusions

The use of DWS as an optical probe of local tracer particle motion provides noninvasive access to the linear viscoelastic properties of a gelling sample at any time during the gelation process. The detailed information about $G'(\omega)$ and $G''(\omega)$, in particular the high-frequency part, provides a new basis to describe local relaxation processes. Thereby a better understanding of the molecular and supramolecular changes underlying the sol–gel transition can be obtained.

Our experimental setup allows simultaneous capture of the fast and slow dynamics of the system, and it thus provides access to the viscoelastic moduli covering a frequency range of up to six decades or more. The superior time resolution and simplicity of the measurement offers important advantages in the study of gelling systems.

Acknowledgements

We thank Pedro Diaz-Leyva for help preparing the manuscript. Financial support from the Swiss National Science foundation is gratefully acknowledged.

References

1. T.G. Mason and D.A. Weitz, *Phys. Rev. Lett.*, 1995, **74**, 1250.
2. J.L. Harden and V. Viasnoff, *Curr. Opin. Colloid Interface Sci.*, 2001, **6**, 438.

3. F. Cardinaux, L. Cipelletti, F. Scheffold and P. Schurtenberger, *Europhys. Lett.*, 2002, **57**, 738.
4. M. Buchanan, M. Atakhorrami, J.F. Palierne, F.C. MacKintosh and C.F. Schmidt, *Phys. Rev. E*, 2005, **72**, 011504.
5. D.J. Pine, D.A. Weitz, P.M. Chaikin and E. Herbolzheimer, *Phys. Rev. Lett.*, 1988, **60**, 1134.
6. G. Maret and P.E. Wolf, *Zeit. Phys. B*, 1987, **65**, 409.
7. C.Y.M. Tung and P.J. Dynes, *J. Appl. Polym. Sci.*, 1982, **27**, 569.
8. F. Chambon and H.H. Winter, *J. Rheol.*, 1987, **31**, 683.
9. H.H. Winter and F. Chambon, *J. Rheol.*, 1986, **30**, 367.
10. J.C. Scanlan and H.H. Winter, *Makromol. Chem. Symp.*, 1991, **45**, 11.
11. R.G. Larson, *The Structure and Rheology of Complex Fluids*, Oxford University Press, New York, 1998.
12. A.B. Rodd, J. Cooper-White, D.E. Dunstan and D.V. Boger, *Polymer*, 2001, **42**, 185.
13. S.B. Ross-Murphy, *J. Texture Stud.*, 1995, **26**, 391.
14. C. Heinemann, F. Cardinaux, F. Scheffold, P. Schurtenberger, F. Escher and B. Conde-Petit, *Carbohydr. Polym.*, 2004, **55**, 155.
15. P. Zakharov, F. Cardinaux and F. Scheffold, *Phys. Rev. E*, 2006, **73**, 011413.
16. F. Scheffold and P. Schurtenberger, *Soft Mater.*, 2003, **1**, 139.
17. H.M. Wyss, S. Romer, F. Scheffold, P. Schurtenberger and L.J. Gauckleret, *J. Colloid Interface Sci.*, 2001, **241**, 89.
18. P. Schurtenberger, A. Stradner, S. Romer, C. Urban and F. Scheffold, *Chimia*, 2001, **55**, 155.
19. S. Romer, F. Scheffold and P. Schurtenberger, *Phys. Rev. Lett.*, 2000, **85**, 4980.
20. Y. Nicolas, M. Paques, D. van den Ende, J.K.G. Dhont, R.C. van Polanen, A. Knaebel, A. Steyer, J.-P. Munch, T.B.J. Blijdenstein and G.A. van Aken, *Food Hydrocoll.*, 2003, **17**, 907.
21. E. ten Grotenhuis, M. Paques and G.A. van Aken, *J. Colloid Interface Sci.*, 2000, **227**, 495.
22. D.G. Dalgleish and D.S. Horne, *Milchwissenschaft*, 1991, **46**, 417.
23. D.G. Dalgleish and D.S. Horne, *Abstract of American Chemical Society*, Vol. 201, 68-COLL Part 1, April 14, 1991.
24. M. Alexander and D.G. Dalgleish, *Langmuir*, 2005, **21**, 11380.
25. T.G. Mason, *Rheol. Acta*, 2000, **39**, 371.
26. P.N. Pusey and W. van Megen, *Physica A*, 1989, **157**, 705.

Chapter 22

Gel and Glass Transitions in Short-Range Attractive Colloidal Systems

Giuseppe Foffi,[1,2] Nicolas Dorsaz[1,2] and Cristiano De Michele[3,4]

[1] INSTITUT ROMAND DE RECHERCHE NUMÉRIQUE EN PHYSIQUE DES MATÉRIAUX IRRMA, PPH-ECUBLENS, CH-105 LAUSANNE, SWITZERLAND
[2] PHYSICS DEPARTMENT, ECOLE POLYTECHNIQUE FÉDÉRALE DE LAUSANNE (EPFL), CH-1015 LAUSANNE, SWITZERLAND
[3] DIPARTIMENTO DI FISICA AND INSTITUTO NAZIONALE PER LA FISICA DELLA MATERIA (INFM), UNIVERSITÀ DI ROMA LA SAPIENZA, PIAZZALE ALDO MORO 2, 00185 ROME, ITALY
[4] INSTITUTO NAZIONALE PER LA FISICA DELLA MATERIA (INFM)—CRS SOFT, UNIVERSITÀ DI ROMA LA SAPIENZA, PIAZZALE ALDO MORO 2, 00185 ROME, ITALY

22.1 Introduction

In the last few decades, the advances in colloidal science have made possible the realization of systems that are characterized by interactions tuneable in range and strength. The size and the shape of the particle can also be finely controlled. One possibility to deal theoretically with the physical properties of these systems is to devise a proper effective interparticle potential by integrating out the solvent (and cosolute) degrees of freedom.[1] Once this step has been carried out, each colloidal particle can be treated as a 'big' atom and the tools of statistical mechanics can be applied.

Interactions in atomic systems are characterized by a range that is long compared to the atomic size and they cannot be adjusted externally. The possibility of controlling the range of the interaction and the particle size opens up a whole range of phenomena that are not observed in atomic systems. Moreover, the larger size of colloidal particles allows scientists to use optical experimental technique such as light scattering and confocal microscopy to investigate the structural and dynamic properties.

Hard spheres are a striking example of the possibilities offered by colloidal systems. A freezing transition was predicted for hard spheres in pioneering work more than 30 years ago.[2] Some 20 years later, a first-order phase transition of purely entropic origin, from a liquid to a crystal, was experimentally detected[3] in a system of colloids with a steep repulsive interaction.

Another interesting case is represented by systems with a short-range attractive interaction – one that is 'short' as compared to the particle size. This is a case that cannot be studied in atomic systems. For example, when polymers with radius of gyration much smaller than the diameter of the colloidal particles are added to a hard sphere system, an effective attractive interaction sets in. In particular, the radius of gyration of the polymer (or, alternatively, a smaller colloidal particle) modulates the range of the attraction. In the short-range attractive colloidal system (SRACS), new thermodynamic behaviour emerges as predicted by theory[4] and simulations,[5] *i.e.*, when the range of the attraction is short enough, the liquid–liquid phase separation becomes metastable with respect to the fluid–crystal phase separation. It is interesting to note that the metastable liquid–liquid critical point still plays an important role owing to the density fluctuations that can favour crystal nucleation, a fact that is of great importance for protein crystallization.[6]

Colloidal systems play an important role also in studying dynamics. For purely hard spheres, for example, a kinetic arrest has been found[7] when a system of sterically stabilized PMMA particles is prepared at very high densities (*i.e.*, at packing fractions above 58%). At these densities the system is frozen into a disordered structure, and it does not relax anymore towards equilibrium, *i.e.*, it is a glass. The existence of this glassy phase has been of great importance to test the predictions of mode coupling theory (MCT),[8] one of the few predictive theories of the glass transition. Indeed, in a set of beautiful experiments, van Megen and Underwood[9] gave the first proof of the predictions of MCT. The phenomenology of the SRACS is even richer. When the range of the system is short, MCT predicts a re-entrant line in the density temperature plane.[10–12] This means that, above a certain density, it is possible to melt the glass on lowering the temperature (*i.e.*, increasing the attraction between particles) and to vitrify it again on further decreasing it. The general phase diagram for the SRACS is depicted schematically in Figure 1. In the high-density region, a re-entrant glass line is generated by the competition between attractive and repulsive forces. Two glassy phases with different dynamic properties, named repulsive and attractive, are generated by these two distinct arrest mechanisms. The phenomenology of the dynamic arrest at large packing fractions is now well established and accepted as a result of many simulations[13–15] and experiments.[16–18] It is also worth mentioning that new phenomena have also been predicted, such as higher-order MCT singularities and an anomalous logarithmic decay in the density correlators.[12]

At low density, the SRACS is known to form an aggregated gel, where a phase-spanning arrested structure leads to a yield stress (see, for example, the recent review by Trappe and Sandkuhler[19] and references therein). When the range of the attraction is reduced enough, MCT predicts that the attractive glass line moves to low density;[11,20] and it has been speculated[11,21] that this line (depicted as the dashed line in Figure 1) is indeed coincident with the gel line in the SRACS.

A useful tool for clarifying these phenomena is computer simulation that allows investigation of the relationship of the interaction to the dynamics. In

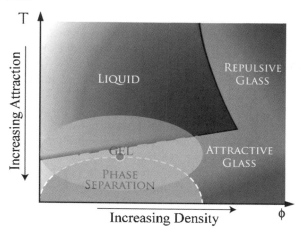

Figure 1 *The general phase diagram for the short-range attractive colloidal system (SRACS)*

this article we discuss results for the SRACS most especially connected to gel formation, a phenomenon ubiquitous in nature, but still far from understood. We always keep open the connection to the glass transition that is present on the high-density side of the phase diagram. Recent results tend to interpret the gel transition in the SRACS as an arrested phase separation where the tendency of the system to form an inhomogeneous structure is frozen in by the tendency of the particle to stick together.[22,23] We shall discuss how the possibility to reach a percolating arrested structure from the one-component fluid (*i.e.*, without any phase separation) seems ruled out by the computer simulations.[24]

22.2 The Gel Transition: Arrested Phase Separation

Within the framework that treats colloidal particles as atoms with a tuneable interaction, the model we shall discuss is a system of hard spheres interacting *via* a square-well potential. We simulate a 1:1 binary mixture of 2000 spherical particles with a size ratio of $d_1/d_2 = 1.2$ which effectively suppresses crystallization. The interaction potential is defined by

$$u_{ij(r)} = \begin{cases} \infty, & r < d_{ij}, \\ -u_0, & d_{ij} < r < d_{ij} + \Delta_{ij}, \\ 0, & r > d_{ij} + \Delta_{ij}, \end{cases} \quad (1)$$

where $d_{ij} = (d_i + d_j)/2$, with d_i being the hard-core diameter of the particle of component i (=1,2) and Δ_{ij} representing the range of the attraction. We shall measure the range of the attraction by the non-dimensional variable $\varepsilon = \Delta_{ij}/(d_{ij} + \Delta_{ij})$ with $\varepsilon = 0.005$. This chosen ε value is arbitrary, but it is representative of all square-well potentials with interaction range smaller than a few percent as far as the thermodynamic and equilibrium properties are concerned.[24,25] For the chosen potential the critical temperature is $T_c = 0.20$.

We choose $k_B = 1$ and set the depth of the potential to $u_0 = 1$. Hence $T = 1$ corresponds to a thermal energy equal to the attractive well depth. The diameter of the smaller species is chosen to be the unit of length, i.e., $d_2 = 1$. Density is expressed in terms of packing fraction $\phi = (\rho_1 d_1^3 + \rho_2 d_2^3)(\pi/6)$ where $\rho_i = N_i/L^3$, L being the box size and N the number of particles of component i. Time is measured in units of $d_2(m/u_0)^{1/2}$.

Newtonian dynamics (ND) has been coded via a standard event-driven algorithm, commonly used for particles interacting with step-wise potentials.[26] Between collisions, the particles move along straight lines with constant velocities. When the distance between the particles becomes equal to the distance where the potential has a discontinuity, the velocities of the interacting particles instantaneously change. The algorithm calculates the shortest collision time in the system and propagates the trajectory from one collision to the next one. Calculations of the next collision time are optimized by dividing the system in small sub-systems, so that collision times are computed only between particles in the neighbouring sub-systems.

Brownian dynamics (BD) has been implemented via the position Langevin equation,

$$\vec{r}_k = \frac{D_0}{k_B T} \vec{f}_k(t) + \vec{r}_k(t), \qquad (2)$$

using the algorithm coding developed by Scala et al.[27] In Equation (2), $r_k(t)$ is the position of particle k, $f_k(t)$ the total force acting on the particle k, D_0 the short-time (bare) diffusion coefficient, and $\vec{r}_k(t)$ a random thermal noise satisfying

$$\left\langle \vec{r}_k(0) \vec{r}_k(t) \right\rangle = k_B T \delta(t). \qquad (3)$$

In this algorithm, a random velocity (extracted from a Gaussian distribution of variance $\sqrt{k_B T/m}$) is assigned to each particle and the system is propagated for a finite time-step, $2mD_0/k_B T$, according to event-driven dynamics. We chose D_0 such that short-time motion is diffusive over distances smaller than the well width.

The phase diagram for the binary mixture is shown in Figure 2. A clear way to indicate the position and shape of the glass line is represented by the equi-diffusivity lines,[14] defined as the loci where D/D_0 is constant. Here D is the diffusion coefficient calculated from the equilibrium mean square displacement (MSD) using the Einstein relation normalized by the ND bare diffusivity, $D_0 \equiv (k_B T d_2^2/m)^{1/2} = v_{th} d_2$, to take into account the differences in the microscopic time because of the different thermal velocity v_{th}. The equi-diffusivity lines already show signs of the re-entrant behaviour that, at high density, characterizes the liquid pocket between the repulsive and the attractive glass; moreover, the extrapolation to the expected glass line is also plotted.[28]

One of the most used models to describe interaction in the SRACS is the Baxter potential[29] that describes the short-attractive interaction in the limit of

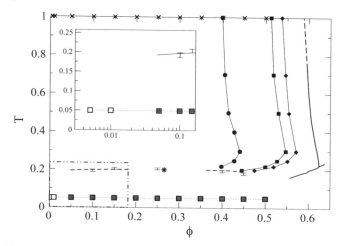

Figure 2 *The numerical phase diagram, temperature T versus particle packing fraction ϕ, indicating the quench analyzed. The crosses are the initial configurations, and the squares the final ones. The squares are shaded if the final structure is a percolating gel. The dashed line and the star are the liquid–liquid coexistence line and the critical point from Miller and Frenkel.[30,42] The continuous lines with symbols are the equi-diffusivity lines, $D/D_0 = 0.01, 0.005$ and 0.001, going from left to right. The bold line (with no symbol) is the extrapolated glass line.[28] The inset shows an enlargement of the low packing fraction region on a logarithmic scale*

vanishingly small range. We shall use this model to estimate the location of the coexistence curve. This potential $V_B(r)$ is defined according to

$$\exp(-\beta V_B(r)) = \theta(r-d) + \frac{d}{12\tau}\delta(r-d), \quad (4)$$

where τ is an adhesive parameter that plays the role of an effective temperature, θ and δ are the Heaviside and the Dirac functions, respectively, and $\beta = 1/k_B T$. The main feature of this model is that it allows us to take the limit of a vanishing interaction range by keeping the second virial coefficient finite, *i.e.*, keeping fixed the overall strength of the interaction. The success of this model is due to the fact that it can be solved analytically within the Percus–Yevick approximation, and that it can be successfully used to interpret experimental data. The adhesive parameter τ is directly related to the virial coefficient by the relation:

$$B_2^* = \frac{B_2}{B_2^{HS}} = 1 - \frac{1}{4\tau}. \quad (5)$$

In this way is it possible to relate the adhesive parameter to the temperature of a finite square-well model by equating the virial coefficients of the two models. Doing this we obtain:

$$1 - \frac{1}{4\tau} = 1 - (e^{\beta u_0} - 1)[(1-\varepsilon)^{-3} - 1]. \quad (6)$$

While the coexistence curve can be calculated analytically with Percus–Yevick theory, the outcome is not reliable. Indeed, the predicted behaviour is thermodynamically inconsistent since the two equations of state obtained by the energy and compressibility route give different results. A more sensible route is to use the results of numerical simulations of Miller and Frenkel,[30,31] who performed accurate Monte Carlo simulations to evaluate the correct Baxter coexistence curve. These data provide an accurate estimate of the coexistence curve of all short-range interacting potentials, via an appropriate law of corresponding states. Indeed, Noro and Frenkel[25] noticed that, when the attraction range is much smaller than a few percent of the particle diameter, the value of the first virial coefficient at the critical temperature is essentially independent of the details of the potential, down to the Baxter limit. Following this prescription we report in Figure 2 the coexistence curve calculated by Miller and Frenkel[30,31] in the ϕ–τ plane after transforming it to the T plane according to Equation (5). To give independent support to the mapping, we calculated explicitly the location of the spinodal line, by bracketing it with the lowest T value at which no phase separation takes place and the highest T value at which we observe a growing spinodal decomposition peak. The calculated spinodal line, also reported in Figure 2, is in very good agreement with Miller and Frenkel estimates.

Figure 2 also illustrates the protocol of our numerical experiment.[23] We prepared several independent configurations equilibrated at $T = 1.0$ at packing fractions ranging from 0.01 to 0.50. These configurations were then quenched at temperature $T_f = 0.05$, deep down in the liquid–liquid phase separating region. The temperature was kept fixed by coupling the system to a thermostat, and the evolution was then followed as a function of time. This same procedure has been followed for both Newtonian and Brownian dynamics.

The first observation is that, if the packing fraction is lower than a certain threshold (the open symbol in Figure 2), the system does not form any percolating structure; whereas above this threshold value (the filled symbols), the system gets arrested in an out-of-equilibrium state. This is an indication that the percolation threshold has been passed. Indeed, it is possible to evaluate the percolation line for this model: it passes close to the critical point, intersects the bimodal, and moves to lower density.[30] Indeed, the transition from a percolating to a non-percolating structure is seen better following the time evolution of the potential energy per particle, U/N.

An important property of the square-well system is the fact that the bonds between the particles are defined with no ambiguity, i.e., two particles are bonded if they are within each other's attractive well. In this way the number of bonds per particle is directly proportional to the energy, i.e., $-2U/(Nu_0)$. In Figure 3, we present the evolution of the energy per particle versus time at the various packing fractions for both ND and BD. This evolution presents three characteristic regimes. At the beginning the energy remains constant for a time that gets longer the lower the packing fraction. This is a reflection of the typical free path for the particles, which increases on lowering the density and regulates the formation of large aggregates. After this initial stage the aggregation process starts to set in. This slows down significantly once a number of bonds of the order of 6 ($U/N \approx -3$) per particle is reached. Two different types of behaviour can be distinguished. At very low density ($\phi < 0.05$) the energy continues to drop and does not seem to reach any stationary value in the simulation time window. For $\phi > 0.05$, however, the time dependence of the energy abruptly stops and the system does not show signs of further evolution. This can be seen as indicating that an arrested structure has formed, consistent with growing evidence, numerical[24,32–35] and experimental,[22] that the gel state results from an arrested phase separation when the phase separation dynamics generates regions of local density sufficiently large to undergo an *attractive* glass transition. If the system were to be quenched to a higher temperature, but one still below the critical point, the system would present a drift in the energy, indicating that the separation process does not arrest during the simulated time.[23]

We report also in Figure 3 the BD results. Despite the different microscopic dynamics, the time dependence of the energy shows a similar trend, and the value of the energy at which the system stops for different densities is indeed very similar to the ND case. Comparing in more detail the two simulations, one notices that the Brownian aggregation dynamics is smoother compared to the Newtonian one. The reason for this difference in behaviour is probably the fact that the ND momentum conservation law requires many-body interactions for cluster aggregation. Indeed, two-body interactions between a pair of monomers cannot produce a bounded dimer state; at least three particles have to interact to bond. Once this happens small clusters start to appear in the system and consequently the aggregation speeds up significantly. As shown in the inset of

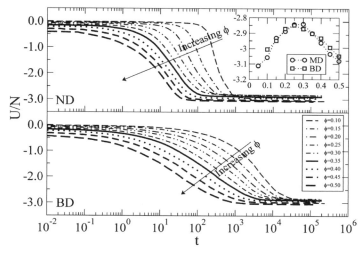

Figure 3 *Evolution of the potential energy per particle, U/N, in Newtonian dynamics (above) and Brownian dynamics (below) as a function of time for different packing fractions above the percolation threshold. The inset shows the energy of the final structure versus packing fraction ϕ*

Figure 3, we can argue that the final energies reached by the two different microscopic dynamics (ND and BD) are equal within numerical error. This is not a trivial issue, since one would expect the microscopic dynamics to play an important role in an out-of-equilibrium situation like the one described above. This suggests that it is possible to use ND to describe the non-equilibrium dynamics of arrested phase separation, with great advantage in term of computational time. This possibility can be exploited when we are interested, as in the present case, in the final arrested structure. But if one is primarily interested in the kinetics of the process (*i.e.*, how the gel structure forms), the issues concerning the microscopic dynamics should be properly addressed.

It is interesting to note that, on the overall range of densities for this type of quench, the final energy presents a maximum close to the critical packing fraction. Under equilibrium conditions, one would expect the energy to grow with the packing fraction; but, in this out-of-equilibrium situation, this is not the case. The origin of this maximum could be related to the strength of the critical fluctuations.

We now turn our attention to the structure of the aggregates. We simulate a one-component system made of 10^4 particles to study large length scales. As before we set $\varepsilon = 0.005$, and configurations equilibrated at $T = 1.0$ were quenched at $T_f = 0.05$. We describe the properties that are obtained when the evolution of the system has stopped, *i.e.*, when the energy has become constant.

Figure 4 shows snapshots of three representative configurations at different packing fractions. In each of the illustrated situations, all the particles belong to a single percolating cluster. This structure clearly resembles a gel with a highly inhomogeneous percolating arrested structure. It is relevant to stress that the percolating structure is formed during the separation process. On increasing the density the percolating structure becomes more uniform and the degree of inhomogeneity is reduced.

One useful concept for gels involves considering the fractal properties of the aggregates.[36] While this concept applies strictly only to extremely low-density aggregates, it is natural to try to understand its degree of applicability to the gels discussed so far. In order to do this we study the integrated radial distribution function

$$n(r) = 4\pi\rho \int_1^r x^2 g(x) dx, \qquad (7)$$

where ρ is the number density, and $g(r)$ the radial distribution function that can be directly evaluated from the simulations. The quantity defined in Equation (7) represents the average number of particles within a distance r of another particle. It can be shown[37] that this quantity can be used to study the fractal dimension since $n(r) \sim (r/r_0)^{d_f}$, where r_0 is a typical length scale and d_f the fractal dimension. The fractal regime is characterized by a typical length scale ξ; for distances larger than ξ, we have $g(r) \sim 1$ and $d_f = 3$; the fractal scaling regime holds for $r_0 \leq r \leq \xi$. Hence, the length scale ξ represents the crossover

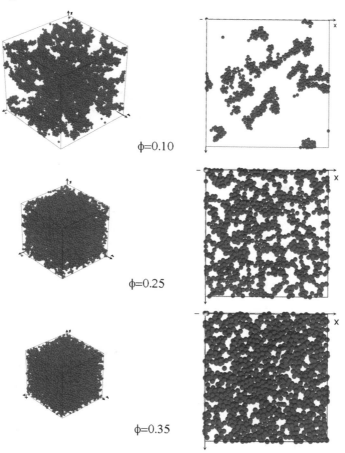

Figure 4 *Snapshots of the final structure for three representative packing fractions. On the left are three-dimensional representations. On the right are two-dimensional slabs*

between the fractal regime and the homogeneous regime, and it can be extracted from the relation[37]:

$$\xi = r_0 \left(\frac{\phi}{n_0}\right)^{1/d_f - 3}. \qquad (8)$$

The quantity n_0 is obtained from the relation $n(r) = n_0 (r/r_0)^{d_f}$ by fitting the numerical results and taking r_0 equal to the radius of the particles. The integrated distribution functions for different packing fractions are shown in Figure 5. For the most heterogeneous case ($\phi = 0.1$), there is a scaling regime characterized by $d_f = 2.2$. From the relation in Equation (8), we obtain a characteristic length scale $\xi = 6.9$. This means that, if we consider the structure of our aggregated systems only up to this length scale, the system is indeed

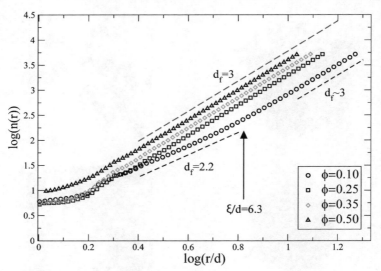

Figure 5 *Log–log plot of integrated radial distribution function n(r) for representative packing fractions. The inferred fractal dimension d_f for the various regimes is indicated, and also the characteristic length scale ξ for $\phi = 0.1$*

fractal. As expected, for $r \geq \xi$, the system exhibits a non-fractal uniform structure. Moving to higher packing fractions we could not find any sign of a fractal structure. Hence it appears that a gel with some fractal structure can be obtained only for a density close to the percolation threshold. When the system is denser, the percolating structure is thicker and compact.

So far we have shown that no fractal structure emerges on increasing the density of the arrested phase separated structure, but we still have to consider whether the system is more homogeneous. To do this we study the static structure factor along the isotherm $T_f = 0.05$. The static structure factor is defined as

$$S_q(t_f) = \langle \rho^*(q,t_f)\rho(q,t_f) \rangle, \qquad (9)$$

where the density variable is defined as

$$\rho(q,t_f) = \sum_{k=1,N} \frac{\exp(i\vec{q}\cdot\vec{r}_k(t_f))}{\sqrt{N}}, \qquad (10)$$

and the time t_f is the final time of the simulation. The structure factor has been studied in ageing in the arrested phase separation regime[23] and also at high density, *i.e.*, in the glass.[38] When the system presents strong inhomogeneities on a length scale larger than the particle diameter, a peak in $S(q)$ emerges at small q. In our case this is due to spinodal decomposition. For $\phi = 0.1$, there is low-q peak that strongly indicates an open structure as also seen from the snapshots of the configuration in Figure 6. The peak moves to higher q on increasing the

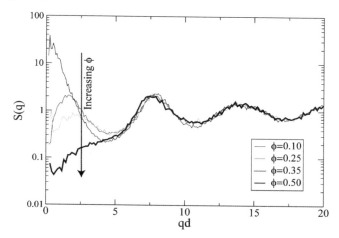

Figure 6 *Static structure factor $S(q)$ of the final configuration at representative packing fractions*

packing fraction, an indication that the typical length scale of the inhomogeneities is decreasing. At the same time the intensity of the peak decreases. Once the packing fraction $\phi = 0.5$ is reached, the static structure factor has no sign of the low-q peak. At such a high packing fraction, the system is so dense that the particles do not have time to diffuse and rearrange, and the homogeneous structure of the quenched liquid is retained. This re-enforces the idea that the gel formed from the SRACS is a spatially inhomogeneous attractive glass (in the MCT fashion[12]), where the inhomogeneity is built by the phase separation process. In this way, we have shown a clear connection between the gel – as an arrested phase separation – and the attractive glass.

Before concluding this section, we discuss the pore distribution function $P(r)$ developed to quantify the porosity of heterogeneous materials,[39] and which has also been used to provide a description of the gel structure obtained by computer simulations.[40] This quantity is defined as the probability, $P(r) \, dr$, that a randomly chosen point inside the pore volume lies between r and $r + dr$ from the pore surface; it is normalized to unity. In Figure 7, we plot the pore distribution function for the initial and final configurations for three packing fractions, $\phi = 0.1, 0.25$, and 0.35. In the initial configurations the particles are homogeneously distributed and the pore-size distribution function presents a peak that reflects the typical inter-particle distance. In the final stage of aggregation, however, the pore distribution decays to zero more slowly. Indeed no peak is present but just a broad distribution of distances that get wider the lower the density. For $\phi = 0.1$, for example, $P(r)$ is spread over more than three diameters, indicating an extremely open structure. For the other two cases this decay takes place on shorter length scale, indicating a more compact structure. In the lower panel of Figure 7, we show the time evolution of the pore distribution function for the case $\phi = 0.25$. To relate the evolution of this

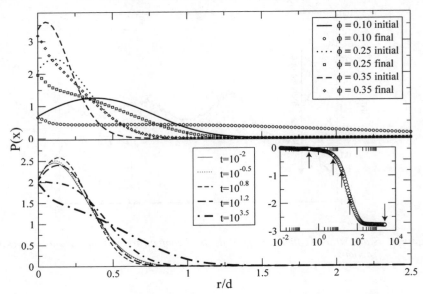

Figure 7 *Upper plot shows pore distribution function P(x) at the initial and final configurations for three representative densities. Lower plot shows evolution in time of pore distribution function for the case $\phi = 0.25$; the energy is plotted versus time in the inset. The arrows indicate the time at which the pore distribution function is evaluated*

function to other observables, we also show in the inset the evolution of the energy. During the first stage, the peak of $P(r)$ does not move. The only change is in the height of the peak, which displays a limited growth, indicating an initial local rearrangement of the particles. In a second stage the peak disappears, following the decrease of the energy, and at same time the pore distribution gets wider, *i.e.*, the pores are growing in size. This confirms that the fast drop in energy and the formation of the open structure proceed simultaneously.

22.3 Dependence on Range of Attraction: Can an Arrested State Emerge from Equilibrium?

In the previous section we discussed how a gel can be obtained by quenching the system into the phase separation region. A question remains open: is it possible to arrest the system without the strong density fluctuations that enhance the formation of open structures?

One of the possibilities that have been explored in the past is the control of the dynamics by varying the range of attraction. Indeed, within MCT there is a critical interaction range below which the attractive glass line takes over the phase separation,[20] and this feature has been used to interpret experimental results for the gel transition.[21] Here we review some results discussed in detail elsewhere.[24]

In general, the MCT overestimates the location of the glass lines and a non-negligible mapping must be applied to theoretical curves before comparing theory with experiment or simulation data.[28,41] Such a mapping has been evaluated for the same square-well potential discussed so far with an attractive range of $\varepsilon = 0.03$. The mapped attractive glass line has been found[35] to end on the right side of the spinodal, hence proving that for this range the attractive glass does not overtake the phase separation.

We have studied the system for different ranges of attraction, ranging from a few percent of the particle diameter down to $\varepsilon = 5 \times 10^{-6}$, which is very close to the Baxter limit. As we discussed before, all the thermodynamic results for short-range systems can be described by the density and the second virial coefficient, i.e., systems with same virial coefficient and at the same density are taken as being in the same thermodynamic state. This can be seen in the lower panel of Figure 8 where the Miller and Frenkel coexisting curve[30] is plotted. As discussed before, all the data from the simulations for the different ranges can be plotted using the same variables, and they agree with the numerical Baxter coexistence curve. In order to quantify the change of the dynamics, we have studied the tagged MSD $\langle r^2(t) \rangle$ that is directly related to the diffusion coefficient by the Einstein relation:

$$\lim_{t \to \infty} \frac{\langle r^2(t) \rangle}{t} = 6D. \qquad (11)$$

We consider two state points, also shown in the lower panel of Figure 8, at two packing fractions, $\phi = 0.2$ and 0.4, to the right and the left of the critical point $\phi_c \sim 0.25$, respectively, and with a virial coefficient value $B_2^* = -0.405$ close to the critical one.

In the upper panel of Figure 8, the MSD is presented for all values of the range investigated in the case of Newtonian dynamics. As we are interested in the structural variations of the diffusion constant, we have to take into account the fact that, in going to the lower limit of the range, we reduce the temperature and consequently the thermal velocity of the monomer. This is achieved by multiplying the time by the bare diffusivity of the single monomer. All the MSD data lie on the same curve, a clear indication that there is no slowing down of the dynamics for either of the two state points considered. Consequently, we have to conclude that it is not possible to obtain a dynamic arrest by reducing the interaction potential range. Similar to what was discussed for the arrested phase separation, this conclusion does not depend upon the microscopic dynamics. In the middle panel of Figure 8, we show the same results for BD. Even if we cannot reach values of the range as short as before, owing to numerical limitations of BD, we can confirm the prediction from ND. We can finally add the equi-diffusivity lines to the phase diagram of Figure 8. In doing so, we have obtained a phase diagram that is universal, not only in its thermodynamics, but also in its dynamics.

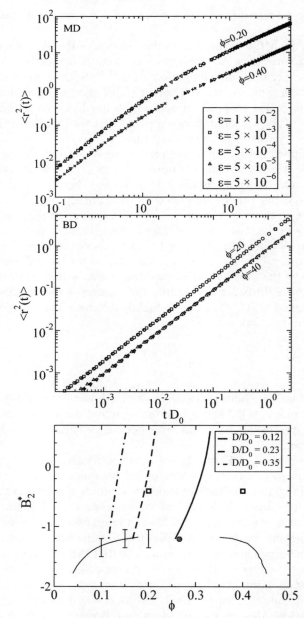

Figure 8 *Mean-square displacement obtained by (a) ND and (b) BD at different values of the range for two packing fractions. (c) The final universal phase diagram. The coexistence line and the critical points are taken from Miller and Frenkel[30,42] and the vertical error bars are our estimations of the spinodal from the simulation. The two squares are the two state points at which we consider the dynamics. The equi-diffusivity lines are also shown*

22.4 Conclusions

We have discussed some of the properties of the gel state as obtained by computer simulations on SRACS. It emerges that the only way to obtain a gel – an open ramified structure that spans the entire simulation box – is through a non-equilibrium quench of the system deep into the phase-separating region. We have characterized this arrested structure by different observables, and we have noticed how these observables are independent of the microscopic dynamics. This feature is non-trivial for a non-equilibrium situation like the one considered here.

Quenching the system to the same final temperature, the structure of the aggregates has been found to depend strongly upon the density at which the quenches are performed. Moving to higher densities, the system loses its open structure and is more homogeneous, becoming *de facto* an attractive glass. The possibility of a glass transition for short enough range of interaction potential is ruled out by the simulations.

Acknowledgement

We thank Prof. F. Sciortino for discussions and comments on the manuscript.

References

1. C.N. Likos, *Phys. Rep.*, 2001, **348**, 267.
2. W.G. Hoover and F.H. Ree, *J. Chem. Phys.*, 1968, **49**, 3609.
3. P.N. Pusey and W. van Megen, *Nature*, 1986, **320**, 340.
4. A.P. Gast, C.K. Hall and W.B. Russel, *J. Colloid Interface Sci.*, 1983, **96**, 251.
5. E.J. Meijer and D. Frenkel, *Phys. Rev. Lett.*, 1991, **67**, 1110.
6. P.R. ten Wolde and D. Frenkel, *Science*, 1997, **277**, 1975.
7. P.N. Pusey, in *Liquids, Freezing and Glass Transition*, D.L.J.P. Hansen and J. Zinn-Justin (eds), North-Holland, Amsterdam, 1991, p. 763.
8. W. Gotze, in *Liquids, Freezing and Glass Transition*, D.L.J.P. Hansen and J. Zinn-Justin (eds), North-Holland, Amsterdam, 1991, p. 287.
9. W. van Megen and S.M. Underwood, *Phys. Rev. Lett.*, 1993, **70**, 2766.
10. L. Fabbian, W. Gotze, F. Sciortino, P. Tartaglia and F. Thiery, *Phys. Rev. E*, 1999, **59**, R1347.
11. J. Bergenholtz and M. Fuchs, *J. Phys., Condens. Matter*, 1999, **11**, 10171.
12. K. Dawson, G. Foffi, M. Fuchs, W. Gotze, F. Sciortino, M. Sperl, P. Tartaglia, T. Voigtmann and E. Zaccarelli, *Phys. Rev. E*, 2001, **63**, 011401.
13. A.M. Puertas, M. Fuchs and M.E. Cates, *Phys. Rev. Lett.*, 2002, **88**, 098301.
14. G. Foffi, K.A. Dawson, S.V. Buldyrev, F. Sciortino, E. Zaccarelli and P. Tartaglia, *Phys. Rev. E*, 2002, **65**, 050802.

15. E. Zaccarelli, G. Foffi, K.A. Dawson, S.V. Buldyrev, F. Sciortino and P. Tartaglia, *Phys. Rev. E*, 2002, **66**, 041402.
16. K.N. Pham, A.M. Puertas, J. Bergenholtz, S.U. Egelhaaf, A. Moussaid, P.N. Pusey, A.B. Schofield, M.E. Cates, M. Fuchs and W.C.K. Poon, *Science*, 2002, **296**, 104.
17. T. Eckert and E. Bartsch, *Phys. Rev. Lett.*, 2002, **89**, 125701.
18. S.H. Chen, W.R. Chen and F. Mallamace, *Science*, 2003, **300**, 619.
19. V. Trappe and P. Sandkuhler, *Curr. Opin. Colloid Interface Sci.*, 2004, **8**, 494.
20. G. Foffi, G.D. McCullagh, A. Lawlor, E. Zaccarelli, K.A. Dawson, F. Sciortino, P. Tartaglia, D. Pini and G. Stell, *Phys. Rev. E*, 2002, **65**, 031407.
21. J. Bergenholtz, W.C.K. Poon and M. Fuchs, *Langmuir*, 2003, **19**, 4493.
22. S. Manley, H.M. Wyss, K. Miyazaki, J.C. Conrad, V. Trappe, L.J. Kaufman, D.R. Reichman and D.A. Weitz, *Phys. Rev. Lett.*, 2005, **95**, 238302.
23. G. Foffi, C. De Michele, F. Sciortino and P. Tartaglia, *J. Chem. Phys.*, 2005, **122**, 224903.
24. G. Foffi, C.D. Michele, F. Sciortino and P. Tartaglia, *Phys. Rev. Lett.*, 2005, **94**, 078301.
25. M.G. Noro and D. Frenkel, *J. Chem. Phys.*, 2000, **113**, 2941.
26. D.C. Rapaport, *The Art of Molecular Dynamics Simulation*, 2nd edn, Cambridge University Press, London, 2004.
27. A. Scala, Th. Voigtmann and C. De Michele, submitted for publication.
28. F. Sciortino, P. Tartaglia and E. Zaccarelli, *Phys. Rev. Lett.*, 2003, **91**, 268301.
29. R.J. Baxter, *J. Chem. Phys.*, 1968, **49**, 2770.
30. M.A. Miller and D. Frenkel, *Phys. Rev. Lett.*, 2003, **90**, 135702.
31. M.A. Miller and D. Frenkel, *J. Chem. Phys.*, 2004, **121**, 535.
32. M.T.A. Bos and J.H.J. van Opheusden, *Phys. Rev. E*, 1996, **53**, 5044.
33. K.G. Soga, J.R. Melrose and R.C. Ball, *J. Chem. Phys.*, 1998, **108**, 6026.
34. J.F.M. Lodge and D.M. Heyes, *Phys. Chem. Chem. Phys.*, 1999, **1**, 2119.
35. E. Zaccarelli, F. Sciortino, S. Buldyrev and P. Tartaglia, in *Unifying Concepts in Granular Media and Glasses*, A.F.A. Coniglio and M.N.H. Herrmann (eds), Elsevier, Amsterdam, 2004.
36. A.H. Krall and D.A. Weitz, *Phys. Rev. Lett.*, 1998, **80**, 778.
37. B.H. Bijsterbosch, M.T.A. Bos, E. Dickinson, J.H.J. van Opheusden and P. Walstra, *Faraday Discuss.*, 1995, **101**, 51.
38. G. Foffi, E. Zaccarelli, S. Buldyrev, F. Sciortino and P. Tartaglia, *J. Chem. Phys.*, 2004, **120**, 8824.
39. S. Torquato, *Random Heterogeneous Materials: Microstructure and Macroscopic Properties*, Springer, New York, 2001.
40. M. Whittle and E. Dickinson, *Mol. Phys.*, 1999, **96**, 259.
41. M. Sperl, *Phys. Rev. E*, 2003, **68**, 011401.
42. M.A. Miller and D. Frenkel, *J. Phys. Condens. Matter*, 2004, **16**, S4901.

Chapter 23

Shape and Interfacial Viscoelastic Response of Emulsion Droplets in Shear Flow

Philipp Erni, Vishweshwara Herle, Erich J. Windhab and Peter Fischer

INSTITUTE OF FOOD SCIENCE AND NUTRITION, ETH ZURICH, CH-8092 ZURICH, SWITZERLAND

23.1 Introduction

Proteins adsorbed at fluid–fluid interfaces are relevant to a number of phenomena in colloidal systems, such as the stability and flow behaviour of food foams and emulsions,[1–3] the mechanics of cell membranes[4] and enzymatic catalysis at liquid interfaces.[5] Unlike small-molecule surfactants, proteins not only reduce the interfacial tension of a lipid–water or air–water interface on adsorption, but they also strongly modify the rheological properties of the interface, under both shear and dilatational/compressional deformations.[6,7] Furthermore, the adsorption of surface-active proteins appears almost irreversible, as has been recently shown in sub-phase exchange experiments:[7–9] only minor desorption and negligible changes in interfacial tension have been observed after a protein solution is replaced with protein-free buffer surrounding a protein-covered pendant drop. These effects have been described for a number of surface-active proteins of both biological and technical relevance, such as β-casein,[10–13] β-lactoglobulin,[13–18] bovine serum albumin[19–21] or lysozyme.[20]

Emulsions, as well as immiscible polymer blends and phase-separated biopolymer mixtures, develop flow-induced morphologies when stresses due to the applied flow overcome the interfacial forces that favour the spherical drop morphology at rest.[22,23] Drops can be subjected to deformation, breakup, and coalescence; all of these processes are associated with characteristic light scattering patterns.[24] Shape anisotropy of emulsion droplets in the micrometre size range can be studied by rheometer-based small-angle light scattering (Rheo-SALS).[25–27] Such measurements have been performed for immiscible polymer blends, where absolute viscosities are very large, and for phase-separated biopolymer mixtures, where interfacial tensions between two different aqueous phases are very small.[25] In both cases, the ratio of interfacial to hydrodynamic stresses is in a range in which drop sizes, timescales and shear stresses are experimentally accessible for rheological and rheo-optical measurements. The flow and interfacial properties of the system can be combined into a dimensionless group, the Capillary number, defined as $Ca = \tau R/\sigma$, *i.e.*, the ratio of hydrodynamic stress (τ) to interfacial stress (σ/R), where σ is the interfacial tension and R is the radius of the undeformed droplet. When Ca exceeds a so-called critical Capillary number Ca_{crit}, droplet break up occurs; values for Ca_{crit} as a function of the ratio of the viscosities of dispersed and continuous phases have been collected for various flow types and materials.[28–31] The breakup behaviour of protein-covered emulsion droplets has been shown to be influenced by the rheological properties of the adsorption layer.[32,33] Recently, the effect of an adsorbed protein layer on the flow-induced deformation of millimetre-sized drops has been demonstrated in an optical flow cell.[34] It was shown that, despite a considerable decrease in interfacial tension, a protein-covered drop is less deformed under identical hydrodynamic stresses due to the viscoelastic properties of the macromolecular adsorption layer.

In this contribution, anisotropy in dilute emulsions under flow is studied by Rheo-SALS and the effects of adsorbed protein layers with interfacial rheology are assessed. Emulsions were prepared with either excess surfactant, sodium dodecyl sulfate (SDS), or a surface-active globular protein, β-lactoglobulin. The SDS was used far above its critical micelle concentration (cmc), and hence the interfacial stress condition of the droplets can be approximated by a pseudo-equilibrium interfacial tension. That is, shear and dilatational interfacial stresses, including interfacial concentration gradients, are absent, or they are balanced on a timescale much faster than our experimental observation time. In contrast, the deformation of emulsion droplets stabilized with β-lactoglobulin is expected to be governed by the solid-like behaviour of the adsorbed protein layer, the properties of which are studied with interfacial rheometry.

23.2 Experimental

23.2.1 Materials

The β-lactoglobulin (80% HPCE, mixture of genetic variants A and B), sodium chloride, sodium dihydrogen phosphate, chloroform (all analysis grade) and SDS (Ultra grade, Mr 288.38, Prod. No. 71725) were purchased from Sigma Chemical (St. Louis, MO) and were used as received. Protein solutions were prepared in phosphate buffer (ionic strength 90 mM, pH 6.7, protein concentration 1 wt.%). For the oil phase, (+)-carvene (Sigma, dynamic viscosity 0.9 mPa s) was used, and the density was adjusted with sucrose acetate isobutyrate (SAIB, Eastman Chemical, USA) at a concentration of 60 wt.%, and giving a dispersed phase viscosity of 32 mPa s. The continuous phase in the single drop experiments was polyethylene glycol (PEG 35000S, Clariant) in aqueous solution at a concentration of 40 wt.% with a viscosity of 2.5 Pa s at 20°C. For the emulsion experiments, the continuous phase was high-viscosity dextrin (BCsweet 01146, Cerestar). Both the dispersed and continuous phases were Newtonian. All water used was from a Millipore Biocel water purification system. Experiments were performed at 20°C. The SDS solutions and buffers were stored at 5°C. Protein solutions were prepared on the days of the experiments and used immediately.

23.2.2 Emulsion Preparation and Characterization

Oil-in-water emulsions were produced in a two-step process. First, a parent emulsion with relatively high oil volume fraction ($\phi = 20$ vol.%) was produced with the aqueous solution (1 wt.%) of protein or surfactant in a laboratory-scale rotor/stator homogenizer (Kinematica Polytron, Switzerland). The volume–area mean droplet diameter d_{32} was in the range 1–10 μm. The parent emulsion was produced with excess emulsifier in the aqueous phase. In the second step, appropriate aliquots of the parent emulsion were diluted with the continuous phase to produce the final dilute emulsions of low phase volume fraction ($\phi = 0.1$ vol.% for SALS and $\phi = 3$–5 vol.% for rheometry) and a sufficiently high continuous phase viscosity to induce detectable droplet deformation. (This latter condition was necessary, since the hydrodynamic stresses attainable at low shear-rates in a low-viscosity environment are several orders of magnitude too weak to overcome the capillary pressure of the droplets in rheometric flows, in particular at small droplet radii.) For the SDS system, the continuous phase additionally contained excess surfactant to yield a final surfactant concentration of 1 wt.% in the dilute emulsion used for the scattering experiments. For the protein system, no additional protein was added to the continuous phase, since we assume that β-lactoglobulin does not significantly desorb from the oil–water interface upon dilution.[8,9]

Particular care had to be taken to avoid inclusion of air bubbles upon mixing of the parent emulsion with the higher viscosity continuous phase. For the experiments shown here, the droplet-free matrices were weighed into centrifuge vials and centrifuged at 5000 g for 5 min to remove air bubbles. Appropriate amounts of the parent emulsion were then slowly added to the continuous phase using a small helix stirrer within the same vials.

To vary the droplet concentration in the final sample, appropriate amounts of buffer or surfactant solution were added to the parent emulsion before mixing. Since the emulsions were prepared in two steps, it should be noted that the final continuous phase ultimately contains water from the parent emulsion, which must be taken into account for the continuous phase rheology. Use of time/concentration superposition methods for the continuous phase moduli and complex viscosity greatly facilitated the experiments when samples with different water concentrations in the dextrin were studied.

Droplet-size distributions were measured by laser diffraction (LS-13320, Beckman Coulter, USA).

23.2.3 Rheo-SALS and Rheology

The Rheo-SALS measurements were performed with the device described by Herle et al.[35] The apparatus is based on the stress-controlled DSR rheometer

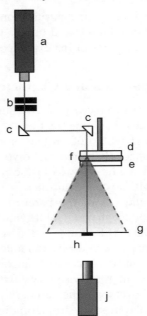

Figure 1 *Arrangement for the Rheo-SALS experiments: (a) He–Ne laser; (b) aperture and neutral density filter; (c) prisms; (d) rotating glass plate; (e) base plate containing circulating heating/cooling water and glass window; (g) translucent screen; (h) beam stop; (j) CCD camera*

(Rheometric Scientific, USA). The light source is a monochromatic 5 mW He–Ne laser (Melles-Griot, USA) with a wavelength of 632.8 nm, guided through the transparent parallel-plate geometry by two prisms (see Figure 1). The plate diameter is 40 mm and the gap between the two quartz glass plates is 1 mm. Disperse phase fractions of $\sim$ 0.1 vol.% were required for the scattering experiments.

Stress-step experiments were performed at nominal shear stresses from 20 to 1500 Pa. Video images of the translucent screen were recorded with a CCD camera (Sony DFW-V 500, Japan) operated at 30 frames per second and mounted vertically below the rheometer. To obtain the calculated Capillary numbers for each experiment, stress values were multiplied by a factor of 0.8, since the laser beam passes through the parallel plate at a radius of 80% of the plate diameter. Scattering patterns were processed with Matlab. Unlike in flow scattering experiments on immiscible polymer blends,[24,26,27] it was unnecessary to pre-shear the samples prior to the scattering experiments, since the initial morphology was produced in the preliminary dispersing step, during which time the droplets were broken down to their final sizes in the homogenizer. Coalescence during the flow was not observed in either the surfactant- or protein-stabilized dilute systems, as was verified by independent droplet-size distribution measurements before and after the experiments.

Rheology (without light scattering) was performed at 20°C in a cell CC16 of concentric cylinder geometry with a Physica MCR300 rheometer (Anton Paar, Germany). The deformation amplitude in the oscillatory tests was set to 20%.

23.2.4 Single Drop Deformation Experiments

To study the effects of an adsorbed protein layer on the deformation of single drops suspended in a sheared continuous phase, we used an optical shear cell with real-time control of the drop position. The device is a modification of Taylor's classic band apparatus and is described in detail elsewhere.[36]

For experiments with protein-covered interfaces, a drop was incubated in a vial containing protein solution ($c = 0.1$ wt.%) for a specific time interval during which the surface-active protein adsorbed to the oil–water interface.[34] As protein adsorption at liquid interfaces is effectively irreversible under the conditions present here, the protein was assumed to behave as an insoluble layer once it was adsorbed.[7–9] Subsequently, the protein-covered drop was carefully immersed in washing buffer to remove unadsorbed protein and transferred into the flow cell *via* a vial containing continuous phase only. Using this procedure, it was possible to perform flow experiments on the identical drop with and without adsorbed protein. The deformation parameter $D = (L - B)/(L + B)$ was used to evaluate the data from the experiments, where L and B are the long and short axes of the drop projected into the plane of velocity *versus* shear gradient.

23.2.5 Interfacial Characterization

The interfacial shear moduli were measured using a biconical disc interfacial rheometer, as illustrated in Figure 2(a) and described in detail elsewhere.[37] A Physica MCR 300 rheometer (Anton Paar, Germany) was adapted for interfacial rheometry. Interfacial tensions at the oil–water interface were measured using the pendant drop method.[38] For tensiometry and interfacial rheometry, all components in contact with the fluids were consecutively cleaned with hot water, ethanol and chloroform, before and after every measurement.

23.3 Results

A particular advantage of the interfacial rheological method used here is the ease with which the complex measuring protocols could be automated. An example is shown in Figure 2 for β-lactoglobulin adsorbed at the oil–solution interface (protein concentration 100 mg L^{-1}, pH 6.7, ionic strength 90 mM). To study the time-dependent rheology of the globular protein films, short frequency sweeps were repeated over several hours. For early film ages, when changes in surface pressure and shear moduli were still occurring fast, the lowest frequency was not tested, since the time necessary to gather sufficient data (about 1/6 of a full oscillation cycle with the device used) was in the range

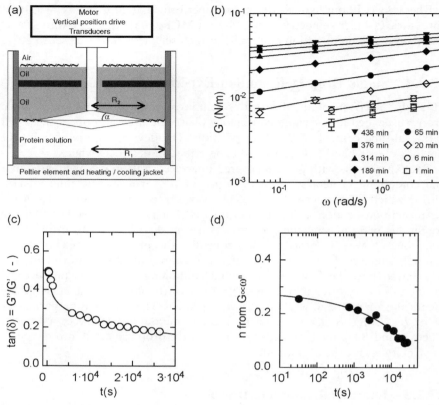

Figure 2 *Interfacial shear rheology. (a) Sketch of the biconical disc rheometer.[37] (b) Combined time and frequency sweep data for an adsorbed layer of β-lactoglobulin at the oil–solution interface. Identical frequency sweep tests were performed at 20°C for different interface ages (c = 100 mg L^{-1}; ionic strength 90 mM; strain γ = 0.75%). (c) Loss tangent G''/G' as a function of the interface age for the same data. (d) Power-law exponent n in fits of $G' \propto \omega^n$ for the same data*

of the timescale of changes in film properties. Figure 2(b) shows the frequency-dependent interfacial storage modulus $G'(\omega)$ at different interface ages. Even at the shortest interface ages studied, the adsorbed β-lactoglobulin films were found to be predominantly solid like ($G' > G''$), indicating that in the concentration range studied here the interface is already densely packed with the globular protein molecules. Figure 2(c) shows the loss tangent $\tan\delta = G''/G'$ at a frequency of $\omega = 0.755$ rad s^{-1} at different interface ages; a decrease in the viscous modulus from ~0.5 to 0.18 G' after 8 h of ageing was observed. The frequency dependence of G' is summarized in Figure 2(d), indicating solidification of the interface as $G' = f(\omega)$ becomes flatter at longer experimental times. For the 1 wt.% SDS system, no significant interfacial shear properties were found in steady, oscillatory or transient shear. Dilatational experiments with

SDS using oscillating pendant drops did not show any measurable interfacial tension response, i.e., the interfacial relaxation process is presumed to occur much faster than our measurements. Static interfacial tensions ($c_{SDS} = 1$ wt.%, 20°C) were 4 mN m^{-1} in aqueous solution (used for the parent emulsion) and 3.9 mN m^{-1} when 40 wt.% PEG was added.

A typical set of results for the frequency-dependent storage and loss moduli of the dilute emulsion in the absence of surface rheological effects is shown in Figure 3. If both the disperse phase and the continuous phase are Newtonian, we observe a 'relaxation shoulder' in the elastic modulus centred above a characteristic frequency related to the shape relaxation of the droplets. The related relaxation timescale is influenced by the interfacial tension, the continuous phase viscosity, the viscosity ratio and the droplet size. For the latter, a mean radius derived from the droplet-size distribution can be used. Emulsion rheological models, such as those by Palierne[39] or Yu et al.[40] can be used to describe emulsion rheological data with shape relaxation. In Figure 3(c) calculated values for $G'(\omega)$ are included (dashed line). The discrepancy above the relaxation shoulder is due to the use of a mean droplet diameter rather than the full size distribution. An alternative representation of these data is shown in Figure 3(d): the relaxation time spectrum $\lambda H(\lambda)$ of the droplets was calculated from the complex modulus $G^*(\omega)$ using standard algorithms.[41] In this representation we can directly relate the relaxation timescale, as indicated by the maximum in the normalized spectrum, to the droplet relaxation process. We note that, in the absence of dispersed droplets, a zero relaxation time spectrum is derived, corresponding to Newtonian scaling with $G' \propto \omega^2$ and $G'' \propto \omega$ (see Figure 3(a)).

In Figure 4 the storage moduli of dilute emulsions ($\phi = 4.5$ vol.%) prepared with excess SDS or β-lactoglobulin are compared. We note the absence of a characteristic relaxation shoulder for interfaces stabilized by the protein. The higher interfacial tension of the protein-stabilized system is unlikely to be responsible for this qualitative difference in the frequency response: if we compare the characteristic timescales calculated according to the Oldroyd equation,[22] a mere shift in the characteristic frequency by about a decade would be observed. If the (static) interfacial tension alone would govern the stress boundary condition between the oil and water phases, we would expect comparable shape responses at comparable Capillary numbers. Therefore, the difference between the surfactant and the protein data appears to be qualitative in nature: it appears that in the protein system the deformation of the micrometre-sized droplets is completely suppressed.

To probe the morphology of the emulsions under flow, we use light-scattering patterns obtained at the maximum deformation rate during an oscillation cycle in the rheometer, i.e., when the oscillatory deformation $\gamma(t)$ passes through zero. In these experiments, the optical technique visualizes the morphology in the velocity–vorticity plane (parallel-plate measuring geometry). For the emulsions stabilized with excess SDS, a characteristic distortion of the light-scattering patterns in the vorticity direction is observed. Such anisotropic scattering patterns have been observed both with phase-separated

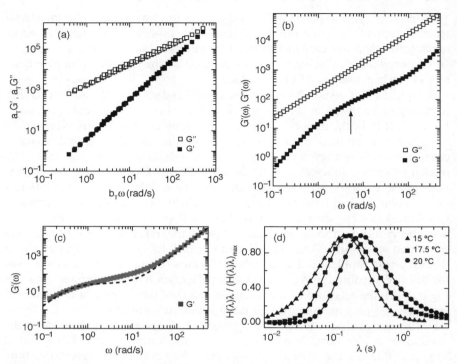

Figure 3 *Frequency-dependent dynamic moduli of emulsions with dominant shape relaxation stabilized by excess SDS. (a) Reduced storage and loss moduli, G' and G'', of the continuous phase. Data are from time/temperature superposition: factor a_T (20, 25, 30°C) = 1, 0.365, 0.232; and factor b_T (20, 25, 30°C) = 1, 1.070, 1.252. (b) G' and G'' at 20°C of an emulsion prepared with SDS (oil phase fraction $\phi = 4.5$ vol.%). The arrow indicates the characteristic relaxation shoulder of the storage modulus, caused by shape relaxation of the deformed droplets. (c) Comparison of experimental $G'(\omega)$ data with calculated[40] values (dashed line) based on measured interfacial tension, individual phase moduli and mean droplet diameter d_{43} from laser diffraction. (d) Normalized relaxation time spectra calculated[41] from $G^*(\omega)$ data at three different temperatures. Maxima indicate characteristic droplet relaxation timescales: reducing the temperature lowers the overall viscosity, thus slowing down droplet retraction*

(bio)polymers[25] and immiscible polymer blends;[24] the value of the Capillary number in our experiment was $Ca \sim 0.1$, which is far below the critical value needed for break up. Therefore, the scattering anisotropy can be assumed to be due to deformation of the droplets in the flow direction. For the emulsion stabilized with β-lactoglobulin, the scattering patterns are isotropic throughout the frequency spectrum. This indicates the absence of any detectable droplet deformation, a result in line with the missing relaxation shoulder in the $G'(\omega)$ curve. We note that other authors[42] have found a markedly different behaviour for immiscible polymer blends containing surface-active agents (the so-called 'compatibilizers'): in those cases, the relaxation shoulder did not disappear in the presence of the compatibilizer, but a second 'slow' relaxation shoulder was

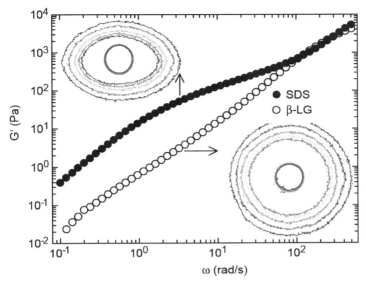

Figure 4 *Comparison of the frequency-dependent storage modulus $G'(\omega)$ for emulsions stabilized with either SDS or β-lactoglobulin (β-LG) (droplet phase fraction ϕ = 4.5 vol.%). Note the absence of a pronounced relaxation shoulder for the protein system. The insets show contour plots of Rheo-SALS patterns obtained for the same emulsions, but with a lower droplet phase volume fraction of $\phi = 0.1$ vol.%. The SALS images were taken at the maximum deformation rate within an oscillation cycle (maximum in $d\gamma(t)/dt$). The innermost ring of the scattering patterns is from the beam stop*

detected. In the latter work, the authors attributed the 'fast' relaxation process to shape relaxation, and the slow one to an interfacial viscoelastic effect. The difference from our results shown here is that in the present case the combined effects of small droplets ($d_{32} \sim$ 2–3 μm) and interfacial rheology of the adsorbed protein do not allow us to detect whether such in-plane interfacial relaxation timescales are present, as they completely suppress deformation of the droplets. We note that, in similar experiments with larger drops in steady shear flow we did find flow-induced anisotropy even for protein-stabilized emulsions; compared to surfactant-stabilized emulsions, these systems showed reduced anisotropy at identical Capillary numbers.[43]

Figure 5(a) shows deformation experiments performed with millimetre-sized drops in the parallel band flow cell. The deformation parameter D is plotted as a function of Capillary number for both a clean drop and a protein-covered drop. The value of Ca is calculated from the static interfacial tension; for the protein layers, we have $\sigma = 11.3$ mN m^{-1}, as measured by pendant-drop tensiometry with a 1-hour-old drop. Although the interfacial tension is lower than the clean interface, the presence of the β-lactoglobulin adsorption layer restricts the ease of deformation, rather than facilitating it. The results are similar to those with drops covered with lysozyme, as reported elsewhere,[34] with the exception that only very minor shape fluctuations (*wobbling*) have been

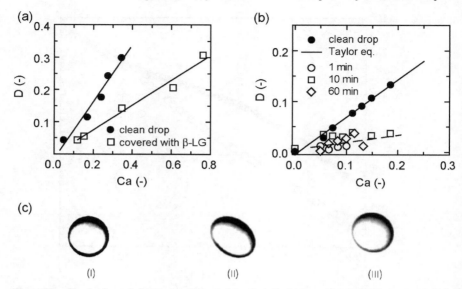

Figure 5 *Single drop deformation experiment. (a) Deformation $D = f(Ca)$ of a oil drop before and after adsorption of β-lactoglobulin (continuous phase viscosity 940 mPa s, drop/continuous phase viscosity ratio 0.096, incubation time of drop in protein solution 60 min). (b) Deformation behaviour for different residence times of the drop in the protein solution. The value of Ca is based on the transient values of the interfacial tension after 1, 10 and 60 min, respectively. (c) Images of the same drop (radius 0.74 mm) with and without the adsorbed protein layer: I, no flow; II, uncovered drop deforming at shear rate of $6.8\ s^{-1}$; III, same drop at same shear rate, but after incubation in the protein solution for 10 min*

observed here. This is likely to be due to the higher continuous phase viscosity in our experiments. In Figure 5(a) the mean deformation values, averaged over at least a second, are reported. We note that, in the static interfacial tension case at the given viscosity ratio, the experimental data collapse onto one line with a slope of approximately one, whereas the values found for the protein-covered interfaces are much lower. All interfacial rheological data measured with β-lactoglobulin showed a strong time dependence (see above).

An obvious question for protein-covered drops is how the interface age influences the small-deformation behaviour. This can be studied by varying the incubation time of the drop in the protein solution. Figure 5(b) shows deformation responses for drops with different interface ages of 1, 10 and 60 min. The magnitude of Ca was calculated from transient values of the interfacial tension measured in the pendant-drop tensiometer at different times: $\sigma = 19$, 16 and 11.3 mN m^{-1} for 1, 10 and 60 min, respectively. In contrast to the uncovered drop, all values for D are lower, but we find that the differences for the individual interface ages are hardly detectable by our method: even at the shortest age of 1 min (which is the shortest realistic timescale accessible with a well-trained experimenter), the protein layers strongly restrict the deformation. This result suggests that the strength of the globular protein layer relevant to the drop deformation is determined at early interface ages, whereas the network properties developed during ageing are of secondary importance. Since the ageing behaviour of a deformed drop under shear hardly correlates with the long-time interfacial shear moduli or viscosity, the role of the dilatational properties should be clarified as well.

The results presented above demonstrate that viscoelastic protein layers adsorbed at the oil–water interface restrict the deformation of emulsion droplets under flow, and that Rheo-SALS is a suitable method to study such effects for droplets in the micrometre size range. At identical Capillary numbers, the flow-induced anisotropy is significantly smaller for protein-covered droplets as compared to drops with an adsorbed small-molecule surfactant. A somewhat unspectacular, but nonetheless important, conclusion can be drawn from Figure 4: for protein-stabilized emulsions with practically relevant droplet sizes ($Ca < 0.1$), scattering will be isotropic even at elevated shear stresses and the flow regime in which deformation occurs will hardly ever be reached. Therefore, protein-covered emulsion droplets, once broken down to their final sizes in the micrometre range, do not deform anymore, especially if the continuous fluid is of low viscosity. Consequently, these droplets should be considered as 'solid' dispersed spheres covered with a charged polymer adsorption layer. Whereas the main contribution to this behaviour is due to the capillary pressure, which is inversely proportional to the radius, Figure 4 shows that interfacial rheological properties enhance the 'effective capillary pressure' and therefore stabilize the droplets at small deformations. We should also keep in mind that all of the above results are valid for small deformations, and consequently for sub-critical Capillary numbers ($Ca < Ca_{\text{crit}}$). The fact that a protein-covered droplet is stabilized in the small-deformation limit does not necessarily mean that it will be more stable against break up at higher deformations. As demonstrated for single drops[32] and later confirmed for emulsions,[33] the role of protein layers in the break up behaviour can even be quite contrary to what we have found for the small-deformation behaviour: Williams et al.[32] found that, whereas β-casein or β-lactoglobulin at low concentrations stabilizes the droplets, higher β-lactoglobulin concentrations result in droplet breakup behaviour essentially controlled by the rigidity of the globular protein network, minimizing the importance of the bulk viscosity. Rigid protein layers can therefore *enhance* droplet break up, even though they stabilize the same droplet at smaller deformations.

Acknowledgements

We thank Bruno Pfister, Rok Gunde, Dani Kiechl, Jan Corsano, Christoph Eschbach and Martina Haug at ETH, and Patrick Heyer, Jörg Läuger and Victor Kusnezov at Anton Paar/Physica. Parts of this project were supported by the Swiss National Science Foundation (SNF project no. 200020-108052) and by ETH Zürich.

References

1. D. Möbius and R. Miller (eds), *Proteins at Liquid Interfaces*, Elsevier, Amsterdam, 1998.
2. B.S. Murray, *Curr. Opin. Colloid Interface Sci.*, 2002, **7**, 426.
3. E. Dickinson, *J. Chem. Soc. Faraday Trans.*, 1998, **94**, 1657.
4. D.E. Discher, N. Mohandas and E.A. Evans, *Science*, 1994, **266**, 1032.
5. L.G. Cascao-Pereira, A. Hickel, C.J. Radke and H.W. Blanch, *Biotechnol. Bioeng.*, 2003, **83**, 498.
6. D.A. Edwards, H. Brenner and D.T. Wasan, *Interfacial Transport Processes and Rheology*, Butterworth-Heinemann, Stonheam, MA, 1991.
7. E.M. Freer, K.S. Yim, G.G. Fuller and C.J. Radke, *J. Phys. Chem. B*, 2004, **108**, 3835.
8. R. Miller, D.O. Grigoriev, J. Krägel, A. Makievski, J. Maldonado-Valderrama, M. Leser, M. Michel and V.B. Fainerman, *Food Hydrocoll.*, 2005, **19**, 479.
9. E.M. Freer, K.S. Yim, G.G. Fuller and C.J. Radke, *Langmuir*, 2004, **20**, 10159.
10. R. Miller, V.B. Fainerman, E.V. Aksenenko, M.E. Leser and M. Michel, *Langmuir*, 2004, **20**, 771.
11. G.B. Bantchev and D.K. Schwartz, *Langmuir*, 2003, **19**, 2673.
12. J. Maldonado-Valderrama, V.B. Fainerman, M.J. Galvez-Ruiz, A. Martin-Rodriguez, M.A. Cabrerizo-Vilchez and R. Miller, *J. Phys. Chem. B*, 2005, **109**, 17608.
13. P. Cicuta and I. Hopkinson, *Europhys. Lett.*, 2004, **68**, 65.
14. E. Dickinson, *Colloids Surf. B*, 1999, **15**, 161.
15. J.T. Petkov, T.D. Gurkov, B.E. Campbell and R.P. Borwankar, *Langmuir*, 2000, **15**, 3703.
16. P. Cicuta, E.J. Stancik and G.G. Fuller, *Phys. Rev. Lett.*, 2003, **90**, 236101.
17. A. Martin, M. Bos, M. Cohen Stuart and T. van Vliet, *Langmuir*, 2002, **18**, 1238.
18. G. Garofalakis and B.S. Murray, *Colloids Surf. B*, 1999, **12**, 231.
19. G. Serrien, G. Geeraerts, L. Ghosh and P. Joos, *Colloids Surf.*, 1992, **68**, 219.
20. J.R. Lu, T.J. Su and R.K. Thomas, *J. Colloid Interface Sci.*, 1999, **213**, 426.
21. L.G. Cascão Pereira, O. Théodoly, H.W. Blanch and C.J. Radke, *Langmuir*, 2003, **19**, 2349.
22. E. Scholten, J. Sprakel, L.M. Sagis and E. van der Linden, *Biomacromolecules*, 2006, **7**, 339.
23. V.E. Ziegler and B.A. Wolf, *Macromolecules*, 2005, **38**, 5826.
24. C.L. Tucker and P. Moldenaers, *Ann. Rev. Fluid Mech.*, 2002, **34**, 177.
25. Y.A. Antonov, P. van Puyvelde and P. Moldenaers, *Biomacromolecules*, 2004, **5**, 276.
26. P. van Puyvelde, P. Moldenaers, J. Mewis and G.G. Fuller, *Langmuir*, 2000, **16**, 3740.

27. J. Vermant, P. van Puyvelde, P. Moldenaers, J. Mewis and G.G. Fuller, *Langmuir*, 1998, **14**, 1612.
28. H.A. Stone, *Ann. Rev. Fluid Mech.*, 1994, **26**, 65.
29. H.P. Grace, *Chem. Eng. Comm.*, 1982, **14**, 225.
30. K. Feigl, S.F.M. Kaufmann, P. Fischer and E.J. Windhab, *Chem. Eng. Sci.*, 2003, **58**, 2351.
31. B.J. Bentley and L.G. Leal, *J. Fluid Mech.*, 1986, **167**, 241.
32. A. Williams, J.J.M. Janssen and A. Prins, *Colloids Surf. A*, 1997, **125**, 189.
33. D.B. Jones and A.P.J. Middelberg, *AIChE J.*, 2003, **49**, 1533.
34. P. Erni, P. Fischer and E.J. Windhab, *Appl. Phys. Lett.*, 2005, **87**, 244104.
35. V. Herle, P. Fischer and E.J. Windhab, *Langmuir*, 2005, **21**, 9051.
36. B. Birkhofer, J.C. Eischen, P. Fischer and E.J. Windhab, *Ind. Eng. Chem. Res.*, 2005, **44**, 6999.
37. P. Erni, P. Fischer and E.J. Windhab, *Langmuir*, 2005, **21**, 10555.
38. R. Gunde, A. Kumar, S. Lehnert-Batar, R. Mäder and E.J. Windhab, *J. Colloid Interface Sci.*, 2001, **244**, 113.
39. J.F. Palierne, *Rheol. Acta*, 1990, **29**, 204.
40. W. Yu, M. Bousmina, M. Grmela and C. Zhou, *J. Rheol.*, 2002, **46**, 1401.
41. T. Roths, D. Maier, C. Friedrich, M. Marth and J. Honerkamp, *Rheol. Acta*, 2000, **39**, 163.
42. E. van Hemelrijck, P. van Puyvelde, S. Velankar, C.S. Macosko and P. Moldenaers, *J. Rheol.*, 2004, **48**, 143.
43. P. Erni, 'Viscoelasticity at liquid interfaces and its effect on the macroscopic deformation response of emulsions', Dissertation no. 16549, ETH Zürich, Switzerland, 2006.

Chapter 24
Enhancement of Stability of Bubbles to Disproportionation Using Hydrophilic Silica Particles Mixed with Surfactants or Proteins

Thomas Kostakis, Rammile Ettelaie and Brent S. Murray

PROCTER DEPARTMENT OF FOOD SCIENCE, UNIVERSITY OF LEEDS, LEEDS LS2 9JT, UK

24.1 Introduction

Incorporation of air bubbles into many food products (fizzy drinks, beer, ice-cream, whipped cream, *etc.*) is an essential aspect of the process by which the desired texture and rheology in these systems are developed. From a thermodynamic point of view, these systems are very unstable and thus it is always a challenge to the food technologist to find the best way to enhance the stability of the bubbles. In the absence of coalescence, the major factor limiting the bubble lifetime is the phenomenon of disproportionation, whereby gas diffuses from smaller bubbles to larger ones as a consequence of differences in the Laplace pressure between bubbles of different sizes.

Recent work has shown[1] that the interfacial elasticity of a wide range of commonly used food proteins is insufficient to slow disproportionation significantly, so that bubbles below ~100 μm diameter dissolve in less than a few hours. On the other hand, it was also found[2] that partially hydrophobic, spherical silica particles (made hydrophobic by chemical grafting of the correct proportion of alkyl chains to their surface) can stabilize bubbles against disproportionation for several days or even months.

One of the key parameters influencing the stabilization of emulsions and foams by solid particles is the size of these particles. There have been many studies in the past where the importance of particle size has been explored in detail. Tang and co-workers[3] examined a wide range of spherical monodisperse hydrophobic silica particles (20–700 nm in diameter). They reported[3] that foam stability decreased with increasing particle size and increased with increasing particle concentration, but without explaining the exact role that the particles played in relation to the stability of the bubbles. Ip and co-workers[4] investigated the dependence of the stability of metallic aluminium foams on silica particle size, wettability and concentration. They concluded[4] that only silica particles with correct wettability could stabilize foams and that stability increased with increasing particle concentration and with decreasing particle size. On the other hand, Hudales and Stein[5] stated that only 'large' (1–10 μm) glass particles in the presence of a surfactant (CTAB) could inhibit thin film rupture and delay film drainage. In addition, theoretical work from Kaptay[6] postulated that solid particles can stabilize bubbles only when the bubbles are not larger than 3–30 μm. In addition, monodisperse hydrophilic silica particles were investigated by Sethumadhavan *et al.*;[7,8] they also found that foam stability was dependent *inter alia* on the size of the particles.

Hydrophobic particles that can adsorb at the air–water (A–W) interface are naturally present in foods, *i.e.*, fat crystals. However, the size, shape and adsorption properties of fat crystals are difficult to control, making it not straightforward to do model experiments with such systems. In contrast, silica particles form good model systems, as their surface-active properties can be easily changed by chemical modification of the surfaces (as demonstrated by Dickinson *et al.*[2]). On the other hand, unmodified hydrophilic silica particles are the only silica particles that could be allowed for use in food systems; but they do not have any tendency to adsorb at an A–W interface. Therefore, a physical method of surface modification must be developed to render such particles hydrophobic enough to adsorb. In addition, any such particles in foods must be of a small size (*e.g.*, <0.1 μm) to avoid complications during digestion and effects on perceived texture and mouthfeel. Such inorganic particles of small size are in fact naturally present in many food stuffs, although in rather small quantities.

In this study, the effect of different types of hydrophilic silica particles on the stability of gas bubbles is described. The particles were composed of either *fumed* or *colloidal* silica, meaning that they were in a powder or liquid form, respectively. The fumed silica particles had a nominal size of 20 nm, although the actual size depends on the state of dispersion of the powder, whereas the colloidal silica samples have a well-defined particle size. The big advantage of colloidal silica as compared to fumed silica is the smaller particle size of the former, as well as the fact that it is highly monodisperse. In this study, however, we focus on a sample available with the smallest mean particle diameter, *i.e.*, 5.5 nm. Adsorption of low-molecular-weight surfactant or protein was used to modify the surface-active properties of the particles. Such systems may act as more realistic models of hydrophobic particles in foods, where both protein and surfactant will compete for adsorption at the A–W interface.

24.2 Materials and Methods

Pure fumed hydrophilic silica particles of nominal diameter 20 nm were specially made by Wacker-Chemie (Munich, Germany). The colloidal silica particles were a gift from Professor Binks (University of Hull). Water from a Milli-Q system (Millipore, Watford, UK), free from surface-active impurities and with a conductivity of less than 10^{-7} S cm^{-1}, was used throughout. Didodecyldimethylammonium bromide (DDAB) (98%), bovine β-lactoglobulin (β-LG, crystallized and lyophilized, with ~90% protein content, lot no. 114H7055), L-α-phosphatidylcholine (lecithin) from frozen egg-yolk (~99%), imidazole and sodium chloride buffer salts were from Sigma (Poole, Dorset). The β-casein was obtained as a freeze-dried sample from the Hannah Research Institute (Ayr, Scotland). The protein gave clear sharp peaks when analysed by fast protein liquid chromatography (FPLC) using a Mono-Q ion-exchange column with a linear gradient.[9] The buffer solution consisted of 0.05 mol dm^{-3} imidazole + 0.05 mol dm^{-3} NaCl, adjusted to pH = 7 by addition of HCl.

Detailed descriptions of the experimental techniques and apparatus to determine bubble stability have been given elsewhere.[2,10,11] A wide range of concentrations of silica particles, DDAB, protein and lecithin were screened in order to test the relative contributions of the particles + amphiphilic molecules to the stability of the bubbles to disproportionation. The experiments involved measuring the change in radius of individual bubbles and also the fraction of surviving bubbles, F, as a function of time at the A–W interface. The quantity F is defined as the ratio of the number of bubbles still visible after a certain time to the number of bubbles present at the start (immediately after their formation). The particle dispersions were prepared, shaken for 15 s to produce the bubbles/foam and then transferred into an observation chamber.[10] The A–W interface was viewed from above through the upper window of the chamber using a microscope, video camera and a video recorder.

The contact angle θ was measured between a drop of DDAB solution and a flat surface formed of fumed silica particles. The flat surface was created by compressing the silica powder in a specially adapted hydraulic press at pressures up to 1250 bar.

24.3 Results and Discussion

24.3.1 Fumed Silica Particles + DDAB

DDAB is a cationic surfactant of very low critical micelle concentration (cmc). It adsorbs strongly to the negatively charged surface of silica.[14] At monolayer coverage it renders the surface hydrophobic by adsorbing with the head group down on the surface.

Figure 1 illustrates the most stable systems obtained at different fumed silica particle + DDAB mixtures. The fraction of surviving bubbles, F, is plotted as a function of time following the creation of the bubbles. When the DDAB concentration was low (1.7×10^{-6} mol dm^{-3}), the particle concentration needed for reasonable bubble stability was 2.2 wt%. The DDAB concentrations giving the most stable systems with 3.0 and 3.5 wt% silica particles were

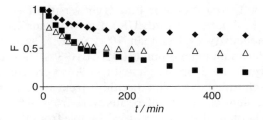

Figure 1 *Fraction F of surviving bubbles as a function of time t at different fumed silica and surfactant concentrations:* ■, 2.2 wt% silica + 1.7 × 10^{-6} mol dm^{-3} DDAB; △, 3.0 wt% silica + 10^{-5} mol dm^{-3} DDAB; ◆, 3.5 wt% silica + 10^{-4} mol dm^{-3} DDAB

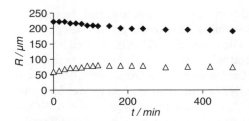

Figure 2 *Radius R of average-sized bubble as a function of time t at different fumed silica and surfactant concentrations:* △, 3.0 wt% silica + 10^{-5} mol dm^{-3} DDAB; ◆, 3.5 wt% silica + 10^{-4} mol dm^{-3} DDAB

10^{-5} and 10^{-4} mol dm^{-3}, respectively. However, this does not take into account the possibility that the bubbles formed in the different systems had different initial sizes, and therefore different intrinsic stabilities.

In Figure 2, the change in the radius of an average sized bubble is presented. The average bubble size in the 10^{-5} mol dm^{-3} DDAB system was much smaller compared to that in the 10^{-4} mol dm^{-3} system. It is well known[1] that small bubbles shrink faster than big ones. So, in part, the difference between results in Figure 1 is due to this difference in average bubble size.

The results described above imply a dependence of stability on the concentration of both the silica particles and the DDAB. That is, it was found that the stability improved with simultaneous increase in the particle concentration and the DDAB concentration. But an increase in concentration of one of the components on its own did not necessarily provide better stability. Contact angle measurements at different DDAB concentrations showed a significant increase in θ from a DDAB concentration of 10^{-6} to 10^{-5} mol dm^{-3} (Figure 3). Nevertheless, for DDAB concentrations above the cmc,[12] the value of θ was found to decrease again. The change in θ explains the change in stability with DDAB concentration. At a DDAB concentration of 10^{-5} mol dm^{-3}, we found $\theta \approx 40°$. The adsorption energy of a particle at the surface of a bubble with such contact angle can be thousands of kT,[10] making particle detachment from the interface almost impossible.

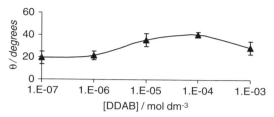

Figure 3 *Contact angle θ of droplets of solutions of DDAB of different concentration on flats of compressed fumed silica particles*

At this point, it is important to distance the contact angle measurements somewhat from the stability data, due to the fact that when the contact angle was measured the surfactant was always in excess, and so the DDAB surface coverage on the silica was in equilibrium with bulk concentration of DDAB. In the bubble stability measurements, however, there could have been a different fractional surface coverage at different particle and surfactant concentrations, because the amount of surfactant was a limiting factor, due to the much higher surface area for adsorption.

24.3.2 Fumed Silica Particles + Protein

In our previous work,[1] it was found that β-LG gave better foam stability against disproportionation than other proteins. But even β-LG-stabilized bubbles shrank slowly and relentlessly with time. So, although β-LG forms highly viscoelastic, gel-like adsorbed layers at the A–W interface,[13] the resulting interfacial elasticity is not adequate to prevent long-term instability of bubbles. A question that arises is whether a mixture of hydrophilic silica particles + β-LG is more effective in providing stability than is β-LG alone. To attempt to answer this question, experiments were performed under exactly the same conditions as those used previously by Dickinson *et al.*[1] for pure β-LG-stabilized bubbles. The protein concentration was 0.05 wt% in pH 7 buffer. Care was taken to generate and study bubbles of the same initial sizes as those analysed in the previous study.[1] The results in Figure 4 show that, when a mixture of silica particles + β-LG was used, the shrinkage of the bubbles was considerably delayed compared to the pure protein system. However, the bubbles were not as stable as for the optimum DDAB + silica systems. Figure 5 compares typical shrinkage kinetics of bubbles of similar initial size for the systems DDAB + silica, β-LG + silica and pure β-LG.

Experiments performed with β-casein, over the same range of protein and particle concentrations as with β-LG, did not give enough stable bubbles for any satisfactory measurements to be made. Thus, the ability of β-casein to stabilize bubbles on its own, while not as great as for pure β-LG, was lost in the presence of the particles. There are two possible explanations for this. Firstly, the silica particles may act as foam-breaking agents in this system. Secondly, the bulk concentration of protein may be reduced by its adsorption to the silica,

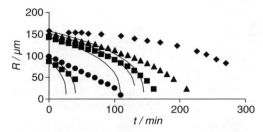

Figure 4 *Comparison of typical bubble radius R versus time t for 1 wt% fumed silica + 0.05 wt% β-LG at pH = 7. The solid lines indicate the expected behaviour of 0.05 wt% β-LG on its own, taken from modelling of previous experimental data in the same bubble-size range[13]*

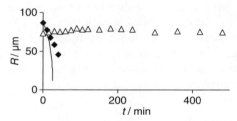

Figure 5 *Comparison of bubble radius R as a function of time t stabilized by fumed silica + DDAB or β-LG at pH = 7: △, 3 wt% silica + 10^{-5} mol dm^{-3} DDAB; ♦, 1 wt% silica + 0.05 wt% β-LG. The solid lines indicate the expected behaviour of 0.05 wt% β-LG[13]*

which means that there is not enough β-casein available to stabilize the bubbles. While it is not clear which of these explanations is the more likely, it is nonetheless noteworthy that there is this pronounced difference in behaviour for two milk proteins.

24.3.3 Fumed Silica Particles + Lecithin

There are few natural cationic surfactants present in food colloids, and to our knowledge no artificial ones are permitted as food additives. Phospholipids are zwitterionic, carrying both a negative and a positive charge at neutral pH. The latter should favour adsorption to a negatively charged surface such as silica. For comparison with DDAB, experiments were therefore conducted with egg-yolk lecithin, as an example of a natural surfactant that might aid hydrophilic particle adsorption at the A–W interface.

Bubbles formed at the A–W interface in a mixed dispersion of silica particles + lecithin were not particularly stable, especially in comparison to the silica + DDAB system. Figure 6 illustrates typical results from several experiments with lecithin + silica where the fraction of surviving bubbles, F, is plotted as a function of time. Whereas with all the other systems studied here the rate of decrease of F was significantly lower than the initial value after two

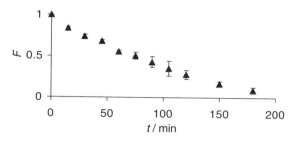

Figure 6 *The fraction F of surviving bubbles as a function of time t for 1 wt% fumed silica + 1.3×10^{-3} mol dm^{-3} lecithin*

or three hours, with lecithin F decreased at approximately the same rate throughout. The relatively poor performance of lecithin compared to DDAB can probably be attributed to its lower adsorption efficiency on silica. The overall net charge on lecithin will be approximately zero under the conditions used. The charge could be made more net positive by decreasing the pH, and so increasing the tendency for adsorption on the negatively charged silica, but this was not tested. Higher lecithin concentrations could also have been tried, but, at the highest concentration used, the solutions were starting to become cloudy, indicating that the limit of solubility had been reached. But this is not to say that variations in temperature or other solution conditions (*e.g.*, ionic strength, concentrations of specific ions) could not be used to enhance the adsorption efficiency of lecithin on the silica.

24.3.4 Colloidal Silica Particles + DDAB

The DDAB concentration was varied from 10^{-7} to 10^{-2} mol dm^{-3} and the colloidal silica particle concentration from 0.5 to 3 wt%. It was found that, for all the particle concentrations in this range, only when the concentration of DDAB was above 10^{-4} mol dm^{-3} was there a significant amount of foam created. Also, for silica contents above 0.5 wt%, the foamability or foam stability did not markedly increase. The effect of DDAB concentration was therefore only investigated in more detail for the case of the particle concentration of 0.5 wt%. The optimum DDAB concentration for maximum stability was found to be 8×10^{-4} mol dm^{-3}. In Figure 7 the F values after 1 h for all the five DDAB concentrations are plotted. As with the fumed silica, the value of F was found to decrease when the DDAB concentration was too high, illustrating the importance of the ratio of surfactant to the particle concentration. Experiments were also carried out to record the evolution of the radius of individual bubbles in the same systems, as shown in Figure 8. Only at the highest concentration of DDAB used (4×10^{-3} mol dm^{-3}) did the bubbles shrink and collapse completely within 60–80 min. This is in agreement with the trend in F in Figure 7: for all the other DDAB concentrations, bubbles were formed that remained stable for many hours.

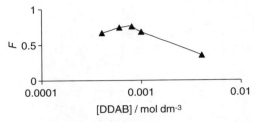

Figure 7 *The fraction F of surviving bubbles after 1 h for 0.5 wt% colloidal silica + DDAB at different surfactant concentrations*

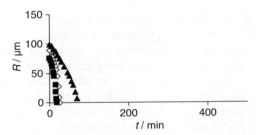

Figure 8 *Typical bubble radius R versus time t for bubbles stabilized by 0.5 wt% colloidal silica + 4×10^{-3} mol dm^{-3} DDAB*

It is clear from the above that, above a certain concentration of silica, there is an optimum ratio of surfactant to particle concentration that gives maximum stability. This makes sense if there is a specific partial coverage of surfactant required on the particles in order to give optimum surface hydrophobicity. Some calculations were therefore carried out to estimate this coverage. The adsorption isotherm for DDAB on silica was not directly determined, but the maximum possible surface coverage was estimated assuming all of the DDAB in the system adsorbs. For this calculation the specific surface area of the 5.5-nm colloidal silica was taken as 500 m^2 g^{-1} assuming spherical particles of density 2.18×10^3 kg m^{-3}. The area per adsorbed DDAB molecule was taken as the area per head group, *i.e.*, 0.38 nm^2.[14] As DDAB is not very soluble in the aqueous phase, but adsorbs strongly to the silica particle surface, this calculation of the maximum possible coverage is expected to be a good estimate of the actual coverage, provided the particles are not significantly aggregated, as aggregation would reduce the actual surface area available for adsorption.

Table 1 shows values of maximum percentage coverage calculated for 5.5-nm hydrophilic silica particles at different values of the surfactant concentration [DDAB]. Table 2 gives the calculated values at different particle concentrations. When the concentration of silica was fixed at 0.5 wt% (Table 1), and comparing the calculated maximum percentage coverage with the [DDAB] giving the most stable bubbles, the optimum surface coverage is 7%. But with [DDAB] fixed at 10^{-3} mol dm^{-3} and the particle content varied (Table 2), the

Table 1 *Theoretical maximum percentage surface coverage of 5.5-nm colloidal silica particles by DDAB molecules, for a 0.5 wt % silica particle concentration at different DDAB concentrations*

[DDAB] (mol dm^{-3})	Surface coverage (%)
4×10^{-4}	4
6×10^{-4}	5
8×10^{-4}	7
1×10^{-3}	9
4×10^{-3}	35

Table 2 *Theoretical maximum percentage surface coverage of 5.5-nm colloidal silica particles by DDAB molecules, for different particle concentrations at 10^{-3} mol dm^{-3} DDAB*

Silica content (wt%)	Surface coverage (%)
0.5	9
1	4
1.5	3
2	2

optimum surface coverage appears to be 3%. Similar considerations show that, at fixed [DDAB] = 4×10^{-3} mol dm^{-3}, the optimum coverage is 11%. Given the possible errors in this calculation due to slightly different states of dispersion of the particles, the range of inferred coverage values, from 3% to 11%, may be considered as reasonably consistent. It is noteworthy that this calculated optimum coverage is quite low. But this is consistent with the contact angle measurements in Figure 3, which suggest that only a relatively small increase in θ (and by implication a small increase in surfactant coverage) is required to impart the necessary degree of hydrophobicity to the particles.

24.3.5 Colloidal Silica Particles + Protein

Measurement of the shrinkage kinetics of bubbles stabilized by silica particles + β-LG was carried out for particle and protein concentrations ranges of 0.5–3 wt% and 0.005–0.5 wt%, respectively. Figure 9 shows the variation of the radii of several individual bubbles in such systems. As with the fumed silica (Figures 4 and 5), it is clear that the shrinkage of bubbles stabilized by the protein + colloidal silica is considerably slower than with the protein alone. On the other hand, Figure 10 shows that the DDAB + silica system is still more effective in stabilization than β-LG + silica. The compositions of the systems in Figure 10 are at the optimum concentrations of particles and amphiphilic molecules giving maximum stability as evidenced by the *F versus* time plots.

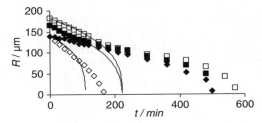

Figure 9 *Typical bubble radius R versus time t for bubbles stabilized by β-LG at pH = 7 in presence of different concentrations of colloidal silica: □, 0.5 wt% β-LG + 1.5 wt% silica; ◇, 0.5 wt% β-LG + 3 wt% silica; ■, 0.05 wt% β-LG + 1.5 wt% silica; ◆, 0.05wt% β-LG + 3 wt% silica. The solid lines indicate the expected behaviour of 0.05 wt% β-LG on its own*[13]

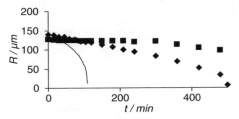

Figure 10 *Comparison of bubble radius R as a function of time t for bubbles stabilized by colloidal silica + DDAB or β-LG at pH = 7: ■, 0.5 wt% silica + 8×10^{-4} mol dm^{-3} DDAB; ◆, 1 wt% silica + 0.05 wt% β-LG*[13]

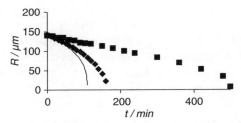

Figure 11 *Comparison of bubble radius R as a function of time t for bubbles stabilized by 0.05 wt% β-LG at pH 7 + different silica particles: ■, 1 wt% colloidal silica; ◆, 1 wt% fumed silica. The solid lines above indicate the expected behaviour of 0.05 wt% β-LG on its own*[13]

In Figure 11 the effect of silica particle size is presented, where the time-dependent radii for bubbles stabilized by 5.5-nm (colloidal) and 20-nm (fumed) silica are compared. The behaviour of a bubble stabilized by pure β-LG is also represented. It is obvious that the smaller colloidal silica particles can enhance the lifetime of the bubbles to a much greater extent than does the fumed silica.

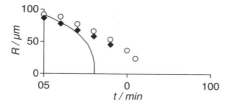

Figure 12 *Comparison of bubble radius R as a function of time t for bubbles stabilized by different silica particles + lecithin or β-LG at pH = 7: ○, 1.5 wt% colloidal silica + 1.3 × 10^{-4} mol dm^{-3} lecithin; ◆, 1 wt% fumed silica + 0.05 wt% β-LG. The solid line indicates the expected behaviour of 0.05 wt% β-LG on its own[13]*

24.3.6 Colloidal Silica Particles + Lecithin

Experiments were conducted with 5.5-nm colloidal silica particles + lecithin. At the same lecithin concentrations as used in the fumed silica + lecithin experiments, there were problems in generating enough bubbles for study, *i.e.*, the foamability of systems was poor. Figure 12 illustrates the typical radius evolution of the few bubbles that could be formed. The bubbles were considerably more stable than the same-size bubbles stabilized by pure β-LG, but they were only slightly more stable than bubbles stabilized by particles + β-LG. Overall the performance of lecithin with the colloidal silica seemed to be slightly worse than with the fumed silica. As discussed above, the reason for this may be poorer adsorption efficiency of the lecithin. Since the specific surface area of the colloidal silica is considerably greater than that of the fumed silica, it may be that the required level of surface hydrophobicity of the particles was not reached.

24.4 Conclusions

It has been shown that hydrophilic inorganic particles, of a size compatible with ingestion, can stabilize air bubbles very effectively against disproportionation, provided their surfaces are physically modified by adsorption of surfactants with positively charged head groups. In general, the smaller (colloidal) silica is more effective, probably because it remains better dispersed. Above a minimum particle concentration, DDAB is very effective, provided the surfactant concentration is not too high or too low. It is necessary to have a low surface coverage of the silica, in the region of 3–11% coverage, and hence only partial particle hydrophobicity. Analogous experiments with lecithin showed that this zwitterionic surfactant was not as effective as DDAB in modifying the surface properties of the silica to enhance bubble stability. Foamability was particularly inefficient with lecithin + silica particles. This is most probably due to the poorer adsorption efficiency of lecithin on silica under the neutral solution conditions considered here.

The use of hydrophilic particles mixed with β-LG gave considerable enhancement in stability compared to the protein on its own. This was not the case with β-casein, however, where hardly any stable bubbles could be formed at all. The increase in bubble stability with β-LG was not as great as with DDAB + silica – but it is not clear how the presence of β-LG might affect the particle hydrophobicity, if at all. As there was no enhancement in stability with β-casein, it is tentatively proposed that the particles may become embedded in the adsorbed β-LG layers around the bubbles, rather than directly adsorbing to the surface of the bubbles. Confocal microscopy has provided some additional support for this hypothesis.[10] So, in the case of β-LG, it is suggested that a protein–particle composite is formed as bubbles shrink, and that this composite ultimately has enough mechanical strength to inhibit disproportionation more effectively than the protein film on its own. Such a mechanism is not possible with β-casein because this disordered protein does not form a strong coherent adsorbed film.

References

1. E. Dickinson, R. Ettelaie, B.S. Murray and Z. Du, *J. Colloid Interface Sci.*, 2002, **252**, 202.
2. E. Dickinson, R. Ettelaie, T. Kostakis and B.S. Murray, *Langmuir*, 2004, **20**, 8517.
3. F.Q. Tang, Z. Xiao, J.A. Tang and L. Jiang, *J. Colloid Interface Sci.*, 1989, **131**, 498.
4. S.W. Ip, Y. Wang and J.M. Toguri, *Can. Metall. Quart.*, 1999, **38**, 81.
5. J.B.N. Hudales and H.N. Stein, *J. Colloid Interface Sci.*, 1990, **140**, 307.
6. G. Kaptay, *Colloids Surf. A*, 2004, **230**, 67.
7. G.N. Sethumadhavan, A.D. Nikolov and D.T. Wasan, *J. Colloid Interface Sci.*, 2001, **240**, 105.
8. G.N. Sethumadhavan, A.D. Nikolov and D.T. Wasan, *Langmuir*, 2001, **17**, 2059.
9. E. Parkinson and E. Dickinson, *Colloids Surf. B*, 2004, **39**, 23.
10. T. Kostakis, R. Ettelaie and B.S. Murray, *Langmuir*, 2006, **22**, 1273.
11. B.S. Murray, E. Dickinson, Z. Du, R. Ettelaie, T. Kostakis and J. Vallet, in *Food Colloids: Interactions, Microstructure and Processing*, E. Dickinson (ed), Royal Society of Chemistry, Cambridge, 2004, p. 259.
12. L. Ramos, T.C. Lubensky, N. Dan, P. Nelson and D.A. Weitz, *Science*, 1999, **286**, 2325.
13. Z. Du, M.P. Bilbao-Montoya, B.P. Binks, E. Dickinson, R. Ettelaie and B.S. Murray, *Langmuir*, 2003, **19**, 3106.
14. J.N. Israelachvili, *Intermolecular and Surface Forces*, Academic Press, New York, 1992.

Chapter 25

Coalescence of Expanding Bubbles: Effects of Protein Type and Included Oil Droplets

Brent S. Murray,[1] Andrew Cox,[2] Eric Dickinson,[1] Phillip V. Nelson[1] and Yiwei Wang[1]

[1] PROCTER DEPARTMENT OF FOOD SCIENCE, UNIVERSITY OF LEEDS, LEEDS LS2 9JT, UK
[2] UNILEVER RESEARCH, COLWORTH LABORATORY, SHARNBROOK, BEDFORD MK44 1LQ, UK

25.1 Introduction

Air may be viewed as a renewable, cheap, 'zero-calorie' replacement for fat in food products, where it can provide the desired bulk volume, texture, stability and appearance. Air bubbles are a traditional structural component of many foodstuffs such as mousses, ice cream, whipped toppings, *etc*. However, apart from instances where the continuous phase of a food foam is solidified (as in meringues and baked products such as bread and cakes), when compared with emulsions and solid dispersions, the disperse phase of a foam is generally not long-lasting. One reason for this transient nature of bubbles is that they are generally more deformable than emulsion droplets (or certainly solid particles) and this makes them more susceptible to coalescence. For instance, the gas bubbles in many foamed products are large enough to be more easily deformed by the hydrodynamic forces operating during processing. Coalescence can also be considerably accelerated by a rapid drop in pressure, for example, in the exit of foam from a mixing or aeration chamber, or in the extrusion of an aerated product from a nozzle, during dispensing, filling, *etc*. This pressure change can lead to significant bubble–bubble coalescence, or to coalescence of bubbles with surface of the product.

Up until now there have been relatively few studies[1-5] that have investigated this problem, either experimentally or theoretically. We reported previously[2] a technique that allows one to observe and measure the behaviour of air bubbles undergoing expansion (or compression and then expansion) due to changes in pressure of the order of a factor of 5 (which seems to be the typical range in many processes). It was shown that such expansions can result in significant coalescence of bubbles that would otherwise be completely stable. Further work suggested that fracture and breakage of interfacial films at critical extents of deformation may also be important in determining stability. Susceptibility to fracture and breakage may be related to the degree of aggregation and cross-linking within the film. Here we outline improved techniques that can be used to examine the behaviour of single bubbles and bulk foams undergoing the same types of expansion. We illustrate the effects of different proteins, mixtures of protein + oil droplets and changes in the interactions between them, on the stability of systems of direct relevance to aerated food products.

25.2 Materials and Methods

25.2.1 Materials

Bovine β-lactoglobulin (β-L, three times crystallized, lyophilized, desiccated, lot no. 21K7079, containing variants A and B), ovalbumin (OA, three times crystallized, lyophilized, desiccated, lot no. 127H7037), glucono-δ-lactone (GDL) (99.0%) and n-tetradecane (99%) were purchased from Sigma-Aldrich (Poole, UK). The commercial whey protein isolate (WPI, BiPro) from Davisco Foods (MN, USA) contained 97.7% protein, 0.3% fat, 1.9% ash and 4.8% moisture. Spray-dried sodium caseinate (SC) (> 82 wt% dry protein, < 6 wt% moisture, < 6 wt% fat and ash, 0.05 wt% calcium) was obtained from DMV International (Veghel, the Netherlands). Water from a Milli-Q system (Millipore, Watford, UK), free from surface-active impurities and with a conductivity of less than 10^{-7} S cm^{-1}, was used throughout. Solutions of pH = 7.0 consisted of 0.05 mol dm^{-3} imidazole buffer + 0.05 mol dm^{-3} NaCl, adjusted to pH = 7.0 by addition of HCl. Invert sugar syrup (82% total solids, 77 ± 1 wt% invert solids, 0.4 wt% ash) was provided by Cadbury Trebor Bassett (Bournville, UK). A 70% corn syrup (LF9, C*Sweet F017Y4) was obtained from Cerestar (Manchester, UK).

25.2.2 Emulsion Preparation and Characterization

In some bubble experiments, a protein-stabilized emulsion was added to the protein solution used to stabilize the bubbles. Oil-in-water emulsions (1 wt% protein, 30 vol% oil) were prepared using a Shields S-500 high-pressure homogenizer. Droplet size-distributions of emulsions were determined using a Mastersizer 2000 multi-angle static light-scattering instrument (Malvern Instruments, UK). The average droplet diameter for each freshly prepared emulsion was d_{43} = 0.4 ± 0.05 µm. In the case of OA, however, it was difficult to obtain stable emulsions in the same droplet size range. It was therefore decided to add some of the SC-stabilized emulsion to the OA solution. Since the volume fraction of emulsion added to the protein solution was low (10^{-3}–1%), and the concentration of the protein solution was typically 1 wt%, it was assumed that the SC added along with the emulsion did not change the bulk protein composition significantly, nor the interfacial properties of the adsorbed OA film.

Some systems were acidified by addition of GDL. To 200 ml of protein solution, or protein solution + emulsion droplets, 0.2 g of GDL was added. The solution was then split into two parts: one was used for the coalescence stability measurements, while the pH of the other was measured simultaneously in the same laboratory (and therefore at approximately the same temperature (18–23°C). At this concentration and temperature, the GDL slowly hydrolyses to lower the pH homogeneously throughout the sample. As the rate of pH lowering was much slower than the timescale of the coalescence stability measurements, in an individual stability experiment the pH could be considered constant (± 0.05 pH).

25.2.3 Single Bubble Layer Experiment

Two techniques were used to measure bubble stability with respect to interfacial expansion. One of these, referred to here as the 'single bubble layer experiment', has been described fully elsewhere.[1]

The basis of the technique is schematically illustrated in Figure 1. Briefly, the equipment consists of a stainless-steel cell containing a flexible square barrier positioned at the air–water interface of the aqueous solution contained within the cell. The barrier sides are 4.2 mm long when undeformed. The barrier shape is maintained by four hinged pins at each corner; these can be moved symmetrically in the directions of the diagonals of the square in order to compress and expand the interface contained within the barrier. The air gap above the interface can also be increased and decreased, to lower and raise the air pressure, by vertical movement of the outer sleeve of the cell. The movement of the sleeve and pins is synchronized, and the geometry of the different parts is such that the relative increase in area of a bubble at the interface due to a fall in pressure is equal to the relative increase in area of the planar interface contained within the barrier. In this way, the individual bubbles at the interface

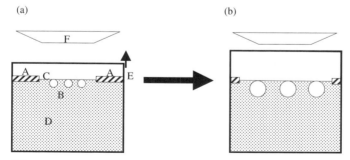

Figure 1 *Schematic representation of the single bubble layer experiments: (a) before expansion; and (b) after expansion. Key: A = flexible square barrier; B = bubbles; C = air–water interface; D = bulk aqueous phase; E = sliding wall of cell, showing direction of motion to increase air gap above the interface; and F = microscope.*

behave as if they are expanding in contact with the surface of a very large expanding bubble. The bubbles at the interface are observed from above *via* a microscope and digital camera, and the images are recorded and stored digitally for subsequent analysis.

Two alternative methods were used to introduce bubbles into the cell. When the aqueous phase was of low viscosity, the bubbles were injected into the cell beneath the barrier with a special syringe.[2] In other experiments where the aqueous phase was viscosified with sugar, it was not possible to create and inject bubbles in this way. Therefore, in these systems, bubbles were pre-formed in the test solution by agitation in a mixer,[1,6] and then a small volume of the aerated viscous solution was pipetted just beneath the planar air–water interface contained by the barrier. In each method, the individual bubbles rose up to the interface, though this took 1–2 min longer in the latter case. The typical size range of the injected bubbles was 10–200 μm. Compared with the bubbles formed by injection, the high viscosity aerated systems contained smaller bubbles, but the size distribution was wider. Around 4–6 min after introducing bubbles beneath the planar interface, the cell was sealed, and then the barrier was expanded while the pressure was simultaneously lowered and the behaviour of the bubbles was recorded.

25.2.4 Foam Stability Test

The second technique used to test bubble stability, called the 'foam stability test', was designed to mimic the behaviour of a real bulk foam undergoing similar rates and extents of expansion as in the single bubble layer apparatus. To facilitate clear observation and quantification of individual bubble stability in the foam, attempts were made to create a monodisperse foam before expansion *via* a special jet employing the technique of hydrodynamic focusing.[7] Figure 2 schematically illustrates the test equipment. The test solution (typically 50 mL) was contained within a rectangular stainless steel cell (capacity 250 mL) possessing two large glass windows in the sides.

The bubble-generating jet consisted of two concentric stainless steel capillary tubes. The external diameters of the outer and inner tubes were 1.82 and 0.50 mm, while their internal diameters were 1.30 and 0.20 mm, respectively. The 'air exit end' of the inner tube was positioned so that its end was inside the outer tube but 10–15 mm from the end of the outer tube. The test solution was circulated from a port in the cell wall and back into the bottom of the cell *via* a peristaltic pump. Air was circulated through the inner tube, *via* another peristaltic pump and tubing, from a port communicating with the top of the air space above the test solution and the other 'air entry end' of the inner tube. To achieve this connection more easily, the 'air entry end' pierces the wall of the outer tubing at a point some 2 cm away from the 'air exit end', so that the connection can be made external to the outer tube. Where the inner tube pierces the outer tube the hole is silver soldered to achieve a perfect seal.

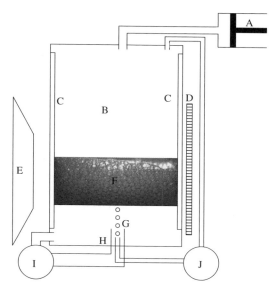

Figure 2 *Schematic diagram of the foam stability apparatus. Key: A = external piston and cylinder for controlling pressure in cell; B = air space; C = glass walls; D = illumination element; E = digital video camera; F = foam; G = foam bubbles issuing from inner capillary tube; H = outer capillary tube; I = peristaltic pump for aqueous phase; and J = peristaltic pump for air phase.*

The whole point of this arrangement (Figure 2) is to enable simultaneous (but independent) concentric flow of the test solution around the exit end of the inner tube, as air is also made to flow out of the inner tube at a different, independently controlled rate. This is the basis of hydrodynamic focusing: that is, under the correct relative flow-rates, the air issuing from the inner tube is made to break up into small monodisperse bubbles, *i.e.*, much smaller than in the absence of the outer liquid flow stream. More complex geometries and flow regimes have been developed and analysed to allow the rapid production of a wide range of perfectly monodisperse bubbles.[7] Nevertheless, the set-up we have employed is robust and reproducible, in that variations of ± 20% in the flow-rate of the air or aqueous phase, or similar changes to the diameters of the capillary tubes used, have little effect on the successful operation of the jet or on the size of the bubbles formed.

The air space in the cell is connected to an external cylinder of larger capacity. This can be pressurized and expanded by a piston driven by a stepper motor under computer control of the rate and extent of the expansion. In the experiments reported here, bubbles were formed at a pressure of 3 bar and the pressure was reduced to 1 bar in 12 s, as indicated by the pressure gauge on the instrument. In this way, the bubbles in the foam could be subjected to similar degrees of expansion/compression as in the single bubble layer experiments.

25.3 Results and Discussion

25.3.1 Effect of Protein Type

Figure 3 shows the results of single bubble layer experiments with solutions of β-L, WPI, SC and OA over a range of bulk protein concentration C_b. In the absence of expansion under the conditions described, there was no significant coalescence over the experimental time scale. As an indicator of stability, the final number of fraction of bubbles coalescing due to the expansion was calculated. This parameter (F_c) does not take any account of differences in stability due to bubble size or to the position of bubbles in the interface relative to the expanding walls, but it turns out in practice (see below) that this is not particularly important in interpreting the results presented here. The rate of expansion was kept constant at $(d\ln A/dt) = 0.07$ s^{-1}, where A is the surface area of the bubbles (or the planar interface), and the extent of expansion was set at $A/A_0 = 3$, where A_0 is the initial area. This corresponds to reducing the pressure in the system from 1 to 0.2 atm in 12 s. The systems contained invert syrup diluted to 40% total solids to give a viscosity of 7.8×10^{-3} Pa s (*i.e.*, approximately eight times higher than without sugar). For every set of conditions, the experiment was repeated at least four times, in each case with typically 50–100 bubbles present before the expansion. The mean values of F_c are plotted, with the error bars representing the standard deviations about the means.

The first key feature to note in Figure 3 is the significant variation in the ability of different proteins to stabilize the bubbles under these conditions, even at quite high values of C_b. Slightly different bubble-size distributions were produced with the different proteins, and detailed analysis has shown[1] that the

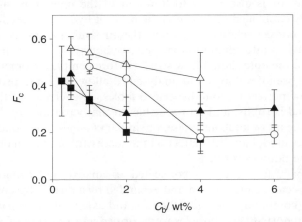

Figure 3 *Fraction of coalescence (F_c) in single bubble layer experiments at different bulk protein concentrations (C_b) at pH = 7, for rate of expansion (d ln A/dt) = 0.07 s^{-1} and extent of expansion (A/A_0) = 3: OA (△); sodium caseinate (○); WPI (▲) and β-L (■). All systems contained 40% invert syrup with a viscosity of 7.8×10^{-3} Pa s*

probability of coalescence is, on average, higher for the larger bubbles. This is probably because of the greater buoyant force acting on larger bubbles, which increases as the cube of the bubble diameter. Nonetheless, this factor does not significantly alter the main trends or the differences between the proteins. For example, for $C_b = 1-4$ wt%, the mean bubble diameter with WPI was 17 μm, while for OA it was 9 μm, even though the value of F_c for OA-stabilized bubbles was higher overall than for WPI-stabilized bubbles, *i.e.*, contrary to the expected trend if stability were to be solely determined by size.

Similarly, the variable tendency for the injected bubbles to cluster together on injection, either in the centre of the barrier, or at its walls, did not seem to have a significant effect on the measured values of F_c. Using a high-speed camera we have confirmed the findings of Liger-Belair[8] that each bubble coalescence event at the planar interface, even at these high protein concentrations, is extremely rapid (complete within <0.5 ms) and locally violent, causing plumes of aqueous phase to be ejected into the air and inducing pronounced (short-term) distortion of the neighbouring bubbles.

By analysing the sequence of individual coalescence events in some bubble clusters in detail, we have shown[1] that coalescence of one bubble (small or large) does not significantly increase the probability of a near neighbouring bubble also coalescing. In fact, in the short term, it actually appears to *reduce* slightly the probability of the neighbour subsequently coalescing. Possibly this is due to a local decrease in the stress in the interface on expansion that triggered the coalescence event in the first place. Alternatively, coalescence of one bubble may raise the local bulk concentration of protein, thereby reducing the local interfacial tension gradients caused by the original expansion. While these effects cannot be easily proved, it seems reasonable to consider them as likely to originate from the generation or existence of local variations in surface composition as the system expands.

Even down to $C_b = 1$ wt%, based on various measurements made elsewhere,[9,10] one expects diffusive adsorption to the newly created interface on expansion to be easily fast enough to replenish regions depleted in adsorbed protein at the expansion rates employed, assuming the decrease in the surface protein load on expansion is evenly distributed throughout the adsorbed layer.

Another key feature of the results in Figure 3 is that, even at the highest protein concentrations studied, the measured value of F_c levels off, but it does not reach zero. Thus, significant instability remains (*e.g.*, $F_c > 20\%$). This would represent, *e.g.*, a significant change in the bubble-size distribution and/or loss of gas from a product. It should be emphasized here that, under all the experimental conditions relevant to Figure 3, the bubbles were completely stable to coalescence for several hours in the absence of expansion. Only on expansion does instability arise, as most of the bubbles that do coalesce do so *during* the expansion; the remainder that coalesce typically do so within 1 min of the expansion ceasing.[1] In other measurements we have shown[1] also that when the expansion rate is reduced by a factor of 10, to 0.007 s^{-1}, the measured values of F_c are similar, although as expected F_c does increase roughly in proportion to the final extent of expansion (A/A_0).

These combined observations lead us to speculate that the main mechanism leading to coalescence in this situation is some sort of catastrophic fracture of the adsorbed protein film at a certain amount of expansion, or a range of interfacial strains. This is substantiated by other recent work.[11–14] For example, measurements of the surface tension γ of adsorbed films of β-L, WPI and SC, when undergoing similar expansions, showed that for the globular proteins there was a marked increase in the maximum change (increase) in γ as the films were aged. This was explained[14] as being due to increased cross-linking within the films with time, which resulted in the films being more liable to fracture on expansion, rather than adjusting to a homogeneous redistribution of protein within the interface. The work of Hotrum *et al.*[11] clearly indicates a greater tendency for film fracture with more rigid protein films, such as those formed from β-L and soy glycinin, in comparison with β-casein. Bos *et al.*[12,13] have also observed such fracture, and have measured yield stresses of protein films in relation to foamability and foam stability.

It seems noteworthy that the least stable bubbles were those formed in the solution of OA, which is well known for its tendency to form highly aggregated films at the air–water interface. The WPI system is more stable than that containing OA, but less stable than that containing pure β-L. Likewise, it has been shown that WPI has a greater tendency to form more highly aggregated interfacial films than pure β-L, which are therefore more likely to fracture on expansion.[14]

One of the clear conclusions from the above is that it seems difficult to find a food protein that, on its own, can prevent bubble coalescence under these conditions of expansion. The only way to reduce F_c close to zero seems to be to increase the aqueous phase bulk viscosity (viscoelasticity) to a point that it is practically in a gelled state. This we have demonstrated[1,15] by incorporating a range of polysaccharide thickeners. Under such circumstances, the bubbles are effectively immobilized in a thin film of bulk elastic gel beneath the planar interface. This is an important result technologically because such thickeners are frequently part of the recipe of foamed products. But such bubble stability results obtained with hydrocolloids present are probably not strictly relevant to the behaviour of a *flowable* foam undergoing a pressure drop.

25.3.2 Effect of Oil Droplets

Another key ingredient of many dairy-based whipped products is the presence of oil droplets. These are often partly crystallized, and their partial coalescence into a network around the surface of bubbles appears to be the major factor influencing the foam stability. Certainly, dispersions of solid particles can be used to form bubbles that are extremely stable to disproportionation,[16,17] but the stability of such systems to systematic pressure variation does not appear to have been tested. As partially crystalline oil droplets are difficult to control in terms of their size and aggregation, we have decided deliberately to test first the effect of completely fluid emulsion droplets on coalescence stability. In this we were also motivated by the desire to produce whippable emulsions with reduced saturated (solid) fat, as part of other ongoing work.[18]

Figure 4 shows some initial results of single bubble layer experiments at pH = 7 with 1 wt% β-L for three different expansion rates and including different volume fractions (ϕ) of oil droplets. These systems did *not* contain added sugar as thickener and the bubbles were formed by the injection method. In Figure 4 (and the subsequent figures), as an aid to clarity in comparing the different data sets, we show the estimated mean error as a single error bar in the top left of the plot. It can be seen that, at very low ϕ the systems containing oil droplets were slightly more stable (lower F_c) compared with no addition of droplets at all, while at high ϕ there was a slight decrease in stability back to that for $\phi = 0$. The trends for SC-stabilized bubbles were similar, but for OA-stabilized bubbles, which were not very stable anyway, all additions of oil droplets decreased the stability further. Because of this, and because of the further complication that the added emulsion had to be stabilized with SC anyway, the OA systems were not studied further. More detailed measurements with β-L and SC were performed at $\phi = 0.25$, since this was found to be approximately the highest value of ϕ giving some increase in stability at neutral pH.

Figure 5 shows the results of single bubble experiments for 1 wt% SC and β-L where the pH of the system was adjusted with GDL. The systems were not thickened with syrup. In all cases the expansion rate was $(d\ln A/dt) = 0.07$ s^{-1} and the expansion ratio was $A/A_0 = 3$. We see that both proteins show a slight decrease in stability (increase in F_c) from pH = 7 to pH ≈ 6. But there is a marked increase in stability, to almost $F_c = 0$, at a pH between 5.4 and 5.6. Below this pH range the curves diverge significantly. For β-L the value of F_c remains low as the pH is reduced further to the lowest pH tested (pH ≈ 4.7), whereas for SC there is a marked decrease in stability again, with the system becoming almost completely unstable below pH ≈ 5. (In fact, few stable bubbles could actually be formed on injection in this pH range.)

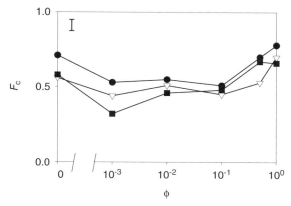

Figure 4 *Fraction of coalescence (F_c) in single bubble layer experiments with 1 wt% β-L at different volume fractions (ϕ) of β-L-stabilized oil droplets, at three different expansion rates (dlnA/dt): 0.007 s^{-1} (●); 0.07 s^{-1} (▽); and 0.35 s^{-1} (■)*

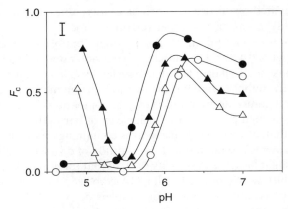

Figure 5 *Fraction of coalescence (F_c) in single bubble layer experiments with 1 wt% protein as a function of pH: β-L (●); β-L + oil droplets (β = 0.25%) (○); sodium caseinate (▲); and sodium caseinate + oil droplets (φ = 0.25%) (△)*

The most obvious interpretation of these trends is as follows. The isoelectric pH (pI) of β-L and SC is pH = 5.3 ± 0.1. This is in the pH region where there is the maximum stability (minimum F_c). Reduction of the electrostatic repulsion between proteins as the pI is approached generally increases the adsorbed amount, and it can also strengthen the net attractive interaction between adsorbed proteins, mechanically strengthening the films.[19] Whereas the β-L remains soluble at or below pI, the SC precipitates. Indeed, in this pH range, the SC solutions became cloudy, and after an hour or so individual precipitates of protein were visible to the naked eye. When the protein becomes so aggregated, the films are likely to be more liable to fracture, and there is not enough free protein to adsorb to freshly formed interfaces; hence the bubbles become very unstable to pressure change. The slight decrease in stability between pH = 6 and 6.5 may be due to reduced electrostatic repulsion between adjacent air–water interfaces, which is not compensated for by any increase in intermolecular strengthening of the film at this intermediate pH. Therefore, this reduced repulsion between interfaces enhances the probability of coalescence.

Figure 5 also shows the results of the corresponding experiments with protein-stabilized *n*-tetradecane emulsion droplets present (φ = 0.25%). There are two main features associated with the inclusion of oil droplets. Firstly, the stability is markedly increased for both proteins, but particularly for β-L, so that at pH values approximately just below pI the bubbles were found to be completely stable ($F_c = 0$). So far we have never observed such high stability under these expansion conditions with any protein on its own at any concentration. Secondly, the added oil droplets have the effect of broadening the stability curves so that the variation with pH is not so marked. For example, with SC there is still a significant increase in F_c at low pH, but this occurs at a lower pH than without droplets and the onset of increased stability occurs at a higher pH than without droplets.

It should be emphasized that in none of these experiments did we observe any oil droplet instability. At this C_b value, and at the rates of interfacial expansion applied, the work of Hotrum et al.[11] suggests that this is indeed what would be expected. Therefore, the droplets did not appear to coalesce, or to enter and spread at the planar air–water interface to form lenses or larger oil droplets. This means that droplet instability factors cannot explain the changes observed; in fact, their occurrence would tend anyway to reduce bubble stability. Thus we propose that the additional stabilizing effect of the added oil droplets is due to the droplets themselves forming a flocculated network as the pH is lowered. Whether this network forms part of the protein network already at the bubble surface – either within it or adsorbed to it – is not yet so clear.

The enhancement in bubble stability on addition of emulsion droplets at pH = 7, where protein flocculation is not expected, suggests that the mere presence of oil droplets at the interface may also be a factor in improving stability. The oil droplets will naturally tend to rise up to the planar air–water interface and so accumulate around the air bubbles located there. Confocal images (not shown) of some aerated (whipped) systems do suggest that the emulsion droplets can become positively associated with the bubbles surfaces even at pH = 7. Partly for this reason, it was also considered necessary to develop a method for looking at bulk foam stability under such expansion conditions.

Figure 6 compares the results from the single bubble layer experiments for SC (Figure 5) with the corresponding results from the foam stability test under the same solution conditions. In the latter case the bubbles were formed at 3 atm and the pressure was dropped to 1 atm in 12 s; but this was still a similar expansion rate to that in the single bubble layer experiments. Note also that in the foam experiments the value of F_c was calculated from the number of bubbles visible at the wall of the cell, and from bubbles coalesced with each

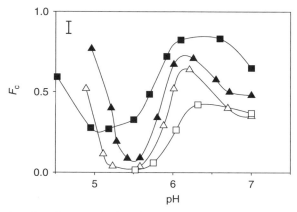

Figure 6 *Fraction of coalescence (F_c) with $C_b = 1$ wt% sodium caseinate as a function of pH in: (a) single bubble layer experiments in the systems sodium caseinate (▲) and sodium caseinate + oil droplets ($\phi = 0.25\%$) (△); (b) foam stability experiments in the systems sodium caseinate (■); and sodium caseinate + oil droplets ($\phi = 0.25\%$) (□)*

other (not just at the surface of the foam). All these factors mean that the magnitude of force driving coalescence is different in the two types of experiment and so differences in the absolute values of F_c in each case might be expected. However, what is most significant is that the overall trends in the F_c data sets are similar for both types of measurements.

In the foam experiment there is a maximum and minimum in F_c at similar pH values as in the equivalent single bubble layer experiment, except that the peak and trough are somewhat broader. In the case of the minimum in F_c (without droplets), this is shifted to a slightly lower pH. Most significantly, in the presence of oil droplets the foam is dramatically stabilized above the pI, compared to without oil droplets, so that F_c is practically zero, within experimental error. Below pH $\approx$ 5.5 the SC system with oil droplets appeared to form a mixed precipitate/gel and stable bubbles could not be formed to perform the expansion test. The results of these experiments on model foams and individual bubbles, as a function of pH and protein type, are in broad agreement with work elsewhere on the whipping of emulsions, in terms of both whippability and foam stability.[18] The results also suggest that the exact structure of the foam (bubble sizes and packing) is not particularly important in determining its stability, but rather the solution conditions and the mechanical properties of the interfacial film or of the intervening aqueous phase surrounding the bubbles.

Finally, since the effect of viscosification of the aqueous phase has been shown to affect bubble stability, some foam stability tests were also performed with sugar syrup present in the aqueous phase. A different syrup was used in this case, the Cerestar corn syrup, which gave an aqueous phase viscosity of 30 ± 5 Pa s. Figure 7 compares the foam stability results from Figure 6 (in the absence of syrup) with those obtained with syrup. Clearly the presence of the syrup alone, in the absence of oil droplets, gives a considerable increase in

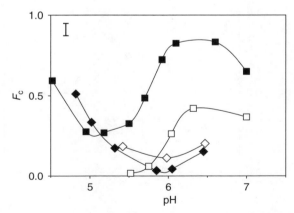

Figure 7 *Fraction of coalescence (F_c) with $C_b = 1$ wt% sodium caseinate as a function of pH in foam stability experiments in the systems: sodium caseinate (■); sodium caseinate + oil droplets ($\phi = 0.25\%$) (□); sodium caseinate + corn syrup (◆); and sodium caseinate + corn syrup + oil droplets ($\phi = 0.25\%$) (◇). Corn syrup viscosity = 30 Pa s*

stability of the foam, which is maximized at pH ≈ 5.9, both with and without droplets. Interestingly, there is the suggestion that the systems are slightly less stable in the presence of oil droplets + syrup, although this difference lies just within the combined experimental scatter in the F_c data. An increased viscosity of the aqueous phase slows down thin film drainage between bubbles, and this is known to enhance the lifetime of quiescent foams. It is not clear, however, how the increased viscosity necessarily operates to improve stability when interfaces are expanded in the manner of these experiments. Certainly, the viscosity is not high enough to slow down the expansion rate: the bubbles are observed to grow in size in response to the pressure change at the same rate as without syrup.

25.4 Conclusions

We have developed two pieces of apparatus for studying the stability to coalescence of bulk foams and individual bubbles when they are subjected to rates and extents of pressure drop typical of many processing conditions involving foamed food products. For the cases considered here, these test methods appear to give complementary results, and they are also highly discriminating in terms of distinguishing the ability of different proteins to stabilize foams under such conditions. The proteins appear unable to stabilize foams completely at neutral pH. For β-L and sodium caseinate, the foam stability is considerably increased as the pH is reduced to the isoelectric point, but below this pH the bubbles in the sodium caseinate systems become highly unstable due to precipitation of the protein. A considerable further increase in stability can be obtained by including a low volume fraction of oil droplets stabilized by the same proteins. This is attributed to the formation of an even more effective cross-linked network of droplets and/or protein + droplets at the air–water interface as the pH is lowered towards the pI.

References

1. B.S. Murray, E. Dickinson, C.K. Lau and E. Schmidt, *Langmuir*, 2005, **21**, 4622.
2. B.S. Murray, I. Campbell, E. Dickinson, K. Maisonneuve, P.V. Nelson and I. Söderberg, *Langmuir*, 2002, **18**, 5007.
3. I. Söderberg, E. Dickinson and B.S. Murray, *Colloids Surf. B.*, 2003, **30**, 237.
4. B.S. Murray, E. Dickinson, Z. Du, R. Ettelaie, K. Maisonneuve and I. Söderberg, in *Food Colloids, Biopolymers and Materials*, E. Dickinson and T. van Vliet (eds), Royal Society of Chemistry, Cambridge, 2003, p. 165.
5. B.J. Hu, A.W. Nienow and A.W. Pacek, *Colloids Surf. A*, 2003, **31**, 3.
6. K. Lau and E. Dickinson, *J. Food Sci.*, 2004, **69**, E232.
7. P. Garstecki, I. Gitlin, W. DiLuzio, G.M. Whitesides, E. Kumacheva and H.A. Stone, *Appl. Phys. Lett.*, 2004, **85**, 2649.

8. G. Liger-Belair, *J. Agric. Food Chem.*, 2005, **53**, 2788.
9. J.A. de Feijter and J. Benjamins, in *Food Emulsions and Foams*, E. Dickinson (ed), Royal Society of Chemistry, London, 1987, p. 72.
10. J.A. de Feijter, J. Benjamins and F.A. Veer, *Biopolymers*, 1978, **17**, 1759.
11. N.E. Hotrum, M.A. Cohen Stuart, T. van Vliet and G.A. van Aken, *Langmuir*, 2003, **19**, 10210.
12. M.A. Bos, K. Grolle, W. Kloek and T. van Vliet, *Langmuir*, 2003, **19**, 181.
13. A.H. Martin, K. Grolle, M.A. Bos, M.A. Cohen Stuart and T. van Vliet, *J. Colloid Interface Sci.*, 2002, **254**, 175.
14. B.S. Murray, B. Cattin, E. Schüler and Z.O. Sonmez, *Langmuir*, 2002, **18**, 9476.
15. B.S. Murray, E. Dickinson, C. Gransard and I. Söderberg, *Food Hydrocoll.*, 2006, **20**, 114.
16. E. Dickinson, R. Ettelaie, T. Kostakis and B.S. Murray, *Langmuir*, 2004, **20**, 8517.
17. T. Kostakis, R. Ettelaie and B.S. Murray, *Langmuir*, 2006, **22**, 1273.
18. K.E. Allen, E. Dickinson and B.S. Murray, *Lebensm. Wiss. Technol.*, 2006, **39**, 225.
19. B.S. Murray, *Curr. Opin. Colloid Interface Sci.*, 2002, **7**, 426.

PART IV
Emulsions

Chapter 26

Role of Protein-Stabilized Interfaces on the Microstructure and Rheology of Oil-in-Water Emulsions

Peter J. Wilde, Alan R. Mackie, Michael J. Ridout, Fiona A. Husband, Graham K. Moates and Margaret M. Robins

INSTITUTE OF FOOD RESEARCH, NORWICH RESEARCH PARK, COLNEY, NORWICH NR4 7UA, UK

26.1 Introduction

Emulsions have applications as wide ranging as food, pharmaceuticals, oil production, printing, agrochemicals and photography. They are incorporated into a broad range of food products, and it is estimated that over 40% of processed foods contain emulsified oils or fats. The emulsified fat is included to impart the desired texture, structure and flavour to foods. Nevertheless, dietary fat intake is a major concern in the Western diet, and it is thought to be a major contributory factor to increasing levels of obesity and related conditions. The sensory perception of fat content in emulsions is poorly understood, and although there are many successful reduced fat products on the market, fat replacement is still seen as a major challenge in the food industry. Therefore, if we can understand more fully the fundamental basis of sensory perception of fat content, we can develop intelligent strategies to control the organoleptic properties of emulsion-based foods. This study is aimed at trying to understand the complex relationship between the physico-chemical and organoleptic properties of oil-in-water (O/W) emulsions.

Proteins are often used to stabilize food emulsions against coalescence, as they possess unique interfacial properties that can confer high levels of long-term stability.[1] Proteins are complex, polyionic, amphiphilic macromolecules, and their unique interfacial properties have been studied for many years.[2,3] Following adsorption at the interface they tend to undergo rearrangement and aggregation processes to form an immobile, elastic interfacial film.[3] Hence, the molecular structure of the proteins can strongly influence their interfacial rheological properties[4,5] and the stability of emulsions and foams.[1,3]

The other important properties of emulsions that are relevant to organoleptic behaviour are microstructure and rheology.[6,7] Apart from the properties of the continuous phase, the main factors which govern emulsion rheology are the properties of the dispersed phase droplets.[6,8] These include the droplet volume fraction, the droplet–droplet interactions, and the droplet viscosity and deformability. Considerable progress has been made in understanding the factors that influence the physical characteristics of model emulsions. It has been shown[9] that there is a sharp transition from liquid to solid-like behaviour as the droplet volume fraction reaches the random packing limit, particularly in monodisperse systems.[10] But our knowledge of the structure and rheology of dispersions containing polydisperse, deformable droplets is still at an elementary stage. Emulsified oil droplets have a fluid internal phase and a flexible interface. The application of stress can cause circulation of the internal phase and may lead to droplet distortion.[11] The deformation of the droplets is influenced by the viscosities of the dispersed and continuous phases, and by the strength of the shear field. Taylor[12] predicted that interfacial tension plays a key role in controlling distortion, and more recently Otsubo and Prudhomme[8] have demonstrated a scaling relation for the viscosity of an O/W emulsion in terms of droplet size, the viscosities of the two phases and the interfacial tension. Even more recently Mason et al.[10,13] showed that the Laplace pressure ($2\gamma/r$), and thus both the interfacial tension γ and the droplet radius r, is an important factor in relation to emulsion rheology, and so should be used as a normalization factor. Evidence that the interfacial tension is not the only characteristic of the interface that may be involved in the deformation of droplets has been supplied recently by the studies of Pozrikidis[14] and Nadim.[15] Pozrikidis[14] studied the deformation of a liquid drop with a constant isotropic surface tension and finite surface viscosity. He found that the surface viscosity acts to suppress the interfacial motion and hence it reduces the magnitude of the droplet deformation. Nadim[15] provided results which suggest that the deformation of droplets and the measured effective viscosity of the emulsion are functions of the surface shear and dilatational viscosities.

Concentrated emulsions stabilized by proteins tend to have greater elastic moduli than emulsions stabilized by low-molecular-weight emulsifiers.[7,16–19] We have already reported[20] that the interfacial structure of simple emulsion systems can have a significant effect on the organoleptic properties of emulsions. We showed that the interfacial composition of test emulsions influences the sensory perception of 'creaminess' and fat content. The main result was that emulsions stabilized by proteins possess an enhanced sensory response to fat content. This was attributed to changes in interfacial composition affecting emulsion rheological behaviour. An enhanced viscosity was found for the protein-stabilized emulsion when compared with a similar emulsifier-stabilized emulsion, but the mechanism underlying the phenomenon was not entirely clear. There have been some theoretical treatments showing how an elastic interfacial layer can influence droplet deformation[15,21] and emulsion rheology,[18] and certainly the elasticity of most globular protein interfaces would induce this effect. Experimental studies of the behaviour of protein-stabilized emulsions have suggested[17] that the interactions between the protein-stabilized droplets may be responsible for the enhanced rheological properties. There have been several studies looking at the interactions between protein-stabilized droplets and surfaces, showing that, although long-range interactions are dominated by electrostatic repulsion,[22–24] short-range interactions can vary for different proteins.[22] Most globular proteins have a large steric repulsion term,[22,24] sometimes due to the formation of multilayers,[22] or adsorbed protein aggregates.[23] Also it has been inferred that non-DLVO hydration forces may exist,[24,25] but the consequences of these interactions have not been fully investigated. It has also been shown[26] that interactions between protein-stabilized emulsion droplets can influence the phase separation behaviour of the emulsion.[26] As far as we are aware, however, there has been no systematic experimental study of the role that an adsorbed protein layer has on the microstructure and bulk rheology of the cream layer of an O/W emulsion.

The aim of the present study was to investigate more thoroughly how the interfacial characteristics of proteins can influence the structure and bulk rheology of O/W emulsions, with a view to understanding how we might be able to manipulate the sensory properties of food emulsions. Our approach is to quantify more precisely the role of interfacial composition by creating emulsions that are as near identical in characteristics as possible, but with varied interfacial rheological properties. This was achieved by first creating an emulsion stabilized by protein (whey protein isolate) possessing an immobile, viscoelastic interface. This interface was then disrupted and displaced by the addition of surfactant[27,28] to produce a near-identical emulsion stabilized by the surfactant. Different surfactant mixtures were used to control the surface charge and zeta potential of the emulsion droplets. The resulting emulsions were then characterized to determine the effect of interfacial composition on the dynamic development of structure and rheology.

26.2 Materials and Methods

O/W emulsions (22 wt% oil) were prepared initially using 0.43 wt% whey protein isolate (WPI) (BiPro, Davisco Foods, MN, USA) with 10 mM citrate buffer at pH = 6.0 as the aqueous phase and 10:90 (by volume) hexadecane:heptane as the oil phase. The emulsion was prepared with a Waring blender using a timed shearing cycle. The particle-size distribution of this 'parent' WPI emulsion was measured. This parent emulsion was then divided into two equal parts and to each sample was added 9.09 g of sodium citrate buffer (pH = 6) with or without 2.7 wt% surfactant. Displacement studies showed that this surfactant concentration was sufficient to remove the protein completely from the oil droplet surface. This produced two separate emulsions with identical oil phase volumes and droplet-size distributions, one stabilized by protein and the other by surfactant. The final continuous phase buffer concentration was 10 mM. The surfactants used were polyoxyethylene-23-lauryl ether (Brij 35) and sodium dodecyl sulfate (SDS) (Sigma-Aldrich, Poole, UK). Droplet-size distribution measurements at the end of the experiments showed no change from the initial values.

Droplet-size distributions of the emulsions were measured using a Coulter LS 230 laser diffraction particle sizer (Beckman Coulter, CA, USA). The data were analysed using an optical model for a fluid with real and imaginary parts of the complex refractive index set to 1.332 and 1.391, respectively. The zeta potential of the emulsions was measured using a Zetasizer 3 (Malvern Instruments, UK) calibrated using the -50 mV standards supplied by the manufacturer. All samples were measured after dilution using the relevant continuous phase (separated by centrifugation).

Microscopy images were acquired using a Biorad 1024 confocal microscope based around a modified Nikon Optiphot microscope. Samples were placed in a specially modified cuvette which had one side replaced with a cover glass. Observations were made using a 60 × oil immersion objective with a numerical aperture of 1.4. Samples were stained using Nile Red, which was added to the sample in powder form and allowed to stand for several days.

The interfacial tension between the aqueous phase of the emulsion and n-hexadecane was measured using the pendant drop technique. Images were acquired using a digital camera (Pulnix, USA) and frame grabber (Matrox, USA) and analysed using in-house software.

The surface and bulk shear rheological measurements were made using a TA Instruments AR2000 controlled stress rheometer (TA Instruments, Crawley, UK) in controlled stress mode for creep and steady-state bulk measurements and in controlled strain mode for oscillatory shear and surface shear measurements. Creaming experiments used a modified cup and bob arrangement with a cup of 150 mm depth and an inner cylinder of 40 mm height. With this arrangement we could monitor the bulk rheology of a sample at a range of heights within a creaming emulsion. A frequency of 1 Hz and a strain of 1% were chosen as the measuring conditions, and samples were monitored over the same time-scale as for the ultrasonic experiments. Rheological measurements of the creamed phase were carried out in a cone-and-plate geometry using a 2° cone in either steady or dynamic mode, as required. Surface rheological measurements at the air–water or oil–water interfaces were undertaken with a polished aluminium biconical disc (6°, 60 mm diameter) over a 2-h time period, again with a strain of 1%, but a frequency of 0.5 Hz.

Oil volume fraction profiles were assessed from time-dependent measurements of ultrasonic velocity through the sample as a function of height within the sample. These data were related to disperse phase volume fraction *via* the Urick equation. Creaming measurements were made at 20°C using an Acoustiscan system (University of Leeds, UK).[29] Readings were taken every 2 mm over the entire height of the emulsion, to give a profile of the dispersed phase volume throughout the emulsion. Measurements continued until the majority of the oil was in the cream layer, and the distribution was approaching equilibrium.

Table 1 Values of zeta potential ζ, mean particle diameter d_{32}, interfacial tension γ, interfacial dilatational modulus $|E|$ and interfacial shear modulus G' for the emulsion systems based on the different emulsifiers

Emulsifier	ζ (mV)	d_{32} (μm)	γ (mN m^{-1})	$|E|$ (mN m^{-1})	G' (mN m^{-1})
WPI (pH 6)	−28.4	6.75	15.8	38.9	0.43
Brij 35	−0.4	6.75	–	–	<0.005
SDS	−79.9	6.75	–	–	<0.005
Brij 35 + SDS	−25.5	6.75	8.4	1.5	<0.005

26.3 Results

The methodology underlying this study was designed to produce emulsions which were as near identical as possible apart from their interfacial compositions, in order to test the influence of interfacial rheology on the bulk structure and rheology of the emulsion cream layers. Mixtures of ionic and non-ionic surfactants were used to modify the surface charge. Table 1 shows the characteristic size distributions, the zeta potentials, and the interfacial properties of the different emulsions. Thus, using the approach described above, two near-identical emulsions were made, one stabilized by a mixture of non-ionic surfactant (Brij 35) and anionic surfactant (SDS) and the other stabilized by whey protein (WPI).

Interactions between emulsion droplets have most impact on the rheological behaviour when they are close-packed in the dense emulsion cream layer. Initial measurements involved sampling the cream layer and measuring various rheological parameters. The WPI emulsions tended to have higher viscosities and showed greater 'yield-stress'-like behaviour than those containing surfactants. As an example, Figure 1 shows the creep–recovery behaviour of two emulsions. The initial creep behaviour is fairly similar for the two samples, although the WPI-stabilized emulsion (solid line) does appear to show an elastic 'jump' when the stress is applied. The WPI-stabilized emulsion also shows significant recovery on removal of the applied stress, again indicating a significant elastic component. In contrast, the surfactant-stabilized emulsion (dotted line) shows hardly any recovery at all, which is more indicative of a viscous sample.

A limitation of the preceding rheological experiments was that the emulsion had to be sampled and then transferred to the rheometer, potentially disrupting any weak structures that may have formed during the creaming process. Thus, a method of measuring the viscoelasticity of the creams *in situ* was developed, as described above. Figure 2 compares the bulk viscoelastic response of the cream layer as it formed in the two systems. The cream layer had more or less fully developed after about 20 h. The moduli for the WPI emulsion were nearly an order of magnitude greater than for the Brij/SDS system. Additionally, in the WPI emulsion the elastic component was greater than the viscous component, whereas in the surfactant emulsion the viscous component was comparable to or slightly greater than the elastic component, confirming the

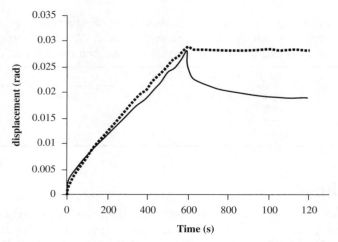

Figure 1 *Creep recovery behaviour of samples of emulsion creams stabilized either by WPI (solid line) or Brij 35 (dashed line). A stress of 0.1 Pa was applied for 600 s and then removed*

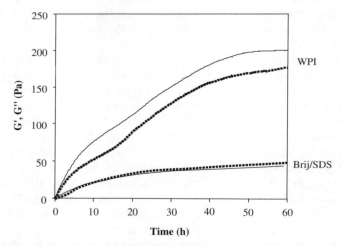

Figure 2 *Storage modulus G' (solid line) and loss modulus G'' (dashed line) for creaming emulsions stabilized by WPI and Brij/SDS*

observations reported in Figure 1. We also note that the rheological behaviour of the creams continued to develop after the cream layer had fully formed, particularly for the protein sample. This indicates that the number and/or the strength of interactions continued to increase over time even after the creaming was assumed to be complete. Another possible contributing factor to the larger elasticity in the protein emulsion is the difference in the interfacial tension for the protein and surfactant systems, and the influence that this has on droplet deformability. Measurements of the interfacial tension of the sub-phases against *n*-hexadecane (Table 1) gave 15.8 mN m^{-1} for the protein-stabilized system, but a much lower value of 8.4 mN m^{-1} for the Brij/SDS system. The subsequent difference in the Laplace pressure would be expected to have a significant effect on droplet deformability, leading to the protein system being more elastic at the higher volume fractions.

The separation of the emulsions under gravity was followed non-invasively using the ultrasonic technique described above. The results are shown in Figure 3 in terms of the disperse phase volume as a function of height for a series of time intervals over a period of 45 h. Both systems show the development of a cream layer at the top of the tube, whose thickness gradually increased with time. The lack of a sharp boundary at the base of the creaming emulsion indicates a broad

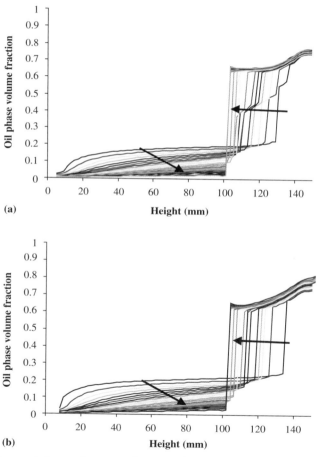

Figure 3 *Dispersed phase volume as a function of height in creaming emulsions stabilized by (a) WPI and (b) Brij35/SDS. Each line represents a set of measurements taken 60 min apart over a period of 40 h. The arrows indicate the progression with time*

size distribution and a lack of flocculation. Because the size distribution of both emulsion systems was identical, the oil droplets moved from the sub-phase at a very similar rate. The region showing the most distinct differences is within the cream layer over the first 8 h of the experiment. In the case of the surfactant-stabilized system [Figure 3(b)], the density of the cream gradually increases from the base to the top of the layer. However, in the case of the protein system, there is a 'pseudo' plateau of lower density at the base of the cream that disappears after about 8 h. This suggests a delay in the rearrangement into a more densely packed layer. After this 8-h period, the ascending droplets seem able to pack into the denser state more directly. This may be a result of the polydispersity of the emulsion, leading to larger droplets ascending into the cream layer first and showing a delay in rearranging. A further difference to note between the behaviour of the two samples is that the final phase volume at the top of the cream layer is some 5% greater in the surfactant-stabilized system than in the protein-stabilized system. This is shown in more detail in Figure 4. Here the oil volume fraction of the top 20 mm of the cream layer is shown as a function of time. The initial rapid creaming seems to be similar for the two samples, but after about 5 h the density of the WPI emulsion equilibrates quickly, whereas the surfactant emulsion continues to increase in density over the course of the experiment. This highlights a difference between the two emulsion systems, as the bulk moduli of the WPI emulsion cream layer are almost an order of magnitude greater than those of the surfactant emulsion cream layer (Figure 2), despite the fact that the former has a lower phase volume (Figure 4).

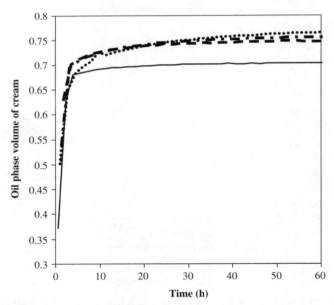

Figure 4 *Dispersed phase volume of upper 20 mm of cream layer as a function of time for emulsions stabilized by WPI (solid line), SDS (dotted line), Brij 35 (dashed line) and SDS/Brij35 (dashed/dotted line)*

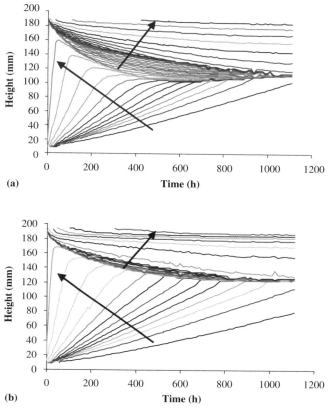

Figure 5 *Dispersed phase volume contour plots for emulsions containing 40% glycerol stabilized by (a) protein and (b) surfactant. Height is plotted as a function of time. The contours are in phase volume steps of 0.02*

In order to investigate these differences more fully, a similar experiment was undertaken in the presence of 40% glycerol, which increased the viscosity of the continuous phase and thus slowed down the creaming by a factor of around 20. The data from this experiment are shown in Figure 5, in the form of contour plots of height against time. The contours represent lines of equal oil phase volume in 2% intervals up to around 80%, depending on the final phase volume at the top of the cream layer. This type of plot allows the determination of the aggregation state of the droplets as they rise. The linearity of the contour lines below the cream layer shows that the droplets have ascended with fixed velocities. This means that the droplets did not aggregate in the sub-cream region. These Stokes velocities can also be used to calculate a size distribution, which was found to be in good agreement with that measured by light scattering at the start of the experiment. Again there was a 'pseudo' plateau at the base of the cream layer in the protein sample [Figure 5(a)]; this persisted until very near the end of the experiment, and it had a depth of some 30 mm at its maximum extent. In this slow creaming experiment, even the surfactant-stabilized emulsion displays the same type of feature [Figure 5(b)], but it is only some 8 mm in depth at its greatest extent. Finally, the phase volume at the top of the cream layer at the end of the experiment is quite significantly different for the two systems, i.e., higher by 0.09 in the surfactant-stabilized system.

One possible explanation for the lower dispersed phase volume of the cream layer in the protein systems is that there was less rearrangement in the layer due to its elastic behaviour. The possible presence of voids or unfilled interstitial spaces in the cream layer was investigated by confocal microscopy. These observations showed that the WPI emulsions exhibited evidence of voids, whereas the surfactant emulsions appeared to show few voids. Image analysis of the phase volumes was in agreement with the ultrasonic analysis. However, one of the most telling differences was not obvious from the static images. In the surfactant-stabilized systems that were investigated, the voids contained small droplets that were undergoing Brownian diffusion. However, in the protein-stabilized systems there was no sign of movement of smaller droplets,[30] demonstrating a clear difference in the close–range interaction for the two systems. Aliquots taken from the base and the top of the two samples were measured in order to determine the particle-size distribution. There were found to be only minor differences between the samples, the main difference being between the top and base of the cream layers (up to a factor of 100 in mean droplet size).

26.4 Discussion

One of the most striking differences between the two systems is in the interfacial rheology (Table 1), particularly the shear rheology. The high moduli values attained by the WPI are typical for interfaces occupied by globular proteins. While the interdroplet interaction range of the two systems was designed to be comparable in terms of the measured zeta potentials, this does not necessarily mean that the short-range interdroplet interaction was similar. Calculation of the local pair potential is difficult to perform accurately for the protein system because of its polyelectrolyte nature. Nevertheless, we can expect substantial differences from the surfactant case.

The objective of the overall project is to determine the importance of the interfacial characteristics of proteins in controlling the bulk rheology of high dispersed phase volume emulsion systems. The systems are chosen to have particular interfacial properties such as high surface elasticity or specific net charge. The two main samples investigated here have similar long-range interaction potentials while having very contrasting interfacial rheological properties. The ionic strength was kept constant, and the two systems had a double-layer thickness of the order of 3 nm. In contrast to previous work using pure β-lactoglobulin as emulsifier,[26] the double-layer thickness was large enough to prevent flocculation in the creaming emulsion. However, once the cream layer becomes compressed, the droplets are forced closer together, and the local interaction potentials for the two systems may be very different, particularly in the light of the differences in surface mobility.

It is known[17] that the rheological properties of dispersions are related to their droplet-size distributions through the Laplace pressure. Because of this, we have kept the droplet size distribution as constant as possible for all the emulsions, and have measured the equilibrium interfacial tension of the systems used to make the emulsions. The surface tensions differ by a factor of two for the protein and surfactant systems, so indicating a predicted twofold variation in the bulk elastic modulus of the emulsions.[9,17] The bulk rheology data of the emulsions (Figure 2) shows that the elasticity is about a factor of 10 larger for the protein-stabilized system than for the surfactant system. As this clearly cannot be explained solely through changes in the Laplace pressure, it suggests that other factors are also involved.

There would seem to be two possible explanations for the difference in the bulk rheological data. First, there is the variability in interfacial rheology as shown in Table 1. These data indicate that, in the concentrated region of the cream, the increased interfacial shear elasticity of the protein system would tend to slow down, and possibly to arrest, the rearrangement of the droplets. This could explain the disparity in the maximum packing fractions for the two systems seen in Figures 3 and 4. Nevertheless, there is clearly no simple linear relationship between interfacial and bulk rheology, as the interfacial shear rheology shows a difference of more than 100-fold between the systems. Further to this, there is also substantial variation in the interfacial dilatational rheology (Table 1). A higher surface elasticity would tend to reduce droplet deformation[15,21] and thus increase bulk elasticity.[18]

The second point of contrast between the protein- and surfactant-coated interfaces is in the close-range interaction. While surfactant-stabilized systems tend to have a repulsive interaction even at close range, it has been shown, using several different methods, such as thin films,[31] magnetic chaining,[23,32] etc., that protein-stabilized interfaces can show a significant degree of adhesion. Thus, when two protein-covered interfaces are brought into close proximity they tend to stick together. This was also suggested[16] by experiments on concentrated emulsion systems which do not spontaneously redisperse on dilution. The cream layer contains droplet interfaces that are forced together by gravity, and the lack of Brownian diffusion observed by confocal microscopy in the cream layer of the protein-stabilized system suggests that the droplets were attracted to each other at close approach, at least more so than for the surfactant-stabilized droplets.

Each of these hypothetical explanations seems equally plausible, and there has been insufficient experimental evidence obtained so far to suggest which is the more important. Thus, for future work, we are designing systems which possess a homogeneous surfactant-like interfacial layer, but also display similar interfacial rheological properties to a protein-stabilized interface. The bulk rheological behaviour of the concentrated dispersion stabilized by this type of system should then provide powerful evidence for determining whether either the interfacial rheology or the close-range interdroplet interaction is the main determining factor.

26.5 Conclusions

We have investigated the rheology and microstructure of emulsion systems in which the interfacial composition was varied, but other physical parameters including the surface charge were kept constant. Significant differences in bulk rheology and creaming behaviour have been found. Emulsions stabilized by protein were found to have a much higher elastic bulk viscoelastic modulus than the corresponding surfactant systems and to have a lower maximum packing fraction. This disparity in behaviour can be attributed to differences in the close range pair–wise interaction, the interfacial rheology, or both of these.

Acknowledgement

We acknowledge the support of BBSRC in funding this work through the Core Strategic Grant to the Institute of Food Research.

References

1. V.N. Izmailova, G.P. Yampolskaya and Z.D. Tulovskaya, *Colloids Surf. A*, 1999, **160**, 89.
2. E. Dickinson, *Colloids Surf. B*, 1999, **15**, 161.
3. M.A. Bos and T. van Vliet, *Adv. Colloid Interface Sci.*, 2001, **91**, 437.
4. E.M. Freer, K.S. Yim, G.G. Fuller and C.J. Radke, *Langmuir*, 2004, **20**, 10159.
5. M.J. Ridout, A.R. Mackie and P.J. Wilde, *J. Agric. Food Chem.*, 2004, **52**, 3930.
6. Th.F. Tadros, *Colloids Surf. A*, 1994, **91**, 31.
7. F. Lequeux, *Curr. Opin. Colloid Interface Sci.*, 1998, **3**, 408.
8. Y. Otsubo and R.K. Prudhomme, *Rheol. Acta*, 1994, **33**, 29.
9. H.M. Princen, *J. Colloid Interface Sci.*, 1983, **91**, 160.
10. T.G. Mason, M.D. Lacasse, G.S. Grest, D. Levine, J. Bibette and D.A. Weitz, *Phys. Rev. E*, 1997, **56**, 3150.
11. R. Cavallo, S. Guido and M. Simeone, *Rheol. Acta*, 2003, **42**, 1.
12. G.I. Taylor, *Proc. R. Soc. (London)*, 1932, **138**, 41.
13. T.G. Mason, *Curr. Opin. Colloid Interface Sci.*, 1999, **4**, 231.
14. C. Pozrikidis, *J. Eng. Math.*, 2004, **50**, 311.
15. A. Nadim, *Chem. Eng. Commun.*, 1996, **150**, 391.
16. T.D. Dimitrova and F. Leal-Calderon, *Langmuir*, 2002, **17**, 3235.
17. L. Bressy, P. Hebraud, V. Schmitt and J. Bibette, *Langmuir*, 2003, **19**, 598.
18. K.D. Danov, *J. Colloid Interface Sci.*, 2001, **235**, 144.
19. M. Oosterbroek and J. Mellema, *J. Colloid Interface Sci.*, 1981, **84**, 14.

20. P.B. Moore, K. Langley, P.J. Wilde, A. Fillery-Travis and D.J. Mela, *J. Sci. Food Agric.*, 1998, **76**, 469.
21. C. Pozrikidis, *J. Non-Newtonian Fluid Mech.*, 1994, **51**, 161.
22. T.D. Dimitrova, F. Leal-Calderon, T.D. Gurkov and B. Campbell, *Langmuir*, 2001, **17**, 8069.
23. T.D. Dimitrova, F. Leal-Calderon, T.D. Gurkov and B. Campbell, *Adv. Colloid Interface Sci.*, 2004, **108–109**, 73.
24. J.J. Valle-Delgado, J.A. Molina-Bolivar, F. Galisteo-Gonzalez, M.J. Galvez-Ruiz, A. Feiler and M.W. Rutland, *J. Phys. Chem. B*, 2004, **108**, 5365.
25. J.A. Molina-Bolivar, F. Galisteo-Gonzalez and R. Hidalgo-Alvarez, *Colloids Surf. B*, 1999, **14**, 3.
26. T.D. Dimitrova, T.D. Gurkov, N. Vassileva, B. Campbell and R.P. Borwankar, *J. Colloid Interface Sci.*, 2000, **230**, 254.
27. J. Chen and E. Dickinson, *J. Sci. Food Agric.*, 1993, **62**, 283.
28. A.R. Mackie, A.P. Gunning, P.J. Wilde and V.J. Morris, *Langmuir*, 2000, **16**, 2242.
29. P.V. Nelson, M.J.W. Povey and Y. Wang, *Rev. Sci. Instr.*, 2001, **72**, 4234.
30. http://www.ifr.ac.uk/science/programme/F1/rheology.html
31. L.G.C. Pereira, C. Johansson, C.J. Radke and H.W. Blanch, *Langmuir*, 2003, **19**, 7503.
32. T.D. Dimitrova and F. Leal-Calderon, *Langmuir*, 1999, **15**, 8813.

Chapter 27

Crystallization in Monodisperse Emulsions with Particles in Size Range 20–200 nm

Malcolm J.W. Povey,[1] Tarek S. Awad,[1] Ran Huo[1] and Yulong Ding[2]

[1] PROCTER DEPARTMENT OF FOOD SCIENCE, UNIVERSITY OF LEEDS, LEEDS LS2 9JT, UK
[2] INSTITUTE FOR PARTICLE SCIENCE AND ENGINEERING, UNIVERSITY OF LEEDS, LEEDS LS2 9JT, UK

27.1 Introduction

A recent search of the literature for papers on crystallization in oil-in-water emulsions with authors who use the term 'nano' has generated 260 references, the first one of which appeared in 1992. The earliest papers were associated with drug delivery and pharmaceuticals. Crystallizing fats were used instead of liquid oils because the solid fat gives up the drug more slowly, providing a controlled release mechanism.[1] An important question is how to crystallize and stabilize the fat. Initially, the emulsions produced were submicron emulsions, sometimes called 'mini-emulsions', with sizes invariably much greater than 100 nm and a nominal size below 1 μm. Usually called 'parenteral' emulsions,[2] they were designed for injection into the blood stream and were normally composed of emulsified soybean oil. The small size was achieved by high-pressure homogenization, and this was necessary to avoid the presence of large oil droplets that might interfere with blood circulation. Initially these parenteral emulsions were produced for intravenous feeding and the prefix 'nano' was not used. However, their use as a vector for drug delivery was clear.

By way of background, it is important to emphasize here that, while many reports can be found of very small particles of surfactant vesicles/micelles (*e.g.*, fractionated egg phospholipids[2]) or swollen micelles, convincing evidence for stable oil particles much below ~100 nm is thin on the ground. For example, particles of size below 100 nm may be produced by dissolution of oil in an organic solvent, followed by dispersion in water; subsequently the oil phase crystallizes. This process has been used to produce cholesteryl acetate particles of ~100 nm;[3] but no reports specify particles much below this sort of size.[4,5] There is no doubt that smaller oil droplets can be produced;[6,7] however, careful examination of the evidence shows that it is doubtful if any of these is an 'emulsion' in the sense of containing distinct, particulate, pure oil particles.

A further difficulty in interpreting evidence in the literature is the lack of essential information on particle sizing techniques, such as the dilution method used in the case of light scattering. Additionally the size characterization methods most frequently used (static and dynamic light scattering) apply only to dilute systems and can give misleading results. This is particularly true with regard to polydisperse systems whose size covers the effective range of the two techniques. A particular issue is the destabilization of the system by Ostwald ripening due to a few large droplets undetectable by dynamic light scattering: the large droplets grow at the expense of the small ones and then cream/sediment out of the system, never to be detected, or their existence in the detected size distribution is rejected as an artefact. Another issue is the disparity between differential scanning calorimetry (DSC), ultrasound and density determinations. For example, the DSC data of Montenegro et al.[8,9] are in striking contrast to the solid content determination from ultrasound, density and X-ray diffraction measurements by other workers, perhaps because of the use of the highly hydrophobic perfluorohexane, which was added to prevent Ostwald ripening of the emulsion, indicating that this nominal 218-nm emulsion was in fact far from monodisperse (size distribution not supplied). The suppression of Ostwald ripening by incorporation of hydrophobic additives is described by Lindfors et al.,[11] but the additive was oil soluble, and therefore Montenegro et al.[8] were no longer studying pure n-hexadecane, but rather a 24:1 mixture of oil + perfluorohexane. There are also many disparities between different sets of DSC data,[7–10] a testament to the huge impact that surfactants and other non-oil components can have on crystallization processes.

We are interested in crystallization in 'nano-emulsions' because it offers a route to large-scale production of small solid particles whose size, shape and size distribution may be extremely well controlled by the properties of the initial emulsion. Once a monodisperse emulsion can be produced and crystallized, then perhaps we shall be able to produce a monodisperse sample of solid particles. All that is necessary then is to reduce the aqueous phase volume by osmotic dehydration, and large quantities of a concentrated nano-sized dispersion may be produced. For example, n-hexadecane may then be a useful and low-cost source of hydrophobic solid particles, for later use as an emulsion stabilizer[12] or foam stabilizer.[13] An additional reason for wanting to work with monodisperse samples is that a high degree of emulsion stability can be achieved with just three components in the system – pure oil, pure surfactant and pure water.

27.2 Materials and Methods

Nano-emulsions may be produced by high-pressure homogenization, ultrasonication and, even under some circumstances, gentle stirring. In the work reported here, emulsions were produced by repeated high-pressure homogenization involving up to 24 passes through the homogenizer. Particle-size polydispersity was minimized by continually recreating new surface until as much of the surfactant as possible had moved to the interface and the surfactant concentration in the bulk was reduced to a minimum. Hydrodynamic stresses within the homogenization process produce a driving force towards a single-sized particle. However, this can take some time to achieve because surface area is proportional to diameter squared and the required energy input is at a minimum proportional to the new surface area produced. So halving the diameter requires at least four times the passes needed to get to the starting point!

We have used three different surfactants and have produced nearly identical size distributions ($d_{32} \sim 130$ nm) with all three: 2 wt.% Tween 20 (Sigma, UK), 1 wt.% Caflon phc060 (ethyl alcoxylate, Uniqema UK) and 20 wt.% polyglycerol ester PGE L-7D (decaglycyl monolaurate, Sakamoto Pharmaceuticals, Osaka, Japan). A coarser emulsion (330 nm) was generated using Caflon as emulsifier by increasing the oil/surfactant ratio by 4 (see Table 1). We have also studied the effects of varying surfactant concentration and number of passes through the homogenizer in the case of Caflon. We have used a Shield homogenizer which is described elsewhere.[20] The emulsions were characterized in terms of (a) three different small amphiphilic surfactants that are quite different chemically and (b) two different mean sizes, $d_{32} = 130$ and 330 nm. In the case of the surfactants, Tween 20 produces micelles of around 15 nm in diameter, which can be measured by dynamic light scattering or by ultrasound scattering. On the other hand, Caflon phc060 micelles could not be detected by either techniques, and it was chosen because it is a much smaller molecule than Tween 20 and a purer chemical substance. PGE was chosen to provide a comparison with the work of Higami et al.[9]

The emulsification principle was to pass the emulsion repeatedly through the homogenizer until the size distribution stopped changing, and to control the mean size by varying the oil/surfactant ratio. The idea was to reduce the surfactant level in the continuous phase (solvent) to a minimum value and continually create new surface until finally a monodisperse emulsion was produced. In all probability the distributions were much narrower. At least 10 passes through the homogenizer were required in all cases.

In detail the preparation steps were as follows. First, it was ensured that both the oil phase and the emulsifier were completely melted to avoid the blockage of the homogenizer and loss of materials. Surfactant was weighed out and made up with water to 95 g. The mixture was stirred at a high enough temperature (60°C) to ensure that all components were liquid. The oil phase was weighed (5 or 20 g) and added gradually to the prepared aqueous phase. The mixture was homogenized at 60°C for 30 min, making 10 separate passes into an empty container while removing the air bubbles after each pass by creaming. The particle-size distribution was measured using a Malvern Mastersizer 2000 static light-scattering apparatus (Malvern, UK). Each sample was degassed in an ultrasonic bath for 10 s, and then loaded into the sound velocity measurement (UVM) cell, stirring the sample using a magnetic stirrer, and removing air

Table 1 Composition and physical property data for experimental emulsions

Sample name	Reference	d_{32} (nm)	ϕ (%)	C_s (wt %)	R_c (Kmin^{-1})	Figure	T_{cryst} (°C)	T_m (°C)	Melting point (°C)
PGE1	This work	130	5	20		Figure 6	2.3	3.3	17.3
Caflon1	This work	123	5	1	2	Figures 2 and 7	1.1	11.1	17.5
Tween 20 (1)	This work	130	20	2	0.013	Figures 3 and 4	3.1	7.2	17.6
Caflon2	This work	330	20	1	—	Figure 8	2.8	10.7	17.6
Tween 20 (2)	Ref. 24	800	20	2	0.06	—	2.5	13	16.5
Tween 20 (3)	Ref. 25	360	20	2	0.07	—	3.5	—	18

Note: d_{32}, average droplet diameter; ϕ, oil volume fraction; C_s, surfactant concentration; R_c, cooling rate during crystallization; T_{cryst}, crystallization temperature; and T_m, melting initiation temperature.

bubbles on the sample surface by a pipette. Temperature was controlled by immersing the UVM cell in a thermostatic bath (Grant, UK). Sound velocity measurements had begun once the temperature inside the cell was at 35°C. More details regarding the sound velocity measurement and the solid content determination can be found elsewhere.[14,15] (The emulsion preparation method is given in detail here because it was found to be essential to follow this procedure to obtain repeatable results.)

We define a sample as 'monodisperse' when it was stable to repeated temperature cycling between 0 and 35°C, where during each cycle the dispersed phase was frozen and thawed. We define 'stability' as the precise reproduction of the sound velocity *versus* temperature curve from cycle to cycle, without any change in apparent particle size between the beginning and end of the experiment.

27.3 Crystal Nucleation Theory

The classical theory of crystal nucleation in lipid systems has been presented in detail elsewhere;[16,17] it will not be repeated here. The concept behind our experiments was to reduce particle size progressively so that volume homogeneous nucleation and volume heterogeneous nucleation were completely suppressed, and either the particles did not crystallize at all or there was surfactant-initiated surface heterogeneous nucleation. It is known that the lauric acid moieties in Tween 20 dissolve into the lipid phase and can nucleate crystallization in the bulk and there is strong evidence that they do this too in cocoa butter and mineral oils.[18–20] Choice of surfactants was also governed by a desire to use different chemical groups (lauric acid in PGE and Tween 20, and a short methylene chain, $-C_{10}$, in the case of Caflon) to influence crystal nucleation. In an earlier experiment[19] we also used sodium caseinate.

It has been suggested[21] that critical fluctuations at the particle interface may play an important role in nucleation. This means that, as particle size is reduced, volume homogeneous nucleation may once again come to play an important part in the crystallization. It is important to recognize that, as the particle size decreases, so the layer of molecules which is in contact with the surface – and which therefore is influenced by it – contributes an increasing proportion of the total volume of the dispersed phase (see Figure 1). As particle size decreases, a reduction in the melting point and the enthalpy change on fusion may also be expected.[22]

There are many potential factors influencing crystal nucleation in particles: diffusion rates, initial reactant distribution, interdroplet collisions, intermicellar exchange of reactants, chemical reactions, Ostwald ripening, the particle-size distribution, autocatalysis, surfactant film flexibility and the critical nucleus size. All of these were accounted for in theoretical modelling work by Tojo and co-workers.[23] One can easily add the following factors: surfactant concentration, type and organization; the application of ultrasound; thermal history; polymorphism (in *n*-hexadecane, as used in our experiments, a rotator–triclinic transformation has been reported[9]); solubility of surfactant in the oil; the

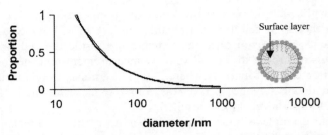

Figure 1 *Proportion of particle volume occupied by a 6-nm deep surface layer as particle size reduces. The value of 6 nm is chosen as the approximate unit cell length for trilaurin*

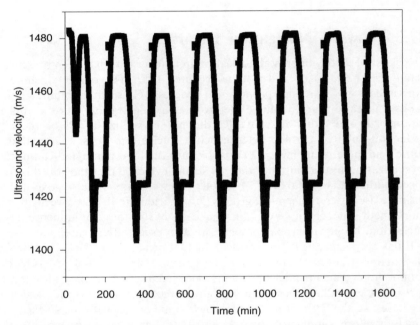

Figure 2 *Ultrasound velocity plotted against time for the 123-nm Caflon-stabilized emulsion (Caflon1). The temperature was cycled at $2°C\ min^{-1}$ between 0 and $40°C$ for 270 h. The size distribution did not change during the experiment*

temperature during crystallization; temperature fluctuations and heat dispersal; the range of interactions between the molecules; depletion flocculation; and interfacial tension.

27.4 Results and Discussion

The stability of our emulsions was assessed by repeated temperature cycling of the emulsion (see, *e.g.*, Figure 2). Data for the 130-nm Tween 20 emulsion are

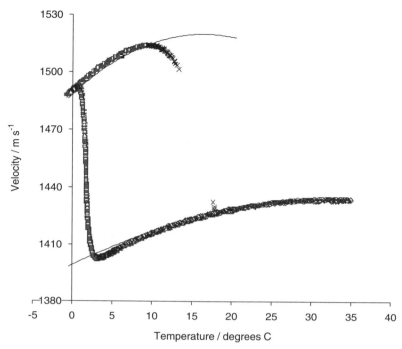

Figure 3 *Sound velocity plotted against temperature for two heating/cooling cycles of the 130-nm emulsion sample (20 vol.% oil) containing Tween 20 (Tween 20 (1)). A heating/cooling rate of $\sim 1°C\ min^{-1}$ was used, but during crystallization and melting the rate was reduced to $0.013°C\ min^{-1}$. All data acquired during the experiment are recorded. The polynomial fits for the temperature dependence of the velocity of sound in the liquid n-hexadecane emulsion and in the fully crystallized n-hexadecane dispersion are shown as lines on the graph ($r^2 > 0.98$)*

shown in Figure 3 and the resultant analysis in terms of crystalline solids is shown in Figure 4. The overall results are summarized in Table 1. Data in Figure 3 are missing during the melting process between 13 and 17.4°C. This is due to attenuation of the sound wave arising from the freezing/melting process.[15] This attenuation process is not normally seen during the crystallization process because the undercooling creates a significant energy difference between the solid and liquid phases, which the thermal fluctuation (~ 300 nK) in the acoustic field is unable to overcome.[15] At temperatures close to the melting point, the energy difference between the liquid and crystalline phases is very small and the thermal fluctuation in the acoustic field is sufficient to tip the system from liquid to solid, and back again, creating an attenuation of the propagating acoustic wave. We have observed this phenomenon in all our experiments.

In Table 1 and Figures 4–8 are presented all our new experimental results together with some earlier data.[24,25] Examination of Table 1 illustrates the general features of the data. First, the crystallization temperature does not vary

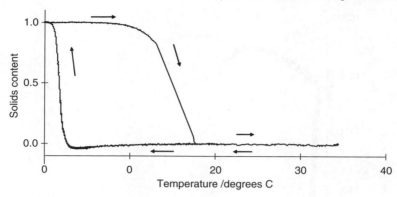

Figure 4 *Crystalline solid content computed for the 130-nm sample (Tween 20 (1)) from the data of Figure 3. Arrows indicate the direction of temperature change*

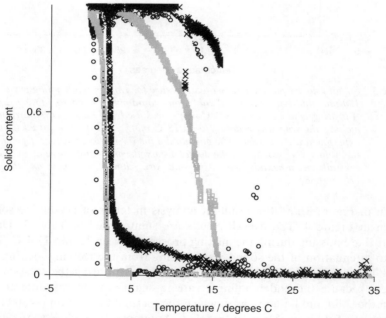

Figure 5 *Solids content plotted against temperature for the three 130-nm samples (PGE1 (□), Caflon1 (○), Tween 20 (1) (■)) and the 330-nm sample (Caflon2, (×))*

significantly. Figure 4 summarizes results for the 130-nm Tween 20 emulsion; these are typical of data we have obtained in the past, *e.g.*, for a 170nm emulsion.[14]

Before examining individual peculiarities of the different emulsions, an attempt has been made to analyse all the data according to classical nucleation

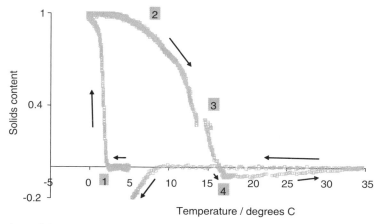

Figure 6 *Solids content plotted against temperature for the 130-nm PGE1 sample. Arrows indicate the direction of temperature change. Label 1 indicates beginning of crystallization; 2, initiation of crystal melting; 3, the point where information was lost due to increased attenuation of the sound; 4, the melted material is different from that at the same temperature on the cooling cycle*

theory using the methods described elsewhere.[17] Examination of Table 1 in Ref. 17 indicates that all types of nucleation can be summarized on a single graph in which $\ln[(1-\phi)/\phi]/T_{abs}$ is plotted against time. Here ϕ is the volume fraction of the oil phase, which comprises crystals, and T_{abs} is the absolute temperature. In this case the slope of the curve is

$$\frac{Nk_B}{h}\exp\left(-\frac{\alpha\Delta S}{R}\right)\exp\left(-\frac{\Delta G^*}{k_B T_{abs}}\right) = \frac{J}{T_{abs}} = \frac{k_x}{L^x T_{abs}} \quad (1)$$

where J is the nucleation rate for crystallization, N the number density of catalytic impurities, k_B Boltzmann's constant, h Planck's constant, $\exp(-\alpha\Delta S/R)$ the probability that a fraction α of the molecules is in the right conformation to crystallize, and R the gas constant. The loss of entropy ΔS on incorporation of material in a nucleus is given by $\Delta S = \Delta H/T_m$, where ΔH is the enthalpy of fusion and T_m is the melting temperature. The quantity ΔG^* is the Gibbs activation energy for the formation of a spherical nucleus. Here we have introduced an arbitrary dimensionality x, which conventionally is 2 for surface nucleation and 3 for volume nucleation, and L is a characteristic length corresponding to the diameter for spheres. A correspondingly appropriate nucleation rate constant k_x is required.

The analysis is complicated by the fact that the temperature is also changing slowly. For this reason the measured temperature is incorporated in Figure 9, while in Figure 10 we plot $\ln[(1-\phi)/\phi]/T_{abs}$ against time. The first thing to notice is that, for the two Caflon emulsions, which were presumably identical in all respects apart from particle size, the slopes whose measured ratio is $0.1001/0.9544 = 9.544$ would have the theoretical ratio $(330 \times 10^{-9})^x/(123 \times 10^{-9})^x$, which for $x = 2$ is 7.2, and for $x = 3$ is 19.3. So we could conclude from this

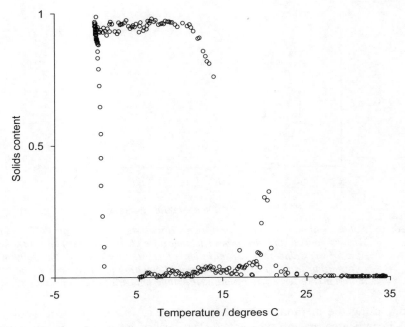

Figure 7 *Solids content plotted against temperature for the 123-nm Caflon1 sample*

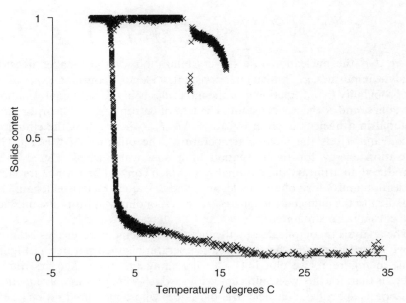

Figure 8 *Solids content plotted against temperature for the 330-nm Caflon2 sample*

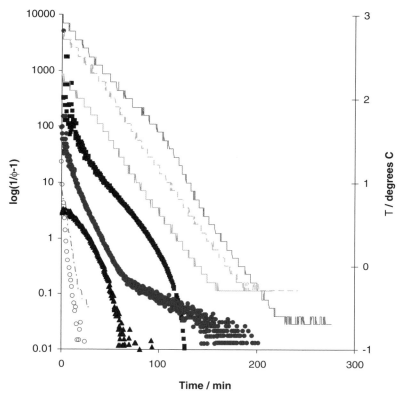

Figure 9 *The quantity ln[(1/ϕ) – 1] versus time (points) for the various sets of data in Figures 4, 6, 7 and 8, together with the corresponding plots of temperature against time (lines). Solid squares and solid line, 130-nm Tween 20; solid triangles and dashed line, 320-nm Caflon2; solid circles and dotted line, 130-nm PGE; open circles and dotted-dashed line, 123-nm Caflon1*

that, at least in the case of the Caflon-stabilized emulsions, nucleation is dominated by surface heterogeneous nucleation. We do not have sufficient data to draw firm conclusions regarding the other surfactants.

Examination of Figure 9 indicates that the nucleation kinetics analysis is insufficient to explain our data. Figure 6 also indicates a very complicated situation where the melting behaviour strongly suggests that the PGE has at least partly dissolved into the bulk oil, modifying both the surface and the bulk properties. Beginning with crystallization at label 1, we note from Figure 9 that there appear to be two stages of crystallization, with an initially faster rate, slowing down below 38% solids until the limit of resolution of the technique is reached. In the case of the other emulsions, the opposite is true with crystallization accelerating until complete. Perhaps the PGE acts as a nucleator and a crystal growth inhibitor, a type of behaviour known for other catalytic nucleators.[18] Label 2 of Figure 6 indicates initiation of crystal melting. As

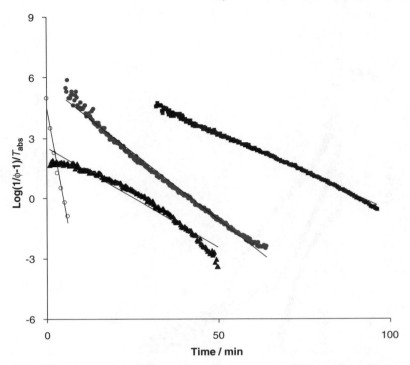

Figure 10 *The quantity $\ln[(1/\phi) - 1]/T_{abs}$ versus time for data in Figure 9. Lines are straight line fits to the data points. Symbols are as for Figure 9. (The data at the beginning and end of the crystallization process have been omitted for simplicity.)*

in previous work,[14] this melting is in the surface layer, but in the case of PGE it is far deeper than that observed with the other surfactants (see Figure 5). Label 3 in Figure 6 indicates the point where information was lost because of increased attenuation of the sound, but this is far less pronounced than for the case of the other surfactants. Label 4 indicates that the melted material is different to that at the same temperature on the cooling cycle. The starting temperature of 40°C was chosen to ensure that whatever phase formed during melting was completely melted before the beginning of another experimental cycle. We note finally that a phase transition is apparent in the cooling liquid just below 10°C. Other workers have seen evidence of such a transition in PGE emulsions; *e.g.*, Sonada *et al.*[7] noted a rotator–triclinic transition around this temperature, but this effect appears far larger in the case of PGE than the other surfactants. Higami *et al.*[9] reported phase transitions on cooling at 8 and 2°C. However, the large amount of PGE present relative to the oil (a 4:1 ratio) does lead us to question the nature of such 'emulsions'.

Our Tween-stabilized emulsions behave here in a manner consistent with all our previous work.[24–26,14–15,17] On the other hand, the two Caflon-stabilized emulsions (Figures 7 and 8) show significant differences between each other and with the other emulsions. First, in the case of the coarser emulsion ($d_{32} = 330$ nm), crystallization begins at the melting point (Figure 8). Secondly, the onset of melting is sudden rather than gradual. Finally, the finer emulsion ($d_{32} = 123$ nm) shows some sort of transition on cooling, just above the melting point, which is associated with a similar but mirror-image transition during the melting (Figure 7).

27.5 Conclusions

Much more work is needed to clarify in detail the processes observed during our experiments. In particular, it seems desirable to carry out X-ray scattering alongside the sound velocity measurements to elucidate the different structures at the surface and within the emulsion particles. We have strong evidence, at least in the case of the Caflon-stabilized emulsions, that the emulsifier is itself nucleating crystallization. In the case of the PGE emulsion, perhaps both nucleation and crystal growth suppression occur. Also the PGE appears to penetrate much more deeply into the bulk of the particle than does either of the two other surfactants. In the case of Tween 20, penetration is little more than one monolayer, as is probably also the case with Caflon.

It is clear from this study that classical nucleation theory requires revision to account for the complex behaviour seen in nano-emulsions. Perhaps this is related to our failure to produce a monodisperse emulsion with mean particle size below 100 nm using standard high-pressure homogenization. One route to smaller sizes is *via* the so-called 'low-energy' emulsification methods,[27] in which the reduction of surface energy is likely to have a profound impact on the crystallization process. An alternative route could be *via* oil-in-water microemulsions.

References

1. K. Westesen, *Colloid Polym. Sci.*, 2000, **278**, 608.
2. B.B. Lundberg, *J. Pharm. Pharmacol.*, 1997, **49**, 16.
3. K. Westesen and T. Wehler, *Colloids Surf. A*, 1993, **78**, 115.
4. B. Sjostrom, A. Kaplun, Y. Talmon and B. Cabane, *Pharm. Res.*, 1995 **12**, 39.
5. H. Bunjes, K. Westesen and M.H.J. Koch, *Int. J. Pharm.*, 1996, **129**, 159.
6. H. Bunjes and M.H.J. Koch, *J. Control. Release*, 2005, **107**, 229.
7. T. Sonada, Y. Takat, S. Ueno and K. Sato, *Cryst. Growth Design*, 2006 **6**, 306.
8. R. Montenegro, M. Antonietti, Y. Mastai and K. Landfester, *J. Phys. Chem. B*, 2003, **107**, 5088.
9. M. Higami, S. Ueno, T. Segawa, K. Iwanami and K. Sato, *J. Am. Oil Chem. Soc.*, 2003, **80**, 731.
10. R. Montenegro and K. Landfester, *Langmuir*, 2003, **19**, 5996.

11. L. Lindfors, P. Skantze, U. Skantze, M. Rasmusson, A. Zackrisson and U. Olsson, *Langmuir*, 2006, **22**, 906.
12. B.P. Binks and C.P. Whitby, *Colloids Surf. A*, 2005, **253**, 105.
13. B.P. Binks and T.S. Horozov, *Angew. Chem.*, 2005, **44**, 3722.
14. M.J.W. Povey, S.A. Hindle, A. Aarflot and H. Hoiland, *Cryst. Growth Design*, 2006, **6**, 297.
15. M.J.W. Povey, *Ultrasonic Techniques for Fluids Characterization*, Academic Press, San Diego, 1997, pp. 54–75.
16. N. Garti and K. Sato (eds), *Crystallization Processes in Fats and Lipid Systems*, Marcel Dekker, New York, 2001.
17. M.J.W. Povey, in *Crystallization Processes in Fats and Lipid Systems*, N. Garti and K. Sato (eds), Marcel Dekker, New York, 2001, p. 251.
18. P.R. Smith and M.J.W. Povey, *J. Am. Oil Chem. Soc.*, 1997, **74**, 169.
19. S.A. Hindle, M.J.W. Povey and K.W. Smith, *J. Am. Oil Chem. Soc.*, 2002, **79**, 993.
20. S.A. Hindle, M.J.W. Povey and K.W. Smith, *J. Colloid Interface Sci.*, 2000, **232**, 370.
21. P.R. ten Wolde and D. Frenkel, *Science*, 1997, **277**, 1975.
22. F. Delogu, *J. Phys. Chem. B*, 2005, **109**, 21938.
23. C. Tojo, F. Barroso and M. de Dios, *J. Colloid Interface Sci.*, 2006 **296**, 591.
24. E. Dickinson, M.I. Goller, D.J. McClements, S. Peasgood and M.J.W. Povey, *J. Chem. Soc. Faraday Trans.*, 1990, **86**, 1147.
25. E. Dickinson, D.J. McClements and M.J.W. Povey, *J. Colloid Interface Sci.*, 1991, **142**, 103.
26. D.J. McClements, E. Dickinson and M.J.W. Povey, *Chem. Phys. Lett.*, 1990, **172**, 449.
27. A. Forgiarini, J. Esquena, C. Gonzalez and C. Solans, *Langmuir*, 2001 **17**, 2076.

Chapter 28
Instant Emulsions

Tim Foster,[1] Alison Russell,[2] Don Farrer,[2] Matt Golding,[1] Roger Finlayson,[1] Anna Thomas,[1] Dan Jarvis[1] and Eddie Pelan[1]

[1] UNILEVER FOOD AND HEALTH RESEARCH INSTITUTE, OLIVIER VAN NOORTLAAN, 3133AT VLAARDINGEN, THE NETHERLANDS
[2] LIFE SCIENCES, COLWORTH PARK, SHARNBROOK, BEDFORDSHIRE MK44 1LQ, UK

28.1 Introduction

Instant food products are well known; some examples are instant topping powders, instant mousses and instant soups. These are generally based upon the principle of either dilution of a stock composition (soup) or dissolving a powder in hot water (soup or dressing). For other food types such as fatty spreads, however, such instant products are not widely known.

Spreads that are generally suitable for use in frying, baking and spreading (on bread, toast or the like) have been available for decades. Almost all emulsion-based food products are sold in a ready-to-use form. With the consequent need to distribute as such, there is a requirement for thermal processing routes (pasteurization/UHT treatment), with the additional need for chilled or frozen distribution to ensure appropriate microbiological stability. As such products are generally offered to the consumer in the form of a final product in a tub or wrapper, they have the disadvantage of having limited storage stability. In particular, storage at elevated temperatures (20–35°C) can lead to phase separation due to the melting of the continuous fat phase. The melting of the fat at such temperatures is required for good eating performance, and is also dictated to some extent by the melting temperature of the triglyceride fat crystals required to structure and stabilize the emulsion during processing and chilled storage.

This paper describes a shelf-stable solution to this problem in terms of a base composition suitable for consumer preparation of a spreadable butter substitute. The base composition is such that the consumer should be able to create the final emulsion by mixing with oil and water in a simple – and preferably manual – operation, thereby allowing the spreadable emulsion to be prepared within a few minutes. The final emulsion should be easy to make and stabilize, and with the potential for generating curiosity and interest in a consumer who has the personal ability to create and observe the change in sensory properties. Contrary to the well-established preparation of products such as soups or dressings from instant powders, the final product described here is not simply obtained by dilution or dissolution of dry material. Nevertheless, starting from a base composition, a stable emulsion structure of high-quality can be obtained with minimal energy input.

Emulsions are commonly employed in food, pharmaceutical and cosmetic industries, where they have the ability to deliver both lipophilic and hydrophilic ingredients in a kinetically stable product. Research in emulsion science and technology has provided understanding about how to manipulate composition, microstructure, (in)stability and texture. Nonetheless, much of the effort in the past has focused on how emulsions can be most efficiently produced, with considerable work done on optimizing emulsion droplet-size distributions through homogenizer head design, membrane emulsification techniques, sonication and so on.

To overcome the need to transport water, and to offer consumers lower-cost alternatives, a range of dried food products has been developed. Improved hydration of powders through agglomeration has led to increased quality of dry food preparations. Additionally, emulsions that can easily be rehydrated, such as powdered toppings, are widely available; these are based upon spray-dried crystallized fat-containing emulsions. Nonetheless, such systems allow little or no control over final emulsion properties, and the spray drying can have a negative impact on the quality of the product, while still requiring sophisticated processing prior to distribution.

The formulation of instant emulsions offers some particularly interesting technical challenges. Interfacial stabilization of oil-in-water (O/W) emulsions is typically provided by milk proteins, but only once the protein has been fully rehydrated. Potentially this may limit the functionality of the protein when used in an instant emulsion. This may be overcome if alternative low-molecular-weight emulsifiers can be used which are delivered from the oil phase. When both types of surface-active ingredients are combined in a mixture of proteins and low-molecular-weight emulsifiers, an understanding of the relative hydration kinetics and surface activity becomes mandatory.

Food emulsions are also structured and stabilized by rheological manipulation of the continuous phase using thickening and/or gelling agents. These are typically hydrated in factory processes, using heat and shear. This is particularly relevant in the case of low-fat emulsions, which increasingly rely upon the structuring of the aqueous phase to provide the required rheological attributes, *e.g.*, low-fat mayonnaise (containing starch) and low-fat spreads (containing gelatin). Hydration rates of thickening and gelling ingredients will also have an effect on the bulk structuring of the aqueous phase of instant emulsions. Furthermore, the competitive hydration of hydrocolloids and surface-active materials may also affect the ability to texturize and stabilize emulsions.

Therefore the issue of how such competing processes affect overall system morphology and properties will be considered here.

Generally, the production of an emulsion requires a high energy input (*e.g.*, using high-pressure homogenization), which adds significantly to production costs. The minimum energy requirement for emulsification can be defined at the simplest level by the thermodynamic equation

$$\Delta G = \gamma \Delta A, \qquad (1)$$

where γ is the interfacial tension, and ΔA the change in surface area of the dispersed phase from a phase-separated system to an emulsion system of specific dispersed phase surface area.

Processing routes such as high-pressure homogenization, which are usually required to achieve effective droplet break-up, are surprisingly inefficient. They typically require an energy input greatly in excess of that given by Equation (1), so that most of the energy supplied is dissipated in the form of heat. Therefore, the ability to develop and design emulsion systems for which the energy input requirement is minimized is of significant scientific and commercial interest.

28.2 Emulsification

Several authors[1–9] have investigated instant or 'spontaneous' emulsions, but not in the field of food science, and not typically utilizing dry powders to aid the structuring and emulsification processes. Water-continuous emulsions from water-in-oil (W/O) microemulsions have been described,[1] as has the spontaneous formation of highly concentrated W/O emulsions,[2,3] and the use of optimized water/oil/surfactant ratios[4] as concentrates to form mini-emulsions upon the addition of a large amount of water.[5] The formulation of spontaneous emulsions has also been suggested[6,7] as an alternative to the use of ultrasonic and high shear devices, and the dynamics of spontaneous emulsification for the fabrication of O/W emulsions has been discussed. Furthermore, some new methodologies[8] and mechanisms[9] have been described to explain the process of spontaneous emulsification.

The work reported here probes the effectiveness of using food emulsifiers to create instant emulsions. Powders for structuring, emulsification, taste and microbiological stability are delivered as a slurry in oil. The cold O/W emulsions can then be formed rapidly with minimal mechanical agitation. The dispersion of hydratable polymers and added solids in the oil serves the same function as mixing a polymer with a non-solvent prior to dissolution in a true solvent: the particles hydrate rapidly and independently in the presence of water on agitation. The conjunction of these two processes results in simultaneous emulsification of the oil and hydration/dissolution of the solids, which thicken and structure the aqueous phase, as well as kinetically trap the dispersed oil phase. Only moderate shear, as can be produced by hand stirring, is required to develop the final 'structured' emulsion. The result is a paste-like O/W emulsion whose properties depend upon thickener levels, oil phase volume, emulsifier type and the nature of the aqueous environment (pH, ionic strength).

Milk proteins are used in many food emulsions because they are seen as acceptable and natural food ingredients. However, delivering the milk powder in an oil slurry means that the proteins have to hydrate before they can provide emulsification of the oil droplets. We have found that, while the preparation of good emulsions is possible, buffer salts are required to reduce the pH so as to provide microbiological stability in the water phase of the final product. The associated disadvantage, however, is that the low pH can cause premature precipitation of the caseinate proteins with subsequent destabilization of the emulsion. The addition of electrolytes (sodium chloride and buffer salts) at the levels required also tends to inhibit or destabilize structured emulsion formation, because of the rapid lowering of the pH on emulsification and increase of the ionic strength of the aqueous phase. These two effects can 'precipitate' the protein before it has developed a stable structure within the emulsion. And local 'flushes' of acidity on initial mixing can cause coating of caseinate granules with precipitated protein before there has been significant dissolution. But emulsions structured in the absence of acid may be post-acidified to well below pH 4.9 without loss of stability. This means that post-acidification with lactic acid or acetic acid is possible.

To overcome the problem of pre-acidification, stable emulsions may be obtained in the presence of a suitable emulsifier such as lecithin. Being soluble in the oil phase, but not to any significant extent in the aqueous phase, lecithin can participate in the early stages of emulsification, as well as can act as a wetting aid for powder dissolution from the oil phase. While some emulsions could be formed in the absence of caseinate, they were found not to be stable to shear. Adopting this approach now leads us to question the activity of caseinate as an emulsifier. Many authors[10–14] have studied the competitive adsorption of caseinate with phospholipid emulsifiers. Heertje[15] has suggested that lecithins are not able to displace proteins completely from the interface in O/W emulsions, even at high phospholipid concentrations. The concentration of caseinate used in their study was 100 times lower than that used here. Additionally they were concerned with displacing caseinate, whereas upon hydration of the caseinate in this case, the protein might attempt to displace the lecithin. Here also we must consider the dynamic nature of the process, where droplets are being created as the caseinate is hydrating. Therefore, the competition is occurring at a virgin interface, and the interfacially active caseinate is also aggregating as a result of the pH decrease, which may kinetically trap the protein at the interface.

To test this hypothesis, lecithin was either fully or sequentially replaced by Tween 20 (a polyoxyethylene sorbitan ester). This nonionic surfactant is much more potent at displacing caseinate from the oil−water interface.[16–19] Stable emulsions could be created with Tween 20 at one quarter of the level of lecithin. However, the emulsions formed were not as viscous. This provides evidence to the fact that the caseinate is not interfacially active in the Tween system, and it is the dual role of emulsifier and structuring agent that caseinate plays in providing the thicker structures when mixed with lecithin. Post-acidification of

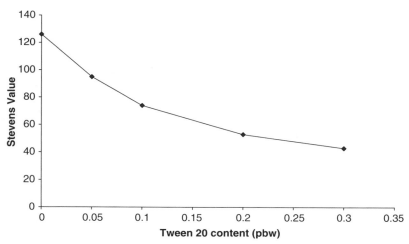

Figure 1 *Effect of the partial replacement of lecithin, with starting level 1 part by weight (pbw) of formulation, by Tween 20 on the texture of an instant emulsion also containing 1 pbw caseinate, 8 pbw starch, 50 pbw sunflower oil, 45 pbw water, 1.2 pbw NaCl and 0.65 pbw buffer salts. Measurements were made using a Stevens Texture Analyser equipped with a standard mesh probe head*

the caseinate + Tween 20 system did not increase the viscosity of the emulsion either, suggesting that Tween 20 may also be interacting with the caseinate, to reduce its structuring potential. Figure 1 shows the effect that sequential substitution of lecithin by Tween 20 has on the 'texture' as measured by a Stevens Texture Analyser.

28.3 Emulsion Stabilization

The example described above serves to show how an acidified protein-containing emulsion can be stabilized. An alternative strategy is to speed up the creation of smaller stable oil droplets. The influences of shear and time were investigated. When the mixture of oil slurry and water is shaken, then the structuring and viscosity increase takes place in the region of 10–20 s. The fast increase in viscosity traps the emulsion and prevents any further 'working' of the system. However, if the mixture is sheared gently, with hand mixing, the system can be agitated for a prolonged period of time. This allows smaller droplet formation, and also additional time for the caseinate to hydrate and participate at the interface. The confocal micrographs in Figure 2 show the effect of shaking and shearing on the oil droplet-size distribution.

Another way to reduce the droplet size is to increase the continuous water-phase viscosity. Grace[20] has shown that droplet break-up can be induced by changing the relative viscosities of the continuous and dispersed phase. The addition of a readily hydratable viscosifier (*e.g.*, guar gum) can increase the aqueous phase viscosity. The effect of increasing the concentration of guar gum

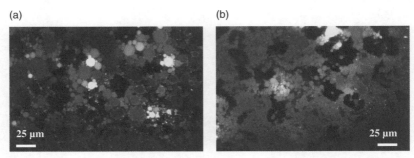

Figure 2 *Confocal micrographs showing (a) the effect of shaking and (b) fork stirring by hand on the microstructure of instant emulsions of the same formulation*

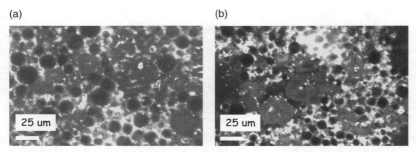

Figure 3 *Confocal micrographs of an instant emulsion containing 1 pbw caseinate, 1 pbw lecithin, 2.75 pbw CWS, 50 pbw sunflower oil, 45 pbw water, 1.2 pbw NaCl and 0.65 pbw buffer salts, as well as (a) 0.5 pbw guar gum or (b) 1.33 pbw guar gum*

on the final oil droplet-size distribution can be seen in Figure 3, where for the two systems identical shearing rates and times were employed. The mechanism by which the final droplet size is formed is probably dominated by droplet break-up in this case. But, if thickening happens too quickly, the emulsion is difficult to form because of too high a difference in the viscosities of the continuous and the dispersed phases.

28.4 Rates of Hydrocolloid Hydration

Hydration of water-soluble polymers on an industrial scale is often aided by pre-dispersion in a non-solvent (*e.g.*, oil). The mixing times are typically tens of minutes at elevated temperatures. Here we are concerned with cold hydration in the time frame of tens of seconds. Efficient viscosifying can be achieved in two ways – namely by swelling of particles (*e.g.*, with starch) or dissolution of polymers (*e.g.*, with xanthan gum, guar gum, *etc.*).

Starches have been designed over the past few decades to serve this purpose, either in the form of pre-gelatinized starch, or cold-water-swelling starch (CWS). Pre-gelatinized starches are particulate materials, as a result of the cooking and drying processes employed in production; they are irregularly shaped 'gelled' particles. The CWS materials are not 'cooked out' as much in the production process, *i.e.*, only to the point of granular swelling, and are subsequently spray dried. The granular integrity of the starch is retained; this is due to cross-linking within the granule, and the fact that many of the CWSs are waxy, *i.e.*, they are low in amylose content. The agglomeration of granules provides readily hydratable powders for dispersion directly into water.

Figure 4 shows the rates of hydration measured in this study, comparing agglomerated and standard CWS with pre-gelatinized starch. The viscosity traces were obtained using a Rapid Visco Analyser (RVA) 4 or Super4, with mixtures of sunflower oil (12.5 g) + starch (4 g) + lecithin (2.5 g) (Bolec, MT), to which water (11.25 g) was added. The mixture was immediately put into the RVA and mixed for 10 min at 25°C at 180 rpm. The relative viscosifying power can be compared at constant starch level, along with the rates of hydration. It seems that delivering these starches from an oil slurry provides no advantage to using the more expensive agglomerated version of the CWS. Indeed, the non-agglomerated sample thickens faster, reaching a plateau viscosity after 40 s, which is reached by the agglomerated sample 20 s later.

Polymeric materials were also followed in a similar way. Xanthan gum is a polyelectrolyte whose solution properties have been well described[21] because it has been the polysaccharide of choice in many food applications since its discovery and commercialization in the 1960s and 1970s. Figure 5 shows the effect of hydrating the xanthan at different concentrations, in the presence or absence of salt (NaCl) and emulsifier (lecithin). What can be clearly seen is that the polyelectrolyte swells and solubilizes quickly in the absence of salt because of charge repulsion. When hydrated in the presence of the emulsifier the onset of the viscosity increase is also enhanced, presumably because small oil droplets

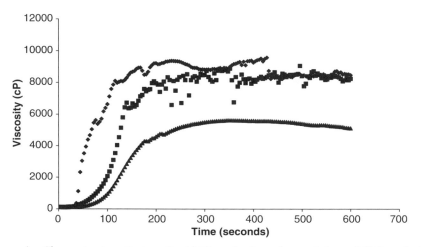

Figure 4 *Change in viscosity in a Rapid Visco Analyser for a mixture of 12.5 g oil, 4 g starch, 0.3 g lecithin + 11.25 g water for cold-water-swelling starch (CWS) (♦), agglomerated CWS (■) and pre-gelatinized starch (▲)*

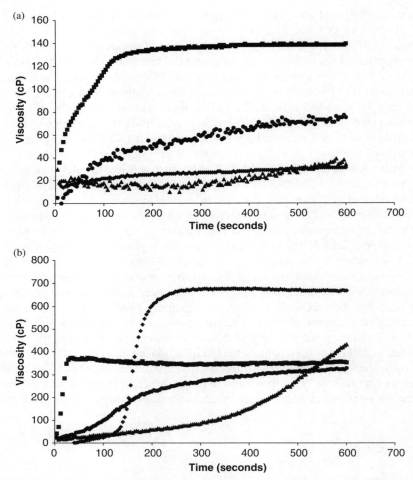

Figure 5 *Change in viscosity in a Rapid Visco Analyser for 12.5 g oil, 11.25 g water and different levels of xanthan/NaCl/lecithin: (a)* ♦, *0.35:1.2:0.3;* ■, *0.35:0:0.3;* ▲, *0.35:1.2:0;* ●, *0.35:0:0. (b)* ♦, *1:1.2:0.3;* ■, *1:0:0.3;* ▲, *1:1.2:0;* ●, *1:0:0*

are formed quickly, increasing the oil–water interfacial area and easing the passage of the polymer particles into the aqueous phase. The same final viscosities are approached for equal polymer and salt contents, but the delay in reaching these values is a function of the speed of hydration. What is also notable is the effect that salt has at the different xanthan concentrations. At the lower xanthan level the salt induces a sevenfold decrease in viscosity, whereas at the higher level only a doubling of viscosity is seen. This can be explained in terms of some earlier work in this area;[21] it is a consequence of the xanthan–xanthan weak gel interactions, which are a function of both polymer and electrolyte concentrations.

28.5 Competitive Hydration

In large-scale food manufacturing, blends of biopolymers are used to supply the requisite product properties. In ice-cream manufacture, mixtures of locust bean gum + κ-carrageenan are used in combination with milk proteins to provide the correct product stability and eating quality. And in the manufacture of low-fat emulsions such as mayonnaise, mixtures of starch with gums such as xanthan are employed effectively to replace the fat-structuring properties. The previous section has highlighted the fast hydration and structuring potential of two different biopolymers. Now let us consider what happens when blends of these two biopolymers are used in this 'instant' structuring route.

Figure 6 shows the effect of changing the level of xanthan in blends with starch in the presence and absence of salt. All experiments were carried out in the presence of an emulsifier (lecithin), to aid the hydration process. The blend with the lowest xanthan content hydrates at slightly later time, but yet it reaches a slightly higher end viscosity when compared to starch alone. As the concentration of xanthan increases, however, it dominates the hydration to such an extent that the swelling and viscosifying potential of the starch is reduced, so that the viscosity of the blend of 0.5 g of xanthan + 3.5 g of starch only achieves the same viscosity as for xanthan alone (see Figure 5). The sample containing 0.25 g of xanthan + 3.5 g of starch hydrates in a two-stage process, whereby the xanthan first enhances the viscosity and afterwards the swelling potential of the starch emerges. In the presence of salt the hydration of the same mixture is delayed, and the onset of the viscosity increase seems to match that of starch alone. The value rises and then falls to a level similar to that of starch alone, indicating a complex competition between xanthan and starch for the

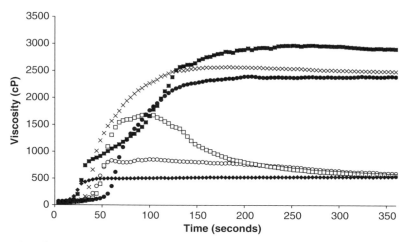

Figure 6 *Change in viscosity in a Rapid Visco Analyser for 12.5 g oil, 11.25 g water and 0.3 g lecithin, and also containing:* ●,○, *4 g CWS;* ×, *0.125 g xanthan + 3.5 g CWS;* ■,□, *0.25 g xanthan + 3.5 g CWS;* ♦, *0.5 g xanthan + 3.5 g CWS. Closed and open symbols indicate absence and presence of NaCl*

early stages of hydration. In this case the starch seems finally to dominate the mix viscosity. It is noteworthy that the starch swells faster – but thickens to a lower degree – in the presence of salt. This could possibly be due to osmotic effects.

What we have demonstrated here is the complexity of instant emulsion formation caused by the presence of different surface-active molecules, and the effect that the aqueous phase constituents can have on the final emulsion structure. The nature of the emulsification process itself also plays a major role in creating structure in the aqueous phase. The complexity of such interrelated factors, particularly when multiple aqueous phase ingredients are being used, warrants further investigation.

References

1. R.W. Greiner and D.F. Evans, *Langmuir*, 1990, **6**, 1793.
2. H. Kunieda, Y. Fukui, H. Uchiyama and C. Solans, *Langmuir*, 1996, **12**, 2136.
3. K. Ozawa, C. Solans and H. Kunieda, *J. Colloid Interface Sci.*, 1997, **188**, 275.
4. K. Bouchemal, S. Briancon, E. Perrier and H. Fessi, *Int. J. Pharm.*, 2004, **280**, 241.
5. S. Lamaallam, H. Bataller, C. Dicharry and J. Lachaise, *Colloids Surf. A*, 2005, **270–271**, 44.
6. N.L. Sitnikova, R. Sprink and G. Wegdam, *Langmuir*, 2005, **21**, 7083.
7. F. Ganachaud and J.L. Katz, *Chem. Phys. Chem.*, 2005, **6**, 209.
8. J.C. Lopez-Montilla, P.E. Herrera-Morales and D.O. Shah, *Langmuir*, 2002, **18**, 4258.
9. J.C. Lopez-Montilla, P.E. Herrera-Morales, S. Pandey and D.O. Shah, *J. Dispersion Sci. Technol.*, 2002, **23**, 219.
10. D.G. Dalgleish, *Food Hydrocoll.*, 2006, **20**, 415.
11. W. van Nieuwenhuyzen and B.F. Szuhaj, *Fett. Lipid*, 1998, **100**, 282.
12. E. Dickinson and G. Iveson, *Food Hydrocoll.*, 1993, **6**, 533.
13. Y. Mine and M. Keeratiurai, *Colloids Surf. B*, 2000, **18**, 1.
14. Y. Fang and D.G. Dalgleish, *J. Agric. Food Chem.*, 1996, **44**, 59.
15. I. Heertje, H. van Aalst, J.C.G. Blonk, A. Don, J. Nederlof and E.H. Lucassen-Reynders, *Lebensm. Wiss. Technol.*, 1996, **29**, 217.
16. G. van Aken, *Colloids Surf. A*, 2003, **213**, 209.
17. E. Dickinson, C. Ritzoulis and M.J.W. Povey, *J. Colloid Interface Sci.*, 1999, **212**, 466.
18. J. Chen, E. Dickinson and G. Iveson, *Food Struct.*, 1993, **12**, 135.
19. J.-L. Courthaudon, E. Dickinson, Y. Matsumura and D.C. Clark, *Colloids Surf.*, 1991, **56**, 293.
20. H.P. Grace, *Chem. Eng. Commun.*, 1982, **14**, 225.
21. W.E. Rochefort and S. Middleman, *J. Rheol.*, 1987, **31**, 337.

Chapter 29

Flavour Binding by Solid and Liquid Emulsion Droplets

Supratim Ghosh, Devin G. Peterson and John N. Coupland

DEPARTMENT OF FOOD SCIENCE, PENNSYLVANIA STATE UNIVERSITY, PA 16802, USA

29.1 Introduction

The concentration of volatile molecules in the head-space is among the most important parameters determining the sensory qualities of a food. The perceived aroma depends on the total concentration of volatiles in the product, and also on the extent to which they are bound by the food. For a system at equilibrium, the distribution of a given volatile between the food and the head-space gas is given by its partition coefficient,

$$K_{ge} = \frac{c_g}{c_e}, \tag{1}$$

where c_g and c_e are concentrations of volatile compounds in the head-space and in the food, respectively. (Formally this should be written in terms of activity, but as concentrations of volatiles are typically very low the difference is negligible.) In a multiphase food, the volatiles will distribute themselves between all of the available phases according to the relevant partition coefficients, and these can be combined into an effective partition coefficient for the food. For example, in a food emulsion, the effective partition coefficient K_{ge} is related to the volume fraction of oil (ϕ_o) and the gas–oil and gas–water partition coefficients, K_{go} and K_{gw}, respectively:

$$\frac{1}{K_{ge}} = \frac{\phi_o}{K_{go}} + \frac{(1-\phi_o)}{K_{gw}}. \tag{2}$$

Equation (2) has proved useful in modelling the effects of changing liquid oil concentration on the head-space concentration of volatiles above an emulsion.[1-3] Nevertheless, the interaction with solid fats is less clear, with some workers arguing[4,5] that solid fat does not interact with volatile molecules, while others suggest[5,6] that there may be some binding of the aroma compounds to the crystalline fat. The goal of this work is to examine in more detail the interactions between solid fat and aroma molecules in a model food emulsion.

29.2 Experimental

Unless otherwise stated, ingredients were purchased from the Sigma Chemical Company (St. Louis, MO). Oil-in-water emulsions (40 wt% oil) were prepared by mixing n-eicosane (Fisher Scientific, Springfield, NJ) with 2 wt% sodium caseinate solution (containing 0.02 wt% thimerosal as an antibacterial agent) using a high-speed blender (Polytron, Brinkmann Instruments, Westbury, NY). The mixture was then recirculated through a twin-stage valve homogenizer (Panda, GEA Niro Soavi, Hudson, WI) for several minutes at different pressures (15–60 MPa) to achieve multiple passes through the valves. The particle-size distribution of the emulsions was characterized by static light-scattering (Horiba LA 920, Irvine, CA) using a relative refractive index of 1.15. The initial emulsion was diluted with deionized water to produce a series of samples with different oil contents, and aliquots (2 mL) were added to 20 mL head-space (HS) vials.

Emulsified alkanes show deep and stable supercooling,[7] and so emulsions of either wholly liquid droplets or wholly solid droplets were prepared by either cooling the vials directly to 30°C, or by cooling to 10°C to induce homogeneous nucleation and then reheating to 30°C. The solid and liquid droplets were stable at this temperature over several weeks – no change in measured particle size, no apparent creaming and no change in the droplet melting or crystallization enthalpy.

A stock aroma mixture was prepared from ethyl butanoate (EB), ethyl pentanoate (EP), ethyl heptanoate (EH) and ethyl octanoate (EO) mixed in a volumetric ratio of 1:2:20:20 and diluted in an equal volume of ethyl alcohol. Aliquots of the stock mixture were added to the vials so that the final concentrations of aroma compounds EB, EP, EH and EO in the emulsion were 10, 20, 200 and 200 μL L^{-1}, respectively. The vials were sealed with poly(tetrafluoroethylene)/butyl rubber septa and equilibrated at 30°C for at least 1 h prior to analysis. Other measurements showed that the values reached after 1 h remained constant for over a week, suggesting the samples had reached equilibrium after this time.

Following equilibration, a sample of the head-space gas (1 mL) was withdrawn using an autosampler (Combi-Pal, CTC Analytics, Carrboro, NC, USA) and injected into a gas chromatograph (Agilent 6890, Agilent Technologies, Palo Alto, CA, USA) operating in splitless mode and equipped with a DB-5 capillary column (30 m × 0.32 mm i.d. with 1-μm film thickness) and a flame ionization detector. The operating conditions were as follows: carrier hydrogen flow-rate = 2.0 mL min^{-1}; inlet temperature = 200°C; detector temperature = 250°C; oven program – 30°C for 1 min, increasing at 35°C min^{-1} to 200°C, and constant at 200°C for 2 min. The static head-space concentrations were determined from peak areas using a standard calibration curve ($R^2 = 0.99$). All measurements are expressed as the mean and standard deviation of at least two full experimental replications.

29.3 Experimental Results and Preliminary Modelling

A cocktail of four volatiles was investigated, and the trends seen were similar for all. Therefore only results for EH will be reported for clarity. The value of K_{ge} was found to decrease with oil volume fraction for model emulsions containing liquid and solid *n*-eicosane droplets (Figure 1), but more rapidly for liquid droplets than for solid ones. Figure 1(a) shows that K_{ge} for liquid droplets depends only upon the volume fraction of oil; indeed an unemulsified

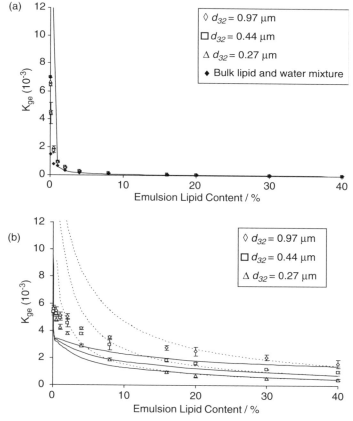

Figure 1 *Effective partition coefficient K_{ge} of ethyl heptanoate (EH) for an n-eicosane emulsion of (a) liquid and (b) solid droplets with various particle sizes. The lines in (a) are calculated from bulk partition coefficients using a volume partitioning model [Equation (2)]. The broken lines in (b) are calculated based on a surface-binding model fit to $\phi_o > 0.1$ ($K^*_{iw} = 4.5 \times 10^{-6}$ m). The solid lines in (b) are based on the combined surface binding/droplet dissolution model described in the text*

mixture of the same ingredients shows the same trend. The value of K_{ge} also decreases with the fat volume fraction for solid droplets [Figure 1(b)]; while this effect is smaller than that seen for liquid droplets, it establishes the existence of some positive interaction between solid fat droplets and aroma molecules. The magnitude of the interaction with solids appears greater for smaller droplets, suggesting it occurs at the surface rather in the bulk of the crystalline droplet.

The variation in K_{ge} was modelled as a function of liquid oil volume fraction using Equation (2) with independently measured bulk partition coefficients ($K_{go} = 6 \times 10^{-6}$, $K_{gw} = 0.0182$). Figure 1(a) shows that the equation fits the data well over the whole concentration range, suggesting that the most important factor is the volume partitioning of the aroma molecules, and that other effects (*e.g.*, Kelvin pressure effects, surface binding of the volatile, *etc.*) are not important enough to be seen in the liquid oil emulsion data.

The particle-size effect seen in the solid fat data cannot be adequately represented solely by a volume-based model. This suggests that interfacial binding may be important. The presence of a binding surface can be incorporated into Equation (2) as an additional two-dimensional phase,[8,9]

$$\frac{1}{K_{ge}} = \frac{\phi_o}{K_{go}} + \frac{(1-\phi_o)}{K_{gw}} + \frac{K_{iw}^* A_s}{K_{gw}}, \quad (3)$$

where A_s is the interfacial area per unit volume of emulsion (expressed as $6\phi_o/d_{32}$), and K_{iw}^* the surface binding coefficient defined as the ratio of the surface excess concentration (volume of aroma compound per unit interfacial area) to the aqueous concentration. As a first approximation, the EH was assumed to interact just at the surface of the solid droplets and not with the bulk of the crystal, *i.e.*, we set $\phi_o = 0$. Equation (3) was then used to model the head-space concentration of EH in equilibrium with the solid droplets. There is a reasonably good fit to the higher volume fraction samples, with the broken lines in Figure 1(b) corresponding to the best fit value of $K_{iw}^* = 4.5 \times 10^{-6}$ m from a statistical fit to the data with $\phi_o > 0.1$. But the model overestimates the head-space concentration at lower oil volume fractions.

As the effect of surface binding is much smaller than the effect of partitioning into liquid oil, we are not able to discount the possibility of similar binding occurring at the liquid oil–water interface. When Equation (3) is applied to the liquid droplet data, a similar value of K_{iw}^* (or even a value several orders of magnitude greater) causes no appreciable change to the predicted curve. For the remainder of the work we assume that surface binding occurs at all lipid–water interfaces with this same binding coefficient regardless of whether the droplets are solid or liquid. However, we only expect to see the importance of surface binding in emulsions where there is no liquid oil present.

29.4 Nature of the Flavour–Binding Interactions

In an effort to better understand the nature of the interactions, a coarse *n*-eicosane emulsion was prepared with a median droplet size of ~10 μm to allow easy observation under the microscope. Various concentrations of solid droplets were mixed with the aroma cocktail and observed under polarized light using an Olympus BX40 microscope (Melville, NY) equipped with a video camera (Sony 3CCD, PXC-970MD, New York, NY) using ×200 and ×1000 magnification lenses. Image manipulation was performed using PAX-it 4.2 software (Franklin Park, NJ). In the absence of added aroma mixture, the droplets appeared as bright specs under polarized light as shown in images (a), (c) and (e) in Figure 2. But, in the presence of the added aroma, many of the solid droplets in the more dilute emulsion had apparently disappeared, suggesting that the crystals had dissolved in the added aroma component [Figure 2(d)].

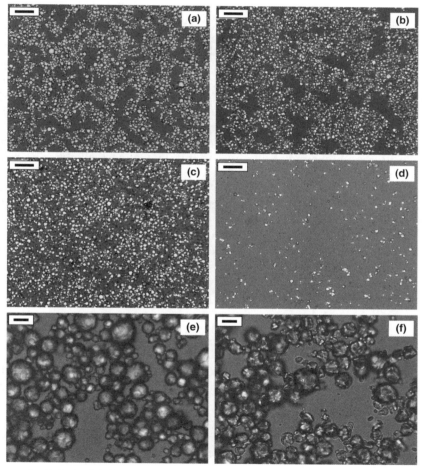

Figure 2 *Polarized light micrographs of coarse solid n-eicosane emulsions: (a) 20% n-eicosane, no added aroma; (b) 20% n-eicosane, with added aroma; (c) 5% n-eicosane, no added aroma and (d) 5% n-eicosane, with added aroma (scale bar = 40 μm). Micrographs (e) and (f) are higher magnification images of the 20% emulsion without and with added aroma (scale bar = 10 μm)*

Clearly, some critical ratio of aroma compound to solid fat is necessary to achieve this, as the same amount of aroma stock added to more concentrated emulsions caused no observable change [Figure 2(b)]. Furthermore, when the individual droplets were observed at a higher magnification (×1000), the aroma-free samples appeared to be solid spheres [Figure 2(e)], whereas in the presence of added aroma an adsorbed layer, presumably composed of aroma molecules, led to the formation of less regular shapes [Figure 2(f)].

It is not possible to make a direct comparison between the large droplets used in the microscopy study and the smaller ones used in the head-space GC analysis, as the smaller droplets could not be adequately observed microscopically. Nonetheless, similar behaviour was observed when differential scanning calorimetry (DSC) was used to compare the melting enthalpies of fine n-eicosane droplets prepared with and without added aroma. Samples (~15 mg) were sealed into aluminium pans and placed in the calorimeter (Perkin-Elmer DSC-7, Norwalk, CT) alongside an empty pan as a reference. The melting enthalpies of n-eicosane in five emulsions ($d_{32} = 0.44$ μm) with different oil contents prepared with or without added volatiles were measured by first cooling to 10°C then reheating at 5°C min^{-1} to 50°C. The ratio of the two enthalpies was used as a measure of the effect of the added aroma on the solid fat content of the droplets. At high oil/aroma ratios (*i.e.*, high oil volume fractions) the melting enthalpy of the n-eicosane was unaffected by the presence of the aroma compound; hence this means that almost all the n-eicosane remained in the solid state. But at low oil/aroma ratios (*i.e.*, low oil volume fractions) the melting enthalpy of the samples containing aroma was much lower than those prepared without added aroma compound, suggesting a much lower solid content of n-eicosane in the former case) (Figure 3). This supports the hypothesis that the n-eicosane can dissolve in the aroma compound if sufficient of it is present and it is adsorbed to the droplet surface. Any successful model must therefore account for both the effect of the aroma on the phase behaviour of the crystalline phase and on the potential binding of the aroma to the solid and liquid phases present.

29.5 Surface Binding and Droplet Dissolution Model

The effects of solid and liquid fat in the droplets on the aroma release properties of the emulsion can be taken into account using a modified form of Equation (3),

$$\frac{1}{K_{ge}} = \frac{\phi_o(1 - \phi_{sf})}{K_{go}} + \frac{(1 - \phi_o)}{K_{gw}} + \frac{K_{iw}^* A_s}{K_{gw}}, \quad (4)$$

where ϕ_{sf} is the solid fat content, *i.e.*, the proportion of the total lipid in the emulsion that is crystalline. We can write a mass balance for the EH, expressed in terms of concentration c and volume V of both the gaseous and the emulsion phases, as denoted by subscripts g and e, respectively:

$$V_{EH} = c_e V_e + c_g V_g. \quad (5)$$

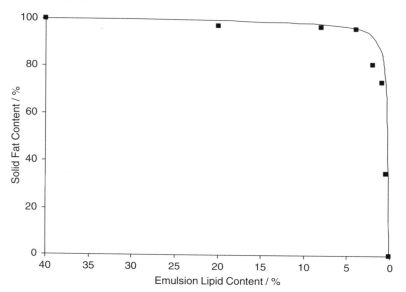

Figure 3 *Solid fat content of n-eicosane emulsions prepared by mixing different concentrations of fully solid droplets with aroma. The line represents a fit from the model described in the text*

Here V_{EH} is the total volume of EH added to the vial. As gas and emulsion are at equilibrium [Equation (1)], we can eliminate c_g from Equation (5):

$$c_e = \frac{V_{EH}}{V_e + K_{ge}V_g}. \tag{6}$$

The EH present in the emulsion system is divided among the liquid oil (subscript o), the aqueous phase (subscript w) and the interphase region (*i.e.*, assuming no direct incorporation into crystalline *n*-eicosane or binding at the oil–fat surface):

$$c_e V_e = c_w [V_e(1-\phi_o) + K_{ow}V_e\phi_o(1-\phi_{sf}) + K_{iw}^* A_s V_e \phi_o]. \tag{7}$$

Rearranging and substituting for the oil/water partition coefficient ($K_{ow}=c_o/c_w$), we have

$$c_o = \frac{K_{ow} c_e V_e}{[V_e(1-\phi_o) + K_{ow}V_e\phi_o(1-\phi_{sf}) + K_{iw}^* A_s V_e \phi_o]}. \tag{8}$$

Finally, by substituting for c_e from Equation (6), we obtain an expression for the concentration of EH in the liquid oil phase:

$$c_o = \frac{K_{ow} V_{EH}}{(V_e + K_{ge}V_g)[(1-\phi_o) + K_{ow}\phi_o(1-\phi_{sf}) + K_{iw}^* A_s \phi_o]}. \tag{9}$$

The quantity K_{ge} can be calculated from Equation (4) and all the other parameters are known. So Equation (9) can be used to calculate the

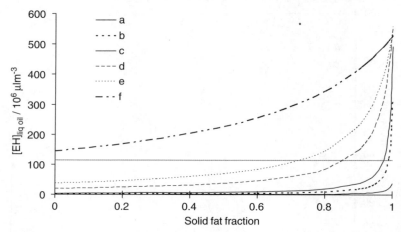

Figure 4 *Concentration of EH in the liquid oil phase of different concentrations of emulsions crystallized to different extents calculated from the bulk partition coefficients and the binding of the aroma to the droplet surfaces ($K_{iw}^* = 4.5 \times 10^{-6}$ m): (a) 40% oil, (b) 10% oil, (c) 5% oil, (d) 1% oil, (e) 0.5% oil and (f) 0.1% oil. The horizontal straight line shows the limiting value of solid fat content from the n-eicosane + aroma phase diagram*

concentration of EH in the oil phase as a function of solid fat content for emulsions of different volume fraction, as shown in Figure 4. As expected, in the completely liquid oil emulsions (zero solid fat fraction), the concentration of EH in the oil phase decreases with oil volume fraction. However, as the droplets crystallize, the EH is progressively confined to a smaller volume of liquid oil, and, while some of it redistributes to other phases, the concentration in the oil phase increases at an increasing rate.

Nevertheless, in a real system there must be equilibrium between the crystalline *n*-eicosane and the liquid *n*-eicosane in solution with the aroma molecules at the temperature of the experiment. The phase diagram of the *n*-eicosane + aroma mixture (Figure 5) has been constructed from the DSC melting points of samples containing different proportions of *n*-eicosane and aroma compound(s). In this case samples were temperature cycled in the DSC from 50°C to -20°C to 50°C at 5°C min^{-1}. At 30°C, the temperature at which all the headspace measurements were made, crystalline *n*-eicosane is in equilibrium with a solution of 46.8% aroma in liquid eicosane (*i.e.*, 1.14×10^8 μL m^{-3} of EH). This limiting value of c_o is also plotted in Figure 4, and the maximum solid fat content ϕ_{sf} for an emulsion of given volume fraction is calculated as the intersection of the line with the corresponding curve. The limiting solid fat content calculated from the phase diagram and the bulk partition coefficients is plotted in Figure 3, and this agrees well with the measured solid fat content of the droplets, suggesting that the dissolution observed should be expected from the phase behaviour of the ingredients. Finally, the gas–emulsion partition coefficients were calculated using these values of ϕ_{sf} in Equation (4); these values are plotted alongside the experimental measurements in Figure 1(b).

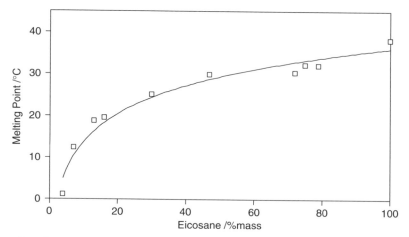

Figure 5 *Phase diagram of oil + aroma system. The melting point of the n-eicosane + aroma mixture is plotted against the weight fraction of n-eicosane*

While the quality of fit is not perfect, the trends are similar to those seen in the experimental data, and the fit is quantitatively better than the original surface-binding model.[9] In particular, the simple surface-binding model predicts an increase at lower oil contents not seen in the experimental data, nor in the refined model that allows dissolution.

It appears that, at a high ratio of fat surface to aroma compound, the volatiles simply adsorb at the droplet surface and can be modelled by a simple surface-binding coefficient. However, if the amount of aroma per unit surface exceeds a certain value, the solid droplets will begin to dissolve in the aroma according to their phase diagram. The additional liquid phase produces a much larger reservoir for the aroma and reduces the amount in the head-space. A simple estimation based on the geometry of the molecules involved suggests that the critical amount of adsorbed material needed to initiate dissolution is much less than the monolayer. The model proposed is based on independent measurements of the bulk thermodynamic properties of the independent phases – gas–liquid oil and gas–water partition coefficients and the aroma + fat phase diagram – with the only adjustable parameter being the surface binding coefficient measured at high oil concentrations where presumably the amount of liquid oil is negligible. The quality of fit between theory and experiment is acceptable, but far from perfect. The region where deviations are substantial is extremely difficult to model because the compound EH has such a high affinity for liquid oil that even a tiny error in the calculated solid fat content leads to a huge difference in the predicted result.

29.6 Conclusions

The distribution of volatile molecules between a liquid emulsion and the head-space can be modelled in terms of the volume partition coefficients. The interactions between solid fat droplets and aroma molecules are weaker but more complex. A small amount of added aroma can bind at the droplet surface, but larger amounts can dissolve the crystalline droplets according to the phase diagram and partition into the liquid oil. The melting and dissolution of droplets in a food emulsion can have a huge effect in the concentration of head-space volatiles. It may provide a way to develop a small-molecule delivery system.

Acknowledgement

This work was partially funded by a grant from the Center for Food Manufacturing, Pennsylvania State University.

References

1. M. Harrison, B.P. Hills, J. Bakker and T. Clothier, *J. Food Sci.*, 1997, **62**, 653.
2. J. Bakker and D.J. Mela, in *Flavour–Food Interactions*, R.J. McGorrin and J.V. Leland (eds), American Chemical Society, Washington, DC, 1996, p. 36.
3. M. Charles, S. Lambert, P. Brondeur, J.-L. Courthaudon and E. Guichard, in *Flavour Release*, D.D. Roberts and A.J. Taylor (eds), American Chemical Society, Washington, DC, 2000, p. 342.
4. P.B. McNulty and M. Karel, *J. Food Technol.*, 1973, **8**, 319.
5. D.D. Roberts, P. Pollien and B. Watzke, *J. Agric. Food Chem.*, 2003, **51**, 189.
6. H.G. Maier, in *Proceedings of International Symposium on Aroma Research*, H. Maarse and P.J. Groenen (eds) Pudoc, Wageningen, 1975, p. 143.
7. G.L. Cramp, A.M. Docking, S. Ghosh and J.N. Coupland, *Food Hydrocoll.*, 2004, **18**, 899.
8. D.J. McClements, *Food Emulsions*, 2nd edn CRC Press, New York, 2004.
9. S. Ghosh, D.G. Peterson and J.N. Coupland, *J. Agric. Food Chem.*, 2006, **54**, 1829.

Chapter 30

Adsorption of Macromolecules at Oil−Water Interfaces during Emulsification

Lars Nilsson,[1] Peter Osmark,[2] Mats Leeman,[3] Céline Fernandez,[2] Karl-Gustav Wahlund[3] and Björn Bergenståhl[1]

[1] DIVISION OF FOOD TECHNOLOGY, LUND UNIVERSITY, P.O. BOX 124, 22100 LUND, SWEDEN
[2] DIVISION OF TECHNICAL ANALYTICAL CHEMISTRY, LUND UNIVERSITY, P.O. BOX 124, 22100 LUND, SWEDEN
[3] DEPARTMENT OF EXPERIMENTAL MEDICAL SCIENCE, MOLECULAR ENDOCRINOLOGY GROUP, LUND UNIVERSITY, 22184 LUND, SWEDEN

30.1 Introduction

Adsorption of macromolecules plays an important role in the formation and stabilization of food emulsions.[1-3] Macromolecules that adsorb and are able to form a coherent interfacial layer can provide stability, and thus the adsorption behaviour is crucial to the understanding of their functionality.

Emulsions are typically made by applying intense mechanical energy to two immiscible liquids and thereby dispersing one in the other.[4] Adsorption during emulsification differs from many other adsorption situations as the surface is being created simultaneously with the adsorption. The adsorption event can typically be described in three steps: (i) transport of molecules to the newly formed interface; (ii) sticking of molecules with sufficient affinity for the interface; and (iii) possible changes in configuration such as unfolding and spreading of the molecules. During first two steps, which occur during very short timescales, recoalescence of the newly formed droplets occurs in parallel with the adsorption, and thus kinetic factors are likely to play a large role and the adsorption rate can be assumed to be critical. High-pressure homogenization is a common way of making an emulsion. In such a device the emulsification occurs in turbulent flow and thus convective mass transport dominates over diffusive transport. To stick at the interface the macromolecule must have sufficient affinity for it; in the case of oil-in-water emulsions, this implies the presence of sufficient hydrophobic groups in the macromolecule.

The macromolecular emulsifying agent used in food emulsions is commonly a protein, as proteins are surface-active and are abundant in most foods. Egg yolk proteins are important in many food applications as they are able to provide stabilizing effects at low pH, which is uncommon for other proteins typically used in food dispersions. This is attributed to the fact that egg yolk contains species with a broad pI range.[5, 6] Proteins in food applications are often a mixture of different species with varying chemical composition. This in turn can give rise to competitive adsorption if the interfacial area is small in comparison to the protein concentration in the bulk solution, which may mean that ultimately only certain protein species are actually adsorbed. In use as emulsifiers and emulsion stabilizers, proteins have some limitations at high ionic strength and also often at low pH. These conditions are, however, common in many foods. And the increasing demand for heat-treated food emulsions further limits their applicability.

The other large group of macromolecules abundant in nature are polysaccharides. Most natural polysaccharides are not sufficiently surface-active to act as emulsifiers on their own.[7] But chemically modified polysaccharides, such as octenyl succinate anhydride starch esters (OSA-starch), do have considerable surface activity. Owing to the ultra-high molar mass and branched structure, adsorbed OSA-starch can provide good steric stabilization in oil-in-water emulsions.[8, 9]

There is a large difference in properties, and a huge difference in molar mass, between egg yolk proteins and OSA-starch. So it is reasonable to assume that their adsorption behaviour will be different. Knowledge of the adsorption behaviour and of the composition of the adsorbed layer is beneficial for understanding the mechanisms of macromolecular adsorption in emulsions. Thus, the purpose of the work presented here is to compare the adsorption behaviour of these two macromolecules with relevance to oil-in-water emulsion formation.

30.2 Experimental

Three OSA-starch samples of waxy barley origin, varying in molar mass and degree of substitution, were provided by Lyckeby Stärkelsen (Kristianstad, Sweden). The properties of the samples are given in Table 1. The protein

Table 1 *Properties of OSA-starch samples*

Sample	Degree of substitution	Degree of branching	Molar mass (10^3 kg mol^{-1})		r_{rms}[a] (nm)
			Before homogenization	After homogenization	
B39-22	0.0224	0.0548	39	12	38
B27-21	0.0213	0.0502	27	8.6	44
B86-10	0.0104	0.0526	86	6.7	41

Source: Modified from ref. 10.
[a] Root-mean-square radius of homogenized samples.

fractions used were the water-soluble plasma proteins from egg yolk, fractionated by the method of McBee and Cotterill.[11] The fraction was characterized with two-dimensional SDS polyacrylamide electrophoresis (Figure 1) and the most abundant proteins were identified with mass spectrometry. These results are given in Table 2.

The emulsions were made with medium-chain triglyceride (MCT) oil by high-pressure homogenization at 15 MPa in a laboratory-scale valve homogenizer. The emulsion surface area was calculated from the Sauter mean diameter of the emulsions as determined by light diffraction and verified with light microscopy. For the three OSA-starch samples, the emulsions were prepared at pH 6.0 with different levels of OSA-starch. For the proteins, two different series of emulsions were prepared. In the first series of experiments the pH was kept constant at 7.0 and the protein concentration was varied. In the second set the protein concentration was kept constant while the pH was varied between 2.8 and 8.0 at constant ionic strength. The particle-size distribution of the emulsions was determined and the amount of adsorbed OSA-starch[13] or protein[14] was obtained *via* serum depletion.

The non-adsorbing protein species were determined using one-dimensional SDS–polyacrylamide electrophoresis.[15] The molar mass and rms radius distributions of OSA-starch were determined using asymmetrical flow-field flow fractionation coupled to multi-angle light scattering and refractive index detection (AsFlFFF-MALS-RI).[16] To allow faster separation, a programmed cross flow rate was employed. The elution was started at an initial cross flow

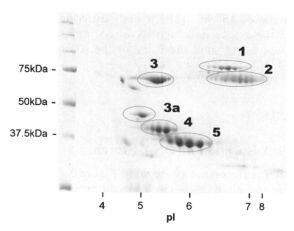

Figure 1 *Two-dimensional polyacrylamide gel of soluble egg yolk plasma protein. The six most prominent protein bands, marked by rings, were identified by mass spectrometry: (1) ovotransferrin or conalbumin; (2) immunoglobulin G; (3) serum albumin or α-livetin; (3a) truncated serum albumin (residues 1−410); (4) yolk plasma glycoprotein YGP42; (5) yolk plasma glycoprotein YGP40. The presence of several spots for most of the proteins is due to differential phosphorylation*
(Modified from Ref. 12).

Table 2 *The molecular weight (MW) and isoelectric point pI of the five dominant protein species in the egg yolk livetin fraction as determined by two-dimensional SDS−PAGE and subsequent mass spectrometry*

Spot	Protein	Observed MW (kDa)	Observed pI	Theoretical MW (kDa)[a]	Theoretical pI[a]
1	Ovotransferrin (conalbumin)	80	6.5−7	75.8	6.69
2	Immunoglobulin G, heavy chain[b]	65−70	6.5−8	60−70	6−7
3	Serum albumin (α-livetin)	65	5−5.7	67.2	5.35
4	Yolk plasma glycoprotein YGP42	40	5.3−5.8	31.4	5.88
5	Yolk plasma glycoprotein YGP40	35	5.5−6.3	31.0	6.16

Source: Modified from ref. 12.
[a] Not taking into account any post-translational modifications, *e.g.*, glycosylation or phosphorylation.
[b] Theoretical values for IgG are only approximate due to the inherent immunoglobulin heterogeneity.

rate of $Q_c(0) = 1.0$ mL min^{-1} and then the flow rate was reduced exponentially with a set time constant, *e.g.*, a half-life of 4 min. After 24 min of elution the cross flow rate was set to zero and the channel flushed without any cross flow for 5 min before the next analysis. The carrier liquid for the separation was 10 mM NaNO$_3$ with 0.002 wt.% NaN$_3$. The molar mass and the rms radius were obtained by Berry's method[17] by fitting a straight line to the data obtained at scattering angles of 43−90°. (The lowest scattering angles of 14−35° were not included as these gave data that were too imprecise.) A dn/dc value of 0.146 mL g^{-1} was used,[18] and the value of the second virial coefficient was taken as negligible.

30.3 Results

30.3.1 Total Adsorption of OSA-Starch

The emulsification performance is shown in Figure 2 as emulsion surface area created (A) *versus* the initial OSA-starch concentration (C_0). These results show that it is possible to create an emulsion with the OSA-starch as the sole emulsifier. If we assume that the amount of available emulsifier limits the effectiveness of the emulsification, then the surface area should increase (*i.e.*, droplet diameter should decrease) linearly with the initial emulsifier concentration. For samples B86-10 and B27-21 the amount of surface area created during emulsification was found to increase strongly with increasing initial OSA-starch concentration, but for sample B39-22 the surface area increase is less pronounced up to a concentration of ~ 1.25 mg mL^{-1}. Replicates of sample B39-22 also exhibited a rather large variation in the surface area

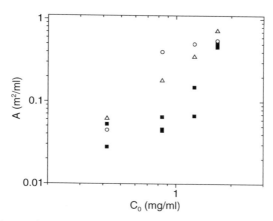

Figure 2 *Emulsion surface area A created by emulsification of 5 wt.% MCT oil with different OSA-starches:* ■, *B39-22;* ○, *B27-21;* △, *B86-10* (Redrawn from Ref. 9).

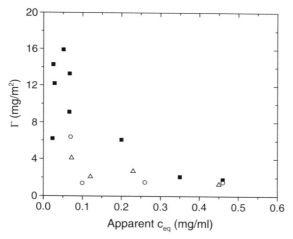

Figure 3 *Adsorption isotherms for the OSA-starches at the oil–water interface:* ■, *B39-22;* ○, *B27-21;* △, *B86-10* (Redrawn from Ref. 9).

created. Microscopy of the emulsions showed no significant flocculation throughout the experiments.

The adsorption isotherm for each of these OSA-starch samples is shown in Figure 3. The surface load Γ is plotted against the apparent equilibrium concentration of OSA starch in the bulk solution (apparent c_{eq}). The surface load for samples B27-21 and B86-10 shows little concentration dependence, and is approximately 1–3 mg m^{-2}, which corresponds quite well to what is expected for a monolayer of an adsorbed macromolecule.[19] However, at lower equilibrium concentrations these two samples show an apparent increase in the surface load. The adsorption isotherm for sample B39-22 shows an even stronger dependence on the apparent equilibrium concentration than the two other samples and the surface load decreases strongly with increasing equilibrium concentration. At low apparent equilibrium concentrations the surface load can become very high (reaching levels of 16 mg m^{-2}) and the variation between replicates was found to be large. These lower bulk concentrations are also where the emulsion surface area is smaller, *i.e.*, the droplet size larger (Figure 2).

The non-Langmuir shape of the adsorption isotherms suggests that the adsorption of the different OSA-starch samples, and in particular B39-22, is a nonequilibrium process. (Hence, we prefer the use of the term 'apparent equilibrium concentration'.) A possibility is that the surface load may depend on polymer availability in relation to the specific surface area during emulsification rather than on the equilibrium concentration. The dependence of the surface load on the bulk concentration of polymer *versus* the specific emulsion surface area has been suggested earlier by Walstra,[4] and experimental results from other authors[20,21] also suggest a correlation. This relation is here termed the 'dynamic surface load', where we are assuming that the adsorption rate controls the adsorbed amount. If the polymers adsorb rapidly, there will not be sufficient time for rearrangement and a thick adsorbed layer will be obtained. In contrast, if the polymers adsorb slowly, full rearrangement is possible and a thin layer will be obtained. The surface load *versus* the dynamic surface load is shown in Figure 4. The results indicate that the surface load increases linearly with dynamic surface load for sample B39-22, while samples B27-21 and B86-10 behave more ideally, *i.e.*, more surface area is created, rather than a greater amount of OSA-starch at the interface.

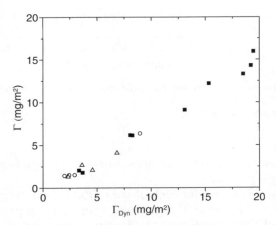

Figure 4 *Emulsion surface load Γ in emulsions with different levels of dynamic surface load (Γ_{dyn}), i.e., the amount of OSA-starch available per specific emulsion surface area:* ■, *B39-22;* ○, *B27-21;* △, *B86-10*
(Redrawn from Ref. 9).

30.3.2 Selective Adsorption of OSA-Starch

To study the molar mass distribution of adsorbed and non-adsorbed OSA-starch, an emulsion with a volatile oil phase (cyclohexane) was prepared with OSA-starch B27-21. In this way the oil phase could be eliminated by flushing with N_2 after emulsification. The molar mass and rms radii distributions could then be determined by asymmetrical flow-field flow fractionation coupled to multi-angle light scattering with refractive index detection (AsFlFFF-MALS-RI). This is a technique that enables us to characterize very large macromolecules such as the OSA-starch.[22-24] The relative molar mass distributions in the upper and lower phase, after emulsion separation, are shown in Figure 5. We can see that the high molar mass components are selectively adsorbed as they are only detected in the upper phase. Consequently, the lower molar mass components are found at higher levels in the lower phase. The concentration of OSA-starch was determined gravimetrically in the whole emulsion, in the lower

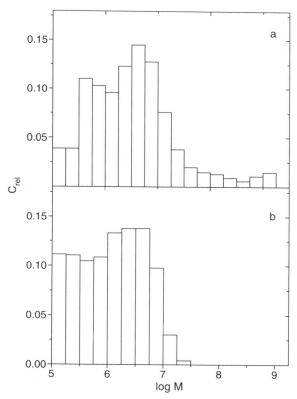

Figure 5 *Molar mass distribution of OSA-starch in (a) upper phase and (b) lower phase. The relative mass concentration of OSA-starch (C_{rel}) is plotted against the logarithm of the molar mass M. The upper phase contained not only the dispersed phase but some aqueous phase as well*
(Redrawn from Ref. 25).

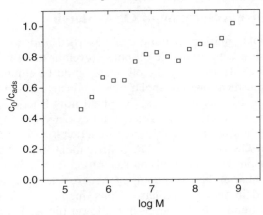

Figure 6 Adsorbed ratio $c_{ads}/c_{whole\ emulsion}$ versus logarithm of the OSA-starch molar mass
(Redrawn from Ref. 25.)

phase, and in the upper phase. The concentration in each molar mass class could then be determined from mass balances using a maximum likelihood estimation.[26] This procedure enables us to plot the adsorbed ratio (adsorbed concentration divided by concentration in the whole emulsion) for each molar mass class, as shown in Figure 6. The plot shows that the adsorbed ratio increases with increasing molar mass, and approaches unity for the largest molecules, while for a molar mass of $\sim 10^5$–10^6 g mol^{-1} the ratio is only 0.5–0.65. Thus, it is clear that there is a preferential adsorption of the ultra-high molar mass components during emulsification.

30.3.3 Protein Adsorption

The adsorption of the protein fraction at pH 7.0 was found to show a typical Langmuir-type behaviour with a plateau value of ~ 1.5 mg m^{-2}, as would be expected for a water-soluble protein fraction adsorbing at a hydrophobic interface.[27] A clear preferential adsorption of the components corresponding to two of the dominant protein bands was observed when the interfacial area was insufficient to accommodate all the available protein. These bands were identified as serum albumin (protein 3) and yolk plasma glycoprotein YGP40 (protein 5).[28]

For experiments in which protein concentration was kept constant and pH was varied, significant degradation of some of the protein was observed below pH 4 (Figure 7), which made it difficult to determine the selective adsorption below this pH. Serum albumin seems to adsorb much more efficiently at pH $\sim$ 4–5, which is slightly below its pI. The glycoprotein YGP40 adsorbs efficiently at any pH, with a slight increase in adsorbed amount at pH $\sim$ 6–8, which is just above its pI. The component IgG adsorbs slightly more at pH 8, which is close to its pI.

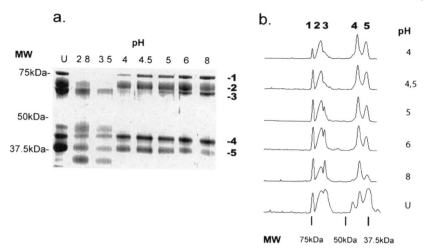

Figure 7 *SDS–PAGE of soluble egg yolk proteins at protein concentration $c_0 = 1$ mg mL^{-1}.*[12] *(a) The bands refer to samples before emulsification (U) and after emulsification at varying pH values (numbers from 2.8 to 8). (b) The integrated lane profiles are labelled as in Figure 1*

30.4 Discussion

Having observed contrasting examples of selective adsorption for two different kinds of macromolecules, we now consider the possible reasons for this selectivity. The adsorption can be described as being controlled by kinetics, affinity, or equilibrium considerations. In what follows, we discuss these mechanisms and their influence on the results obtained in this study.

30.4.1 Kinetically Controlled Adsorption

The experiments for OSA-starch gave very high surface loads. In a high-pressure homogenizer the adsorption occurs under turbulent flow conditions, and the transport to the interface is dominated by convection rather than by diffusion, which would be the case under quiescent conditions. Owing to the convective mass transport, larger molecules are transported more rapidly to the interface, which is the opposite to what applies for diffusional mass transport. The adsorption process can be described as a collision between particles in turbulent flow. By invoking the theory of isotropic turbulence, and assuming that the particles are not smaller than the turbulence microscale, known as the Kolmogorov scale, the collision frequency can be represented by[29,30]

$$N = 1.4\pi(r_p + r_d)^{7/3}\varepsilon^{1/3} n_p n_d, \quad (1)$$

where r_p and r_d are the radii of the polymer and the emulsion droplets, and n_p and n_d are the number concentrations of these species. We can express the flux of macromolecules to the interface as

$$\frac{dc_i(t)}{dt} = -k A c_i(t) \left[1 - \sum_i \theta_i(t)\right], \quad (2)$$

where

$$k = \frac{1.4(r_{p,i} + r_d)^{7/3} \varepsilon^{1/3}}{4r_d^2}, \quad (3)$$

and c_i is the macromolecule concentration for molar mass class i in the solution. The quantity $1 - \sum_i \theta_i(t)$ describes the saturation of the interface, where $\theta_i = (\Gamma_i/\Gamma_{max,i})$. The saturation term can be expressed as

$$\theta_i(t) = [c_i(0) - c_i(t)](a_i/A), \quad (4)$$

where A is the emulsion interfacial area; the quantity a_i is the projected area of a macromolecule in class i,

$$a_i = \frac{\pi r_{p,i}^2 N_A}{M_i}, \quad (5)$$

where N_A is Avogadro's number and M_i is the molar mass of class i. We can then model the adsorption in the molar mass range investigated experimentally. This gives us a set of differential equations that can be solved numerically:

$$\left\{\begin{array}{c} \frac{\partial c_n}{\partial t} \\ \frac{\partial c_i}{\partial t} \\ \frac{\partial c_1}{\partial t} \end{array}\right\} = \left\{\begin{array}{c} -k_n(r_{p,n}, r_d) c_n(t) \left[A - \sum_{i=1}^{n} (c_i(0) - c_i(t)) a_i\right] \\ -k_i(r_{p,i}, r_d) c_i(t) \left[A - \sum_{i=1}^{n} (c_i(0) - c_i(t)) a_i\right] \\ -k_1(r_{p,1}, r_d) c_1(t) \left[A - \sum_{i=1}^{n} (c_i(0) - c_i(t)) a_i\right] \end{array}\right\}. \quad (6)$$

The results from the modelling are compared with the experimental results in Figure 8 in terms of adsorbed ratio. The data show that the ultra-high molar mass components adsorb preferentially and that the adsorption can be described theoretically as collisions between particles in turbulent flow. Thus the high surface loads obtained experimentally may be because of an over-representation of the ultra-high molar mass components at the interface, which as such may give rise to higher surface loads than expected. Because kinetics are important, owing to the short timescales for adsorption during emulsification,[9,31] it is likely that nonequilibrium structures and jamming will arise at the interface as the macromolecules will not have sufficient time to optimize their configurations from a thermodynamic point of view. Furthermore, the surface loads may be determined by Apollonian packing,[32,33] which means that late arriving molecules find fewer available adsorption sites and so have to 'squeeze in' between the already adsorbed molecules. This results in a type of close packing which resembles Apollonian packing, i.e., a space-filling packing circles or spheres (see Figure 9). To describe this, Douglas et al.[33] have proposed the adaptive random sequential adsorption (ARSA) model, where the molecules are

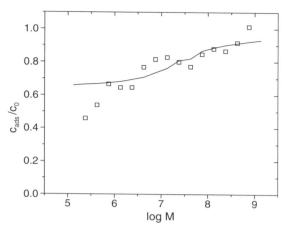

Figure 8 *Comparison of experimental data (open squares) and predictions from the model described in Equation (6) shown as adsorbed ratio plotted against the logarithm of the molar mass*
(Redrawn from Ref. 25).

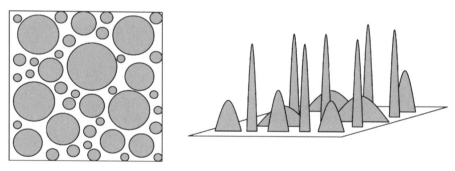

Figure 9 *Schematic illustration of an adsorbed layer of macromolecules closely resembling Apollonian packing (left) and the layer obtained from the ARSA model, where the macromolecules are treated as deformable droplets with a constant volume (right)*
(Redrawn from Ref. 33).

treated as deformable 'droplets' with a constant volume. Late arriving molecules thus become elongated to stick to the remaining adsorption sites, which results in very thick layers (referred to by the authors[33] as a matted carpet) and correspondingly high surface loads, as illustrated in Figure 9.

30.4.2 Affinity-Controlled Adsorption

Adsorption of the water-soluble plasma proteins exhibits a pronounced selectivity with a preference for YGP40 (protein 5) and serum albumin (protein 3).

The simplest interpretation would be that the most hydrophobic protein should adsorb selectively. However, these proteins are all water soluble and they do not display any significant differences in overall hydrophobicity. A second possible explanation is that a higher molecular weight would give more potential sticking possibilities (per molecule), which should favour adsorption. However, the results indicate favoured adsorption for one high-molecular-weight and one low-molecular-weight species, depending on pH, seemingly excluding the possibility that the molecular weight has an important overall influence on the adsorptive competitiveness in the system. A third alternative explanation could involve the kinetics. In this system with comparatively small molecules adsorbing, we might expect diffusion to determine the adsorption rate. However, this hypothesis would also imply that the competitiveness should follow the molecular weight, when it does not. The fourth possibility would be that the competitiveness is controlled by the distribution of hydrophilic and hydrophobic domains in the protein molecule. Both YGP40 and serum albumin contain long contiguous stretches of low hydrophilicity, as can be seen from the Kyte–Doolittle plots[34] in Figure 10. Therefore, a coherent hydrophobic block in the amino acid sequence might be an important property in the adsorption selectivity, as proteins once adsorbed can unfold and spread out at the interface.[35–37]

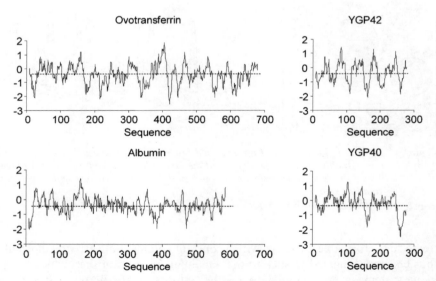

Figure 10 *Kyte–Doolittle plots of four of the five most abundant proteins in the egg yolk livetin fraction with a window of 15 amino acids.[34] The positive direction of the y-axis implies increasing hydrophobicity. No plot is shown for immunoglobulin G due to the large sequence variation. IgG tends to consist of alternating hydrophobic and hydrophilic stretches, as it is the case for ovotransferrin. To facilitate comparison, a horizontal dashed line has been added at –0.4, representing the glycine residue in the Kyte–Doolittle algorithm*
(Redrawn from Ref. 12).

The emulsifying activity of proteins shows a weak positive correlation with their protein surface hydrophobicity.[38] The emulsifying activity is also strongly related to the ability to unfold at the interface, so that partially unfolded proteins often have higher emulsifying activity that is related to the higher surface hydrophobicity and the greater flexibility.[39,40] The amount of spreading and its timescale depend on the protein structure, the solubility, and the nature of the interface itself. The timescale for interfacial spreading ranges from ~ 10 s for flexible proteins such as β-casein up to $\sim 10^3$ s for globular proteins[41] (although values as small as 10^{-2} s have been reported for bovine α-lactalbumin[42]).

The importance of the hydrophobic block character for the adsorption of other surface-active macromolecules such as triblock copolymers has been reported.[43,44] The results of these studies show that a longer coherent block of monomers with a high affinity for the interface tends to give greater surface-activity and higher surface loads. Griffiths et al.[45] has studied the role of copolymer architecture on the adsorption at interfaces and found that a cyclic triblock copolymer (in which the two hydrophilic blocks were covalently bound together) had surprisingly similar adsorption behaviour to a linear triblock copolymer of the same composition. This is indeed an interesting result that could to some extent be related to the efficiency of the adsorption of proteins. Furthermore, the heterogeneity in the distribution of hydrophobic substituents in modified celluloses has been shown to influence the clouding behaviour of such derivatives.[46] Thus the distribution of hydrophobic groups plays a large role in determining the amphiphilic character and surface activity of water-soluble macromolecules, and we suggest that the Kyte–Doolittle plots in Figure 10 reflect this concept.

30.4.3 Equilibrium-Controlled Adsorption

What we mean by 'equilibrium-controlled adsorption' is the situation in which the rates of adsorption and desorption are equal, e.g., as described by the Langmuir isotherm. Our results for the adsorption of OSA-starch show that the highest molar mass species are preferentially adsorbed at the interface while the lower molar mass species reside to a larger extent in the bulk solution. The interfacial displacement of low molar mass polymers by higher molar mass polymers, in dilute solutions at equilibrium conditions, is well known and can be well described for homopolymers by the Scheutjens–Fleer (SF) theory.[19] The shorter chain polymers adsorb more rapidly at the bare surface because of their higher diffusion coefficients, but they are then displaced by the slower diffusing long chains because the entropy loss upon adsorption is lower for the latter whereas the gained adsorption energy is roughly the same. Naturally, this interpretation assumes reversible adsorption, but this is very unlikely to be the case when the adsorption is due to the hydrophobic effect and the macromolecules are large. Care should therefore be taken when interpreting experimental results from equilibrium theories on systems like these with strong evidence for nonequilibrium effects.[47]

30.5 Conclusions

This study has considered two different adsorption situations. The first situation involving competitive adsorption is the case of hydrophobically modified starch, for which the relative adsorption follows the size of the molecule. We suggest that the composition of the adsorbed species is controlled by the sequence of adsorption, which is then controlled by the flux towards the interface. The second situation involves the egg yolk protein fraction. The competitive adsorption is described according to a difference in the amino acid sequence of the proteins. We explain these variations in adsorbed amount for different species in terms of differences in the affinity for the interface, *i.e.*, the adsorption strength.

Acknowledgement

Financial support from the Centre for Amphiphilic Polymers (Lund, Sweden) is gratefully acknowledged.

References

1. E. Tornberg, *J. Sci. Food Agric.*, 1978, **29**, 867.
2. E. Dickinson and G. Stainsby, *Food Technol.*, 1987, **41**(9), 74.
3. B. Bergenståhl, in *Gums and Stabilisers for the Food Industry*, Vol. 4, G.O. Phillips, D.J. Wedlock and P.A. Williams (eds), IRL Press, Oxford, 1988, p. 363.
4. P. Walstra, in *Encyclopedia of Emulsion Technology*, Vol. 1, P. Becher (ed), Marcel Dekker, New York, 1983, p. 57.
5. W. Ternes, *J. Food Sci.*, 1989, **54**, 764.
6. L. Nilsson, P. Osmark, C. Fernandez and B. Bergenståhl, unpublished.
7. E. Dickinson, *Food Hydrocoll.*, 2003, **17**, 25.
8. S. Tesch, C. Gerhards and H. Schubert, *J. Food Eng.*, 2002, **54**, 167.
9. L. Nilsson and B. Bergenståhl, *Langmuir*, 2006, **22**, 8770.
10. L. Nilsson, M. Leeman, K.G. Wahlund and B. Bergenståhl, *Biomacromolecules*, 2006, **7**, 2671.
11. L.E. McBee and O.J. Cotterill, *J. Food Sci.*, 1979, **44**, 656.
12. L. Nilsson, P. Osmark, C. Fernandez, M. Andersson and B. Bergenståhl, *J. Agric. Food Chem.*, 2006, **54**, 6881.
13. P. Åman, E. Westerlund and O. Theander, in *Methods in Carbohydrate Chemistry*, J.N. BeMiller, D.J. Manners and R.J. Sturgeon (eds), Vol. 10, Wiley, New York, 1994, p. 111.
14. M.M. Bradford, *Anal. Biochem.*, 1976, **72**, 248.
15. U.K. Laemmli, *Nature*, 1970, **227**, 680.
16. B. Wittgren and K.-G. Wahlund, *J. Chromatogr A.*, 1997, **760**, 205.
17. G.C. Berry, *J. Chem. Phys.*, 1966, **44**, 4450.

18. J. Brandrup, E.H. Immergut and E.A. Grulke (eds), *Polymer Handbook*, 4th edn, Wiley, New York, 1999.
19. G.J. Fleer, M.A. Cohen Stuart, J.H.M.H. Scheutjens, T. Cosgrove and B. Vincent, *Polymers at Interfaces*, Chapman & Hall, London, 1993.
20. E. Tornberg, A. Olsson and K. Persson, in *Food Emulsions*, S.E. Friberg and K. Larsson (eds), Marcel Dekker, New York, 1997, p. 279.
21. S. Tcholakova, N.D. Denkov, D. Sidzhakova, I.B. Ivanov and B. Campbell, *Langmuir*, 2003, **19**, 5640.
22. M. Andersson, B. Wittgren and K.-G. Wahlund, *Anal. Chem.*, 2001, **73**, 4852.
23. S. Lee, P.-O. Nilsson, G.S. Nilsson and K.-G. Wahlund, *J. Chromatogr. A*, 2003, **1011**, 111.
24. G. Modig, L. Nilsson, B. Bergenståhl and K.G. Wahlund, *Food Hydrocoll.*, 2006, **20**, 1087.
25. L. Nilsson, M. Leeman, K.G. Wahlund and B. Bergenståhl, submitted for publication.
26. A.J. Dobson, *An Introduction to Generalized Linear Models*, Chapman & Hall/CRC, Boca Raton, FL, 2002.
27. E. Dickinson and G. Iveson, *Food Hydrocoll.*, 1993, **6**, 533.
28. J.-I. Yamamura, T. Adachi, N. Aoki, H. Nakajima, R. Nakamura and T. Matsuda, *Biochim. Biophys. Acta*, 1995, **1244**, 384.
29. M.A. Delichatsios and R.F. Probstein, *J. Colloid Interface Sci.*, 1975, **51**, 394.
30. L.A. Spielman, in *The Scientific Basis of Flocculation*, K.J. Ives (ed), Sijthoff & Noordhoff, Alphen aan den Rijn, 1978, p. 63.
31. F. Innings, L. Fuchs and C. Trägårdh, submitted for publication.
32. B.B. Mandelbrot, *The Fractal Geometry of Nature*, Freeman, New York, 1983.
33. J.F. Douglas, H.M. Schneider, P. Frantz, R. Lipman and S. Granick, *J. Phys., Condens. Matter*, 1997, **9**, 7699.
34. J. Kyte and R. Doolittle, *J. Mol. Biol.*, 1982, **157**, 105.
35. M.E. Soderquist and A.G. Walton, *J. Colloid Interface Sci.*, 1980, **75**, 386.
36. W. Norde and A.C.I. Anusiem, *Colloid Surf.*, 1992, **66**, 73.
37. C.P. Wertz and M.M. Santore, *Langmuir*, 1999, **15**, 8884.
38. A. Kato and S. Nakai, *Biochim. Biophys. Acta*, 1980, **624**, 13.
39. A. Kato, Y. Osako, N. Matsudomi and K. Kobayashi, *Agric. Biol. Chem.*, 1983, **47**, 33.
40. S. Damodaran, in *Food Chemistry*, O.R. Fennema (ed), Marcel Dekker, New York, 1996.
41. P. Walstra, *Physical Chemistry of Foods*, Marcel Dekker, New York, 2003.
42. M.F.M. Engel, C.P.M. van Mierlo and A.J.W.G. Visser, *J. Biol. Chem.*, 2002, **277**, 10922.
43. J.A. Shar, T.M. Obey and T. Cosgrove, *Colloid Surf. A*, 1998, **136**, 21.
44. C.A. Prestidge, T. Barnes and S. Simovic, *Adv. Colloid Interface Sci.*, 2004, **108–109**, 105.

45. P.C. Griffiths, T. Cosgrove, J. Shar, S.M. King, G.-E. Yu, C. Booth and M. Malmsten, *Langmuir*, 1998, **14**, 1779.
46. H. Schagerlöf, M. Johansson, S. Richardson, G. Brinkmalm, B. Wittgren and F. Tjerneld, submitted for publication.
47. B. O'Shaugnessy and D. Vavylonis, *J. Phys. Condens. Matter*, 2005, **17**, 63.

PART V
Texture, Rheology and Sensory Perception

Chapter 31

Tribology as a Tool to Study Emulsion Behaviour in the Mouth

D.M. Dresselhuis,[1,2] E.H.A. de Hoog,[1,3] M.A. Cohen Stuart[2] and G.A. van Aken[3]

[1] WAGENINGEN CENTRE FOR FOOD SCIENCES, P.O. BOX 557, 6700 AN WAGENINGEN, THE NETHERLANDS
[2] LABORATORY OF PHYSICAL CHEMISTRY AND COLLOID SCIENCE, WAGENINGEN UNIVERSITY AND RESEARCH CENTRE, P.O. BOX 8038, 6700 EK WAGENINGEN, THE NETHERLANDS
[3] NIZO FOOD RESEARCH B.V., P.O. BOX 20, 6710 BA EDE, THE NETHERLANDS

31.1 Introduction

Fat is both appreciated and disliked for its presence in food – appreciated for the unique sensation it gives to various food systems, but disliked for the relatively high amount of calories it provides to the body. Therefore, much interest exists in the development of low-fat products, which upon eating still give rise to the desired sensation. This sensation is often referred to as 'creamy', a term indicative of richness, high quality, and a pleasant sensation.[1,2]

In general, food products are roughly developed following the lower part of the scheme depicted in Figure 1. A product is engineered, which includes optimizing the composition and the processing conditions in order to obtain a physically and chemically stable food product. Subsequently, this product is sensorially evaluated, which often leads to re-adjustment of the formulation and/or processing conditions.

In the development of a food emulsion, one faces the challenge of retaining the appreciated creamy sensation even when the product contains only a small amount of fat. This development process can be rather time-consuming, since it typically involves a trial-and-error approach. Generally, the influence of oral processing on the food product is not considered, although the actual perception of food takes place in the mouth, *i.e.*, during and after oral processing (see Figure 1). By gaining more knowledge about the mechanisms involved in fat perception, as well as about the physico-chemical processes taking place in

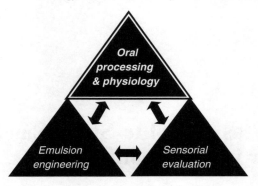

Figure 1 *General product development scheme for a food system*

the mouth, we should be better able to engineer food emulsions with the desired creamy in-mouth sensation.

31.2 Sensory Perception of Emulsions

What determines the positive creamy sensation of an oil-in-water emulsion? There has been a lot of study and controversy on this matter. Many studies have attempted to relate fat-related attributes like creaminess and fattiness to chemical and rheological properties of emulsions. It has been shown that creaminess perception is related to emulsion viscosity, thin film rheology, and release of fat-associated flavours, and that these factors are interrelated.

Kokini[3] investigated several commercial products and indicated that there is a mathematical relationship between the perceived thickness, smoothness, and the creaminess of a product. This suggests that the perception of creaminess is partly related to the viscosity of a product. Moore et al.[4] confirmed such a relationship, but pointed out that there are also additional features of emulsified fat, which contribute to its perception. Frost and co-workers[5] tried to enhance the fat perception of a low-fat milk by adding thickener, whitener, aroma, and by varying the droplet size. Addition of a thickener increased the viscosity but it did not significantly increase the perception of creaminess. In another type of study, Richardson et al.[2] compared ratings of overall liking of food emulsions with creaminess ratings; they also could not find a relationship between viscosity and creaminess. On the other hand, Akhtar et al.[6] did conclude that viscosity was of utmost importance to creaminess perception.

Evidence that viscosity alone cannot explain a creamy and fatty sensation of food emulsions was assessed from an unusual type of food research. By monitoring neuron signals of the brain of a rhesus macaque, the researchers showed[7–9] that viscosity and fat can be perceived independently. Researchers studying the effect of volatiles on emulsion perception elaborated this finding. They indicated that not only mouthfeel, including the sensation of thickness, is of importance for creamy sensations, but also flavour release.[1–2,8,9,10] The data of Kirkmeyer and Tepper[11] support the notion that flavour and mouth-feel attributes are highly integrated in the creaminess perception, and they suggest that humans are incapable of easily separating these sensations. Our own work[12] also confirms that flavour is of importance for the perception of creaminess. Our sensorial quantitative descriptive analysis has shown that applying different fat phases in oil-in-water emulsions resulted in a difference in creaminess ratings. Creaminess was closely related to the rated perception of thickness and smoothness, but also to flavour intensity and creamy flavour.[12] This suggests that both the viscosity and the flavour release are of importance.

Nevertheless, taking into account the effect of flavour still does not fully explain the fact that rheological measurements cannot be directly related to creaminess sensations. The inability to make this link is partially due to the complexity of oral processing. During this processing, a food product is rubbed and squeezed between tongue and palate, it is mixed with saliva, and the temperature of the product rises. In contrast to the simple shear applied in standard rheological equipment, the food in the mouth is subject to a complex shear flow pattern, which mainly occurs in the thin film present between tongue and palate. Several authors have indicated[13-15] that it is not the bulk behaviour of an emulsion as studied with standard rheology that predicts the creamy sensation of a product, but the thin film rheology as studied in tribology. In the present work, we focus on studying the ability of tribology to be a tool in understanding in-mouth behaviour of emulsions and consequently in understanding their sensory perception.

31.3 Tribology and Sensory Science

Tribology is the science of the study of friction, lubrication processes, and wear phenomena caused by surfaces sliding over one another. Tribology is widely studied in relation to high-pressure production processes, *e.g.*, in the metals industry. Here friction and wear should be minimized, implying that the friction forces must be reduced by means of lubrication of the surfaces. We can distinguish three lubrication regimes. In the hydrodynamic regime the lubricating layer is thicker than the maximum height of the surface asperities, resulting in a complete separation of the two opposing surfaces. In boundary lubrication the lubricating layer is thinner than the surface roughness, typically only a few molecular layers thick. A third regime is the mixed regime in which both boundary lubrication and hydrodynamic lubrication occur.[16]

Several authors[13-15] have applied this tribology approach to the low-pressure oral processing of food. Upon swallowing a food product, the tongue is pressed against the palate, giving rise to a friction force with the food product acting as a lubricant. From the combination of sensorial evaluation of food systems and tribological measurements conducted on these same systems, they have inferred[13-15] an inverse relation between lubricational properties of the system and the sensorially perceived creaminess and fattiness.

The amount of friction between two sliding surfaces is predominantly determined by the nature of the sliding surfaces, the load applied, and the speed of shearing. So, for example, the friction forces are considered to be determined by the roughness, the wetting characteristics, and the deformability of the two sliding surfaces. Furthermore, the ability of a surface to interact, by chemical or physical means, with the opposing surface and with the lubricant confined between the surfaces will also influence the friction force. In order to reduce the friction between a pair of sliding surfaces, a lubricant should be able to wet the surfaces, preferably forming a thick viscous layer on each of the surfaces. When a viscous layer with a thickness exceeding the mean asperity height is present, a hydrodynamic pressure can build up within this viscous layer at high sliding speeds (hydrodynamic lubrication) resulting in low friction forces. In the case of oral processing, it is likely that the oral mucosa is 'mixed lubricated', implying that both the viscosity of the food and the surface characteristics determine the in-mouth friction.

In their experiments designed to relate sensory perception of emulsions to tribological data, Malone et al.[14] used oil-in-water emulsions of similar viscosity but varying in fat content. The emulsion droplets were stabilized by a polyglycerol ester. Tribological measurements on the emulsions were conducted with a steel ball on a silicone rubber disk. Malone and co-workers[14] concluded that the measured difference in frictional properties could be correlated to the objective perception of fat. De Wijk and Prinz[15] used a home built set-up consisting of a rotating rubber band and a metal cylinder to test the lubricational behaviour of various fat-containing products. Like Malone et al., they concluded[15] that friction is closely associated with fat-related texture attributes.

Lee and co-workers[13] used tribological measurements to predict the difference in perception among chocolate samples varying in particle size, composition, and processing. In the 'pin-on-disk' set-up used to measure the lubricational behaviour of the various chocolate samples, Lee et al.[13] used different pins and disks in order to study the effect of deformation of the surface on lubricational behaviour. In analogy with findings from high-pressure tribological measurements, the elastic modulus of the surfaces was also found to be crucial for measurements at relative low pressure.[13]

The various experiments described above were all designed to predict the *in-mouth* sensation of an emulsion. Yet, all of the experiments were conducted with artificial surfaces. The question then remains: if measurements are performed with surfaces so different from the oral mucosa, can they really mimic in-mouth friction and thereby predict sensory perception?

31.4 Importance of Surface Characteristics

Depending on whether it is liquid, solid, or semi-solid, the food goes through several stages before being swallowed and digested. Emulsions are considered to be semi-solid food systems. There are three phenomena that are thought to relate to perception taking place in the mouth upon oral processing of an emulsion:[17] (i) the interaction between emulsion components and/or saliva, (ii) the interaction between emulsion components and the mucous layer covering the tongue, and (iii) the rubbing and squeezing of the emulsion against the palate. This rubbing of the product against the palate generates a friction force which can be sensed by mucosal mechanoreceptors.[18] As surface characteristics are of great relevance in high-pressure processing, it is likely also that they are of importance in the low-pressure in-mouth processing of food. In our work we have characterized a pig's tongue surface and have compared it with polydimethylsiloxane (PDMS), a surface often used as a mouth-mimicking surface. Pig's tongue was chosen as the surface mimicking human tongue since humans consume roughly the same type of diet as pigs.[19] Based on observations made during preparation of the pig's tongue, we expect that there is no longer a mucous layer present on the pig's tongue samples used in our measurements.[19] We have characterized the surfaces with respect to roughness, deformability, and wetting characteristics.

The roughness of the pig's tongue was determined using stereomicroscopy and confocal laser scanning microscopy (CLSM). The filiform papillae, giving the tongue its roughness, are on average 300 times longer than the average asperity height on the PDMS surface we used (see Figure 2). The roughness of the PDMS surface is likely to be similar to the roughness of the steel, PTFE, Zirconia, and rubber surfaces used by Malone, de Wijk and Lee.[13–15] The mean asperity height on PDMS was around 5 nm, allowing hydrodynamic lubrication by a food product to occur. This suggests that, based on solely the difference in the observed roughness between tongue and PDMS, the friction forces in the mouth are of larger magnitude than those measured in a tribometer with, say, rubber and steel as the tribopair.

As well as roughness, the elastic moduli of the two sliding surfaces are of importance in friction force measurements. Friction depends on the contact pressure, which is related to the reduced elastic modulus of the two sliding surfaces. In our work the elastic modulus of PDMS, a relatively soft rubber, is being compared with pig's tongue tissue. We have shown[19] that the elastic modulus of PDMS is two orders of magnitude larger than that of pig's tongue. These results indicate that most of the oral simulating tribological experiments using rubber surfaces are conducted at a contact pressure that cannot be reached in the mouth.

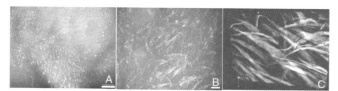

Figure 2 *Illustration of macroscopic roughness of pig's tongue: (A) stereomicroscopy image, scale bar = 1000 μm; (B) stereomicroscopy image, scale bar = 200 μm; and (C) 3-D CLSM image, 300 × 500 μm²*

The wetting characteristic of the surface is another aspect that can influence the capacity of an emulsion to act as a lubricant and thus consequently its capacity to reduce the friction. We have found[19] that a water droplet spreads fast on a slightly moist pig's tongue which is not covered by a mucous layer. Drying the tongue tissue with pressurized air results in a 'dry' surface. On such a surface a water droplet scarcely spreads at all, indicating that the tissue is then more hydrophobic, and hence comparable to PDMS.[19] In the human mouth the oral mucosa is covered by a gel-like mucous layer, containing mainly mucins (glycoproteins) and water. Our results show that the oral mucosa is intrinsically hydrophobic, but it is covered by a mucous layer which gives it the characteristics of a hydrophilic surface.

In summary, then, we can say that a pig's tongue not covered by a mucous layer has a similar hydrophobicity to PDMS. In the mouth, however, a mucous layer covers the surface; so PDMS fails to mimic the tongue in terms of its wetting characteristics. Considering the elastic modulus and the roughness, PDMS differs greatly from pig's tongue; the synthetic polymer surface is much smoother and it has an elastic modulus that is two orders of magnitude larger.

To what extent does the behaviour of an emulsion upon shearing between artificial surfaces deviate from in-mouth emulsion behaviour? To identify these differences we used a home-built experimental arrangement called the optical tribological configuration (OTC).[19] In this arrangement (Figure 3) a sample of emulsion was sheared between a pair of surfaces, either PDMS/glass or tongue/glass, under a certain load F_z. During each experiment the lower plate oscillated for a number of cycles at a certain speed. The friction force F_x was measured,

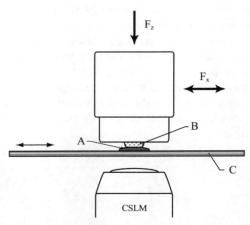

Figure 3 *The OTC. Emulsion sample A is confined between an upper surface B (e.g., pig's tongue or PDMS) and a glass surface C. A normal force F_z is applied and the friction force F_x is measured while surface C is oscillating. The sample is observed by CLSM*

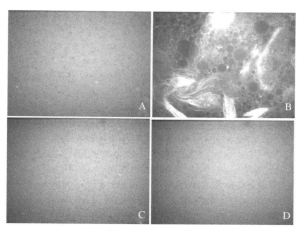

Figure 4 CLSM images of an oil-in-water emulsion (1 wt% WPI, 40 wt% sunflower oil). Sample sheared between pig's tongue and glass: (A) before applying a load; (B) after applying a load and shearing, still under compression. Sample sheared between PDMS and glass: (C) before applying a load; (D) after applying a load and shearing, still under compression. Image size = 144 × 108 μm^2

and simultaneously the emulsion droplet behaviour under shear was observed by CLSM, making it necessary to shear the sample against an optically transparent glass plate. Shearing an emulsion sample between PDMS and glass results in a friction force that is much lower than the force measured upon shearing the same emulsion between pig's tongue and glass. In fact, the friction force is so low that it cannot be determined accurately with the set-up used.[19] Owing to the fact that PDMS is much smoother than pig's tongue, it seems likely that a hydrodynamic layer is formed on the surface upon shearing, resulting in a very low friction force.

The behaviour of the lubricating emulsion was observed by confocal microscopy during shearing in the OTC. The black spheres in Figure 4 are emulsion droplets and the white structures in Figure 4(B) are tongue papillae. Shearing with the pig's tongue [Figure 4(A) and 4(B)] resulted in coalescence of the emulsion, whereas shearing with PDMS [Figure 4(C) and 4(D)] gave no observable coalescence. A possible reason for this difference in coalescence behaviour is the dissimilarity between PDMS and pig's tongue with respect to deformability and roughness of the surfaces. The emulsion droplets are, in the case of shearing with the pig's tongue, confined between the papillae. This can cause the droplets to be subject to an extensive shear flow regime in the confined space between the papillae hairs. Another explanation is that the droplets are sheared against each other between the papillae surfaces. In the latter case the micro-roughness of the tongue surfaces is of importance,[19] as well as the deformability of the surface. In contrast to the pig's tongue, PDMS has a completely smooth surface with a much higher elastic modulus. Therefore, upon shearing with PDMS, no coalescence was observed.

31.5 In-Mouth Emulsion Behaviour and Perception

The observed differences in coalescence behaviour (Figure 4) and in measured friction force show that surface characteristics are of importance in low-pressure processes. The intriguing question now is: what are the implications of these results for using tribology as a predictive tool for sensory perception of emulsions? We need to investigate whether a difference in coalescence in the mouth can influence the perception. To this purpose, emulsions varying in their sensitivity to coalescence were engineered and then sensorially evaluated.

In our study[12] we have prepared oil-in-water food emulsions, stabilized by 1 wt% WPI (Davisco) and containing 10 wt% sunflower oil (Cargill, Amsterdam) with either small droplets ($d_{32} \sim 1.5$ μm) or large droplets ($d_{32} \sim 2.5$ μm). The larger droplets were expected to be more sensitive to coalescence due to the lower Laplace pressure. A second group of emulsions was engineered with solid fat as the fat phase instead of the liquid sunflower oil. Solid fat in emulsion droplets is partly crystalline, causing those droplets to be more sensitive to partial coalescence. With increasing droplet size, there is an increase in the probability that fat crystals are able to penetrate the interfacial layer stabilizing the emulsion droplet.[20] Furthermore, as mentioned before, the Laplace pressure stabilizing the droplet is lower for larger droplets. A sensory experimental panel, trained specifically on liquid emulsions, evaluated the emulsions in three separate sessions. The panel consisted of eight panellists, who evaluated the emulsions on odour, taste, mouth-feel, and after-feel attributes.[12]

The emulsions containing sunflower oil, and differing solely in particle size, showed no significant difference in the perception of creaminess, thickness, and coating after-feel. These attributes are also suggested to relate to friction. When using solid fat as the dispersed phase we did find an effect of the droplet size on the perception of these attributes.[12] Panellists rated emulsions with larger droplets containing solid fat as more creamy, and as giving a more coating after-feel. These results suggest that the expected difference in the occurrence of partial coalescence between smaller and larger droplets containing solid fat is indeed sensorial evaluated as more creamy.

As already mentioned in Section 31.2, viscosity has been suggested to relate to the perception of creaminess. Figure 5 shows that the viscosity of our samples was roughly the same for all droplet sizes, which assures us that the differences in the perception of creaminess between the samples was not due in this case to any differences in viscosity. CLSM observations of the sheared emulsions did not show conclusive evidence, however, that samples having initially larger droplets were more vulnerable to coalescence than the smaller droplets. This is due to the fact that the droplets enlarged by coalescence cream out of the focal plane of the microscope.[19]

Figure 6 shows the friction force data of the emulsions sheared between pig's tongue and glass. The differences in friction forces between the samples are small, and not in all cases significant, but they do show a trend. (The variability between the different tongue samples introduces an error, which is almost equal in magnitude to the difference in friction force between the samples.) The

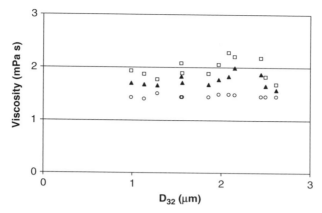

Figure 5 *Viscosity of emulsions stabilized by 1 wt% WPI, containing 10 wt% sunflower oil or 10 wt% solid fat, as function of the mean droplet size d_{32} at three shear-rates: □, 51.3 s^{-1}; ▲, 90.1 s^{-1}; and ○, 1030 s^{-1}*

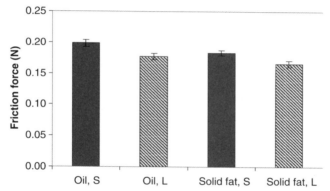

Figure 6 *Friction force measurements of emulsions sheared between pig's tongue and glass. The emulsions contained either 10 wt% sunflower oil, having small (Oil, S) or large (Oil, L) droplets, or 10 wt% solid fat, having small (Solid fat, S) and large (Solid fat, L) droplets. Error bars indicate calculated standard error*

friction force data show that the samples with solid fat give rise to a lower friction force than the samples containing sunflower oil (Figure 6). This indicates that the samples with solid fat are better lubricants than the samples with sunflower oil. The samples with the larger droplets give a lower friction upon shearing. These results are explained by the difference in sensitivity to coalescence and subsequent oil release.

Studies relating friction of emulsions to sensory perception are up until now limited. So far most friction results have been obtained with thickened oil-in-water emulsion systems such as custards. De Wijk and Prinz[15] found an opposite relation between particle size, friction force, and creaminess in their custard systems as compared with our results. However, a system in which emulsion droplets are entrapped in a gelled matrix is likely to behave differently from one having a non-gelled continuous phase. Furthermore, de Wijk and Prinz used artificial surfaces in their shearing experiments, and we have already established that such surfaces differ greatly from the pig's tongue surfaces used in our experiments.

From our combined friction data and sensory data, we conclude that the presence of solid fat instead of sunflower oil in emulsion droplets enhances the perception of creaminess, and that larger droplets give a more creamy sensation and a lower friction force upon shearing.

31.6 Summary

There is a great interest in gaining more knowledge on the relation between fat-related attributes such as creaminess and food emulsion properties. Various characteristics such as viscosity, release of fat flavour, and tribological behaviour are suggested to be relevant for perception. In this study we have focused on studying the ability of tribology to be a tool in understanding emulsion in-mouth behaviour and consequently in understanding the perception of an emulsion. Tribological results are to a large extent determined by the properties of the sliding surfaces. It is therefore crucial to use surfaces with relevant properties. To mimic human oral surfaces we have used pig's tongues, which we characterized in terms of roughness, wetting properties, and deformability. The effects of these surface characteristics on emulsion behaviour were identified by comparison to an artificial surface (PDMS). Large differences in friction forces and coalescence behaviour were observed between measurements conducted with pig's tongue and PDMS. Sensorial analyses confirmed that a difference in the sensitivity of an emulsion to coalescence indeed influences the fat perception of fat-related attributes. Furthermore, emulsions more sensitive to coalescence also gave rise to a lower friction force, indicating that tribology can be one of the tools in understanding emulsion perception provided that mouth-relevant surfaces are used. Nevertheless, the measured differences in friction force between the samples are small, suggesting that tribology data should be supported by other confirmatory methods in order to be a useful tool.

Acknowledgements

The authors thank Franklin Zoet and Jan Klok for their help in, respectively, preparing the samples for the sensory tests and performing the OTC measurements. We thank Monique Vingerhoeds and Rene de Wijk for analysing the sensory data.

References

1. D. Kilcast and S. Clegg, *Food Qual. Pref.*, 2002, **13**, 609.
2. N.J. Richardson-Harman, R. Stevens, S. Walker, J. Gamble, M. Miller and A. McPherson, *Food Qual. Pref.*, 2000, **11**, 239.
3. J.L. Kokini, *J. Food Eng.*, 1987, **6**, 51.
4. P.B. Moore, K. Langley, P.J. Wilde, A. Fillery-Travis and D.J. Mela, *J. Sci. Food Agric.*, 1998, **76**, 469.
5. M.B. Frost, G. Dijksterhuis and M. Martens, *Food Qual. Pref.*, 2001, **12**, 327.
6. M. Akhtar, J. Stenzel, B.S. Murray and E. Dickinson, *Food Hydrocoll.*, 2005, **19**, 521.
7. I.E. de Araujo and E.T. Rolls, *J. Neurosci.*, 2004, **24**, 3086.
8. E.T. Rolls, *Physiol. Behav.*, 2005, **85**, 45.
9. J.V. Verhagen, E.T. Rolls and M. Kadohisa, *J. Neurophysiol.*, 2003, **90**, 1514.
10. C. Yackinous and J.X. Guinard, *J. Food Sci.*, 2000, **65**, 909.
11. S.V. Kirkmeyer and B.J. Tepper, *Chem. Senses*, 2003, **28**, 527.
12. D.M. Dresselhuis, E.H.A. de Hoog, M.A. Cohen Stuart and G.A. van Aken (unpublished).
13. S. Lee, M. Heuberger, P. Rousset and N.D. Spencer, *Tribol. Lett.*, 2004, **16**, 239.
14. M.E. Malone, I.A.M. Appelqvist and I.T. Norton, *Food Hydrocoll.*, 2003, **17**, 763.
15. R.A. de Wijk and J.F. Prinz, *Food Qual. Pref.*, 2005, **16**, 121.
16. H.-J. Butt, K. Graf and M. Kappl (eds), *Physics and Chemistry of Interfaces*, Wiley-VCH, 2003.
17. G.A. van Aken, M.H. Vingerhoeds and E.H.A. de Hoog, in *Food Colloids: Interactions, Microstructure and Processing*, E. Dickinson (ed), Royal Society of Chemistry, Cambridge, 2005, p. 356.
18. M.R. Heath, *Food Qual. Pref.*, 2002, **13**, 453.
19. D.M. Dresselhuis, E.H.A. de Hoog, M.A. Cohen Stuart and G.A. van Aken (submitted for publication).
20. P. Walstra, in *Food Chemistry*, 3rd edn, O.R. Fennema (ed), Marcel Dekker, New York, 1996.

Chapter 32

Saliva-Induced Emulsion Flocculation: Role of Droplet Charge

Erika Silletti,[1,2] Monique H. Vingerhoeds,[1,3] Willem Norde[2,4] and George A. van Aken[1,5]

[1] WAGENINGEN CENTRE FOR FOOD SCIENCE (WCFS), P.O. BOX 557, 6700 AN WAGENINGEN, THE NETHERLANDS
[2] LABORATORY OF PHYSICAL CHEMISTRY AND COLLOID SCIENCE, WAGENINGEN UNIVERSITY, P.O. BOX 8038, 6700 EK WAGENINGEN, THE NETHERLANDS
[3] AGROTECHNOLOGY AND FOOD SCIENCES GROUP B.V., P.O. BOX 17, 6700 AA WAGENINGEN, THE NETHERLANDS
[4] UNIVERSITY MEDICAL CENTRE, UNIVERSITY OF GRONINGEN, P.O. BOX 196, 9700 AD GRONINGEN, THE NETHERLANDS
[5] NIZO FOOD RESEARCH, P.O. BOX 20, 6710 BA EDE, THE NETHERLANDS

32.1 Introduction

Human saliva is a complex heterogeneous biological fluid composed of proteins, electrolytes, small organic compounds and water.[1–3] It is involved in several functions in the oral cavity and is responsible for maintaining oral health and for protection of the teeth and mucosal surfaces.[4] Mixing with saliva plays an important role in facilitating food manipulation and bolus formation in the oral cavity.[5] For example, the enzymatic activity of salivary α-amylase is responsible for the initial breakdown of starch-based foods before swallowing.

The human perception of food products is becoming increasingly important for the food industry in relation to product design and evaluation. Often the perception of foodstuff – as, for example, an emulsion – cannot be directly related to the texture of the product before consumption. Moreover, several authors have reported[6–9] the influence of saliva properties, such as flow, composition and lubrication, on sensory perception and flavour release. It is becoming evident that knowledge on the interaction between saliva and food is of primary importance for understanding the oral processing of food.

Therefore investigations of the dynamic processes occurring in the mouth which affect food structure are needed.

Recently, the physical/chemical effect of saliva on protein-stabilized food emulsions was reported.[10,11] Saliva was shown to induce flocculation in emulsions stabilized by several emulsifiers, including sodium caseinate and whey protein isolate. Depletion flocculation has been indicated to be responsible for droplet flocculation in the emulsion + saliva mixtures. The salivary mucins, among other proteins and protein complexes present in saliva, are thought to be responsible for this behaviour. Mucins are a family of polydispersed molecules which are the main constituents of the slime layers that cover the mucous epithelia throughout the body. They are composed of disulfide-linked monomers that contain heavily glycosylated domains carrying a net negative charge due to the presence of sialic acid residues and sulfate groups.

In this chapter, we provide an overview of the role of emulsion droplet charge on saliva-induced flocculation, meant as a first clue to explain the oral behaviour of emulsions. The results are described in more detail elsewhere.[12] The second part of this chapter aims to gain insight into the flocculation mechanism of positively charged emulsions. Lysozyme was chosen as a model for studying the interaction between saliva and positively charged emulsifiers both in solution and adsorbed on the emulsions droplets. We suggest that complex formation between lysozyme and saliva plays a major role in the flocculation of positively charged emulsions upon mixing with saliva.[13]

32.2 Materials and Methods

32.2.1 Materials

Freeze-dried β-lactoglobulin (β-lg) purified according to the previously described method[14] was provided by Wageningen Centre for Food Sciences (WCFS, Wageningen, the Netherlands). Lysozyme from chicken egg white (L6876) was obtained from Sigma-Aldrich Chemie (Zwijndrecht, the Netherlands) and used without further purification. Polyoxyethylene sorbitan monolaurate (Tween 20) was purchased from Quest International (Zwijndrecht, the Netherlands), sodium dodecylsulfate (SDS) from Merck (Schuchardt, Germany), and cetyltrimethylammonium bromide (CTAB) from BDH Chemicals (Poole, England). Sunflower oil (Reddy, Vandemoortele, the Netherlands) was purchased from a local retailer. The BCA™ Protein Assay Kit was from Pierce Biotechnology (Rockford, IL, USA) and sodium azide was from Merck (Schuchardt, Germany).

32.2.2 Saliva Collection and Handling

Unstimulated whole human saliva was collected for 30 min each from 11 healthy non-medicated volunteers from 8.30 to 10.30 a.m. The collection procedure is extensively described elsewhere.[12] Briefly, after optionally having breakfast and brushing their teeth, donors refrained from eating and drinking, with the exception of water, for 2 h before donation. Saliva was collected with closed lips for a couple of minutes and then expectorated into ice-chilled vessels. The samples were kept constantly on ice during both donation and handling. Saliva was pooled, centrifuged to remove cellular debris, frozen in liquid nitrogen and stored at $-80\,°C$. This pooled saliva supernatant is indicated in the text simply as saliva.

Storage did not affect the pH of the saliva, which ranged from 6.8 to 7.1. Protein content determination of the saliva was performed according to the BCA method,[15] with bovine serum albumin as standard. The determined protein content varied from 1.2 to 1.6 mg mL^{-1}.

32.2.3 Preparation of Emulsions and their Mixtures with Saliva

The surfactants SDS, Tween 20 and CTAB were dissolved in demineralized water for 2 h at room temperature. The proteins β-lg and lysozyme were dissolved overnight at 4°C in demineralized water and in 10 mM NaCl, respectively. Pre-emulsions were prepared using an Ultra-Turrax T25 Basic (IKA-Werke) and subsequently homogenized at room temperature with 10 passes through a Delta Lab-scale homogenizer (Delta Instruments) operating at a pressure of 70 bar or (in the case of the lysozyme emulsion) 100 bar. The different pressure was used to obtain similar mean droplet sizes. Oil-in-water emulsions stabilized by SDS, Tween 20, CTAB and β-lg at pH = 6.7 contained 40 wt% sunflower oil and 1 wt% emulsifier. The β-lg emulsion at pH = 3.0 was prepared by lowering the pH after homogenization. The lysozyme emulsion (pH = 6.7, 1 wt% protein, 10 mM NaCl) contained 20 wt% sunflower oil.

Droplet-size distributions were obtained as averages of three measurements performed by laser diffraction with the Mastersizer Hydro 2000S (Malvern Instruments, Southborough, UK). Dilution with water was applied during analysis. Zeta potential measurements of emulsions were carried out using an Electroacoustic Spectrometer model DT 1200 (Dispersion Technology, Bedford Hills, NY, USA).

The emulsion + saliva mixtures containing 10 wt% oil phase and 0.6 mg mL^{-1} of salivary proteins were prepared at room temperature by adding saliva to diluted emulsions with 10 mM NaCl in the continuous phase. The pH of the mixtures was around 6.8 for emulsions made at pH = 6.7 and around 4.3 for the β-lg emulsion made at pH = 3.0. Unless specified otherwise, the pH in the text refers to the pH of the emulsions.

32.2.4 Characterization of the Flocculation

Flocculation was studied by light microscopy, particle-size analysis and rheology. Light microscopy images were taken using an Olympus BX 60 microscope equipped with an Olympus DP 70 camera (Olympus Nederland, Zoeterwoude, the Netherlands). Size-distribution analysis of the emulsion + saliva mixtures was conducted by laser diffraction as indicated above. The rheological behaviour of emulsions upon addition of saliva was characterized by measuring the shear-rate dependent viscosity. Shear viscosity measurements were carried out in duplicate at 20°C with 10 min recovery time using a Paar Physica MCR 300 rheometer with cone-and-plate geometry CP75-1 of angle 1° (0.0175 rad) and gap-width 0.05 mm (at the tip). The shear-rate was initially increased from 1 to 1500 s^{-1} and then decreased again to 1 s^{-1}.

32.2.5 Characterization of Complex Formation between Lysozyme and Saliva

For the detection of complex formation between saliva and the positively charged protein emulsifier, a 10 mg mL^{-1} lysozyme protein stock solution was prepared in 10 mM NaCl. The mixture with saliva contained 5 mg mL^{-1} of lysozyme stock solution and 0.6 mg mL^{-1} of salivary proteins. Confocal scanning laser microscopy (CSLM) was performed with Leica TCS SP microscope used in the single-photon fluorescence mode. The set-up was configured with an inverted microscope (model Leica DM IRBE) and an Ar/Kr laser. The objective lens used was a 63X/UV/1.20NA/water immersion/PL APO. Oregon Green (OG) was used to covalently stain the lysozyme. With stirring for 2 h at room temperature, 2.5 mg Oregon Green™ 488 was dissolved in 249.5 mg of dimethylformamide (DMF), and then 4 mg of the OG-DMF was added to 2 g of the sample to be labelled. The sample solution was gently stirred overnight at 4°C avoiding exposure to light. After mixing with saliva, the sample was excited at 488 nm and the emission fluorescence was detected between 500 and 560 nm.

32.3 Results and Discussion

One of the key parameters influencing the saliva-induced emulsion flocculation, which is expected to play a role in the in-mouth perception of the emulsions, is the charge on the emulsion droplets. Therefore the saliva-induced flocculation of emulsions with different droplet charges was examined.[12] A summary of the flocculation behaviour in relation to the ζ-potentials on the emulsions droplets is shown in Table 1.

Microscopy and laser diffraction analysis were carried out to characterize the effect of charge on the morphology and size of the flocs upon mixing an emulsion with saliva. The effect of droplet charge on the surface-weighted mean diameter d_{32} of the emulsions before and after mixing with saliva is shown in Figure 1. While the d_{32} values of the original emulsions varied minimally (between 0.6 and 1 µm), a large variation (from 0.6 to 45 µm) was found for the mixtures with saliva. For negatively charged and neutral emulsions stabilized by SDS, β-lg (pH = 6.7) and Tween 20, the d_{32} value remained unchanged before and after mixing with saliva. Microscopy pictures of the mixtures reveal that for the SDS-stabilized emulsion, with $\zeta = -90$ mV, no flocculation had occurred. But flocculation was observed for the β-lg and Tween 20 emulsions

Table 1 *The ζ-potential values of the different emulsions and the corresponding flocculation behaviour of the emulsions (10 wt% oil) mixed with salivary proteins (0.6 mg mL^{-1})*

Emulsifying agent	pH	ζ (mV)	Type of flocculation
SDS	6.7	−90	None
β-lg (neutral)	6.7	−44	Reversible[b]
Tween 20	6.7	−2	Reversible[b]
β-lg (acidic)	4.3[a]	+21	Irreversible[b]
Lysozyme	6.7	+30	Irreversible[b]
CTAB	6.7	+74	Irreversible[b]

[a] Measured ζ-potential referring to pH of mixture with saliva.
[b] With respect to dilution.

upon mixing with saliva. The light-scattering results indicate that this flocculation behaviour was reversible upon dilution with water, in line with depletion as driving force for flocculation. The saliva protein composition suggests a dominant role of the large negatively charged salivary mucins in the depletion mechanism, as seen for other non-adsorbing biopolymers, *e.g.*, methylcellulose, dextran or exocellular polysaccharide.[10,16–18] In case of the SDS emulsion + saliva mixture, however, the strong electrostatic repulsion between the emulsion droplets presumably overcomes the combined van der Waals attraction and depletion forces. This explains the non-flocculation behaviour, as similarly found by Blijdenstein *et al.*[18] in emulsions containing dextran at concentration lower than 1 wt% and at low ionic strength (NaCl concentration ≤ 10 mM).

For positively charged emulsion droplets, stabilized by β-lg (pH = 3.0), lysozyme and CTAB, strong flocculation, irreversible upon dilution, was observed (Figure 1). This flocculating behaviour resulted in an increase in the d_{32} value of the mixtures with increasing ζ-potential. The microscopy pictures reveal the presence of heterogeneous flocs, containing both spherical and long thread-like structures, easily visible for emulsions with $\zeta = 30$ mV. The large size of these flocs (> 200 µm) was not detected by laser diffraction. In fact, both d_{32} and the particle-size distribution (not shown) did not indicate the presence of such large flocs. It is possible that the flocs seen by microscopy are composed of smaller ones, which fell apart during the particle-size measurements. In addition, both the model used for the calculation of the sizes (the particles are considered as spheres) and the broadness of the distribution would affect the determined size. We hypothesize that complex formation between the emulsifier on the droplet surface and the salivary proteins – for example, the negatively charged mucins – could be responsible for the observed flocculation behaviour.[12]

In Figure 2 the width of the particle-size distribution is depicted as a function of the ζ-potential. The width, expressed as the ratio $(d_{0.9} - d_{0.1})/d_{0.5}$, where $d_{0.1}$, $d_{0.5}$ and $d_{0.9}$ are the sizes for 10, 50 and 90% of the sample, is compared for the emulsions before and after saliva addition. In line with expectations, the

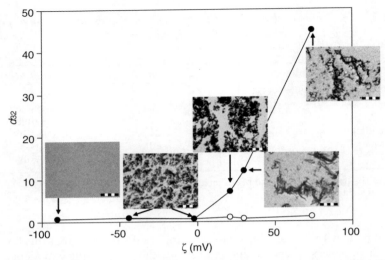

Figure 1 *Effective mean droplet diameter d_{32} values of the emulsions ($\bigcirc$) and the emulsion + saliva mixtures ($\bullet$) as a function of the ζ-potential. Also shown are some light microscopy pictures of the corresponding mixtures. Scale bar = 200 μm*

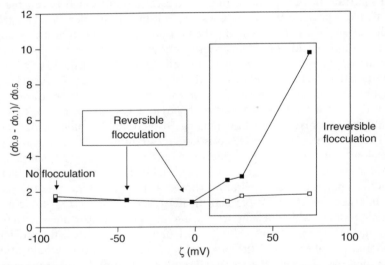

Figure 2 *The width of the particle-size distribution, defined as $(d_{0.9} - d_{0.1})/d_{0.5}$, for the differently charged emulsions before mixing ($\square$) and after mixing ($\blacksquare$) with saliva. Existence and reversibility of flocculation is indicated*

distribution width showed no changes upon saliva addition for emulsions stabilized with SDS, β-lg (pH = 6.7) and Tween 20 ($\zeta < 0$ mV). On the other hand, for positively charged emulsion droplets stabilized with β-lg (pH = 3.0), lysozyme and CTAB, the width of the distribution increased after mixing with saliva. The increased polydispersity of the above mentioned emulsion + saliva mixtures also affects the d_{32} values. This, together with the previously mentioned considerations, might explain why the measured d_{32} values (e.g., 45 μm for the CTAB-stabilized emulsion + saliva mixture) are far from the sizes observed by microscopy (> 200 μm).

Flocculation affects the viscosity of emulsions, typically leading to shear-thinning behaviour.[19,20] Viscosity measurements can therefore provide useful information about the state of flocculation and hence about the flocculation mechanism. Higher viscosities, even at high shear-rates, are obtained in samples where droplets are strongly and irreversibly bound to each other. Conversely, aggregates in emulsions flocculated by depletion fall apart at high shear-rates, resulting in similar viscosities to non-flocculated emulsions. A convenient way to evaluate different flocculating emulsion + saliva mixtures is to calculate the ratio η_{mix}/η_{emul} in which the viscosity of the mixture (η_{mix}) is normalized by the viscosity of the corresponding emulsion (η_{emul}). In this way, the viscosity increase due to flocculation caused by saliva is independent of the original emulsion viscosity. As expected, emulsion + saliva mixtures exhibited shear-thinning behaviour (Figure 3). Higher positive ζ-potential values of the emulsions resulted in a stronger increase in the viscosity after saliva addition, as indicated by the greater values of η_{mix}/η_{emul}. The results indicate that the flocs of positively charged emulsions in the presence of saliva are not completely disrupted by shear over the applied shear-rate range, which is in line with the irreversibility of the flocculation towards dilution (Figures 1 and 2).

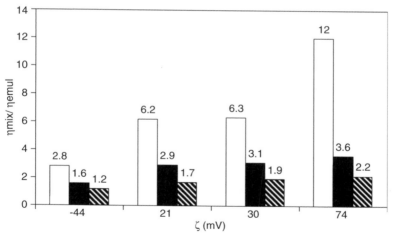

Figure 3 *Viscosity of the emulsion + saliva mixtures normalized by the viscosity of the emulsions calculated at 10 s^{-1} (white columns), 100 s^{-1} (black columns) and 900 s^{-1} (shaded columns) for different ζ-potentials: −44 mV for β-lg (pH = 6.7); +21 mV for β-lg (pH = 4.3) (as in the mixture of saliva); +30 mV for lysozyme; and +74 mV for CTAB*

The above results underline the important role of electrostatics in causing the enhanced viscosity of the positively charged emulsion + saliva mixtures. These combined findings seem to indicate the presence of an internal structure where the droplets are strongly bound together. Two mechanisms, which might occur simultaneously, are proposed to explain the droplet flocculation. Firstly, the droplets can be connected to each other *via* the binding of salivary protein to the emulsifier adsorbed on the droplet surfaces. This mechanism is known as bridging flocculation. It has been described for other food-related systems, *e.g.*, β-lg emulsions + carboxymethylcellulose at pH = 3,[21] and bovine serum albumin emulsions + ι-carrageenan.[22] Secondly, a network of salivary proteins and emulsifiers in the continuous phase could be formed, thereby entrapping the emulsion droplets. Since the strongly negatively charged mucins constitute up to 30% of the proteins in unstimulated saliva, a complex between these salivary proteins and the adsorbed emulsifier layer on the droplet surface or in the continuous phase is likely to be readily formed. This hypothesis is supported by the previously described[23] strong interaction between pig gastric mucins and polymer solutions of cationic gelatin and chitosan at pH = 5.5.

To support the hypothesis that salivary components directly interact with positively charged polymers adsorbed to the droplet surfaces, we have studied by CSLM the effect of saliva addition to a lysozyme protein solution. Lysozyme was chosen as the model component for the study because it could easily be covalently labelled with the fluorescent dye Oregon Green. Furthermore, the mixing of the lysozyme protein solution with saliva did not change the pH from the saliva's physiological value, as did occur on mixing with β-lg emulsions made at pH = 3.0.

Macroscopic pictures were taken of saliva and its mixture with the lysozyme protein solution. Whereas saliva is a transparent liquid, the mixture was turbid due to the formation of complexes, which were clearly visible with confocal microscopy (Figure 4). This confirms that complex formation can indeed occur between lysozyme and salivary components in solution. It is also feasible that this interaction could also occur at the oil–water interface between saliva and lysozyme (as well as with other positively charged emulsifying agents).

Most food proteins form complexes with anionic hydrocolloids in the pH region where the two macromolecules carry opposite net charge.[24] Complexes of whey protein, β-lg and lysozyme with gum arabic and/or carrageenan, as well as other mixed food biopolymer systems, have been extensively investigated.[25–28] In these systems electrostatic attraction is considered to be the main driving force for complexation. Our findings suggest that a similar complexation mechanism takes place involving saliva and various food components.

32.4 Conclusions

This study has aimed to improve understanding of the flocculation behaviour of emulsions as induced by saliva. We have focused on the role of the charge on the emulsion droplets and the contribution of electrostatics to the flocculation

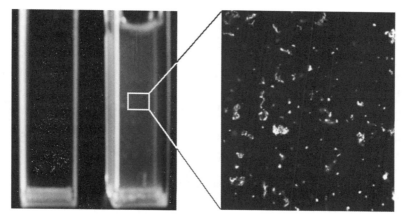

Figure 4 *Saliva (0.6 mg mL^{-1}) before mixing (left cuvette) and after mixing (right cuvette) with protein (5 mg mL^{-1} lysozyme solution) at pH = 7. On the extreme right is a CSLM image (79 × 79 μm) showing complex formation between saliva and OG-lysozyme*

mechanism of emulsion + saliva mixtures. Different scenarios have been shown to occur depending upon the measured ζ-potential of the droplets. No flocculation was found for highly negatively charged emulsions, which is ascribed to electrostatic repulsion preventing close approach of the droplets. Reversible flocculation was found for weakly negatively charged and neutral emulsion droplets, probably induced by salivary mucins *via* depletion interactions. Instead, for positively charged emulsion droplets, we propose a bridging mechanism induced by electrostatic attraction between salivary proteins and the surface of the droplets. Furthermore, we have demonstrated that saliva can form complexes in solution with a positively charged polymer such as lysozyme, and this is likely to occur in the continuous phase of emulsions as well. It reinforces the idea that the electrostatic interaction between salivary components – for example, the negatively charged mucins – and positively charged emulsifiers, both in the adsorbed state and in the continuous phase, can have an influence on emulsion behaviour in the oral cavity.

References

1. A. Zalewska, K. Zwierz, K. Zólkowski and A. Gindzienski, *Acta Biochim. Pol.*, 2000, **47**, 1067.
2. A. van Nieuw Amerongen, E.C.I. Veerman and A. Vissink, *Speeksel, speekselklieren en mondgezondheid*, Bohn Stafleu Van Loghum, Houten, 2004, p. 384.
3. S.P. Humphrey and R.T. Williamson, *J. Prosthet. Dent.*, 2001, **85**, 162.
4. A. van Nieuw Amerongen and E.C.I. Veerman, *Oral Dis.*, 2002, **8**, 12.

5. J.F. Prinz and R.A. de Wijk, in *Flavour Perception*, A.J. Taylor and D.D. Roberts (eds), Blackwell, London, 2004, p. 40.
6. M.E. Malone, I.A.M. Appelqvist and I.T. Norton, *Food Hydrocoll.*, 2003, **17**, 763.
7. M.E. Malone, I.A.M. Appelqvist and I.T. Norton, *Food Hydrocoll.*, 2003, **17**, 775.
8. L. Engelen, R.A. de Wijk, J.F. Prinz and F. Bosman, *Physiol. Behav.*, 2003, **78**, 165.
9. K.B. de Roos, *Int. Dairy J.*, 2003, **13**, 593.
10. M.H. Vingerhoeds, T.B.J. Blijdenstein, F. Zoet and G.A. van Aken, *Food Hydrocoll.*, 2005, **19**, 915.
11. G.A. van Aken, M.H. Vingerhoeds and E.H.A. de Hoog, in *Food Colloids: Interactions, Microstructure and Processing*, E. Dickinson (ed), Royal Society of Chemistry, Cambridge, 2005, p. 356.
12. E. Silletti, M.H. Vingerhoeds, G.A. van Aken and W. Norde, *Food Hydrocoll.*, in press.
13. E. Silletti, M.H. Vingerhoeds, G.A. van Aken and W. Norde (manuscript in preparation).
14. H.H.J. de Jongh, T. Gröneveld and J. de Groot, *J. Dairy Sci.*, 2001, **84**, 562.
15. P.K. Smith, R.I. Krohn, G.T. Hermanson, A.K. Mallia, F.H. Gartner, M.D. Provenzano, E.K. Fujimoto, N.M. Goeke, B.J. Olson and D.C. Klenk, *Anal. Biochem.*, 1985, **150**, 76.
16. R. Tuinier and C.G. de Kruif, *J. Colloid Interface Sci.*, 1999, **218**, 201.
17. C.K. Reiffers-Magnani, J.L. Cuq and H.J. Watzke, *Food Hydrocoll.*, 2000, **14**, 521.
18. T.B.J. Blijdenstein, W.P.G. Hendriks, E. van der Linden, T. van Vliet and G.A. van Aken, *Langmuir*, 2003, **19**, 6657.
19. E. Dickinson and M. Golding, *Colloids Surf. A*, 1998, **144**, 167.
20. E. Dickinson and M. Golding, *J. Colloid Interface Sci.*, 1997, **191**, 166.
21. T.B.J. Blijdenstein, A.J.M. van Winden, T. van Vliet, E. van der Linden and G.A. van Aken, *Colloids Surf. A*, 2004, **245**, 41.
22. E. Dickinson and K. Pawlowsky, *J. Agric. Food Chem.*, 1997, **45**, 3799.
23. E.E. Hassan and J.M. Gallo, *Pharm. Res.*, 1990, **7**, 491.
24. E. Dickinson, *Food Hydrocoll.*, 2003, **17**, 25.
25. F. Weinbreck, R. de Vries, P. Schrooyen and C.G. de Kruif, *Biomacromolecules*, 2003, **4**, 293.
26. F. Weinbreck, H. Nieuwenhuijse, G.W. Robijn and C.G. de Kruif, *J. Agric. Food Chem.*, 2004, **52**, 3550.
27. C. Schmitt, C. Sanchez, A. Lamprecht, D. Renard, C.M. Lehr, C.G. de Kruif and J. Hardy, *Colloids Surf. B*, 2001, **20**, 267.
28. F. Weinbreck and C.G. de Kruif, in *Food Colloids, Biopolymers and Materials*, E. Dickinson and T. van Vliet (eds), Royal Society of Chemistry, Cambridge, 2003, p. 337.

Chapter 33

Surface Topography of Heat-Set Whey Protein Gels: Effects of Added Salt and Xanthan Gum

Jianshe Chen, Eric Dickinson, Thomas Moschakis and Kooshan Nayebzadeh

PROCTER DEPARTMENT OF FOOD SCIENCE, UNIVERSITY OF LEEDS, LEEDS LS2 9JT, UK

33.1 Introduction

Surface topographic properties have a huge impact on the appearance and physical characteristics of food (and non-food) materials. The visual appearance and surface texture of a food product are primarily determined by its surface roughness, its surface moistness, and its surface oiliness, as well as by its geometric shape and colour. The oral perception of smoothness, grittiness, and juiciness of the food is greatly influenced by the contact that the food makes with the surface of the tongue and palate.

The surface of a man-made material can be created by either a physical process or a chemical process. In the former case, a mechanical force is applied to remove the external layer of the material and so to create a surface with designed characteristics (roughness/smoothness, glossiness, *etc.*). Materials in this category are often hard and non-deformable, such as steel, glass, wood, *etc.* The tribological properties and characterization of such materials have been studied extensively and are well reported in the literature.[1] Surface creation by a chemical process involves those cases where the surface of the material is created by assembly of molecules or colloidal particles through mechanisms of surface adsorption, deposition, interaction, or aggregation. The properties of such a surface are influenced not only by the chemical composition at the surface, but also by the interactions between the surface components. Assembled particle systems have received considerable attention in recent years with the growing interest in nanoparticle materials. But so far there has been limited research on the surface creation of assembled particle systems.

The lack of knowledge about the surface creation of assembled particle systems is largely due to the lack of suitable techniques for their surface characterization. Self-assembled particle systems are often mechanically weak, and so a small surface load may cause significant network deformation and surface distortion. Additionally, the presence of surface moisture makes some surface characterization techniques (*e.g.*, scanning electron microscopy) not feasible for these systems. The contribution of the associated hydrodynamic flow of surface water also makes it more complicated for such systems in interpreting results from traditional tribology techniques. Nevertheless, there have been reported successes of using classical surface friction techniques on surface investigation of wet polymer gels. For instance, Gong and Osada[2,3] have characterized the surface properties of a number of synthetic polymer gels by measuring surface friction as a function of surface load and sliding speed. And Baumberger *et al.*[4] have successfully used a modified tribometer to assess the surface slipperiness properties of gelatin gels.

Recently, we developed a simple surface friction device and used it for surface characterization of wet protein particle gels.[5] The technique gave effective and reliable differentiation of the surfaces of heat-set whey protein gels. Complementary confocal laser scanning microscopy (CLSM) was used[6] to establish characteristic topographical features of the same protein particle gels.

Particle gels differ significantly from polymer gels both in terms of the gelation mechanisms and in their mechanical properties.[7,8] A particle gel is a network of particle strings or clusters formed as a result of the interaction and aggregation of colloidal particles, while a polymer gel arises from the cross-linking of polymer chains at high enough concentration. A polymer gel typically has a transparent visual appearance with a relatively large linear regime (up to several hundred percent strain) and shear-hardening mechanical behaviour.[7] A particle gel may be visually opaque or rather translucent depending on the size of the individual particles and the particle clusters. Mechanically, a particle gel has a much shorter linear regime, and it often fractures at small deformations (a few percent).[8] The microstructural features of some protein particle gels have been thoroughly investigated in relation to their mechanical and physical properties.[9] The aim of the present work is to understand the surface characteristics of these particle gels. The model system studied in this investigation is a heat-set whey protein gel with salt and xanthan gum added to control the surface texture and microstructure.

33.2 Materials and Methods

Commercial whey protein isolate (WPI) (Lacprodan DI-9224) was provided by Arla Foods (Videbaek, Denmark). The product contained > 93.5 wt% protein, < 0.2 wt% fat, <0.2 wt% lactose, and < 6 wt% moisture. Sodium chloride and polysaccharide xanthan gum (99% purity) were purchased from Sigma-Aldrich (UK). Milli-Q (Millipore, Bedford) purified water was used for the preparation of all solutions.

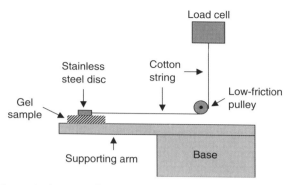

Figure 1 *A schematic diagram showing the experimental arrangement for the surface friction measurements. The friction force is recorded by the load cell and the sliding speed and distance are controlled by a stepper motor connected to the load cell*

Protein solutions (14 wt %) were prepared with or without addition of salt (0.2 M NaCl). The required amount of xanthan was added to the protein solution using a stock xanthan solution. A certain volume of protein solution (50 ml of protein solution for the salt effect experiments, and 30 ml of degassed protein solution for the xanthan effect experiments) was carefully transferred into an aluminium dish (diameter 81 mm, height 11 mm). Each dish was covered with a lid to stop moisture loss and was transferred to a 90°C waterbath and left for 1 h in order to produce a firm gel. The level of the dish was carefully checked using a spirit-levelling device.[5] The xanthan was labelled covalently with a fluorescent dye (FITC) following the procedure given by Tromp *et al.*[10]

The experimental arrangement for the surface friction measurements is shown in Figure 1. The friction force between the gel and a stainless steel disc was measured as a function of sliding speed of the disc (from 10 μm s^{-1} to 10 mm s^{-1}) and as a function of surface load (up to 600 Pa).[5,11] The surface microstructure was observed using a Leica CLSM (Leica Microsystems, Germany) and a Philips XL30 environmental scanning electron microscope (ESEM). The procedures adopted for the microstructural observations and image analysis can be found elsewhere.[6]

33.3 Results and Discussion

33.3.1 Effect of Added Salt

Figure 2 shows the surface friction forces for heat-set WPI gels containing no salt and 200 mM NaCl. Large differences in the surface characteristics of the two gels are apparent. Figure 2(a) shows surface friction as a function of sliding speed, and Figure 2(b) shows the surface friction as a function of surface load. We can see that the presence of added salt during heat-induced gelation makes

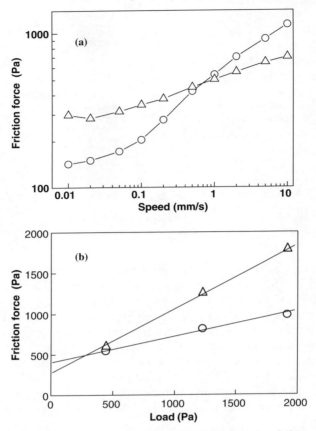

Figure 2 Surface friction force as a function of (a) sliding speed and (b) surface load for heat-set gels: ○, no added salt; △, 200 mM NaCl

the surface friction of the WPI gel much less speed-dependent but much more load-dependent. The surface friction of the protein gel containing no added salt was found to increase by almost 10 times as the sliding speed increased from 0.01 to 10 mm s^{-1}, while the friction force of the gel containing 200 mM NaCl gave a much smaller increase over the same range of sliding speeds. The friction forces for both gels show a linear dependence on the surface load. However, the slope for the salt-containing gel is $2\frac{1}{2}$ times higher than that for the protein gel containing no salt.

Classical friction theory interprets the surface friction as the work required to overcome the energy barrier caused by surface asperities. The magnitude of the energy barrier depends on the roughness of the surface and on the load. Therefore, the friction force for every surface has a characteristic relationship with the surface load. The rougher the surface, the greater is the load dependence. The results in Figure 2(b) suggest a smooth surface for the heat-set WPI gel without salt, but a relatively rough surface for the gel containing 200 mM

NaCl. Also, according to the classical friction theory, the friction force should remain constant at various sliding speeds (if no acceleration is involved) and it should be zero for zero surface load. However, the experimental data in Figure 2 contradict these predictions: the friction force increases with increasing sliding speed and there is a non-zero friction force at the extrapolated zero surface load for both protein gels. This contradiction can be attributed to the existence of surface water and the presence of adsorbing polymer chains.[2] For wet protein gels, it has been shown[11] that contributions to surface friction can be attributed to three different sources: surface roughness, surface hydrodynamic flow, and surface adsorption of polymer chains. The contributions from hydrodynamic flow and polymer chain adsorption can be clearly seen from the friction force and sliding speed relationships in Figure 2(a). The contribution of polymer chain adsorption increases with the sliding speed and reaches a maximum at some medium speed. At very high sliding speeds, this contribution will become negligible because there is not enough time for polymer chains to adsorb to the substrate.[2] In this case, the friction force will show a linear dependence on the sliding speed. Therefore, in theory, it is possible to estimate the thickness of surface water based on the speed-dependence of surface friction when polymer chain adsorption becomes minimal. Unfortunately, the maximum speed of our device was only 40 mm s^{-1}; but we have noticed that even at this speed the gel sample tends to deform because of the high level of friction and its relatively weak mechanical strength. Examining the friction force data in Figure 2(a), we do not see a linear relationship between force and sliding speed. We conclude that the contribution from hydrodynamic flow is still not large enough to dominate. But based on the fact that the salt-containing gel has lower speed-dependence (Figure 2(a)), one can infer that this gel has a thicker layer of surface water. This agrees with the load-dependence test results: a rough surface would have more void spaces to hold water.

Figure 3 shows CLSM surface images for the two protein gels with (a) no salt and (b) 200 mM NaCl. We can see that the former has a much smoother and flatter surface. There is hardly any measurable surface irregularity for this gel. In contrast, the salt-containing gel has a rough surface with numerous clearly identifiable peaks and valleys. It is plausible to assume that the peaks are composed of protein aggregates and the valleys are the void spaces containing surface water. The CLSM images have been further examined for surface roughness estimation. Figure 4 shows surface profiles of the two protein gels. The surface roughness has been quantified in terms of the root-mean-square roughness, R_q. The gel with no salt addition has $R_q \approx 0.2$ μm, but the gel containing salt has $R_q \approx 2.4$ μm, *i.e.*, more than 10 times larger.

It is well known[11] that the presence of salt leads to a coarser microstructure and weaker mechanical strength for heat-set WPI gels. Our observation of a much rougher surface for the gel containing 0.2 M NaCl seems consistent with existing observations of bulk microstructure. However, it is worth noting that the peaks and valleys at the surface of the gel in Figure 3(b) appear to be oriented towards the surface. While this directional preference for the protein aggregates at the surface has not been observed in the bulk microstructure of whey protein gels,[9] it is unclear yet what causes this surface anisotropy. One possible

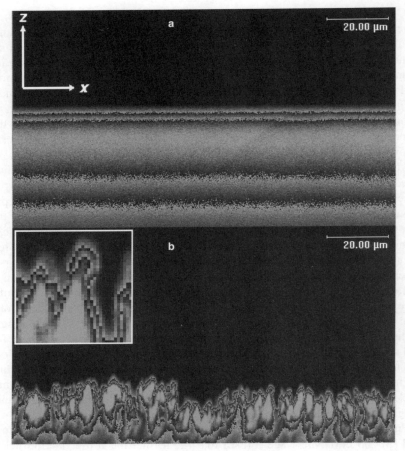

Figure 3 *CLSM images of the surface of heat-set whey protein gels: (a) no added salt; (b) 200 mM NaCl. The inset in (b) is an enlargement of the tips of the surface asperities*

explanation is surface evaporation. The upward mass transport of water may cause protein aggregates to become oriented. But this seems unlikely, considering the fact that the gelation was carried out in well-sealed dishes with a small head space. Another possible explanation would be a different assembly mechanism of the protein aggregates at the surface. While protein clusters may diffuse and aggregate in any direction in the bulk phase, protein aggregation at the surface is only possible in lateral directions and from beneath.

33.3.2 Effect of Added Xanthan Gum

The presence of a small amount of the hydrocolloid, xanthan gum, produces a dramatic effects on the topographical features of the heat-set whey protein gels.

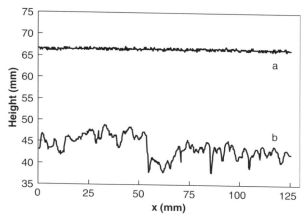

Figure 4 *Surface height profiles (in z-direction) determined from CLSM images as function of position in x-direction for heat-set whey protein gels. (a) No added salt; (b) 200 mM NaCl*

Figure 5 shows (a) the speed dependence and (b) the load dependence of the surface friction for whey protein gels (14 wt% protein, 200 mM NaCl) containing different amounts of xanthan gum (0, 0.05 and 0.25 wt%). It is noticeable that the addition of the polysaccharide causes a lowering of surface friction over the whole range of sliding speed. The friction force for the gel containing 0.25 wt% xanthan has the highest speed-dependence over the tested speed range; this can probably be attributed to the additional contribution of adsorption/desorption of xanthan chains at the surface. However, the gel containing no xanthan becomes highly speed-dependent at high sliding speeds (> 1 mm s^{-1}), indicating probably the great importance of hydrodynamic flow at the surface under these conditions. The load-dependence of the surface friction becomes reduced in the presence of xanthan. These data suggest that the xanthan has a significant 'surface smoothing' effect on the heat-set whey protein gels.

The surface smoothing effect of the added hydrocolloid is also clearly evident from the 3-D surface images obtained by CLSM observation. Figure 6 shows protein-stained surface profiles for gels containing (a) no added xanthan, (b) 0.05 wt% xanthan, and (c) 0.25 wt% xanthan. It should be noted that these images have the same length scales for the *x*- and *y*-coordinates, but slightly different magnifications in the *z*-direction. The surface of the gel containing no xanthan appears to be the most rough, with high peaks and deep valleys spread out over the whole surface. In contrast, the surface for the gel containing 0.25 wt% xanthan appears much smoother. The calculated values of the average surface roughness, R_q, are (a) 3.8, (b) 3.0, and (c) 1.5 μm. This estimated surface roughness is based on the surface profile of just the protein aggregates, *i.e.*, with the distribution of the polysaccharide not included in the calculation. Even so, it is obvious that the presence of the xanthan causes a dampening effect on protein aggregate morphology at the surface, thereby leading to a smoother surface.

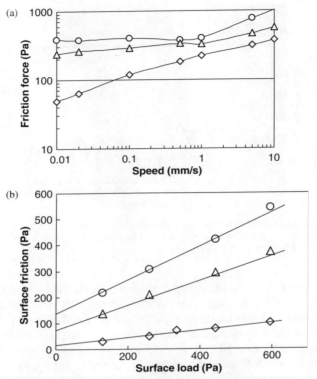

Figure 5 *Surface friction as a function of (a) sliding speed (at constant surface load of 482.5 Pa) and (b) surface load (at a constant sliding speed 1 mm s^{-1}) for heat-set whey protein gels containing different amounts of xanthan:* ○, *none;* △, *0.05 wt%;* ◇, *0.25 wt%*

Separate CLSM observations with both the protein and the polysaccharide labelled separately have also been made to examine the surface distribution of the xanthan. Figure 7 shows the CLSM images of the surface of the gel containing 0.05 wt% xanthan. In Figure 7(a) we can see the polysaccharide distribution is oriented towards the surface in the form of highly concentrated stripes. Peaks of protein aggregates are clearly identifiable in Figure 7(b), with the open valleys appearing just at the positions where xanthan stripes are located. The combination of Figures 7(a) and 7(b) gives the dual-stained image shown in Figure 7(c). Now we clearly see a rather dense surface layer with segregated protein and xanthan regions. It appears that the outer part of the hydrocolloid regions protrudes slightly further out from the surface than does the protein.

Figure 8 shows the ESEM surface images for the gel containing (a) no xanthan and (b) 0.05 wt% xanthan. We observe completely different topographical microstructure for the two gels. For the gel with no xanthan addition, its surface layer was composed of rather regularly spaced protein aggregates separated by holes and void spaces (Figure 8(a)). We presume that these holes and void spaces extend well into the bulk phase of the gel and that they

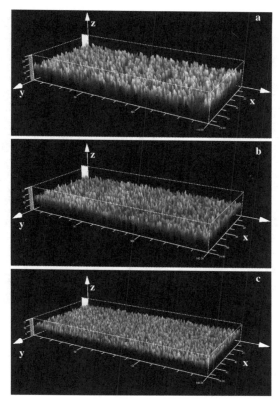

Figure 6 *Topographic CLSM images of protein distribution at the surface of heat-set whey protein gels containing different amounts of xanthan: (a) none, (b) 0.05 wt%, and (c) 0.25 wt%. The scales for the x- and y-coordinates are 0–150 µm and 0–75 µm for all three images. The scales for the z-coordinates are (a) 0–22.5 µm, (b) 0–20 µm, and (c) 0–15 µm*

represent surface water. In contrast, the gel containing 0.05 % xanthan appeared smooth and almost fully covered by a layer of polymeric material (Figure 8(b)), with no observable holes across the whole surface. This picture is consistent with the one shown by CLSM in Figure 7(c). It seems reasonable to believe, therefore, that a very thin layer of xanthan covers the outer surface of the heat-set WPI gel. This polysaccharide layer closely 'seals off' the surface and so makes the biopolymer gel surface much smoother. We should point out that the presence of such a low xanthan content has no significant effect on the bulk microstructure of heat-set whey protein gels. That is, ESEM images indicated[12] little difference between the microstructures of a bulk gel without added polysaccharide and a bulk gel containing 0.05 wt% xanthan.

The effects of xanthan on heat-set WPI gels have been previously reported in terms of the mechanical properties and the microstructure of the bulk gel state. Byrant and McClements[13] showed that, in the presence of salt, mixtures of xanthan + heat-denatured whey proteins are thermodynamically incompatible,

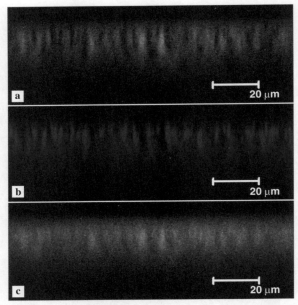

Figure 7 *Surface profile of the heat-set whey protein gel containing 0.05 wt% xanthan by CLSM observation: (a) xanthan stain, (b) protein stain, and (c) dual stained*

leading to xanthan- and protein-rich regions. Turgeon and Beaulieu[14] also observed phase incompatibility between whey protein and polysaccharides and postulated that such incompatibility could lead to spherical inclusion of whey protein by polysaccharides. These observations seem broadly consistent with our observations of the surface distribution profiles of whey protein and xanthan in Figure 7: the protein and hydrocolloid appear to reside in separate local unmixed domains at the surface. It would seem reasonable that one consequence of thermodynamic incompatibility between protein aggregates and xanthan would be exclusion of the latter from the aggregated protein network into an outer polysaccharide-rich layer.

33.4 Concluding Remarks

This work has demonstrated that the topographical properties of heat-set WPI gels are strongly influenced by the addition of either salt or hydrocolloid. Large differences can be clearly seen from both surface friction force measurements and microstructural observations. The presence of 0.2 M NaCl makes the protein gel surface much rougher and wetter, which is consistent with the coarse microstructure of the bulk gel. However, the anisotropy in the orientation of the protein aggregates at the surface would seem to indicate a different mechanism of surface creation from that of bulk network formation.

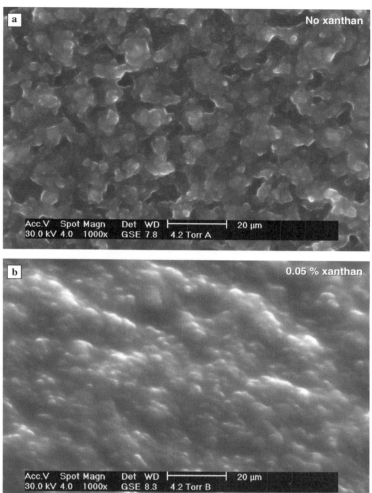

Figure 8 *ESEM surface images of heat-set whey protein gels containing (a) no xanthan and (b) 0.05 wt% xanthan. The scale bar represents 20 μm*

The presence of a very small concentration of xanthan gum makes the protein particle gel surface much smoother. This surface smoothing effect appears to be reasonably consistent with other microstructure observations previously reported in the literature. Although surface exclusion or surface deposition of incompatible xanthan could be a possible mechanism for this smoothing behaviour, further detailed investigation is needed to prove this hypothesis definitively. A general conclusion from this investigation is that it should be possible to produce protein particle gels with surface mechanical properties and functionalities designed for specific applications.

Acknowledgements

J.C. thanks Mr. Phillip Nelson for the construction of surface friction device and Dr. Luis Pugnaloni for useful discussions.

References

1. N.K. Myshkin, A.Y. Grigoriev, S.A. Chizhik, K.Y. Choi and M.I. Petrokovets, *Wear*, 2003, **254**, 1001.
2. J. Gong and Y. Osada, *J. Chem. Phys.*, 1998, **109**, 8062.
3. J. Gong, G. Kagata, Y. Iwasaki and Y. Osada, *Wear*, 2001, **251**, 1183.
4. T. Baumberger, C. Caroli and O. Ronsin, *Euro. Phys. J. E*, 2003, **11**, 85.
5. J. Chen, T. Moschakis and P. Nelson, *J. Texture Stud.*, 2004, **32**, 493.
6. J. Chen, T. Moschakis and L.A. Pugnaloni, *Food Hydrocoll.*, 2006, **20**, 468.
7. A.H. Clark, in *Food Polymers, Gels and Colloids*, E. Dickinson (ed), Royal Society of Chemistry, Cambridge, 1991, p. 322.
8. J.A. Yanez, E. Laarz and L. Bergstrom, *J. Colloid Interface Sci.*, 1999, **209**, 162.
9. M. Langton and A.-M. Hermansson, *Food Hydrocoll.*, 1996, **10**, 179.
10. R. Tromp, F. Velde, J. Riel and M. Paques, *Food Res. Int.*, 2001, **34**, 931.
11. M. Verheul and S.P.F.M. Roefs, *J. Agric. Food Chem.*, 1998, **46**, 4909.
12. K. Nayebzadeh, J. Chen, E. Dickinson and T. Moschakis, *Langmuir*, 2006, **22**, 8873.
13. L.L. Lowe, E.A. Foegeding and C.R. Daubert, *Food Hydrocoll.*, 2003, **17**, 515.
14. S.L. Turgeon and M. Beaulieu, *Food Hydrocoll.*, 2001, **15**, 583.

Chapter 34

Mechanisms Determining Crispness and Its Retention in Foods with a Dry Crust

Ton van Vliet,[1,2] Jendo Visser,[1,2] Wim Lichtendonk[1,4] and Hannemieke Luyten[1,3]

[1] WAGENINGEN CENTRE FOR FOOD SCIENCES, P.O. BOX 557, 6700 AN WAGENINGEN, THE NETHERLANDS
[2] WAGENINGEN UNIVERSITY AND RESEARCH CENTRE, P.O. BOX 8129, 6700 EV WAGENINGEN, THE NETHERLANDS
[3] AGROTECHNOLOGY AND FOOD INNOVATIONS, P.O. BOX 17, 6700 AA WAGENINGEN, THE NETHERLANDS
[4] TNO NUTRITION AND FOOD RESEARCH, P.O. BOX 360, 3700 AJ ZEIST, THE NETHERLANDS

34.1 Introduction

Various popular food products consist of a crispy crust and a non-crispy moist inner part. Well-known examples are many bread types and deep-fried battered snacks like fried fish. For the first type the crust has originally the same chemical composition as the rest of the food product, while for the second type the crust stems from a coating with another (composite) material.[1] For both types there are changes in the local composition during the manufacturing process – especially the concentration of water and its distribution. The crust can be characterized as a cellular solid material. After production, the crust loses its crispy behaviour at various rates. This process takes several hours for most types of bread; but for many deep-fried battered snacks it happens much faster, often within half a hour.[2,3]

It is generally accepted that the crispy behaviour of a food is related to its fracture behaviour. The characteristic requires brittle fracture accompanied by acoustic emission[1,4–6] Not all brittle food products are crispy. For instance, hard candies and chocolate bars may fracture in a brittle way; but, though it may differ between countries, they are mostly not called 'crispy'. In other cases products may be considered to be more hard than crispy. To be crispy (or crunchy) the consumer must perceive multiple fracture events and the work of fracture during mastication must be relatively low. These requirements set clear restrictions on the morphology and on the mechanical properties of the solid matrix of crispy foods.[1,7] This will be worked out further below. In general, for a product to be perceived by consumers as crispy, it must fulfil certain requirements at the molecular, mesoscopic and macroscopic scales.[1] (Since the difference between 'crispness' and 'crunchiness' is not well established,[1] we will only use the first term here.) We will discuss the mechanisms acting at the various length-scales. The emphasis will be on the requirements for crispy behaviour at the mesoscopic scale, as well as on processes at this length-scale causing loss of crispness. But also the relevant aspects at molecular and macroscopic scale will be discussed briefly.

34.2 Materials and Methods

Samples of dry biscuit (Knappertje, Verkade) and toasted rusk rolls (echte beschuit, Bolletje) were used as models for dry crispy foods. Both these foods are cellular solids. The products were bought in a local supermarket and stored at $\sim 23°C$ under a relative humidity of $\sim 30\%$. For each experimental session a new package was opened, and the first and last biscuits were discarded. Model deep-fried snack crusts were prepared in a special mould in which a set amount of batter could be deposited on a hydrophobic sieve (mesh size = 200 μm) giving a batter film of 1–2 mm thick above a silica gel with a water activity of 0.8. The whole was deep fried for 2.5 min in oil at $180°C$. It resulted in flat crispy products with a structure similar to the crust of deep-fried 'chicken nuggets'.[8]

Fracture experiments were performed using a Texture Analyser (Stable Micro Systems TA.XT.plus). In one type of experiment the biscuits were fractured using wedge penetration (wedge angle 30°) at various speeds. In another type of experiment, the fracture behaviour and acoustic emission were determined by cutting the rusk rolls at a controlled speed in the range 0.2–40 mm s^{-1} using a thin stainless steel razor blade fitted to the Texture Analyser.[9] In this way, at low cutting speed (<1 mm s^{-1}), the force drops and acoustic emission related to the fracture of individual beams or lamellae forming the cellular sponge structure could be recorded. Analogue data from the Texture Analyser were sent to a Brüel & Kjær Pulse front-end system, where they were converted into a digital signal (65 kHz). All tests were done in an acoustically insulated room, and the motor of the Texture Analyser was acoustically insulated.

Sound emission was recorded using a 1/2-inch, type 4189 free-field Deltatron microphone (Brüel & Kjær, Naerum Denmark) with a frequency band of 6.3 Hz to 20 kHz and a sensitivity of 50 mV Pa^{-1}. The analogue signal was converted to a digital one using the same front-end system as for the force data. Recording, replay and basic signal analysis were performed using Brüel & Kjær pulse Labshop software, and more detailed signal analysis using Brüel & Kjær Sound Quality type 7698 software. Shortening of the time between two sound events was done using Sound Forge (Sony Pictures Digital, USA).

34.3 Mechanism Acting at the Molecular Scale

In the literature the subject of 'crispness' is often exclusively linked to processes acting at the molecular scale. It is clear that, without the right material properties at the molecular scale, a product cannot be crispy. The main requirement for the material properties is related to the observation that the material must exhibit brittle fracture accompanied by acoustic emission. This means solid-like behaviour and high crack-growth speeds. Figure 1 shows results obtained from the slow cutting of a rusk roll.[9]

To generate sound emission the crack growth speed should be above 1/4 to 1/2 of the maximum speed of stress waves in a material, the Rayleigh speed, as defined by[10]

$$v_{\text{Rayleigh}} = \sqrt{\frac{E}{\rho}}, \qquad (1)$$

where E is Young's modulus and ρ the density of the solid material. For most dry crispy foods, the modulus of the solid material is ~ 1–2×10^9 N m^{-2},[11,12] and the density is ~ 1–1.5 kg m^{-3}. This implies a Rayleigh speed of 1000–1500 m s^{-1} and a crack growth speed of 300–400 m s^{-1}.[10] The crack growth speed in the crispy biscuit has been found to be too high to measure by a high-speed camera (>100 m s^{-1}).[2]

Another requirement for a product to be perceived as crispy is that the work of mastication should be relatively low. The product should not be too hard. This implies a relatively low fracture force. Since the modulus (stiffness) of solids is high, a low fracture force can only be achieved if the fracture strain is

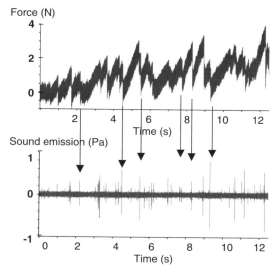

Figure 1 *Measured force and accompanying sound emission as a function of time during cutting of a toasted rusk roll at a speed of 0.4 mm s^{-1}*

low, but this is not enough. The modulus of solids is typically $\sim 10^9$–10^{10} N m^{-2}. Even for a fracture strain below 1%, this implies a fracture stress that is about equal to that of human teeth, which is around 2×10^8 N m^{-2}. To have a fracture stress clearly below that of the teeth, the solid material of the food must contain rather large defects,[13] or it must be built of rather thin beams, struts and films, with a sponge- or foam-like structure. Indeed, most dry crispy products have a sponge-like structure.

The main reason for loss of crispness is a change in properties of the solid matrix as a result of water sorption from the environment or water transport from other parts of the product.[14–16] This process results in an increase in the free volume of the product, allowing a greater mobility of the molecules in the system. The original idea was that loss of crispness, as a result of water acting as a plasticizer, is governed by its effect on the glass transition.[17,18] At a water content below the glass transition, the molecular mobility of the (macromolecular) components responsible for material stiffness is close to zero. Around the glass transition the segments of macromolecules become mobile and start moving randomly. At a still higher water content the (whole) molecules become mobile resulting in the gel-like behaviour of cross-linked polymer systems. The water content at which the glass transition occurs depends strongly upon the temperature. A similar transition as described above for increasing water content may occur with increasing temperature. This transition is normally characterized by its mid-point and is denoted as the glass transition temperature T_g.

Loss of crispness occurs over a water content range and a plot of it against the 'equilibrium' water activity has a sigmoid shape.[19] Relating loss of crispness to the mid-point of the glass transition is not very useful for several reasons. Most food products are composed of several ingredients, each characterized by its own glass transition range. Moreover, it has often been observed that crispness is already affected at water contents below T_g.[20–25] Mobility changes below T_g have also been observed for various enzymes.[26] Another complication is that other components such as small solutes like sugar molecules may act as plasticizers. Moreover, at low concentrations, an increase in the content of water[25,27] or fructose[28] may lead to an increase in the stress required for deformation. These various observations do not fit with the concept that T_g is the only determining factor for a product being crispy or not.

The above-mentioned aspects can be understood quite well by recognizing that fracture consists of two processes:[29,30] fracture initiation and fracture propagation. Fracture initiation occurs when the local stress in the material exceeds the breaking stress of the bonds between the structural elements, giving the solid-like properties to the material. Subsequently, the small crack so formed may grow spontaneously if certain conditions are fulfilled. This so-called fracture propagation condition is governed by the net energy balance:

$$W = W' + W'' + W_f \qquad (2)$$

where W is the amount of energy supplied to the material during deformation, W' the part of the deformation that is elastically stored, W'' the part that is dissipated and W_f is the fracture energy.[1,31] Fracture propagation occurs when the value of W' is so high that the differential energy that is released due to stress relaxation in the material in the vicinity of the formed crack surpasses the differential energy required for crack growth.[31,32] Energy dissipation processes other than that resulting from fracturing (W'') directly affect the amount of energy available for crack growth and hence the speed of crack propagation. Crack growth speed may become high enough to get sound emission if the released energy available for crack growth is sufficiently high, *i.e.*, if energy dissipation is low and the volume in which stress relaxation occurs is large enough. Our hypothesis is that even a small increase in energy dissipation (*e.g.*, due to an increase in water content) will result in (i) an increase in the amount of energy required to get fracture, implying an anti-plasticizing effect of small amounts of water (higher fracture stress); and (ii) a delay of the transition from crack initiation to crack propagation, implying fewer acoustic emission events and maybe another pitch of the emitted sound.

The molecular approach has great benefits in understanding crispy behaviour and the loss of it during storage under deteriorating conditions. However, there are also observations that cannot be explained based solely on a molecular approach. For instance, it is hard to explain in molecular terms why the crust of most deep-fried snacks loses its crispy behaviour much faster than the crust of baked bread-type products (Figure 2). Such differences can be explained, however, by also considering mechanisms determining crispness acting on longer length-scales, *i.e.*, mesoscopic and macroscopic length-scales. Vincent[33,34] and Luyten *et al.*[2,7] point to the importance of the structure at the mesoscopic length-scale for determining the number and size of the force drops and therewith crispness perception. Moreover, oil adsorption by the cellular structure after deep frying of snacks may be an important factor in the fast decrease in crispness of these snacks after frying.[35]

34.4 Mechanism Acting at the Mesoscopic Scale

34.4.1 Estimation of the Structural Length-Scale for Crispness

As mentioned above, for a food product to be perceived as crispy its fracture behaviour should be characterized by multiple fracture and sound events while the work of mastication should be relatively low. To get acoustic emission during fracture of a material, the crack speed has to accelerate from zero to the required speed of 300–400 m s^{-1}. This already sets a minimum for the required length of the crack involved. Beams and lamellae that are thinner than this minimum length will fracture at a low force, but there will be no acoustic emission.

Fracture will start at small defects (cracks or pores) in the beams and lamellae forming the cellular structure. Vincent[33] has reported a defect size in crisps of ~10 µm, which is comparable to the size of the smallest pores in biscuits (Figure 3) and other crisp products. On the other hand, larger pores may cause a growing crack to stop or to slow down temporarily; in any case a

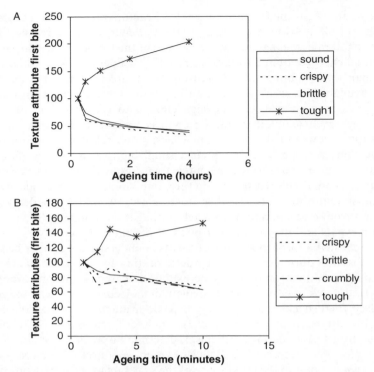

Figure 2 *Sensory texture attributes (indicated) of model bread (A) and of model snack (B) as a function of ageing time*

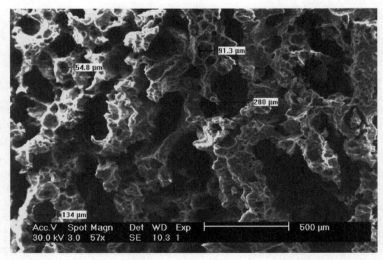

Figure 3 *Scanning electron microscope image showing the cellular structure of a crispy biscuit. (Picture taken by B. Heijns and M. Wevers, KU Leuven, Belgium.)*

pore may lead to a blunting of the crack tip, resulting in a lower local stress concentration. This will involve energy dissipation.

The force–deformation curve of a crispy cellular solid food product has a jagged shape. We propose that the size of the force drops during fracture of a cellular solid food is an important parameter in determining final crispness perception. By using a force and sound sampling rate of 65 kHz, the fracture and accompanying acoustic emission of individual beams and lamellae could be determined in toasted rusk rolls and Cracotte crackers.[8,9] It could be concluded that there is only a very low acoustic emission (or none at all) for a force drop smaller than about 0.3 N; but for force drops larger than 3–6 N, products could be graded as being more hard than crispy. It is likely that the exact numbers will depend upon the product involved, as the boundary between when a product is called primarily 'crispy' or 'hard' will be subjective and dependent upon the product involved and what the consumers are used to. Nevertheless, based on the numbers given above for the fracture force of an individual beam or lamella, it is possible to make a rough estimate of the thickness of the beams involved in an optimum crisp product.[7]

Let us assume that the weakest parts of the structure of the crisp product are square rectangular beams of length $\sim$200–500 µm, fixed at both ends, which are deformed in bending. The relationship between the fracture stress and the beam characteristics is[36]:

$$\sigma_{\text{fracture}} = E\varepsilon = \frac{3}{4}\frac{FL}{bd^2}. \quad (3)$$

Here, σ_{fracture} is the fracture stress, E the Young's modulus of the beam, ε the fracture strain, F the fracture force, L the beam length, b the beam width and d the beam thickness. For $E \approx 2 \times 10^9$ N m^{-2} and $\varepsilon \approx 0.01$,[11,12] the calculated beam thickness is in the range 130–380 µm. Such a thickness is, however, likely to be an overestimate for two reasons. Firstly, in using Equation (3), we neglect any shear contribution to the resistance against bending of the beam, which is not (fully) allowed for in view of the small length/diameter ratio of a large proportion of the beams.[36] Secondly, the assumption that the applied force acts on the middle of the beam is also a rough approximation. If loading would be at one quarter of the end of the beam, the calculated beam thickness would be about 30% less for the same fracture force. So, a better estimate of the lower limit of the beam thicknesses calculated from Equation (3) is probably in the range 50–100 µm. A similar calculation made for the lamellae between gas cells results in similar film thicknesses. Vincent[34] has mentioned a value of 0.05 N as the minimum force drop for a fracture event detectable by humans. Using this number in a calculation according to Equation (3) gives a minimum beam thickness of $\sim$70 µm, in accordance with the range just mentioned.

34.4.2 Need for Crack Stoppers

The conclusion that the crack has to grow at a speed $>$300 m s^{-1} for acoustic emission during the fracture of a crispy product has at first sight unexpected consequences. When eating a crispy product the sound emission usually lasts for a few hundred milliseconds; moreover, several sound pulses are heard. Humans can only hear single sound pulses as individual sound events when the time interval between the pulses is at least 3–5 ms.[37,38] A short passage in piano music is perceived as a single acoustic event when an artificial pause between two parts is shorter than 10–15 ms.[39] For a minimum of three individual sound events to be heard, this implies that the crack must run at high speed for at least 10–15 ms, which implies a minimum size of the fracturing food of 3–5 m. This is, of course, unrealistic. So the conclusion is that, during the eating of a crispy product, the fast growing cracks must be initiated several times at different places, and they must stop after some time. So the presence of 'crack stoppers' is essential for a product to be perceived as crispy.

Large pores in the cellular structure are likely to act as crack stoppers. The minimum size of these crack stoppers must be such that the time interval between two sound pulses is about 5 ms. During the biting of a product, the cracks are initiated by the teeth penetrating the product at a speed of 30–40 mm s^{-1}. The required minimum distance between two points of crack initiation is then $\sim$150–200 µm. This is the combined size of the beam or lamella that is fracturing and the pore size, resulting in a minimum pore size of $\sim$100 µm. Pores of this size are generally present abundantly in crispy foods (*e.g.*, see Figure 3). The thickness of crispy products varies roughly between 2 mm (*e.g.*, the crust of bread, or a battered deep-fried snack) and 10 mm (*e.g.*, roasted bread), implying that crack initiation and stopping will occur several times during biting, even if the crack propagation proceeds along a line perpendicular to the outside.

34.4.3 Need for Overlapping Sound Events

During the fracturing of a single beam, the amplitude of the single sound event is relatively low. For instance, the measured sound pressure of the emitted sound during the fracture of a single beam in toasted rusk roll has been found to range from $\sim$0.2 to $\sim$0.5 Pa (or 80–88 dB) (Figure 1).[9] (Sound pressure in decibels can be calculated from the sound pressure P in Pa using the relation: 1 dB$\equiv$20 log $P/20 \times 10^{-6}$). The sound pressure emitted during biting, or cutting at the speed of biting, of a toasted rusk roll is up to 100 dB.[9] These data show that one needs some overlap of sound signals from individual fracture events in order to reach the sound pressures normally observed during eating of these crispy products.

On average the sound signal emitted during fracture of a single beam lasts for at most 1 ms.[7,9] To produce a sound signal with 10–12 dB the sound pressure should be increased by a factor of 3–4 times, and so there must be 3–4 fracture events per millisecond – in fact, there must be rather more, since sound energy is only high during about half a millisecond. The highest speed during biting at which the teeth approach each other is 30–40 mm s^{-1}. Assuming that a product that is eaten is on average 4 cm wide, we can calculate a maximum average size

Table 1 *Minimum and maximum sizes of the structural elements at the mesoscopic length-scale in a dry cellular crispy food as estimated from sound emission properties and mechanical and physiological constraints*

Structural element	Minimum size (µm)	Reason	Maximum size (µm)	Reason
Beam	50–100	For fast fracture	~400	To prevent too hard texture
Pore	~100	For distinguishable events	450–650	To produce overlap of sound events

for the gas cells that allow us to get 3–6 fracture events per millisecond, taking into account that each gas cell is surrounded by 6–8 lamellae. This rough calculation gives a maximum pore size of ~450–650 µm. Vincent[33] also concluded that fracture events in a crispy food have to overlap in time for the reason that otherwise the force drop on fracture would be too small to be detected during biting. Another aspect that should be taken into account is that the cellular structure of the product is typically heterogeneous. Depending upon the thickness of the beam or lamella involved, the force drop and emitted sound pressure on fracture will vary. Presumably it is only the relatively large events that are noticed during consumption of a crispy product.

An overview is given in Table 1 of the minimum and maximum sizes of the structural elements in a dry cellular crispy food required for an optimum crispy product as estimated from sound emission properties and mechanical and physiological constraints. Of course, the precise numbers will depend to some extent on the particular product involved.

34.4.4 Frequency Spectrum of Emitted Sound

As mentioned above, the emitted sound signal on fracture of a single beam or lamella in a crispy product only lasts for at most 1 ms. A common way to analyse the frequency distribution of sound is by fast Fourier transform (FFT) analysis.[40] In view of the constraint on the duration of a single sound event, there can be no contribution to the signal with frequency below $(1/10^{-3}$ s), *i.e.*, below 1 kHz. A frequency calculation involving low frequencies is still possible if the whole sound signal consists of many exact repetitions of the sound event, so producing a long periodic signal, although this is not precisely the case for the complete sound signal for crispy products. However, for a not too heterogeneous product, probably one can deal with the complete sound signal as a periodic signal. Several researchers (*e.g.*, de Belie *et al.*[41,42]) have indicated that the low frequency part contributes to crispness. The question is which product properties determine the low frequency part of the signal.

At least part of the set of low frequencies calculated by FFT analysis for a long periodic sound signal will be determined by the periodicity, *i.e.*, the distance between the successive signals.[40] We have investigated if this effect also plays a role in determining the low frequency part of the sound signals emitted on fracture of a crispy product.[7] To do this, we have manipulated the measured sound signal recorded during the cutting of a toasted rusk roll at 0.2 mm s^{-1}. The signal contained 30 separated sound events in a time interval of 4.5 s. Random noise was removed using Sound Quality software. Next, using Sound Forge software, the silent periods between the sound events were shortened without affecting the properties of these sound events. It resulted in a sound signal of 100 ms containing the same 30 separated sound signals. The FFT analysis was done on both the 4.5 s and the 100 ms sound signals. The results in Figure 4 show that the overall shape of the frequency spectrum at higher frequencies (roughly above 1 kHz) was not affected, but the shortening of the interval time clearly caused an increase in peaks at lower frequencies (below 1 kHz) and the appearance of peaks at the harmonic frequency of the occurrence of the sound events (300 Hz). This shows that the calculated frequency spectrum below 1 kHz depends primarily on the number of separate sound events and much less on the properties of the single sound events. This implies that, for cellular solid crispy foods, the low frequency part is primarily determined by the morphology and architecture of the product, and only to a lesser extent by the solid matrix properties.

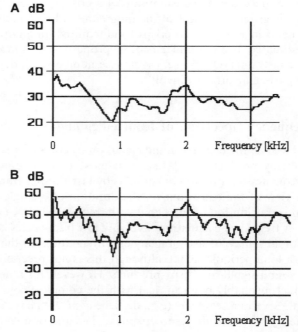

Figure 4 *Frequency spectrum of sound emitted during slow cutting at 0.2 mm s^{-1} of a fresh toasted rusk roll: (A) sequence of 30 fracture events in 4.5 s; (B) same sound with reduced silent periods in between the sound events (30 events in 100 ms)*

34.4.5 Effect of Oil Presence after Deep Frying

A problem for the texture of deep-fried snack products with a crispy crust and a soft interior is that often the crust rapidly loses its crispy character – typically within 3–20 min after frying. In contrast, crispy loss is much slower for bread with a crispy crust. In general, the decrease in crispness of the crispy crust is attributed[1] to plasticizing of the solid matrix of the crust as the result of the uptake of water transported from the moist inside to the crust or taken up from the environment. However, this cannot explain the large difference in loss of crispness between deep-fried products and baked products with a similar overall structure. An important difference between the two types of products is that, after frying, the adhering oil is sucked into the deep-fried products during cooling down.[43]

To study the possible role of the suction of oil into a dry crispy food, test pieces of toasted rusk rolls ($\sim 5 \times 5 \times 1$ cm) were dipped for 5 s in hot (180°C) or cold (35°C) arichidonic oil. The fracture behaviour was studied by cutting at a speed of 1 mm s^{-1}. Also the fracture behaviour was determined for model deep-fried snack crusts (2.5 min in oil at 180°C) by wedge penetration at a speed of 40 mm s^{-1}. From a portion of the snacks, the adhering oil was removed from the sample as soon as possible after frying by wiping with tissue paper.[35]

Measured force and sound pressure curves *versus* cutting time are shown in Figure 5 for rusk rolls without oil and after dipping in cold oil. The force *versus*

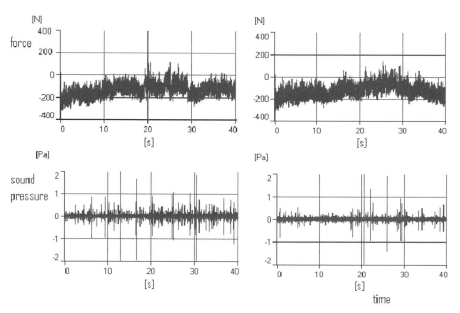

Figure 5 *A typical example of the force and sound pressure plotted against time for cutting experiments on toasted rusk rolls without oil (left) and after dipping in cold oil (right)*

Table 2 *Number of force peaks and sound peaks per unit area of material during a cutting experiment of toasted rusk roll with a razor blade. In brackets are standard deviations (n = 3)*

Sample treatment	Number of force peaks per cm^2	Number of sound peaks per cm^2
Dry, no oil	79 (9)	100 (20)
Soaked with oil at room temperature	81 (6)	42 (7)
Soaked with oil at 180°C	87 (12)	66 (2)

time curves look about the same (the small differences are not significant). The same applies to the data for the number of mechanical peaks per unit area of cut cross section of the rusk roll (Table 2). In contrast, however, the curves of sound pressure against time differ considerably. The average size of the peaks after oil treatment was smaller, indicating a lower acoustic energy. Furthermore, the number of acoustic peaks was much lower for the rusk rolls that were dipped in oil (Table 2), the effect being larger for the samples soaked in oil at room temperature. A possible explanation for the smaller effect on sound emission of dipping the rusk roll in hot oil is that more oil was already drained from the sample before the cutting test because of the much lower oil viscosity at 180°C. The average take-up after dipping toasted rusk roll in oil was 286 ± 20 wt% for cold oil and 226 ± 33 wt% for hot oil ($n = 5$).

Similar results were obtained for the effect of wiping oil from the model crust directly after frying. In Table 3 the numbers of force and sound peaks are given that are larger than the threshold values of 1 N and 1.5 Pa, respectively. While the number of force peaks does not differ significantly between the standard and wiped snacks, the sound peak number is significantly higher for the wiped snacks. Moreover, the sound pressure level of the sound signals was on average 5 dB higher for the wiped snacks (data not shown).

So the conclusion is that sucking of oil into the cellular sponge structure of a crispy product has no significant short-term effect on its mechanical properties. But it strongly affects the sound emitted on fracture. Uptake of oil will result in the presence of oil in the pores or oil droplets in the larger holes.[44-46] This means that at least part of the sound that is emitted near the crack tip will have to pass an oil layer before it is transferred to the surrounding air. In principle, then, there are a few acoustic mechanisms that may explain this effect. The main ones are damping of the sound when it travels through the air or oil and reflection and refraction of the sound at the oil–air interface. Damping of sound in air is very weak.[47] In oil it is greater because of the higher viscosity of the oil. But, since the oil layers involved are very thin (≤ 1 mm) the attenuation of sound during its transport through the oil should be very low. In contrast, reflection of sound at the oil–air interface will be very large because of the large difference in acoustic impedance between air and oil. The acoustic impedance Z is defined as the product of the density and the acoustic velocity c in the

Table 3 *Number of force peaks and sound peaks larger than threshold values of 1 N and 1.5 Pa, respectively, that can be distinguished in the force and sound versus deformation curves as a function of ageing time for deep-fried model crust snacks, with and without wiping to remove liquid oil. In brackets are standard deviations*

	Number of force peaks		Number of sound peaks	
Ageing Time (min)	Standard	Wiped	Standard	Wiped
5	21.3 (3.3)	22.4 (2.7)	75 (18)	111 (24)
7	23.3 (2.0)	25.3 (1.7)	69 (21)	107 (25)
10	20.3 (2.6)	23.7 (2.3)	46 (13)	77 (15)
12	19.1 (1.6)	18.7 (2.1)	38 (16)	69 (17)
15	17.6 (2.4)	15.2 (1.6)	42 (15)	59 (19)

material;[47] values for air and oil are $\sim 4 \times 10^2$ and $\sim 1.3 \times 10^6$ kg m^{-2} s^{-1}, respectively. The reflection coefficient R for sound is given by

$$R = \frac{(Z_2/Z_1) - \sqrt{1-(n-1)\tan^2 \alpha_i}}{(Z_2/Z_1) + \sqrt{1-(n-1)\tan^2 \alpha_i}}, \qquad (4)$$

where $n = (c_2/c_1)^2$ and α_i is the angle of incidence of the sound wave. Symbols with subscript 1 refer to the material on the side from which the sound comes. For an oil–air interface this results in a reflection coefficient of -1, implying total reflection.

In spite of total reflection of sound at an oil–air interface, one can still hear a sound signal on fracture of crispy solids containing oil, although it is less loud than for the same material containing no oil or less oil. We assume that the oil penetrating into the cellular solid only covers part of the solid–air interface, thereby allowing part of the acoustic emission during fracture of the solid material to pass directly into the air without passing through an oil layer. Moreover, new surfaces formed on fracture are unlikely to be covered by oil.

It was also noticed that there was an effect of the type of oil for the rusk roll experiments.[35] Oil that became solid at room temperature had less effect on the sound than liquid oil. The acoustic impedance for a solid fat is likely to be somewhat higher than for a liquid oil, and so a difference in reflection cannot explain this observation. Our tentative explanation is that, on solidification of the oil, small cracks are formed in the fat layer which allow passage of the sound.

34.4.6 Change in Mode of Fracturing with Increasing Water Content

Another point that should be considered is that, on deterioration of a crispy product, the mode of fracturing changes on the mesoscopic scale, as depicted schematically in Figure 6 for the case of fracture due to penetration by the tip of

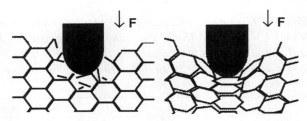

Figure 6 *Schematic representation of the response on penetration by the tip for a rounded cylinder of a dry crispy cellular solid (left) and of the same product after storing at high (80%) relative humidity (right)*

a rounded cylinder.[8] During the transition from 'crispy' to 'non-crispy', the fracturing mechanism of a cellular solid material will change from brittle fracture of 'single' beams and lamellae at relatively low forces to ductile fracture initially involving the bending of groups of beams and lamellae, which finally may rupture together at a much larger overall force. The latter will be the case as long as the water content does not become too high, even though the fracture force of the individual beams and lamellae may already have decreased due to the weakening of the solid matrix. For a toasted rusk roll, an increase in the cutting force has been observed from < 1 N up to 6–7 N at a cutting speed of 0.4 mm s^{-1}.[9]

34.5 Effects at the Macroscopic Scale

At the macroscopic scale, the difference in mechanical properties between the crust and the inner moist part of a product is of importance. Depending upon the ratio of the stiffness of the crust to that of the moist layer below it, the crust may behave in different ways when the product is subjected to deformation by a wedge-like structure (*e.g.*, a tooth). When the stiffness of the two parts of the product is of the same order, the wedge will penetrate the crust layer; but when the stiffness of the layer below it is clearly lower, the crust layer will deform more in a bending mode, the extent of deformation depending upon the ratio of the stiffnesses. In extreme cases, the transition from a weak to a stiff sub-layer will cause the same change as going from a three-point bending test to the wedge penetration of a crust layer situated on a solid foundation. In the latter case the fracture process generally will take longer for the same deformation speed. Fracture of a crust in three-point bending occurs mostly as one main fracture event, sometimes preceded by smaller fracture events, while in wedge penetration there are more fracture events of comparable size. Three-point bending involves more of a snapping type of fracture, whereas wedge penetration is more of a crushing type of fracture. So, in principle, the ageing and drying out of the crumb during bread storage may also induce a change in crispness perception of the crust. In practice, however, other processes that can cause a change in crust crispness, such as the uptake of water, will tend to proceed faster.

Other macroscopic properties affecting crust fracture behaviour are the presence of sides supporting the crust and the curvature of the crust. The presence of stiffer sides will support wedge penetration above three-point bending of the crust during biting as long as the distance between the sides is of the order of 10 cm or less, depending upon the mechanical properties and thickness of the crust and the mechanical properties of the layer below. A crust with convex curvature to the outside will also bend less easily than a flat crust. Maybe this is a main reason why breads characterized by a crispy crust often have a rounded shape – at least in one direction.

34.6 Conclusions

The crispy behaviour of dry cellular solids is determined by various mechanisms acting at molecular, mesoscopic and macroscopic length-scales. On the mesocopic scale it is especially the morphology at length-scales of 50–500 µm that is of greatest importance for the crispness of the product. The minimum and maximum sizes of the structural elements on the mesoscopic scale have here been estimated from sound emission properties and mechanical and physiological constraints. For a product consisting of a crispy crust and a non-crispy moist inner part, the variation in mechanical properties over macroscopic distances (>1 mm) can also be important.

Acknowledgements

We thank EefJan Timmerman for performing experiments on the effect of oil uptake by rusk rolls. The study was funded by the Wageningen Centre for Food Sciences, an alliance of major Dutch food industries, Wageningen University and Research Centre, Maastricht University, and TNO Nutrition and Food Research, with financial support from the Dutch government.

References

1. H. Luyten, J.J. Plijter and T. van Vliet, *J. Texture Stud.*, 2004, **35**, 445.
2. H. Luyten, W. Lichtendonk, E.M. Castro, J.E. Visser and T. van Vliet, in *Food Colloids: Interactions, Microstructure and Processing*, E. Dickinson (ed), Royal Society of Chemistry, Cambridge, 2005, p. 380.
3. G. Dijksterhuis, H. Luyten, R. de Wijk and J. Mojet, *Food Qual. Pref.*, 2007, **18**, 37.
4. B. Drake, *J. Food Sci.*, 1963, **28**, 233.
5. Z. Vickers and M.C. Bourne, *J. Food Sci.*, 1976, **41**, 1158.
6. L. Duizer, *Trends Food Sci. Technol.*, 2001, **12**, 17.
7. H. Luyten and T. van Vliet, *J. Texture Stud.*, 2006, **37**, 221.
8. J.E. Visser, W. Lichtendonk, R.J. Hamer and T. van Vliet, submitted for publication.

9. E.M. Castro-Prada, H. Luyten and T. van Vliet, in *Proceedings of the 4th International Symposium on Food Rheology and Structure*, P. Fischer, P. Erni and E.J. Windhab (eds), ETH Zürich, Switzerland, 2006, p. 389.
10. J. Fineberg and M. Marder, *Phys. Rep.*, 1999, **313**, 1.
11. A. Baltsavias, A. Jurgens and T. van Vliet, *J. Cereal Sci.*, 1999, **29**, 235.
12. A.-L. Ollett, R. Parker and A.C. Smith, in *Food Polymers, Gels and Colloids*, E. Dickinson (ed), Royal Society of Chemistry, Cambridge, 1991, p. 537.
13. T. van Vliet, in *Food Macromolecules and Colloids*, E. Dickinson and D. Lorient (eds), Royal Society of Chemistry, Cambridge, 2005, p. 447.
14. E.E. Katz and T.P. Labuza, *J. Food Sci.*, 1981, **46**, 403.
15. M. Peleg, *Crit. Rev. Food Sci. Nutr.*, 1997, **37**, 491.
16. M. Harris and M. Peleg, *J. Texture Stud.*, 1996, **73**, 225.
17. L. Slade and H. Levine, in *Water Relations in Foods*, L. Slade and H. Levine (eds), Plenum, New York, 1991, p. 29.
18. S. Abblett, A.H. Drake, M.J. Izzard and P.J. Lillford, in *The Glassy State in Foods*, J.M.V. Blanshard and P.J. Lillford (eds), Nottingham University Press, Loughborough, 1993, p. 89.
19. M. Peleg, *J. Texture Stud.*, 1994, **25**, 403.
20. R.J. Hutchinson, S.A. Mantle and A.C. Smith, *J. Mat. Sci.*, 1989, **24**, 3956.
21. G.E. Attenborough, A.P. Davies, R.M. Goodband and S.J. Ingman, *J. Cereal Sci.*, 1992, **16**, 1.
22. R.J. Nichols, I.A.M. Appelqvist, A.P. Davies, S.J. Ingman and P.J. Lillford, *J. Cereal Sci.*, 1995, **21**, 25.
23. M. Le Meste, G. Roudaut and S. Davidou, *J. Therm. Anal.*, 1996, **47**, 1361.
24. I. Fontanet, S. Davidou, C. Dacremont and M. Le Meste, *J. Cereal Sci.*, 1997, **25**, 303.
25. G. Roudaut, C. Dacremont and M. Le Meste, *J. Texture Stud.*, 1998, **29**, 199.
26. R.M. Daniel, R.V. Dunn, J.L. Finney and J.C. Smith, *Annu. Rev. Biomol. Struct.*, 2003, **32**, 69.
27. Y.P. Chang, P.B. Cheah and C.C. Seow, *J. Food Sci.*, 2000, **65**, 445.
28. M. Peleg, *J. Texture Stud.*, 1996, **73**, 712.
29. A.G. Atkins and Y.-M. Mai, *Elastic and Plastic Fracture*, Ellis Horwood, Chichester, 1985.
30. T. van Vliet, H. Luyten and P. Walstra, in *Food Polymers, Gels and Colloids*, E. Dickinson (ed), Royal Society of Chemistry, Cambridge, 1991, p. 392.
31. A.A. Griffith, *Phil. Trans. Roy. Soc. (London)*, 1920, **A221**, 169.
32. T. van Vliet, H. Luyten and P. Walstra, in *Food Colloids and Polymers*, E. Dickinson and P. Walstra (eds), Royal Society of Chemistry, Cambridge, 1993, p. 175.
33. J.F.V. Vincent, *J. Sci. Food Agric.*, 1998, **78**, 162.
34. J.F.V. Vincent, *Eng. Failure Anal.*, 2004, **11**, 695.
35. T. van Vliet, J.E. Visser and H. Luyten, *J. Food Eng.*, submitted for publication.

36. W.C. Young, *Roark's Formulas for Stress and Strain*, McGraw-Hill, New York, 1989, p. 101, p. 200.
37. W.M. Hartman, *Signals, Sound and Sensation*, Springer, New York, 1997.
38. K. Krumbholz and L. Wiegrebe, *Hearing Res.*, 1998, **124**, 155.
39. A.S. Galembo and A. Askenfelt, *Acoust. Phys.*, 2006, **52**, 144.
40. B.P. Marchant, *Biosyst. Eng.*, 2003, **85**, 261.
41. N. de Belie, V. de Smedt and J. de Baerdemaeker, *Postharvest Biol. Technol.*, 2000, **18**, 109.
42. N. de Belie, M. Sivertsvik and J. de Baerdenmaeker, *J. Sound Vibr.*, 2003, **266**, 625.
43. M. Mellema, *Trends Food Sci. Technol.*, 2003, **14**, 364.
44. C.M. McDonough and L.W. Rooney, *Cereal Foods World*, 1999, **44**, 342.
45. J.M. Aguilera, D.W. Stanley and K.W. Baker, *Trends Food Sci. Technol.*, 2000, **11**, 3.
46. P. Bouchon, P. Hollins, M. Pearson, D.L. Pyle and M.J. Tobin, *J. Food Sci.*, 2001, **66**, 918.
47. D.T. Blackstock, *Fundamentals of Physical Acoustics*, Wiley, New York, 2000, p. 108.

Subject Index

Absorption
 kinetics of, 5, 19–21
 of colloidal particles, 25
 of glucose, 3–4, 9
 of lipids, 10–11, 22–6
 of nutrients, 19–20
 of protein, 9–10,
Acid milk gels, 257–302
 xanthan gum in, 289–302
Acoustic emission, from crispy food, 486–97
Acoustiscan, 388
Activity coefficient/factor, 130–3, 211
Adhesive hard-sphere model, 332–3
Admul WOL, see polyglycerol polyricinoleate
Adsorbed protein film(s)/layer(s)
 adhesion of, 395
 cross-linking in, 376
 displacement of, 252, 387, 416
 effect of ageing on, 352–3
 effect on bubble/foam stability, 177–8, 188–92, 357, 361–2, 365–8, 376, 378
 effect on droplet deformation/break-up, 344–53
 effect on emulsion properties, 169–75, 385–97
 effect of polysaccharides on, 195–208
 effect of surfactant on, 209–25
 fracture of, 370, 376, 378
 gel-like character of, 189, 361
 mixed, 169–75
 structure of, 167, 203–6, 216
 surface rheology of, 203–6
 thermodynamic model of, 210–2
 thickness of, 214, 215–6, 218, 236, 237

Adsorption
 affinity-controlled, 443–5
 competitive, 168, 246, 434, 444
 diffusion-controlled, 201–3, 212–3, 223, 444
 during emulsification, 433–48
 from protein + surfactant solutions, 209–26
 kinetically controlled, 441–3
 kinetics of, 196, 199–203, 212–3, 423, 438, 442
 of lecithin, 363, 367
 of polyelectrolyte, 145–8
 of protein, 167–75, 196, 209–25, 245, 343, 347, 440, 443–5
 of protein + polysaccharide, 195–208
 of starch, 436–40
 of surfactant(s), 209–26, 245–6, 359, 361
 random sequential, 442–3
 thermodynamics of, 210–2
Adsorption energy, particle, 360
Adsorption isotherm(s)
 from theory, 249–52
 of OSA-starch, 437–8
Ageing
 effect on crispness, 490, 495, 497–8
 of protein layer, 352–3
Aggregates
 fibrillar, 57–8, 64–6
 protein, 35–56, 177–94, 296, 477–8
 rod-like, 38, 57, 63–6
 shape of, 59–64
 structure of, 35–68, 334–6

Subject Index

Aggregation (*see also* flocculation)
  computer simulation of, 36, 37, 269–87
  diffusion-controlled, 35–7
  irreversible, 35–7
  kinetics of, 157–61, 180, 184–5
  of casein(ate)(s), 155–66
  of casein micelles, 10, 195, 262–6, 293–4
  of droplets, 379
  of globular proteins, 35–56
  reaction-controlled, 36, 180
  thermodynamics of, 58–64
Albumin, *see* serum albumin
Amphiphile(s), *see* surfactant(s)
α-Amylase, 119, 123–5, 463
Amylopectin, 117, 122
Amylose, complexation with ligands, 117–25
Appollonian packing, 442, 443
Arabic gum, *see* gum arabic
Aroma compounds, in emulsions, 423–32
Atomic force microscopy (AFM), 38, 229, 231, 233, 238, 240–1
Attractive colloidal systems, 327–42

Baxter potential model, 330–2, 339
Beeswax, 104–6, 110
Biconical disc interfacial rheometer, 347–8, 388
Bile salts, 22–4
Binding
  of calcium ions, 133–4
  of flavour in emulsions, 423–32
  of protein to polysaccharide, 196, 197, 200
  of whey protein to casein, 257–66, 273–80
Bioactive molecules in milk, 8–11
Bioavailability
  definition of, 3, 4
  of casein, 9–10
  of glucose, 9
  of lipids in milk, 10–11
  of neutraceuticals, 88
Biopolymer network, probe particle in, 306–7
Biopolymer solution, microrheology of, 319–26
Biopolymer + surfactant layers, 245–56
Birefringence, 118, 122

Biscuit(s), acoustic/fracture properties of, 486–94
Bovine serum albumin (BSA), 40, 46, 50, 465, 470
Bragg diffraction peaks, 92–100
Bread, crispness of, 489–90, 495, 498–9
Brewster angle microscopy (BAM), 229, 231, 233–41
Bridging
  by polyelectrolytes, 145–8
  by salivary protein, 470, 471
Bridging flocculation, 171, 197, 308, 470
Brij 35 surfactant, 388–92
Brittle fracture, 485, 487, 498
Brownian dynamics (BD), 249, 270, 330–3
Brownian motion, 305, 313, 316, 320
Brush-like layer, 294
Bubble(s)
  clustering of, 375
  coalescence of, 191, 369–82
  disproportionation of, 191, 357–68, 376
  effect of oil droplets on, 376–81
  effect of pressure change on, 369–81
  imaging of, 183–4, 372
  size distribution of, 372, 375
  shape analysis of, 199
  stability of, 191–2, 357–82
Bubble pressure, maximum, 213–4

Caflon surfactant, 401–11
Calcium ions
  binding of, 133–4, 157–66
  effect on caseins, 157–66
Calcium phosphate, 155–66
  amorphous, 163, 165
  precipitation of, 160
Calorimetry, *see* differential scanning calorimetry
Capillary number, 344, 346, 349–53
Capillary pressure, 353
Capsules, leakage from, 103–15
Carbon dioxide, solubility of, 131, 133
Carboxymethylcellulose, 470
ι-Carrageenan, 470
κ-Carrageenan, 421
Casein(ate)(s),
  as emulsifier, 308, 311–6, 370, 374–81, 416

Subject Index

as thickening agent, 416–8
bubble/foam stabilization by, 374–81
complexation with surfactant, 246
digestion of, 9–10
hydration of, 416–7
interactions of, 155–66
precipitation of, 380–1, 416
self-assembly of, 161–6
α_s-Casein, 156, 157–62
β-Casein, 156, 161–2, 228–41, 254, 353, 361–2, 368, 445
κ-Casein, 10, 161–2, 257–8, 264–6, 294
complexation with whey protein, 258, 264–6
Casein micelle(s)
aggregation/coagulation of, 10, 195, 257, 262–6, 293–4
calcium ions in, 157
interaction with pectin, 195
interaction with whey protein, 258, 263–6, 269–86, 294–5
stability of, 257, 262–6
structure of, 161–6
Caseinate nanoparticles, 308
Cellular structure, of crispy foods, 489–93, 496, 499
Cellulose, modified, 445
Cetyltrimethylammonium bromide (CTAB), 464–9
Charge distribution, on adsorbed molecules, 229–30
Charge regulation, 143, 149–51
Charge stabilization, *see* electrostatic stabilization
Charged patches, 149
Chitosan, 470
Cholesterol, 22, 24, 29, 88
Chymosin, 10, 257, 262
Citrate/citric acid, 110, 158, 159
Clustering, *see* aggregation, flocculation
Coacervate capsules, 104–5, 108, 112–3
(*see also* complex coacervates)
Coagulation, *see* aggregation
Coagulation time (CT), 158–60
Coalescence
at surface, 457
in emulsions, 458, 459
of bubbles, 191, 369–82
of droplets, 346, 379, 458
partial, 376, 458

Coarsening of foams, 191, 192 (*see also* disproportionation)
Cold gelation, 37, 51
Cold-water-swelling starch (CWS), 418–21
Collapse
monolayer, 231–40
network, 316–7
Collision frequency, 441
Colloid stability, DLVO theory of, 156–7, 161
Competitive adsorption/displacement, 168, 210, 239–40, 246, 416, 434, 444
Competitive hydration, 414–5, 421–2
Complex(es)
amylose–lactone, 118–25
amylose–lipid, 125
κ-casein–whey protein, 257–66
electrostatic, 467–71
inclusion, 117–26
lactoferrin–β-lactoglobulin, 170, 171
lysozyme–saliva, 464, 466–71
protein–polyelectrolyte, 148–52
protein–polysaccharide, 50–1, 195–208
protein–surfactant, 211, 246
Complex coacervate(s), 108, 112–3, 196, 197
Complex modulus, 306, 321
Computer simulation,
Brownian dynamics, 249, 270, 330–3
Monte Carlo, 130–53, 332
Newtonian dynamics, 330–3, 339
of aggregation, 36, 37, 269–87
of gelation, 269–88, 328–41
of macromolecular solutions, 130–53
Conalbumin, 435–6
Confocal laser scanning microscopy (CLSM), 305
of emulsions, 308–16, 388, 394, 418, 456–8
of pig's tongue surface, 455
of protein aggregates, 38, 51
of protein gel(s), 49, 286, 474–82
of protein + polysaccharide mixture(s), 51, 290, 292–3
of saliva + lysozyme, 466, 470, 471
particle tracking with, 308–16
Contact angle, 359, 360–1
Convective mass transport, 441

Copolymer(s), adsorption of, 248, 254, 445
Crack growth, speed of, 487–91
Crack stoppers, 491–2
Cream layer(s), 388–95
Creaminess, 386, 451–3, 458, 460
Creep, rheological, 389, 390
Crispness, of foods with dry crust, 485–501
Critical capillary number, 344, 353
Critical micelle concentration, 59, 213, 359
Critical point, liquid–liquid, 328
Critical surface pressure, 232, 235
Cross-linking, of adsorbed protein, 376
Crunchiness, *see* crispness
Crust, mechanical properties of, 485–501
Cryo-TEM, 27, 38, 45–5, 70–80
Crystalline fat, flavour binding of, 423–32
Crystallization
  in emulsions, 399–412, 423–32
  of starch, 117–25
Crystals, fat, 358, 413, 423–32
Cubic phase(s), 23–9, 70–84
Cubosomes, 27, 28, 70–84
Cutting, sound emission during, 493–6

Darwinian selection, 5, 7, 13
Debye (screening) length/parameter, 62, 131, 156
Debye–Hückel (DH) approximation/ theory, 130, 131–9, 153
Deep fried snack crust, 486, 489, 495–7
Deformation
  of droplets, 343–53, 386, 391
  of simulated gel, 283–5
Delivery
  from liquid crystalline phases, 23–30, 70, 75–84
  of essential nutrients, 3–5
Denaturation, protein, 65–6, 179–80, 183–6, 285–6, 294
  kinetics of, 179–80, 184–5
Density profile, of polymer layer, 253–4
Depletion flocculation, 307–8, 317, 464, 467, 469, 471
Depletion force/interaction, 292, 294, 467
Deposition, layer-by-layer, 197
Detection threshold of particles, 104–5

Didodecyldimethylammonium bromide (DDAB), 359–68
Dielectric continuum model, 129–30
Dielectric permittivity, of protein, 135–6
Diet-related diseases, 1–4
Differential scanning calorimetry (DSC), 90–99, 119, 122–5, 400, 428
Diffusing wave spectroscopy (DWS), 257–68, 319–26, 295
  theory of, 258–60, 321
Diffusion
  from capsule, 106
  in biopolymer matrix/network, 306–7, 323–4
  in Newtonian liquid, 309–10
  in emulsion, 311–5, 394, 395
  of gas between bubbles, 177–8, 191
Diffusion coefficient
  in capsule wall, 106, 113
  of adsorbing species, 201–3, 212–3, 220–1
  of complexes, 201–3
  of particle(s), 181, 259, 306, 330, 339
Diffusion-controlled adsorption, 201–3, 212–3, 223, 444
Diffusion-limited (cluster) aggregation (DLCA), 35–7, 180, 186, 187, 282
Digestion
  effect of food structure on, 1–15, 20–1
  of emulsions, 21–6
  of lipids, 21–2, 109, 113–4
  of milk, 8–11
  of starch, 118
  of wax, 109, 113–4
Digestive system, 19–20
Diglycerol monooleate (DGMO), 70, 82–3
Dil(at)ational elasticity/modulus, 199, 203–5, 214, 389
Dipalmitoylphosphatidyl choline (DPPC), 228–34, 238–41
Discontinuous micellar cubic structure(s), 87–102
Displacement, competitive, 239–40, 246, 387, 416
Disproportionation, 191–2
Disulfide bonds/bridges/links, 37, 286, 464
DLVO theory, 156–7, 161
Double emulsion, 78, 104

Double-layer repulsion, 139, 146
Double-layer thickness, 394
Drainage, of foam, 183, 190–2
Droplet(s)
  aggregation of, 379
  coalescence of, 346, 379, 458
  deformation of, 343–53, 386, 391
  disruption of, 344, 353, 417–8
  effect on bubble stability, 369–82
  in shear flow, 343–56
  spreading of, 379
Droplet charge, effect on flocculation, 463–72
Droplet dissolution model, 428–31
Droplet shape, in shear flow, 343–55
Droplet size (distribution) (*see also* particle size)
  determination of, 308, 388, 400–1, 424, 435, 465
  effect on aroma partition, 425–6
  effect on crystallization, 401–11
  effect of protein content on, 169, 170
  relationship to flocculation, 466–9
  variation of, 399–401, 414, 417–8
Dynamic elastic modulus, 182
Dynamic light scattering, 181–2, 188, 198, 199, 202, 321, 401
Dynamic surface behaviour, *see* surface rheology
Dynamic surface tension, 213–23

Egg yolk protein(s), 434–6, 440–1, 443–5
Einstein relation, 330, 339
Elastic(ity) modulus, 231, 235–6, 386, 455, 456 (*see also* rheology, surface rheology)
Electric birefringence, 65
Electrical conductimetry, 183
Electron microscopy
  cryogenic, 27, 38, 45–5, 70–80
  environmental, 475, 480–3
  scanning, 38, 49–50, 490
  transmission, 23, 27, 28, 38, 50
Electrolyte solutions, theories of, 129–54
Electrophoretic mobility, 181–2, 186–7 (*see also* zeta potential)
Electrostatic interactions
  in emulsions, 167–76, 197, 466–70
  in macromolecular solutions, 129–54
  in polyelectrolyte solutions, 129–54
  in protein + phospholipid films, 227–43
  protein–polysaccharide, 195–208
Electrostatic stabilization, 257 (*see also* DLVO theory)
Ellipsometry, 214, 215
Emission of sound, from crispy food, 486–97
Emulsification (*see also* homogenization)
  adsorption during, 433–48
  in human stomach, 22
  methods of, 400–1, 411, 415–7
  of microemulsions, 70–84
Emulsifier(s), *see* surfactant(s)
Emulsion(s)
  behaviour in the mouth, 451–61
  creaming of, 308, 317, 388, 391–4
  crystallization in, 399–412, 423–32
  digestion of, 21–6
  double/duplex, 78, 104
  flavour binding in, 423–32
  flocculation of, 307, 315–7, 379, 463–72
  formation of, 106–8, 345
  instant, 413–22
  monodisperse, 399–412
  particle dynamics in, 309–16
  rheology of, 307, 310–1, 317, 344–53, 385–97, 466, 469
  sensory attributes of, 386, 451–3, 458, 460
  shelf-life of, 307, 316
  spontaneous, 415
  structure of, 312–7, 385–97, 418, 457
  viscosity of, 452, 458–9, 466, 469
  whippable, 377, 380
Emulsion droplets
  break-up of, 344, 353, 417–8
  effect on bubble stability, 369–82
  in shear flow, 343–55, 386
  spreading of, 379
Emulsion gel, 317, 460
Emulsion stability, effect of saliva on, 463–72
Encapsulation with wax, 103–16
Environmental scanning electron microscope (ESEM), 475, 480–3
Enzymatic degradation, of starch, 117–8, 123–5, 463
Essential nutrients, 1–5

Ethyl heptanoate (EH), 424–6, 428–31
Excluded volume, 131

Fast Fourier transform (FFT) analysis, 493–4
Fat crystals, 358, 413
  flavour binding by, 423–32
Fat digestion, 22
Fat globule(s), in milk, 10–11
Fattiness, perception of, 386, 452, 453
Fibre, 20, 118
Fibrillar aggregation, 57–8, 64–6
Flavour binding by emulsion droplets, 423–32
Flavour perception, 452–3
Flocculation (see also aggregation)
  acid-induced, 379
  bridging, 171, 197, 308, 470
  by saliva, 463–72
  depletion, 307–8, 317, 464, 467, 469, 471
  of emulsions, 307, 315–7, 379, 463–72
Flory–Huggins χ-parameter, 156, 248
Flow fractionation, 435, 439
Fluorescein isothiocyanate (FITC), 106–11, 308, 475
Foam(s)
  coarsening of, 191–2
  drainage of, 183, 190–2
  formation of, 182–3
Foam (in)stability
  effect of oil droplets on, 376–81
  effect of particles on, 178, 191, 361, 376
  effect of pressure change on, 379–81
  effect of protein on, 177–8, 189–92
  effect of surfactant on, 246
  effect of viscosity on, 381
Foam stability test, 372–3
Foamability, 189
Foamscan, 183, 189
Food structure, importance to nutrition, 1-15, 20
Form factor, 259
Fourier analysis/transform, 493–4
Fractal aggregate(s), 35–54, 57–8, 334–6
Fractal dimension(ality), 36, 40–54, 57–8, 62–3, 180, 279–81, 334–6
Fractal gel (model), 52–3, 260
Fractal scaling, 35–6, 52–3, 335

Fracture
  of adsorbed protein film, 370, 376, 378
  of crispy foods, 485–99
Fracture energy, 489
Fracture force, 487, 491, 498
Fracture strain/stress, 487–9, 491
Frequency spectrum, of sound, 493–4
Friction
  at protein gel surface, 474–84
  at tongue surface, 455–7
  between tongue and glass, 458–9
  measurement of, 475
  relationship to sensory properties, 453

Gastrointestinal tract, 19–20
Gel(s)
  acid (skim) milk, 257–66, 289–302
  definition of, 265, 319–20
  heat-set, 473–84
  hybrid, 279
  β-lactoglobulin, 38–53
  particle, 473–83
  rheology of, 51–3, 261–5, 269–88, 294–8
  structure of, 43–54, 289–302
  surface topography of, 473–84
Gel point, 263–5, 320
Gelatin, 108, 112, 175, 305, 320, 323–4, 470
Gelation
  acid-induced, 257–66
  cold, 37, 51
  computer simulation of, 269–88, 328–41
  heat-induced, 35–56, 475
  of biopolymer solutions, 319–26
  of gelatin, 323–4
  of globular proteins, 35–56, 473–84
  of starch, 122
Genomics, 11–13
Glass transition, 327–41, 488
Globular proteins, self-assembly of, 35–56
Glucono-δ-lactone (GDL), 260–1, 264, 290–1, 295–6, 370–1
Glucose, absorption of, 3–4
Glycemic index, 4
Glycoprotein(s)
  egg-yolk, 435, 436, 440
  in the mouth, 456

Subject Index

Guar gum, 417–8
Gum arabic, 108, 112, 195, 197, 470

Hairy layer, 257, 262, 264, 294
Hard spheres, 35–6, 327–8
Head-space analysis, 423–6
Heat treatment of milk, effect on gelation, 257–302
Helix formation, amylose, 118, 121–5
Henderson–Hasselbalch equation, 62
Hexosomes, 27, 28, 70–83
Hofmeister series, 64
Homogenization, 399–401, 411, 415, 424, 433, 465
Hydrocolloids
　hydration of, 414, 417, 418–20
　stabilization mechanism of, 317, 376
Hydrodynamic diameter/radius, 181, 188, 198, 199, 203
Hydrodynamic flow, at surface, 474, 477, 479
Hydrodynamic focusing, 372–3
Hydrodynamic layer, 457
Hydrodynamic lubrication, 453, 454
Hydrodynamic thickness, 257
Hydrogen bonding, 229–30
Hydrophilic particles, 357–68
Hydrophobic interaction(s), 60–4, 210, 229
Hydrophobicity
　of adsorbing species, 248, 444–5
　of capsule wall material, 112
　of particles, 357–8, 364–5, 367
　of tongue surface, 456
Hydroxyapatite, 163, 165

Image analysis, 182–3, 292, 299, 309, 316
Immunoglobulin G (IgG), 435–6, 440, 444
Inclusion complexes, amylose, 118–25
Incompatibility, *see* thermodynamic incompatibility
Instant emulsions, 413–22
Interaction(s)
　adhesive, 331–2
　aroma–fat, 423–32
　bridging, 145–8
　Coulombic, 210, 211
　depletion, 292, 294, 467
　effect of multivalent ions on, 139–40
　effect of titrating groups on, 140–5
　electrostatic, 129–54, 156, 195–208, 227–44
　excluded volume, 313
　hard core/sphere, 131
　hydrodynamic, 270
　hydrogen bonding, 229, 230
　hydrophobic, 60–4, 210, 229–31
　lysozyme–saliva, 464, 466–71
　many-body, 333
　of casein(ate), 155–66
　of casein micelles, 262–6, 292–4
　of aggregating particles, 37
　of charged hard spheres, 129–30
　of charged macromolecules, 136–48
　of particles in simulation, 270–2
　particle–protein, 307
　protein–phospholipid, 227–44
　protein–polyelectrolyte, 148–52
　protein–polysaccharide, 195–208
　protein–protein, 140–5
　protein–surfactant, 209–23
　screened Coulomb, 136–9
　short-range, 327–42
　square-well, 329–32
　steric, 175, 262, 266, 294
Interfacial elasticity, 199, 203–5, 214, 389
Interfacial rheology/viscoelasticity, *see* surface rheology
Interfacial tension, *see* surface tension
Interfacial viscosity, 386
Iodine binding, 119, 123–5
Ion–ion correlations, 139

Jet homogenizer, 308

Kinetics
　of adsorption, 196, 199–203, 212–3, 423, 438, 442
　of aggregation, 157–61, 180, 184–5
Kyte–Doolittle plot(s), 444–5

α-Lactalbumin, 149–52, 445
Lactation, evolutionary role of, 6–8
Lactoferrin, 167–76
β-Lactoglobulin
　aggregates of, 38–53, 64–6, 177–94
　bubble stabilization by, 361–2, 365–8, 374–8

complex/interaction with pectin, 198–206
denaturation of, 285–6
droplet stabilization by, 344–51
foaming properties of, 177–94
gelation of, 38–53
heating with charged cosolutes, 177–94
in emulsions, 167–76, 464–70
interaction with lactoferrin, 167–76
interaction with polyelectrolyte(s), 149–51
Lactones, complexation with amylose, 118–25
Lactose, digestion of, 9
Lag time, during adsorption, 200–2
Langmuir adsorption/isotherm, 440, 445
Langmuir–Blodgett film, 229
Langmuir trough, 228
Laplace equation, 182
Laplace pressure, 357, 386, 391, 395, 458
Latex particles, 178
Layer-by-layer deposition, 197
Leakage from capsules, rate of, 103–15
Lecithin, 359, 362–3, 367, 416–21
Light microscopy (see also confocal laser scanning microscopy)
of aggregates, 38
of bubbles, 359, 371–2
of emulsions, 427–8, 465–70
of gels, 292, 299
Light scattering (see also diffusing wave spectroscopy)
from aggregates, 39–48, 180–8
from emulsions, 344–6, 349–53
from polymers, 435–6, 439, 440
Limonene, 80–2
Lipase(s), 22–3, 109
Lipid digestion, 21–2, 113–4
Liquid crystalline phase(s), 69–85
during digestion, 23–5
Livetin(s), 435, 437, 444
Locust bean gum, 421
Loss tangent, 348
Lubrication at tongue surface, 453–60
Lycopene, 88–9, 97–100
Lysozyme, 143, 145–6, 149–52, 196
adsorption of, 210, 213–23
interaction with saliva, 464, 466–71

Macromolecular adsorption, during emulsification, 433–48
Macromolecular solutions, electrostatics in, 129–54
Maltese cross, 118, 120, 121
Maltodextrin(s), 117
Marangoni effect, 245
Mayonnaise, 317
Mean free path, of photons, 259–65, 321
Mean-square(d) displacement (MSD), 260–5, 306–15, 321, 330, 339–40
Membrane, milk fat globule, 11
Mesh size, of biopolymer network, 307, 320
Micellar calcium phosphate, 155, 161–6
Micellar cubic structures (cubosomes), 79–80, 84–100
Micellar structures, solubilization in, 87–102
Micelle(s), 23, 401, 84–100
Microcapsules, leakage from, 103–15
Microcrystallites, calcium phosphate, 161–5
Microemulsion(s), 23–7, 69–84, 97
Microencapsulation, 103
Microrheology, 305–26
Microscopy, see atomic force microscopy, light microscopy, etc.
Microspheres, tracking of, 308–14
Microstructure, see structure
Milk
acidification of, 257–302
bioactive molecules in, 8–11
composition of, 6, 8
heated, 258, 263–6, 269–70, 290
Milk fat globule(s), 10–11
Mini-emulsions, 399
Mode coupling theory (MCT), 328, 338–9
Monodisperse emulsions, 399–412
Monoglycerides, self-assembly of, 23–7, 69–84
Monolayer collapse, 231–40
Monte Carlo (MC) simulation, 130–53, 332
Mouth, emulsion behaviour in, 451–61
Mucins, salivary, 456, 464–71
Mucous layer, 455, 456
Multilayers, 175, 220, 238, 387

Subject Index

Multiple emulsion, *see* double emulsion
Multivalent ions, effect on interactions, 139–40

Nanoclusters, calcium phosphate, 161–6
Nano-emulsions, 399–400, 411
Nanoparticles, 88–9, 399–400
Network
   diffusion in, 306–7, 323–4
   of salivary proteins, 470
   of oil droplets, 316–7, 379, 381
   percolating, 282–4
Neutron reflectivity, 167, 237
Newtonian dynamics (ND) simulation, 330–3, 339
Nile Red, 309, 312, 388
Nucleation
   heterogeneous, 403, 409
   homogeneous, 403, 424
   theory of, 403, 406–7, 411
Nucleation and growth, 38, 65
Nutraceuticals, solubilization of, 87–102
Nutrition research
   colloid science in, 29–30
   food structure in, 1-15
   historical pespective of, 1–4

Octenyl succinate anhydride starch esters (OSA-starch), 434–40, 441, 445
Oil droplet(s), *see* droplet(s), emulsion droplet(s)
Oil presence, after frying, 495–7
Oligosaccharides, in milk, 6
Optical microrheology, 319–26
Optical tribological configuration (OTC), 456–7
Oral mucosa, 454, 456
Oral processing, 451–60
Oregon Green, 466, 470
Oscillatory rheology, 53, 261, 291, 295, 298, 320, 347–51, 388
Osmotic pressure gradient, 103, 104, 293–4
Ostwald ripening, 400 (*see also* disproportionation)
Ovalbumin
   adsorption of, 203, 370
   aggregation/gelation of, 40–7
   bubble/foam stabilization by, 374–7

Ovotransferrin, 435–6, 444
Oxidation, protection against, 89

Partial coalescence, 376, 458
Particles
   absorption of, 25–6
   effect on bubble stability, 178, 191, 357–68
   hydrophilic, 357–68
   hydrophobic, 357–8, 364–5, 367
   starch, 418–9
   tracking of, 305–18
Particle/particulate gel(s), 49–50, 53, 473–83
Particle size (*see also* droplet size (distribution))
   of (nano)emulsions, 399–412
   of protein aggregates, 180–8, 296
   threshold for oral detection, 104–5
Particle tracking, 305–18
Partition coefficient(s), 103–4, 108, 113, 423–32
Pectin, 175, 195, 198–206
Pendant drop technique, 182, 189, 214, 388
Peptides, electrostatic interactions of, 136–9
Perception of texture (*see also* sensory perception)
   of crispy foods, 485–6, 490–1
   of emulsions, 385–6, 452–3, 463
   of heat-set protein gels, 473–83
Percolation, 36, 282–4, 332–6
Percus–Yevick theory, 331–2
Permittivity, of protein, 135–6
Pharmaceutical formulation, 20
Phase contrast light microscopy, 292, 299
Phase diagram
   of complex food structures, 21
   of monoglyceride + water system, 71
   of oil + aroma system, 430–1
   of short-range attractive colloid, 327–9
Phase separation
   during aggregation, 37
   in acid milk gels, 289–301
   in adsorbed layer, 246, 249, 252
   microscopic, 311–7
   of amylose + amylopectin, 122, 125
   of emulsion, 307, 311–7, 413
   of liquid crystalline system, 71

of protein aggregates, 183
of starch, 117, 122, 125
protein/polysaccharide, 289–301, 481–2
Phase transition(s)
first-order, 232, 251–2, 255, 327–8
in colloidal systems, 327–42
in interfacial layers, 232, 235, 238–41, 245–56
second-order, 235
Phosphate, interaction with casein, 157–66
Phosphatidylcholine, see lecithin
Phospholipid(s)
as emulsifier, 416–21
caseinate displacement by, 414
effect on bubble stability, 362–3, 367
in films with protein, 227–43
Phosphorylation, 156, 161–5, 435
Phosphoserines, in casein(s), 161–3
Photon correlation spectroscopy (PCS), see dynamic light scattering
Photon transport mean free path, 259–65, 321
Phytosterols, 29, 30, 88–9, 94–100
pK_a of titrating group(s), 135–6
Pig's tongue, 455–7, 460
Plant metabolites, secondary, 5
Plasticizing effect, 488, 489, 495
Plateau border, 190
Poisson–Boltzmann equation, 130, 135
Polarized light microscopy, 119–25, 427
Polyacrylamide electrophoresis, 169, 172, 174, 435, 436, 441
Polydimethylsiloxane (PDMS), 455–7, 460
Polydispersity, 36, 392, 400, 401
Polyelectrolyte solutions, theory of, 129–54
Polyglycerol ester (PGE), 401–11
Polyglycerol polyricinoleate (PGPR), 104, 106–7, 110
Polymer gel, 474
Polysaccharide(s) (see also pectin, xanthan, etc.)
as emulsifier, 434
effect on bubble stability, 376
effect on emulsion stability, 197, 311–7
effect on protein gel surface, 478–83

effect on protein gelation, 50–1, 289–301, 478–83
interaction with protein, 195–208, 289–301, 481–2
Polysorbate, see Tween 20
Polystyrene particles, 308, 320
Pore-size distribution
of crispy food, 489–93, 496
of simulated gel, 337–8
Potato starch, 118–23
Power-law scaling, 35–6, 40, 52–3, 335
Pressure change, effect on bubbles/foam, 369–81
Primitive model of electrolyte solutions, 129–31
Protease(s), digestion by, 9–10
Protein(s)
adsorption of, 167–75, 196, 209–25, 245, 343, 347, 440, 443–5
aggregation on heating, 64–6, 180–8
bubble stabilization by, 178, 191, 357–68
dielectric permittivity of, 135–6
denaturation of, 65–6, 179–80, 183–6, 285–6, 294
self-assembly of, 35–68, 227–44
Protein adsorbed layer(s), see adsorbed protein film(s)/layer(s)
Protein–polysaccharide complexes/interactions, 50–1, 195–208

Q_L phase, 88–100

Radial distribution function, 163–5, 334–6
Radio-labelling, 214, 217
Radius of gyration, 35–47, 294, 328
Random sequential adsorption, 442–3
Rate-limited cluster aggregation (RLCA), 180, 186, 187
Reflectivity, monolayer, 229, 232–3, 236–7
Reflection, of sound, 496–7
Refractive index
in ellipsometry, 215–6
in light scattering, 262
Relaxation
of bonds, 283
of droplet shape, 349–51
of microscopic domains, 311–3, 316
Relaxation spectrum, 320

Relaxation time, 259, 265, 316, 349
Rennet-induced casein gels, 257, 286
Retrogradation, 118
Reversed hexagonal phase, 23, 27
Rheology
   computer simulation of, 283–85
   of acid-induced milk gels, 261–5, 269–88, 294–8
   of biopolymer network, 306–7, 317
   of cream layers, 388, 389–95
   of crispy food, 487–8, 491
   of emulsions, 307, 310–1, 317, 344–53, 385–97, 466, 469
   of gelatin solution, 324–5
   of globular protein gels, 51–3
   shear-thinning, 307, 310–1, 469
Rheometer-based small-angle light scattering (Rheo-SALS), 344–6, 349–53
Rhodamine B, 108, 111
Root-mean-square displacement, 271 (see also mean-square displacement)
Rotational diffusion coefficient, 271
Rubber surface, friction at, 454, 455–7

Salad dressing(s), 317
Saliva, 453, 455, 464–5
   effect on emulsions, 463–72
Sauter mean diameter, 435
Scanning electron microscopy (SEM), 38, 49–50, 490 (see also environmental scanning electron microscope)
Scattering, structure from, 38–49 (see also light scattering, X-ray scattering)
Scheutjens–Fleer theory, 248, 445
SDS–PAGE, 169, 172, 174, 435, 436, 441
Second virial coefficient, 180–1, 331, 339
Secondary emulsion, 174
Self-assembly
   effect of electrostatic interactions on, 227–44
   of casein + calcium phosphase, 161–6
   of liquid particles, 69–86
   of monoglycerides, 23–7, 69–84
   of particle systems, 474
   of protein(s), 35-68
   of protein + phospholipid films, 227–43
   of starch spherulites, 117–26

   of surfactants, 57–68
Self-consistent-field (SCF) theory, 245–56
Self-similar structure, 35–6, 40–1
Sensory perception
   of creaminess, 386, 451–3, 458, 460
   of crispness, 485–6
   of emulsions, 452–3, 454, 458–60, 463
   of fat, 385–6, 451
Sequestrant, 157–8
Serum albumin, egg-yolk, 435–6, 440, 443–4 (see also bovine serum albumin)
Shear modulus, see rheology
Shear (flow)
   effect on droplets, 343–55
   effect on gelation, 320
Shear-thinning, 307, 310–1, 469
Shearing between surfaces, 456–7
Shelf-life, emulsion, 308, 317
Short-range attractive colloidal systems (SRACS), 327–42
Shrinkage, of bubbles, 361–5
Silica (nano)particles, 178, 357–68
Silicone rubber, 454 (see also polydimethylsiloxane)
Simulation, see computer simulation
Skim milk powder, 290
Small-angle light scattering, in rheometer, 344–6, 349–53
Small-angle X-ray scattering (SAXS), 70–83, 90–100
Sodium caseinate, see casein(ate)(s)
Sodium dodecyl sulfate (SDS), 169, 175, 210, 212–9, 344–5, 348–51, 388–92, 464–8 (see also SDS–PAGE)
Soft matter, definition of, 155
Solid (fat) content, of emulsion droplets, 403–11, 423–31, 458–9
Solid fat, effect on crispness, 497
Solubility, of carbon dioxide, 131, 133
Solubilization
   in liquid crystalline phases, 69–70, 75–84, 87–102
   of neutraceuticals, 87–102
Solvent quality, 156, 161
Sound emission, from crispy food, 486–97
Sparging, 182–3
Spherulites, starch, 117–26
Spinodal decomposition, 332, 336

Spontaneous emulsification, 22, 415
Spray drying, 414
Spreading
  of droplets, 379
  of protein, 445
Spreads, fatty, 413
Square-well potential, 329–32
Stability
  microbiological, 413, 416
  of bubbles/foams, 177–8, 357–82
  of casein aggregates/micelles, 156–61, 195, 262–6
Stability ratio, 180, 185
Starch
  as thickener, 417–9, 421–2
  as emulsifier, 434–40, 441, 445
  cold-water-swelling, 418–21
  dispersion(s) of, 118–25, 419–22
  hydration of, 419–22
  pre-gelatinized, 419–20
  resistant, 118
Starch granule(s), 117–8, 123, 419
Starch spherulites, complexation with ligands, 117–26
Steric interaction/repulsion, 175, 262, 266, 294
Steric stabilization, 171, 246, 257, 294, 434
Sticking probability, 180, 185
Sticky sphere model, see adhesive hard-sphere model
Stokes–Einstein equation, 181, 198, 259, 260–3, 306, 311, 321
Stokes velocity, 393
Storage/loss moduli, 53, 261–5, 291, 294–8, 319, 324–5, 348–51, 390
Stress/strain behaviour, simulated, 283–5
Structure(s)
  effect of shear flow on, 349–51
  micellar, 60, 79–80, 84–100
  of acid milk gels, 289–302
  of adsorbed layer(s), 203–6, 216, 233–41, 253–4
  of aggregates, 35–68, 334–6
  of biopolymer network, 307
  of casein micelle, 161–6
  of cream layer, 391–5
  of crispy food, 489–93, 496
  of emulsions, 312–7, 385–97, 418, 457
  of heat-set gels, 43–54, 473–84
  of pig's tongue, 455
  of self-assembled particles, 69–85
  of self-assembled proteins, 35–67
  of starch spherulites, 117–25
  percolating, 332–6
  relationship to elasticity/rheology, 51–3, 297, 312–7
  rod-like, 41–2
  supramolecular, 227–8
  surface, 473–84
Structure factor, 40, 43–6, 259, 336–7
Structure–function relations, 11–13
Sulfhydryl–disulfide exchange, 192
Sulfhydryl groups, 178
Surface activity, see adsorption, surface tension
Surface anisotropy, of protein gel, 477, 482
Surface binding (coefficient), 426, 428–32
Surface coverage,
  calculated, 249–52
  of cationic surfactant, 364–5, 367
  protein, 169, 172–4, 217, 219, 252
Surface creation, 473–4, 482
Surface denaturation, 209
Surface dil(at)ational modulus, see dil(at)ational elasticity/modulus
Surface elasticity
  effect on bubble stability, 191
  of protein layer, 189–90, 376, 378
Surface equation of state, 210–1
Surface friction, 453, 455–7, 474–84
Surface heterogeneous nucleation, 409
Surface hydrophobicity, 445
Surface load, of OSA-starch, 438
Surface pressure (see also surface tension)
  of protein, 210–1, 234–6
  of protein + polysaccharide, 200–3
  of protein + surfactant, 211, 228, 238–41
Surface pressure isotherm, 217, 219–20, 228, 231–5
Surface rheology, 182, 189, 199, 344, 347–8, 388
  effect on bubble/foam stability, 190–1
  of complexes, 197, 203–6
  relationship to bulk rheology, 394–6
Surface roughness/smoothness, 455–7, 473, 476–83
Surface shear modulus, 234, 389, 394
Surface shear viscosity, 386

Surface tension (*see also* adsorption, surface pressure)
  effect on emulsion droplets, 391, 395
  measurement of, 182, 198–9, 213–4, 347, 376, 388, 391
  of protein + surfactant, 210, 213–23
Surface texture, 473
Surface viscoelasticity, *see* surface rheology
Surface water, 474, 477
Surfactant(s)
  adsorption of, 209–26, 245, 246, 359, 361
  as emulsifier, 401–11
  effect on bubble stability, 357–68
  in mixed layers with biopolymers, 245–56
  self-assembly of, 57–68
Surfactant + protein mixtures, 209–44, 416

Temperature cycling, 403, 404
Tensiometry, 182, 189, 198, 213–4, 347
Texture Analyser, 417, 486
Texture perception, 385–6, 452–3, 463, 473–86, 490–1
Thermodynamic incompatibility, 246, 248, 481–2
Thermodynamics
  of adsorption, 210–2
  of aggregation, 58–64
  of emulsification, 415
  of polyelectrolyte solutions, 129–54
  of self-assembly, 58–64
Titrating groups, effect on protein interactions, 140–5
Tongue surface, 455–60
Trajectories, of probe particle(s), 309, 311–4, 316
Transmission electron microscopy (TEM), 23, 27–8, 38, 50
Tribology, 451–61, 473–4
Turbidimetry, 46–8, 160
Turbulent flow, 433, 441–2
Tween 20
  as emulsifier, 401–11, 416–7, 464–8
  hydrolysis of, 109, 113–4

Ultrasound
  attenuation of, 405, 410
  crystalline emulsion content from, 400–11
  monitoring creaming with, 388, 391–4

Ultra-turrax, 108–9, 465
Urick equation, 388

Vesicles, 72, 76
Viscoelastic response of emulsion droplets, 343–55
Viscoelasticity, *see* rheology
Viscosity
  from particle tracking, 314–6
  of emulsions, 308, 310–6, 417–22, 452, 458–9, 466, 469
  of starch dispersions, 419–22
  of xanthan solutions, 308, 311–2
Volatiles, distribution in emulsions, 423–32

Wax
  digestibility of, 109, 113–4
  encapsulation by, 103–15
Wetting, of tongue surface, 456
Whey protein(s) (*see also* β-lactoglobulin, *etc.*)
  complex/interaction with casein(s), 257–86, 294–5
  gels of, 473–84
  heat denaturation of, 258, 263, 294
  interaction with polysaccharide, 195
Whey protein isolate (WPI)
  as emulsion stabilizing agent, 387, 389–94
  bubble stabilization by, 374–6
Whippable emulsions, 377, 380
Winter–Chambon criterion, 320, 324–5

Xanthan (gum), 109
  hydration of, 419–21
  in acid milk gels, 289–302
  in emulsion systems, 109, 307–17, 419–21
  in heat-set protein gels, 473–84
X-ray diffraction, 119, 121, 165
X-ray scattering, 40–6, 69–83, 90–100, 411

Yield stress, 317, 328, 376, 389
Young's modulus, 487, 491

Zeta potential, 169–73, 203, 215, 257, 388–9, 465–71
Zwitterionic surfactant, 229, 362